Frontiers of Nano-Optoelectronic Systems

NATO Science Series

A Series presenting the results of scientific meetings supported under the NATO Science Programme.

The Series is published by IOS Press, Amsterdam, and Kluwer Academic Publishers in conjunction with the NATO Scientific Affairs Division

Sub-Series

I.   **Life and Behavioural Sciences**   IOS Press
II. **Mathematics, Physics and Chemistry**   Kluwer Academic Publishers
III. **Computer and Systems Science**   IOS Press
IV. **Earth and Environmental Sciences**   Kluwer Academic Publishers

The NATO Science Series continues the series of books published formerly as the NATO ASI Series.

The NATO Science Programme offers support for collaboration in civil science between scientists of countries of the Euro-Atlantic Partnership Council. The types of scientific meeting generally supported are "Advanced Study Institutes" and "Advanced Research Workshops", and the NATO Science Series collects together the results of these meetings. The meetings are co-organized bij scientists from NATO countries and scientists from NATO's Partner countries – countries of the CIS and Central and Eastern Europe.

Advanced Study Institutes are high-level tutorial courses offering in-depth study of latest advances in a field.
Advanced Research Workshops are expert meetings aimed at critical assessment of a field, and identification of directions for future action.

As a consequence of the restructuring of the NATO Science Programme in 1999, the NATO Science Series was re-organized to the four sub-series noted above. Please consult the following web sites for information on previous volumes published in the Series.

http://www.nato.int/science
http://www.wkap.nl
http://www.iospress.nl
http://www.wtv-books.de/nato-pco.htm

Series II: Mathematics, Physics and Chemistry – Vol. 6

Frontiers of Nano-Optoelectronic Systems

edited by

Lorenzo Pavesi
INFM and Department of Physics,
University of Trento, Italy

and

Eugenia Buzaneva
Kyiv National Taras Shevchenko University,
Ukraine

Kluwer Academic Publishers

Dordrecht / Boston / London

Published in cooperation with NATO Scientific Affairs Division

Proceedings of the NATO Advanced Research Workshop on
Frontiers of Nano-Optoelectronic Systems: Molecular-Scale Engineering and
Processes
Kiev, Ukraine
May 22–26, 2000

A C.I.P. Catalogue record for this book is available from the Library of Congress.

ISBN 0-7923-6745-6 (HB)
ISBN 0-7923-6746-4 (PB)

Published by Kluwer Academic Publishers,
P.O. Box 17, 3300 AA Dordrecht, The Netherlands.

Sold and distributed in North, Central and South America
by Kluwer Academic Publishers,
101 Philip Drive, Norwell, MA 02061, U.S.A.

In all other countries, sold and distributed
by Kluwer Academic Publishers,
P.O. Box 322, 3300 AH Dordrecht, The Netherlands.

Printed on acid-free paper

TABLE OF CONTENTS

Part III. Biomolecular Technologies

Part IV. Molecular-Scale Engineering: polymers and carbon-based nanostructures

Part V. New technologies for producing nano- systems

PREFACE

Since their discovery, low dimensional materials have never stopped to intrigue scientists, whether they are physicists, chemists, or biochemists. Investigations of their nature and functions have always been and still are numerous and as soon as a solution is found for a given question, another one is raised. The coupling of nano-materials with photonics, i. e. nano-photonics, has produced a boiling pot of idea, problems, discovery and applications. This statement is abundantly illustrated in the present book.

The interest in nano-optoelectronic materials and systems is very widespread, what gives a really international and multicultural flavour to nano-optoelectronic meetings. One of them was organized by our-self in May 2000 in Kiev as a NATO Advanced Research Workshop and EC-Spring School. The arrival of the new millennium provides an obvious transition point at which many aspects of nano-science and nano-engineering of nano-photonic systems can be assessed with respect to the research progresses made in the preceding decades and to the challenges that lie ahead in the coming decades. This book was planed to mark this with the objective of presenting a collection of papers from experts, which provide broad perspectives on the state-of-the-art in the various disciplines of nano-science and nano-engineering and on the directions for future research. The content of this book provides a snapshot, at a particular point in time, on such different subjects such as photonic band gap crystals, semiconductor quantum dot and wire lasers, silicon optoelectronics, carbon-based nanostructures physics, polymer based nano-composite and quantum wires, DNA nano-technology and silicon bio-compatibility, nano-scale optical characterization, spray and cluster deposition, self-assembly, imprint technology, quantum computing and quantum dots based computation. While the snapshot archived in the following pages is inevitably ephemeral in view of the rapid pace of regrouping within novel nano-optoelectronic systems, it is hoped that the perspectives offered here would significantly influence nano-sciences and nano-engineering for decades to come.

It should be noted that the success of our multidisciplinary meeting and of this Book was guaranteed in advance by the enthusiastic work of the organizing commitee: L. P. Biro (Hungary), N. Garcia (Spain), S. Ossicini (Italy), P. N. Prasad (USA), C. M. Sotomayor Torres (Germany) and G. Stegeman (USA). We thank all the authors who responded eagerly to this demanding task and promptly delivered high quality manuscripts. The timely publication of this book would have not been possible without the efficient and cheerful help of our editorial assistants, Dr. Viviana Mulloni and Dr. Olexandr Gorchinsky. We are most grateful to them for all their efforts.

We gratefully acknowledge the generous support of the NATO Scientific Affairs Division, of the Directorate-General Information Society of the European Commission, and of our Universities which made the workshop and this resulting book possible. We also express our gratitude to all those have contributed to the workshop, and to Kluwer for efficient publication.

Lorenzo Pavesi
Trento, Italy, August 2000

Eugenia Buzaneva
Kiev, Ukraine, August 2000

Photographs of the participants to the NATO ARW in Kyiv (May 2000)

INTRODUCTION TO FRONTIERS OF NANO-OPTOELECTRONIC SYSTEMS

L. PAVESI
INFM and Dipartimento di Fisica, Università di Trento, Italy
E. BUZANEVA
Kyiv National Taras Shevchenko University, Ukraine

A workshop on the frontiers of nano-optoelectronic systems was held in Kyiv at the end of May 2000. The topic chosen is emerging as one of the most exciting realms of science to-day. Nano-optoelectronics (also named nano-photonics) is defined as the study of optical phenomena and technology occurring at the nanoscale, it deals with processes and structures, which are spatially localized in domains smaller than the wavelength of visible radiation. It is interesting to note that actual device applications of nano-photonics are already on the market. Some of them are dominating the laser market: i. e. the quantum well lasers. During the workshop several different approaches to nano-photonics were presented: fundamental aspects of nano-scale science, low dimensional III-V semiconductors, low-dimensional silicon, molecular scale engineering, bio-molecular technologies, polymers and carbon based nanostructures, low dimensional photonic crystals, novel architectures of quantum opto-electronic systems and nano-scale characterization of photonic materials. In this Introduction, we report an excerpt of the various round-tables held during the meeting among the participants on the frontiers of nano-optoelectronic as well as the trends of its future development. We refer to the various section of the book, where the concepts here delineated are discussed in more details.

In the field of low dimensional III-V semiconductor based structures,[1] concepts are emerging which are looking not only to the spatial confinement of carriers but also to the spectral confinement of radiation as in optical microcavities. Indeed, by coupling these two concepts, record external quantum efficiencies of more than 70% in light emitting diodes have been reached. Concepts which are emerging in this field are the control of the atomic-scale growth processes which permits to obtain ordered arrays of self-aggregated quantum dots, or the growth of quantum wires on V-shaped substrates, ... All these processes are based on the understanding of the role of the surface potentials (strain, capillary forces and entropy of mixing) on the growth. These are the driving forces, which have to be understood and mastered. The key word with this respect is self-assembly which should reduce the cost of production and should increase the homogeneity of the produced nanostructures. Quantum dots and quantum wires are used in many different optical devices, such as lasers or modulators. Nanostructures should demonstrate dramatic improvements with respect to more-common systems (quantum wells) otherwise they will not make a real breakthrough. In the last years no big improvements have been reported. On the other hand, external facts

---

[1] See part I

claim for a real need of them: e.g. optical interconnect and the scaling down predicted by the Moore' law. The status of silicon-based optoelectronics is slowing down after ten years of excitement.[2] Even the research on Er-doped materials that has some more impetus shows very slow progress. At the workshop there was the first-world report on optical gain observation in silicon nanostructures. People are thinking that this could re-start the field if confirmed. Another issue still open for applications of silicon nanostructures is the problem of carrier injection into them and the carrier transport from the electrode to a single quantum dot. All these aspects need to be improved if one is hoping to put silicon as an optically active medium into the nano-photonics arena.

Spectacular results on the non-linear optical properties of polymers, interpreted as quantum wires have been presented.[3] A conjugated polymer, the backbone of which consists of multiple sequential carbon single, double and triple bonds, forms the quantum wire. The facile movement of the π-electrons results in the large polarizabilities (and hyperpolarizabilities) observed. The integration of organic/inorganic materials (not only semiconductors but also metals) can be used to enhance the non-linear optical properties of the organics. Here the key word is integration among the various materials. Another way to form nanostructure is the encapsulation of metals into carbon nanocages. New aspects of materials science rises: (1) The encapsulate consists of a small number of atoms that is measurably confined by the size of the carbon cage; (2) collective physical properties of these clusters are novel and can be studied as a function of the number of the atoms providing insight into size-dependent properties. In the nanometer size regime, the number of atoms in the bulk of the particle is close to that in the surface region. By varying the size of the particle, the effects of the bulk vs. surface region can be studied independently. Control at the atomic level of nano-structure is performed with the cluster deposition.[4] Experimental demonstration of soft-landing of size-selected clusters stimulates the move to large-scale production of size-selected clusters for opto-electronic applications. Other formation techniques have been discussed: sol-gel process, spin-on processes, and self-assembled monolayer dipping and imprint techniques. All these have pros and cons.

A new boost was provided by the statement that biology is nanotechnology that works.[5] Biologists proposed that DNA could be used as the construction material at the nanoscale. The advantages of DNA with this respect are i) the predictable intermolecular interaction, ii) the availability of convenient modifying enzymes and of automated chemistry, iii) the existence of locally stiff polymers, and iv) the fact that they are externally readable. The immediate challenges in this area consist of extending the system to three-dimensions, to combine the chemistry to scaffold other nanoscale systems, such as carbon nanotubes or proteins, to generalize the action of nanomechanical devices to programmability, to extend the system from periodicity to algorithmic assembly and to investigate the value of generalized complementarities in linking motifs. The concepts of self-assembly can be used for molecular scale electronics. In fact two approaches for nano-technology are pursued: scaling-down and scaling-up. Microelectronics relays on the scaling down of circuit features. While

---

[2] see part II
[3] see part IV
[4] see part V
[5] see part III

molecular chemistry relays on the scaling-up: molecular structures are constructed from smaller components. The idea here is to combine DNA which has poor electronic properties but excellent self-assembly properties with electronic materials which have excellent transport properties but poor self-assembly properties. One has to consider that the operational principle of bio-molecular devices is certainly different from that of present electronics. The main advantage is the possibility of exploiting the complexity of the biological materials, for example the easy of self-organizing/positioning nanometer scale objects with nanometer scale accuracy by using the self-recognizing of DNA. Also novel applications of nanostructured silicon to form smart and active biomaterials are emerging. Here the vision is towards a hybrid approach where silicon electronics and biology are working side-by-side to form sophisticated artificial organs. Neuron-silicon chips for signal transductions are currently persecuted. Many new concepts such as complexity and self-assembly or self-organization are emerging from biology and spilling over the more mature semiconductor technology. An interesting issue is whether the semiconductor physics will help biology more than what biology will help semiconductors. The most appealing aspects are those emerging at the interfaces among nano-photonics, nano-electronics and nano-biology. A new field is emerging named nano-clinics where the interaction between the human body and the therapeutic agent is controlled and occurs at the nanoscale (e. g. micelle for localized drug delivery).

Improvements in first principle calculations towards large-scale and real dynamics computing have been discussed.[6] There was a common hope that real processes such as recombinations or luminescence could be tackled positively in the near future.

Another interesting field is that of the control or manipulation of electrons and photons at the single particle level.[7] At this scale one needs to understand the statistics and the electron/photon coupling. The transfer of concept from wave-optics to materials science as that of interference, phase control of optical trapping is opening many new perspectives. Many hopes of improved photonic devices are based on the development of true photonic band-gap crystals. These are three-dimensional dielectric lattices, which modify the photon dispersion. Photonic crystals offer prospects for numerous applications: low threshold lasing, optical power limiting, chemical and bio-sensing, and optical switching.

A prediction has been made that when the ultimate limit of micro-electronic down scaling (Moore' law) will be reached the new word/concept will be quantum physics. The logic behind the new computational algorithms no longer will be binary logic but quantum mechanics (combinatorial logic).[8] Also the principles ruling the operations of devices will be true quantum mechanics. The benefits of entering this field will be a gain in complexity and the easy of down scaling. The use of some different physics may imply either a *different* implementation of the *same* rules or the implementation of *novel* rules. Along the first route we have witnessed the transition from mechanical to electromechanical and to electronic computers. For the latter, a recent proposal concerns the application of quantum mechanical rules for digital information processing. Appropriate quantum systems will have to be designed for such new applications. A number of options are being investigated: among them semiconductor nanostructures (e.g. electronic quantum boxes) might seem to be first choice.

---

[6] see part II
[7] see part I
[8] see part VII

However, the difficulty to isolate the desired electronic degrees of freedom from the rest leads to decoherence times, which are much shorter than can be obtained in other implementations. In addition, the inhomogeneous size distribution causes disorder in the energy schema, which might imply unpredictable behaviours of the quantum gates. Clearly the quantum computers are breakthrough concepts. The question, which remains open, is: can these theoretical concepts be mapped into technological applications?

In recent years much efforts have been focused on studying nanoscale optical interactions[9] and preparing nanostructured optical materials[10]. Due to the fact that near-field optical microscopy overcomes the diffraction limit, it is now possible to probe the optical interactions at nanoscopic region. High spatial resolution techniques, specifically SMT and AFM, can be used to isolate regions as small as 1-10 nm which leads to focus on one isolated nanostructure and avoid the problem of inhomogenieties. Here one is interested in knowing the ultimate limit of these techniques. Also one is arguing whether one can exploit them, i. e., modify the properties of a quantum dot, or even of a single chemical bond. Furthermore, the high degree of precision and resolution offered by near-field optical microscopy also makes it a potential candidate in the photo processing and photo fabrication. Recently, nanosize control of the local structure has also emerged as a frontier area of materials research, which provides an opportunity to manipulate excitation dynamics as well as to influence energy transfer. Rare earth doped glasses and organic-inorganic hybrid photo refractive nanocomposites are two examples in which nanostructures have been used to control the electric, luminescence, and non-linear optical properties. This new frontier of nanophotonics offers numerous opportunities for both fundamental research and application. From theoretical and experimental perspectives, further studies can be extended to other non-linear optical processes, such as optical wave mixing and coherent anti-stokes Raman scattering (CARS). The information obtained can provide a better understanding of nanoscopic electrodynamics. From application perspective non-linear near-field microscopy holds promise for high-contrast, high-resolution and site specific imaging. Nanofabrication and nanoscale patterning using nanoscale photochemistry will receive increasing attention.

In conclusion, nano-scale physics is very healthy field of research with a lot of promising applications and developments foreseeable for opto-electronics.

---

[9] see part VI
[10] see part V

NANOPHOTONICS: NANOSCALE OPTICAL SCIENCE AND TECHNOLOGY

PARAS N. PRASAD , YUZHEN SHEN, ABANI BISWAS, JEFF WINIARZ
Photonics Research Laboratory, Institute for Lasers, Photonics and Biophotonics, Departments of Chemistry, Physics, Electrical Engineering, and Medicine, State University of New York at Buffalo, Buffalo, NY 14260

Abstract

Nanophotonics defined as nanoscale optical science and technology is a new frontier, which includes nanoscale confinement of radiation, nanoscale confinement of matter, and nanoscale photophysical or photochemical transformation. Selected examples of our research work in each of these areas are presented here. Nonlinear optical interactions involving nanoscale confinement of radiation is both theoretically and experimentally studied using a near-field geometry. The effort in nanoscale confinement of optical domains is focused to control excitation dynamics and energy transfer as well as to produce photon localization using nanostructured rare-earth doped glasses and novel inorganic-organic photorefractive nanocomposites. Spatially localized photochemistry using a near-field two-photon excitation is being developed for nanofabrication and nanoscale memory.

1. Introduction

Emerging as one of the most exciting realms of science today is the field of nanophotonics. Defined as the study of optical phenomena and technology occurring at the nanoscale, nanophotonics deals with processes and structures which are spatially localized in domains smaller than the wavelength of visible radiation. Given the broad definition of nanophotonics, it is convenient to divide it into three main parts, each of which is represented through work conducted in our laboratory: (i) nanoscale confinement of radiation by using a near-field optical probe or photon localization such as in a photonic crystal, (ii) nanoscale confinement of matter by using nanostructured materials, and (iii) nanoscale photophysical or photochemical transformation for photoprocessing.

In recent years much more efforts have been focused on studying nanoscale optical interactions and preparing nanostructured optical materials. Due to the fact that near-field optical microscopy overcomes the diffraction limit, it is now possible to probe the optical interactions at nanoscopic region. Furthermore, the higher degree of

1

L. Pavesi and E. Buzaneva (eds.), Frontiers of Nano-Optoelectronic Systems, 1–10.
© 2000 *Kluwer Academic Publishers. Printed in the Netherlands.*

precision and resolution offered by near-field optical microscopy also makes it a potential candidate in the photoprocessing and photofabrication. Recently, nanosize control of the local structure has also emerged as a frontier area of material research, which provides an opportunity to manipulate excitation dynamics as well as to influence energy transfer. Rare-earth doped glasses and organic-inorganic hybrid photorefractive nanocomposites are two examples in which nanostructures have been used to control the electric, luminescence, and nonlinear optical properties.

Our approach utilizes nanoscale confinement of radiation, nanoscale confinement of matter and nanoscale photophysical or photochemical transformation to investigate nanoscale matter-radiation interactions. Selected examples of our studies in each of these areas include: (1). nonlinear optical interactions involving nanoscale confinement of radiation are both theoretically analyzed and experimentally probed using a near-field geometry, (2). nanoscale confined optical domains to control excitation dynamics and energy transfer and to produce photon localization are illustrated by examples of nanostructured rare- earth doped glasses and photorefractive nanocomposites, and (3).spatially localized photochemistry using a near-field two-photon excitation is utilized for nanofabrication and nanoscale memory.

2. Theoretical modeling of nanoscale second-order optical interactions

We have developed a theoretical description of nanoscale nonlinear optical processes in a near-field geometry[1] using multiple multipole (MMP) model[2], which is a compromise between a purely analytical and a purely numerical approach. This model can be used for solving Maxwell's equations for arbitrarily shaped, isotropic, linear and piecewise homogeneous media. With the MMP model, the electromagnetic field $f \in \{E, H\}$ within individual domains D_i is expanded by analytical solutions of Maxwell's equations

$$f^{(i)}(r, \omega_0) \approx \sum_j A_j^{(i)} f_j(r, \omega_0). \qquad (1)$$

The basic function $f_j(r, \omega_0)$ satisfies the wave vector equation for the eigenvalue q_j

$$-\nabla \times \nabla \times f_j(r, \omega_0) + q_j^2 f_j(r, \omega_0) = 0. \qquad (2)$$

The parameters $A_j^{(i)}$ of the series expansions are numerically matched on the interfaces. The main feature of the nonlinear optical processes in a near field mode is that the phase matching requirement is relaxed, as the domains are much smaller than the coherence length. The theoretical calculation shows that the near-field distribution from a confining tip has two components: (i) for a central angular distribution, the radiation has a real wave vector, and corresponds to the allowed light in the sample, (ii) the outer angular distribution represents an imaginary wave vector, which produces the forbidden light in the sample. Figure 1 shows the near-field intensity distribution of the second-harmonic generation (SHG) from NPP[3] crystals. The field intensity of the SH wave is highly localized in both the lateral and the vertical dimensions within the probe center, i.e. about 50nm×50nm. Unlike the forbidden light, the allowed light is more localized at the center of the tip, and contributes to better resolution.

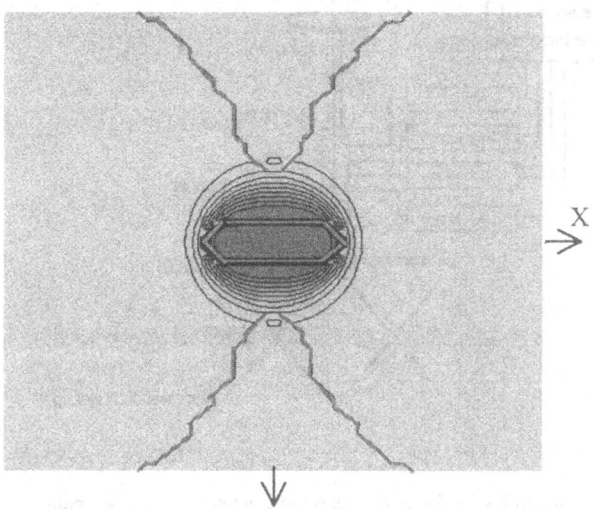

Figure1. Optical near-field intensity distribution of Second-harmonic generation (n_1=1.47, λ=800 nm, d=30 nm, d_{21}=200*10^{-9} esu, d_{22}=70*10^{-9} esu)

3. Nanoscale second-harmonic generation and two-photon excitation

Second-harmonic generation (SHG) is effective and a remote sensing technique for second-order optical nonlinearities. With the development of near-field scanning microscopy (NSOM)[4-8] and photon scanning tunneling microscopy (PSTM)[9-13], it is possible to probe SHG at nanometer scale. We have extended SHG studies to PSTM. The schematic diagram of experimental setup is shown in Figure 2. A Q-switched Nd: YLF laser is used as an excitation source at 1.047 μm with an average power of 250 mW. The laser beam is focused by a lens and illuminates the sample that is mounted with an index matching oil on a fused silica prism under total internal reflection. The measurement is performed on NPP crystals. Figure 3 shows the SH image of isolated NPP nanocrystals, in which the full width at half maximum (FWHM) of the optical intensity profile is 360 nm. Since SH intensity is proportional to the square of effective susceptibility d_{eff}, the nonlinear optical contrast in SH image is related to the local variation of d_{eff}, and the gray scale is a quantitative measure of local d_{eff}. The local orientation of nanocrystals under excitation indicates high local degree of molecular order. The molecules are orientated in such a way that a net transition dipole moment is exhibited in the plane of crystal film, which corresponds to a local polar axis in the nanocrystals. The random orientation of local crystals observed in SH images also suggests that the phase of these microscopic units are unrelated, which may lead to isotropy in the bulk crystals.

4

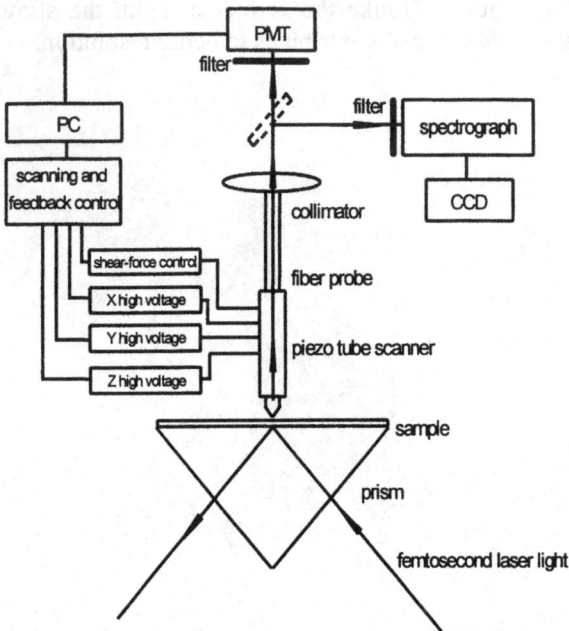

Figure 2. Schematic of PSTM for second-harmonic generation and two-photon excitation

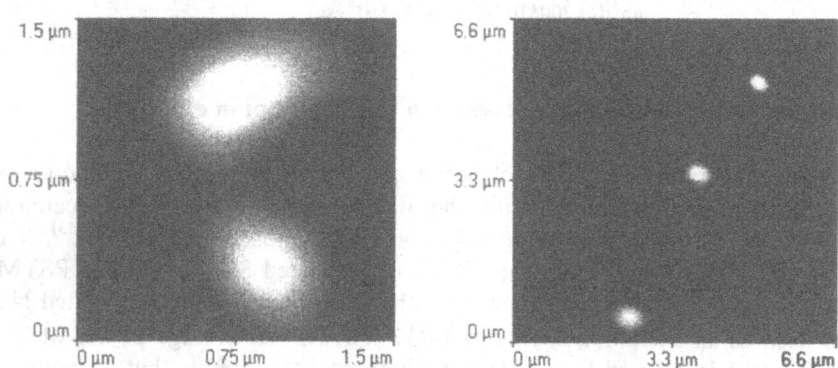

Figure 3. Second-harmonic image   *Figure 4.* Two-photon fluorescence image

Two-photon excitation microscopy is a nonlinear optical imaging technique, which has the advantages of effective rejection of background, reduced volume of photobleaching, and depth discrimination. Some of these advantages can benefit PSTM. We have extended femtosecond two-photon excitation to PSTM, and demonstrated that two-photon PSTM provides advantages in signal-to-noise improvement and in system alignment compared to two-photon NSOM. The schematic diagrams of two-photon PSTM is the same as in Figure 2. A self mode-locked Ti:Sapphire laser is used as an excitation source with an average power of 12 mW for two-photon PSTM. Since the dispersion length is only 5 mm, pulse broadening and nonlinear effects are negligible. The measurement is performed on a two-photon chromophore (AF-390) that was synthesized at the Polymer Branch of the U.S. Air Force Research Laboratory. Figure 4 shows the near-field two-photon fluorescence image, and the FWHM of the optical intensity profile is 340 nm. It is evident that the emission patterns in the fluorescence image are oriented in the same direction for all isolated particles. Such an effect is mostly due to the polarization-dependent dipolar interaction. Since the incident light is linearly polarized, once excited, the molecules act as electric dipoles and dipole orientations approximately follow the excitation polarization. Therefore, the spatial fluorescence feature indicates the high degree of molecular order in the isolated nanoparticles.

4. Nanostructured materials

4.1. RARE-EARTH NANOCOMPOSITES IN LAF$_3$ LATTICE

Rare earth doped transparent glass ceramics has been developed by sol-gel method in our laboratory. Rare earth doped lanthanum halide powders prepared by a sol-gel method coupled with HF reaction atmosphere is found to be highly efficient emitter at 1.3μ m [14]. The glass ceramic is densified at 950^0C. Fluorination was carried out in-situ. The absorption spectrum in the wavelength range of 200-2000nm of Er^{3+}-Yb^{3+} co-doped

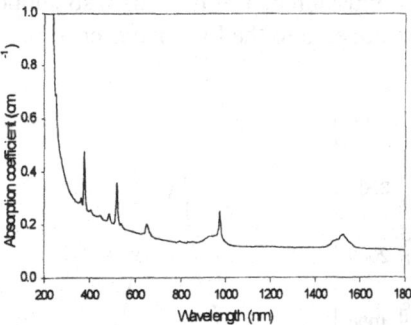 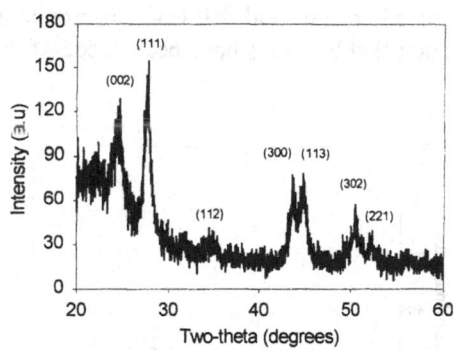

Figure 5. Absorption spectrum of sol-gel ceramics at 900^0C   *Figure 6.* XRD pattern of the sample at 900^0C

sample is shown in Figure 5. All the absorption peaks are corresponding to the electronic transitions of Er^{3+} and Yb^{3+} ions and are similar to what has been observed in glass ceramics prepared by melt quenching [15]. Figure 6 shows XRD spectrum of the glass ceramics. Sharp diffraction peaks are observed which are superimposed on the characteristics broad halo of silica glass. The peaks at around 2θ = 24, 27, 35, 44 and 50^0 are identified as reflections from the (002), (111), (112), (300), (113), (302) and (221) planes of hexagonal LaF$_3$ crystals. Both TEM and XRD studies indicate that LaF$_3$ nano-crystals have been precipitated within the silica matrix.

Upconversion fluorescence was observed using a diode laser emitting at 973 nm. Figure 7 shows the upconversion spectrum of the sample co-doped with Er^{3+} and Yb^{3+}. Several peaks in the UV and VIS region have been observed corresponding to transitions of Er^{3+} excited ions to the ground state [16]. The peaks at 379, 407, 450, 489,

520, 540 and 660 nm are due to $^4G_{11/2} \to {}^4I_{15/2}$, $^2H_{9/2} \to {}^4I_{15/2}$, $^4F_{5/2} \to {}^4I_{15/2}$, $^4F_{7/2} \to {}^4I_{15/2}$, $^2H_{11/2} \to {}^4I_{15/2}$, $^4S_{3/2} \to {}^4I_{15/2}$ and $^4F_{9/2} \to {}^4I_{15/2}$ respectively. The emission lines at 540 and 660 nm are visible with naked eye even with ~27 mW of pump power. Upconversion emission from Er^{3+} in Er^{3+}/Yb^{3+} co-doped matrix under 973nm excitation is predominantly due to the resonant energy transfer from Yb^{3+} to Er^{3+} [17]. The multi-phonon relaxation rate of upconverted emissions is very sensitive to the phonon energy of the matrix [18] and thus, efficient upconversion luminescence is facilitated in low phonon energy matrices. The emissions from ^4F_J (J = 5/2, 7/2) states, i.e., the bands at 450 and 489 nm are hardly observed in high phonon energy matrix. The energy gap to the next lower levels from these ^4F_J levels is between 1000 to 1500 cm^{-1}. Therefore, de-excitation via multi-phonon relaxation to lower energy state is highly likely even in fluoride glasses that quenching of these emissions occurs. The presence of emission from ^4F_J levels in the present glass ceramics indicates that the Er^{3+} ions are present in a site with very low phonon energy of the LaF_3 lattice. IR emission spectrum of the sample in the wavelength range of 1400-1700nm is shown in Figure 8. The spectrum shows maxima at 1530nm with a broad shoulder at longer wavelength side. The emission is due to the $^4I_{13/2} \to {}^4I_{15/2}$ transition of Er^{3+} ions and is similar to that has been reported in alumina-silica matrix [19]. The FWHM of the 1.5µm band is about 65nm, which is even larger than that from aluminosilicate glasses [20]. Due to the size difference between La^{3+} and Er^{3+} ions, the Er^{3+} ions in LaF_3 matrix are experience a distorted crystal field that leads to broadening of emission at 1.5µm. This also supports the fact that Er^{3+} ions have been successfully partitioned into the LaF_3 nano-crystals.

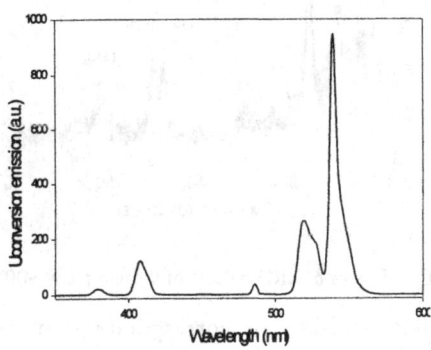

Figure 7. Upconversion spectrum of the sample

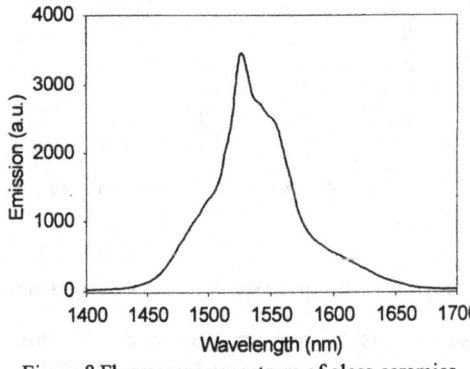

Figure 8. Fluorescence spectrum of glass ceramics

4.2.   NANOCOMPOSITES OF CDS:PVK FOR PHOTOREFRACTIVE EFFECTS

Doping of photoconductive (PC) or PR polymer composites with semiconductor nanocrystals, also known as quantum dots, leads to a new class of organic-inorganic hybrid materials.[21-24] Due to the fact that the physical properties of semiconductor nanocrystallites are dominated by quantum confinement, proper control of particle size is critical in any investigation involving these materials. In the current investigation, the surface of the nanocrystallites has been capped through the covalent addition of thiols, which not only serve as a source of sulfide ions but also as growth moderators.[25] In this case, a relatively small amount of nanocrystals are used (~ 1 wt.%) to photosensitize the polyvinylcarbazole (PVK) matrix, thus ensuring that the nanocrystals are isolated from each other and are responsible for charge generation while the polymer is responsible for charge transport.[26,27] Based on this knowledge, it can be assumed that the photoconductive mechanism is initiated with the absorption of a photon by a CdS cluster, which in turn gives rise to an electron-hole pair which is rapidly trapped on the surface of the particle. Given that the electron affinity of CdS exceeds that of PVK, the transfer of a hole from the nanocrystal to the polymer is energetically favorable.

In order to draw a quantitative comparison between the photocharge generation efficiency, Φ, of the PC polymer doped with CdS and the same polymer doped with C_{60}, Φ of this process for these two composite materials is charted as a function of electric field in figure 1. Buckminsterfullerene is used in this experiment because it is considered to be an exceptional photosensitizer for PVK.[28] However, it is evident in Figure 1 that when suitably sized CdS nanocrystals are employed, the photocharge generation quantum efficiency of the PVK:CdS-nanocomposite can be fashioned such that it significantly exceeds that of the PVK:C_{60} composite (above 0.2 wt % C_{60}, phase separation was observed). It is noted that for the aforementioned composite, the dark current density is approximately 0.05 % of the photocurrent density.

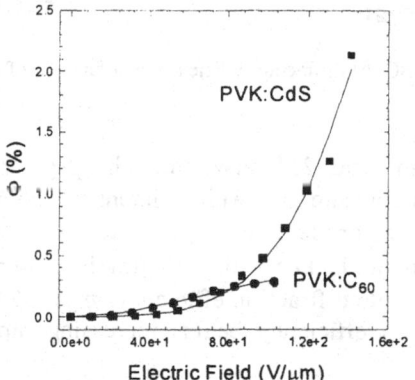

Figure 9. Photocharge generation quantum efficiency as a function of applied electric field at 514.5 nm for the PVK:CdS-nanocrystal (1 wt%) composite (squares) and the PVK:C_{60} (0.2 wt%) composite (circles).

The photorefractive properties of the PVK:NPP:TCP:CdS composite samples were studied via two wave mixing (TWM) and degenerate four wave mixing (DFWM) techniques using an oblique experimental setup. Figure 10 depicts the data obtained in the TWM portion of the experiment to verify intensity exchange between laser beams and quantify Γ for the PVK:TCP:NPP:CdS composite. For this experiment the writing beams, I_1 and I_2, with p-polarization and intensities of 227 mW/cm^2 and 83 mW/cm^2, were used while an external dc electric field E_0 = 107 V/μm was applied to a 168 μm thick sample at time t = 70 s and turned off at t = 190 s. A TWM gain coefficient of Γ =39.5 cm^{-1} was calculated for the data depicted in the figure; however, it is noted that when the sample was subjected to an external electric field of 119 V/μm, a TWM gain coefficient of Γ = 59.5 cm^{-1} was measured just prior to the sample experiencing dielectric breakdown. From a practical point of view, the optical amplification, Γ, must exceed the absorption loss, α, of the photorefractive sample in question.[29] In this case the optical absorption of the sandwiched sample (glass/ITO/polymer composite/ITO/glass) at 514.5 nm was measured to be α = 8.7 cm⁻¹, yielding a net gain coefficient, Γ - α, of 30.8 cm⁻¹ at 107 V/μm and 50.8 cm⁻¹ at 119 V/μm.

In the DFWM experiment the writing beams, I_1 and I_2, were s-polarized and had

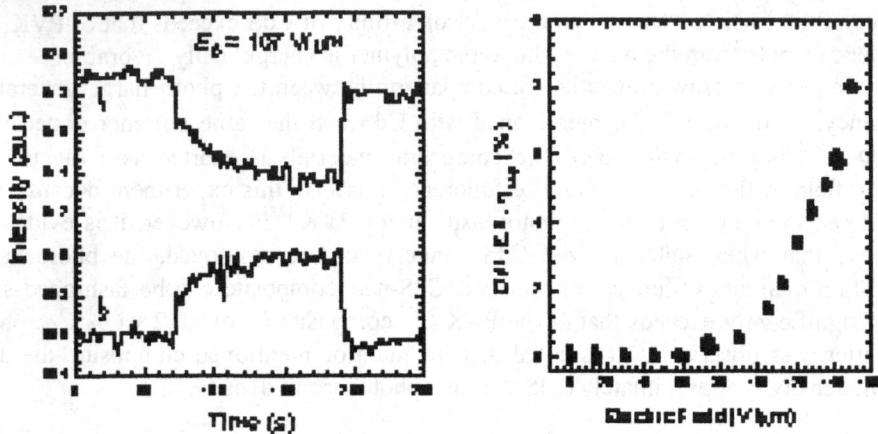

Figure 10. a) Asymmetric TBC, b) diffraction efficiency as a function of applied electric field in a DFWM experiment

intensities of 218 mW/cm² and 211 mW/cm². The p-polarized reading beam, I_r, propagated in the direction opposite to I_1 with an intensity of 5.6 mW/cm². The electric field dependence of the photorefractive DFWM steady state diffraction efficiency obtained in a sample with a thickness of 146 μm is presented in figure 2b. The maximum obtained steady state diffraction efficiency, η_{ssp}, is 8% at E_0 = 137 V/μm. It is noted that the diffraction efficiency begins increasing rapidly only after E_0 has exceeded 80 V/μm.

5.   Nanoscale optical memory with two-photon excitation

Nanoscale optical storage has been carried out on magneto-optic and phase-change media recently.[30-32] In our experiments, a spatially localized photochemical reaction, induced in a near-field geometry, is demonstrated with both one-photon and two-photon excitation on a nanometer scale and used for nanoscale optical memory and nanofabrication. The same self-mode-locked Ti:Sapphire laser is used as the excitation source at 800nm with an average power of 5.5 mW for two-photon excitation and frequency doubled at 400 nm with an average power of 3 mW for one-photon excitation. A dye-doped polymer film (AF-380/PVP) is prepared as the recording medium. Nanoscale domains result when the irradiated dyes are photobleached after being exposed for longer than 10 seconds. These areas show up black due to the loss of fluorescence, and are clearly distinguishable from the surrounding fluorescent areas.

This difference in fluorescence intensity is easily detectable and can be used to represent 1 and 0 in recording.

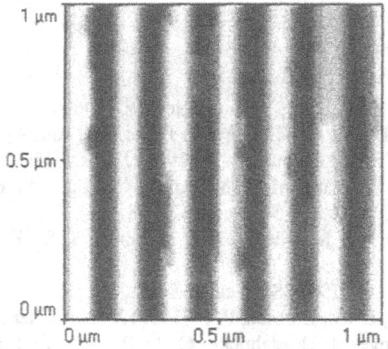

Figure 11. Fine grating structure fabricated with near-field two-photon excitation

The spatial localization of photobleaching depends on the probe size and the intensity distribution of the local field. With one-photon excitation, photobleaching is linear and occurs in the entire zone of illumination. Consequently, relatively large data bits form. However, this can be improved by nonlinear photobleaching with two-photon excitation, because the quadratic dependence of two-photon excitation on light intensity limits the effective excitation to a small volume and enhances the local field confinement in the vicinity of the probe. Figure 11 shows the image of the stripes recorded with two-photon excited photobleaching. The width of a written line is 80 nm, with adjacent line spacing of 160 nm. This result implies that a fine grating ruled with 6250 lines/cm is available with near-field optical technique.

Acknowledgement

We thank Dr. G. Maciel for his help. This work was supported by the Directorate of Chemistry and Life Science of the Air Force Office of Science Research through a University of Southern California MURI Program and in part by the Polymer Branch of the Air Force Research Laboratory at Dayton, Ohio.

Outlook

This new frontier of nanophotonics offers numerous opportunities for both fundamental research and application. From theoretical and experimental perspectives, further studies can be extended to other nonlinear optical processes, such as optical wave mixing and coherent anti-stokes Raman scattering (CARS). The information obtained can provide a better understanding of nanoscopic electrodynamics. From application perspective nonlinear near-field microscopy holds promise for high-contrast, high-resolution and site specific imaging. Nanofabrication and nanoscale patterning using nanoscale photochemistry will receive increasing attention. Photonic crystals offer prospects for numerous applications: low threshold lasing, optical power limiting, chemical and biosensing, and optical switching.

References

1. Jiang, Y.; Jakubczyk, D.; Shen, Y.; Swiatkiewicz, J.; Paras, P. N. (2000) Opt. Lett. 25, 640
2. Novotny, L.; Pohl, D. W.; Regli, P. (1994) J. Opt. Soc. Am. A 11, 1768.
3. Zyss, J.; Nicoud, J. F. and Coquillay, M. (1984) J. Chem. Phys. 81, 4160.
4. Phol, D. W.; Denk, W.; Lanz, M. (1984) Appl. Phys. Lett.. 44, 651.
5. Kopelman, R.; Tan, W. (1993) Science 262, 778.
6. Betzig, E.; Trautman, J. K.; Harris, T. D.; Weiner, J. S.; Kostelak, R. L. (1991) Science, 251, 1468.
7. Xie, X. S.; Dunn, R. C. (1994) Science, 265, 361.
8. Reid, P. J.; Higgins, D. H.; Barbara, P. F. (1996) J. Phys. Chem. 100, 3892.
9. Paesler, M. A.; Moyer, P. J.; Jahncke, C. J.; Jhonson, C. E.; Reddick, R. C.; Warmack, R. J.; Ferrell, T. L. (1990) Phys. Rev. B 42, 6750.
10. Courjon, D.; Sarayeddine, K.; Spajer, M. (1989) Opt. Comm. 71, 23.
11. Goudonnet, J. P.; Bourillot, E.; Adam, P. M.; Defornel, F.; Salomon, L.; Vincent, P.; Neviere, M.; Ferrell, T. L. (1995) J. Opt. Soc. Am. A 12, 1749.
12. Zhang, P.; Haslett, T. L.; Douketis, L.; Moskovits, M. (1998) Phys. Rev. B 57, 15513.
13. Shen, Y.; Jakubczyk, D.; Xu, F.; Swiatkiewicz, J.; Prasad, P.N.; Reinhardt, B. A. (2000) Appl. Phys. Lett. 76,1.
14. J. Ballato, R. E. Riman and E. Snitzer, (1997) Opt. Lett. 22, 6 91
15. W. Xu, G. Chen and J.R. Peterson, (1994) Chem. Phys. Lett. 224, 56
16. M. J. Weber, (1967) Phys. Rev. 157, 262
17. F. Auzel, (1993) Proc. IEEE 61, 758.
18. L. Wetenkamp, G.F. West and H. Többen, (1992) J. Non-Cryst. Solids 140, 35.
19. X. Orignac, D. Barbier, X.M. Du, R.M. Almeida, O. McCarthy and E. Yeatman, (1999) Opt. Mater. 12, 1.
20. B.J. Ainslie, (1991) IEEE. J. Lightwave Tech. 9, 220.
20. Wang, Y.; Herron, N. (1992) Chem. Phys. Lett. 200, 71.
21. Herron, N.; Wang, Y. US Patent 5,238,607.
22. Wang, Y. (1996) Photoconductive Polymers" in Kirk-Othmer Encyclopedia of Chemical Technology, Fourth Edition Wiley, New York.
23. Winiarz, J. G.; Zhang, L.; Lal, M.; Friend, C. S.; Prasad, P. N. (1999) J. Am. Chem. Soc.,121, 5287
24. Nosaka, Y.; Yamaguchi, K.; Miyama, H.; Hayashi, H. (1988) Chem. Lett. 605.
25. Wang, Y.; Herron, N. (1992) Chem. Phys. Lett. 92, 71.
26. Wang, Y. (1996) Pure and Appl. Chem. 68, 1475.
27. Moerner, W. E.; Silence, S. M. (1994) Chem. Rev, 94, 127.
28. Prasad, P. N.; Cui, Y.; Swedek, B.; Cheng, N.; Kim, K. S. (1997) J. Phys. Chem. B. ,101, 3530.
29. Betzig, E.; Trautman, J. K.; Wolfe, R.; Gyorgy, E. M.; Finn, P. L.; (1992) Appl. Phys. Lett. 61, 142.
30. Hosaka, S.; Shintani, T.; Miyamoto, M.; Hirotsune, A.; Terao, M.; Yoshida, M.; Fujita, K.; Kammer, S.; (1996) Jpn. J. Appl. Phys. 35, 443.
31. Terris, B. D.; Mamin, H. J.; Rugar, D.; (1996) Appl. Phys. Lett. 68, 141.

THREE-DIMENSIONAL NANOSTRUCTURES WITH ELECTRON AND PHOTON CONFINEMENT

S. V. GAPONENKO
Institute of Molecular and Atomic Physics
National Academy of Sciences, Minsk, 220072 Belarus
gaponen@imaph.bas-net.by

Abstract

We consider evolution of matter from isolated nanocrystals to quantum dot solids and from microcavities to photonic solids. A possibility of simultaneous electron and photon confinement in mesoscopic structures is considered, e.g. quantum dot in a microcavity and quantum dot in a photonic crystal. Colloidal crystals with self-organisation on nanometer to micrometer scale are shown as the suitable mesoscopic structures to trace these effects experimentally and to design artificial matter with engineering of optical and electronic properties.

1. Introduction

Electron states in bulk solids and nanostructures differ from those relevant to free space because of the translational symmetry of a given crystal lattice and of the specific boundary conditions determined by confinement geometry. Therefore electron density of states and optical transition probabilities can be controlled in nanostructures by means of quantum confinement effects. Propagation of electromagnetic waves in media other than vacuum can be significantly altered as well by means of periodic spatial modulation of the dielectric constant and by spatial restriction. This leads to a concept of photonic band engineering including photonic crystals and photonic quantum boxes (microcavities). Because of the large difference in the characteristic length scales relevant to electrons and photons, i. e. an electron de Broglie wavelength in solids (~10 nm) and an optical photon wavelength (~500 nm), electron and photon densities of states can be engineered separately within the same mesostructure. Furthermore, since spontaneous emission of photons by quantum systems is proportional to the density of photon states, it is possible to create mesoscopic structures with complete control of spontaneous emission spectrum and decay rate by means of the proper electron and photon confinement geometries. The examples of such structures are nanocrystals in a

11

L. Pavesi and E. Buzaneva (eds.), Frontiers of Nano-Optoelectronic Systems, 11–22.

12

microcavity and nanocrystals embedded in a three-dimensional photonic crystal. In this paper we consider properties of isolated and assembled semiconductor nanocrystals and advances in synthesis and studies of mesoscopic systems (microcavities and photonic crystals) with simultaneous electron and photon confinement.

2. Isolated Nanocrystals

In nanostructures, quantum size effect results in size-dependent modification of band structure. Among nanostructures, nanocrystals with three-dimensional confinement are most interesting objects. Nanocrystals possess intermediate properties between small-atomic clusters and bulk crystals. Their electronic and optical properties can be described in a large range of sizes (approximately 2...50 nm) in term of size-dependent relationships. Because of the important role of quantum-size effects and three-dimensional confinement, nanocrystals are often referred to as *quantum dots*. The latter are considered as small bits of matter whose properties smoothly evolve toward parent bulk solid with increasing size. Optical and electron properties of quantum dots are the subject of several recently published books[1,2,3,4,5].

Confinement effects on electron properties of nanocrystals include size-dependent energy spectra, probabilities of optical transitions, specific non-radiative recombination dynamics, many-body interactions and optical gain spectrum. Because of the finite number of electron-hole pair excitations in every nanocrystal, the concepts of electron-hole gas and plasma are not valid and optical absorption, emission and gain processes are to be described in terms of creation and decay of many-particle states.

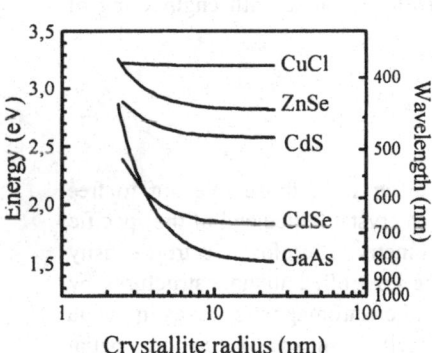

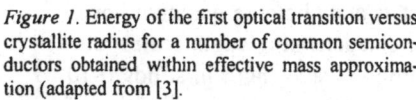

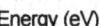

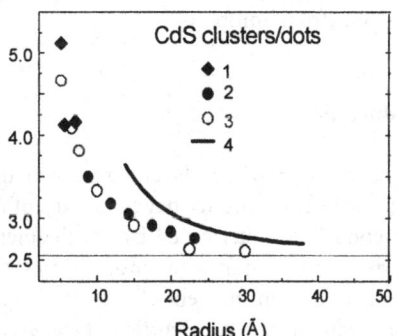

Figure 1. Energy of the first optical transition versus crystallite radius for a number of common semiconductors obtained within effective mass approximation (adapted from [3].

Figure 2. Energy of the first optical transition versus radius for CdS clusters and nanocrystals by means of (◆)CNDO/S-technique; (●)tight-binding approach; (O)pseudopotential method + perturbational approach; solid line – effective mass approximation (adapted from [3]).

Representative data on size-dependent resonant optical absorption (HOMO–LUMO transition) for a number of common semiconductors are given in Figs. 1,2. For semiconductor materials with smaller band gap (and larger excitonic Bohr radius) like, e.g. GaAs, CdTe, CdSe variation in size results in a tunable absorption onset over the whole visible spectrum.

3. From Isolated Nanocrystals to Quantum Dot Solids

Nanocrystals can be arranged in spatially organized ensembles by means of various techniques[3,6,7]. In case of dense ensembles of nanocrystals it is reasonable to ask: Why should not we try to fabricate artificial materials using nanocrystals as building blocks instead of atoms and molecules? Crystallization of matter on the supramolecular level is known and being actively investigated in physics, chemistry and material science. Self-organized crystal structures built of colloidal particles are referred to as *colloidal crystals*. Since the first identification of natural colloidal crystals, namely a specific type of virus, the features of colloidal crystals were found to be inherent in a number of natural and artificial objects (ref. [8] and refs. therein). With respect to nanocrystals self-organized to form a macroscopic colloidal crystal, a term *"quantum dot solid"* has been introduced. The latter concept proposed by Murray et al. [9] implies a kind of condensed matter with spatial organization on a scale comparable to the electron de Broglie wavelength. The basic properties of dense quantum dot ensembles might be expected to reproduce the features inherent in conventional solids, i.e. formation of energy bands in a perfect lattice and a coexistence of localized and delocalized electron states in a disordered quantum dot structures (Fig. 3). By analogy with multiple quantum wells which form a planar superlattice one can consider quantum dot solid as three-dimensional superlattice formed by quantum dots.

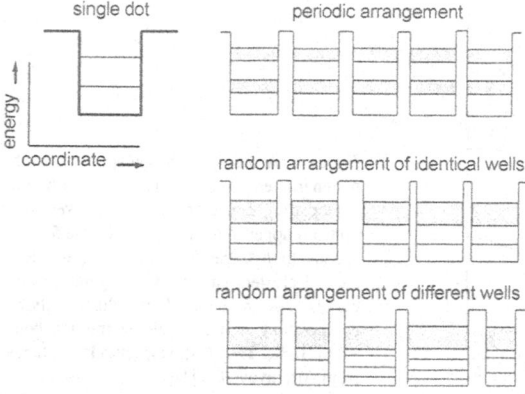

Figure 3. A sketch of electron levels in an isolated quantum well and in periodically and randomly arranged ensembles of identical and different wells.

14

Quantum dot solid is a new type of mesoscopic solid state structure whose electron and optical properties are controlled by electron confinement in every nanoparticle and collective effects from spatial organization of nanocrystals. The systematic analysis of quantum dot aggregates and superlattices in terms of an evolution from isolated nanocrystals to a quantum dot solid has not been performed to date. Studies of electronic energy transfer in CdSe quantum dot solids[9] revealed long-range resonance transfer of electronic excitation from the more electronically confined states of the small dots to the higher excited states of the large dots. Foerster theory for long-range resonance transfer through dipole-dipole interdot interaction was used to explain electronic energy transfer in these close-packed nanocrystal structures. However absorption spectrum of these structures remains unchanged.

It is reasonable to consider build-up of collective electron states over a macroscopically large number of nanoparticles in close-packed quantum dot structures. Recently we found that the absorption spectrum of a quantum dot ensemble exhibits a systematic modification with dot concentration from a set of discrete subbands inherent in isolated nanocrystals to a smooth band-edge absorption similar to that of bulk semiconductors[10]. The results are interpreted in terms of an evolution from individual (localized) to collective electron states delocalized within at least a finite number of nanocrystals.

We investigated ensembles of smaller CdSe nanoparticles of 1.6 nm diameter capped with organic groups. Concentration of particles varied by means of dissolution in a polymer film. Optical absorption spectra of the samples investigated show reversibly a successive evolution from a cluster-like spectrum with finite number of discrete absorption subbands to a smooth nearly structureless absorption spectrum. The onset exhibits a monotonous shift to the low-energy side (Fig. 4).

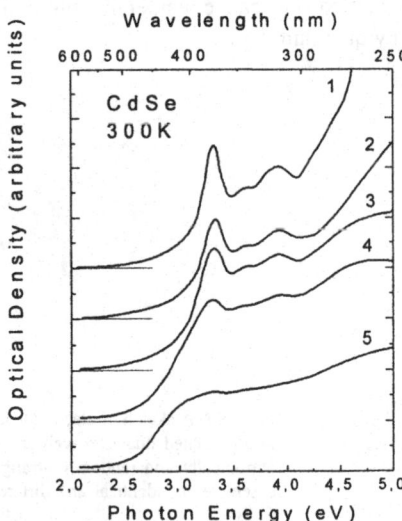

Figure 4. A set of absorption spectra of CdSe nanoparticles, average radius is 1.6 nm. Nanoparticle concentration progressively rises up from upper to lower curves. Curve 5 corresponds to a CdSe film consisting of close-packed clusters capped with organic groups, curves 1-4 correspond to cluster/polymer composition, polymer volume fraction being 37%, 18%, 3%, 1%, respectively. Adapted from Artemyev et al [10].

A systematic modification has been also observed in photoluminescence excitation (PLE) spectra. Not only PLE spectral shape evolves with concentration but also inhomogeneous broadening inherent in diluted quantum dot systems diminishes[11]. Note that reversible long-wave shift of absorption spectrum with growing concentration was reported but not interpreted by other authors [12].

The effects of systematic modification of absorption and excitation spectra with increasing concentration of nanocrystals gain a reasonable explanation in terms of collective phenomena in a system of quantum dots. In the case of a perfect one-, two-, or three-dimensional periodic lattice of identical quantum dots, one can expect the formation of mini-subbands from a discrete set of electron (and hole) levels. This effect is similar to the formation of minibands in one-dimensional superlattice of quantum wells. Noteworthy, in the case of quantum dots a three-dimensional superlattice can be developed. Such a superstructure has no analog among the other nanostructures and offers, in principle, a new type of solids with a three-dimensional band structure on a superatomic scale. Because no long-range order occurs in the structures investigated, we consider the electron properties of disordered solids as a relevant example. In disordered solids, at certain degree of disorder a transition occurs from localized electron states to coexisting delocalized and localized ones separated by a mobility edge. The effect is known as the Anderson transition since 1958. It is extensively examined in disordered solids[13] and is being investigated with respect to classical waves like electromagnetic and acoustic waves.

Coexisting discrete (localized) and band-like (delocalized) states in randomly arranged identical and different wells is depicted schematically in Fig.3 for the case of a one-dimensional arrangement. In the case of a one-dimensional system such a structure possesses localized electron states only. However, in two- and three- dimensional systems, and for certain parameters the Anderson transition occurs giving rise to delocalization of the electron wave function and conductivity within a percolation cluster. This consideration remains valid for quantum dots and can explain the concentration dependence of the absorption spectrum. A development of electron (and hole) bands result in appearance of wide absorption band related to higher delocalized electron and lower hole states along with discrete absorption lines due to transition to lower (localized) electron from higher hole states. At very high concentration, a contribution from localized states may be negligible similar to impurity semiconductors and a bulk-like structureless absorption spectrum develops.

This qualitative consideration has been proved by numerical calculations[10] which show that delocalization more readily occurs in close-packed smaller crystallites whereas in the case of larger dots, even packing of particles up to a volume fraction of 0.5 does not result in a noticeable fraction of delocalized states. This result provides an explanation, for the lack of observation of any modification of absorption spectra with increasing concentration in Ref. [9] where larger CdSe nanocrystals were examined. Taking into account the presence of an organic shell in the experiments, even close-packed ensembles of large crystallites may not satisfy the delocalization condition. Therefore, close-packed large CdSe quantum dots are more likely molecular solids with resonant long-range energy transfer. However, for smaller dots a noticeable delocalization occurs at concentration far from close-packing. For dot radius of 1 nm, 75% of electron states were indeed found to be delocalized at volume fraction of 0.1. On the contrary, a fraction of 0.52 and 0.74 was delocalized for close-packed solid spheres in a simple and face-centered cubic lattice, respectively. Therefore, close-packed small quantum dots reproduce, to a large extent, the properties of atomic solids, including not only energy and charge transfer but also the formation of collective energy states.

These findings are believed to pose a number of issues related to electron properties of quantum dot superlattices like, e.g., electron band structure of a quantum dot supercrystal, renormalized electron and hole effective masses within a superstructure, modified electron-hole interaction (super-exciton) and other. One can foresee new prospects in synthesis of mesoscopic structures whose electron and optical properties are controlled by three-dimensional confinement of electrons within every nanocrystal and by collective phenomena due to spatial organization of nanocrystals.

4. Photonic Crystals

The concept of quantum dot solid is relevant to the case of a three-dimensional super-structure with individual particles of size comparable to the de Broglie wavelengths of an electron and a hole. Even more interesting with respect to its optical properties it seems to be a three-dimensional superstructure consisting of particles of the order of the photon wavelength, i.e. 10^2-10^3 nm when the optical range is considered.

Many properties of solids which do not imply electron-electron interactions are due to the wave properties of quantum particles. These include band gap formation, weak localization, Anderson localization, effective mass and others. These phenomena should occur in case of classical waves as well. With respect to electromagnetic waves this consideration leads to the concept of *photonic solids*. Disordered dielectric structures with high concentration of particles possess multiple wave scattering and exhibit weak localization of light waves known as coherent back scattering. High concentration of particles with high refraction index can lead to a complete localization of optical waves similar to Anderson localization of electrons [14].

Mesoscopic structures with a *periodic* space modulation of dielectric function with a period of the order of photon wavelength are referred to as *photonic crystals*. Light propagation and photon density of states are significantly modified when alternating layers, needles or balls are arranged in a one-, two-, or three-dimensional lattice, respectively. One can see that one-dimensional photonic crystal is nothing else but a well-known dielectric mirror with alternating $\lambda/4$ layers. Because of the Bragg condition, for some wavenumbers (namely, $k_n = 2\pi n/\lambda$) the standing waves form and thus they are reflected backward because they cannot propagate throughout the medium.

This is shown schematically in Fig. 5. In a vacuum or in a homogeneous medium, the dispersion relation $E(p)$ for photons follows the straight line (Fig. 5(a)). In an inhomogeneous, e.g. stratified medium $E(p)$ relation is non-linear. The dispersion curve of photons in a periodic medium with alternating layers of the same thickness, $a/2$ but different refraction indices, n_1 and n_2, has a forbidden gap at the wavenumber

$k_m=m\pi/a$, where m is an integer. At these values standing waves occur and no propagating waves are possible. The larger is the n_1/n_2 ratio the wider is the forbidden gap. Because of the periodicity of the medium, wavenumber values differing by $2m\pi/a$ are equivalent and $E(p)$ can be plotted in a reduced form (Fig. 5 c). The interval of wavenumbers $\left[-\frac{\pi}{a};+\frac{\pi}{a}\right]$ is the first Brillouin zone for photons.

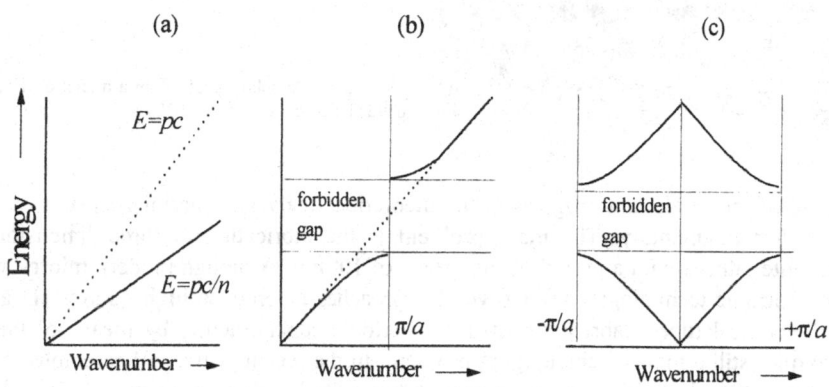

(a) (b) (c)

Figure 5. Formation of photonic bands in a one-dimensional periodic structure. (a) Dispersion curve of photons in vacuum (dashed line) and in a continuous medium (solid line); (b) Dispersion curve of photons in a periodic medium with alternating layers; (c) Reduced dispersion curve within the first Brillouin zone.

Photonic band gap structures offer a possibility to control the spontaneous emission rate by means of the control of the photon density of states. This becomes possible since the spontaneous emission of light by atoms, molecules and solids is not an intrinsic property but the result of their interaction with the zero field electromagnetic oscillations. The rate of spontaneous emission is directly proportional to the photon density of states. In free space, the photon density of states follows an ω^2 dependence. This corresponds to the electromagnetic vacuum. When free space is replaced by an inhomogeneous medium, the photon density of states changes resulting in a modified spontaneous emission rate and spectrum. The phenomenon has been predicted for the first time in 1947 by E.M. Pursell[15] and is currently and extensively investigated in a number of artificial and natural mesoscopic structures: microcavities, dielectric slabs, interfaces, water-in-oil micelles, phospholipid bilayers, and biological membranes. In the context of the topic under consideration, the experiments on semiconductor microcavities should be outlined. These includes three-dimensionally confined semiconductor microcrystals (*photonic quantum dots*) [16], semiconductor quantum wells in a planar Bragg microcavity[17], quantum dots in a planar microcavity[18], and quantum dots in a dielectric microsphere [19].

V.P. Bykov [20] was the first to outline the possibility of the frozen "excited atom + field" state when a periodic medium offers no propagating mode at the resonance

18

frequency. This was followed by a systematic development of the concept of photonic crystals[21,22,23,24] and experimental investigations in this field.

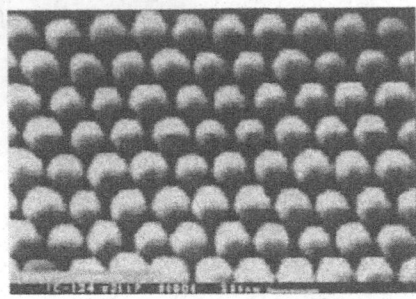

Figure. 6. Microphotograph of an artificial opal sample[31]. Particle size is about 250 nm.

In spite of the significant progress in the theoretical analysis, experimental studies are still rather fragmentary. The main problem is the fabrication of three-dimensional solid-state lattices with a period of the order of 10^2 nm. Although modern micro- and nanoelectronic technologies offer several approaches to produce high-quality 1D and 2D dielectric lattices, fabrication of a 3D periodic superstructure by means of these techniques still remains a challenging problem. In the recent years, self-assembled colloidal structures have been proposed as prototypes for 3D photonic crystals [25]. Solid-state silica colloidal crystals known as gem artificial opals have been considered then as promising prototype structures for the synthesis of solid state photonic crystals for the visible and near-infrared range[26,27]. These findings stimulated extensive studies of optical properties of colloidal crystals, their use as templates for fabricating their high-refractive 3D replicas, and investigations of the spontaneous emission of dye molecules and semiconductor nanoparticles embedded in colloidal crystals and their replicas[Ref. 28 and refs therein].

Gem opals are known to consist of spatially arranged silica microspheres organized either in cubic or hexagonal lattices[29,30]. Iridescent properties of natural and artificial opals are due to interference in a spatially periodic structure.

Typical microphotograph of the artificial opal is presented in Fig. 6 [31]. Close-packing of monodisperse submicron globules in a face-centered cubic lattice is evident. Noteworthy, every globule has its internal structure. It consists of smaller SiO_2 nanoparticles. All the lattices show a dip in the optical transmission spectrum, with the spectral position depending on the lattice period and propagation direction. The nature of spectrally selective transmission of a disperse medium with vanishing dissipation is nothing else but multiple scattering and the interference of light waves. This can be intuitively understood in terms of the Bragg diffraction of optical waves. In other words, formation of a pronounced stop band is indicative of a reduced density of photon states within the sample, and thus can be classified as a photonic pseudogap phenomenon. By analogy with solid crystals, the structure under consideration resembles features of a gapless semiconductor or a semimetal. The gap does exist but it is not omnidirectional.

The observed spectrally-selective characteristics of closed-packed SiO$_2$ microspheres were succesfully reproduced by means of the multiple wave scattering theory using quasi-crystalline model. The model implies a finite number of equally spaced monolayers, each consisting of identical close-packed spherical particles[31]. On the assumption of the statistical independence of the individual monolayers, it is possible at first to find the scattering amplitude of a single monolayer taking into account multiple scattering of the particles within the layer and then to account for the irradiation between the different monolayers within the samples under consideration.

The voids between close-packed microspheres form a sublattice which can be impregnated with a liquid or polymer. This provides a possibility to increase the band gap due to enhancement of the refraction index modulation. In a similar manner, resonant species like molecules and nanocrystals can be embedded as well. The first experi-

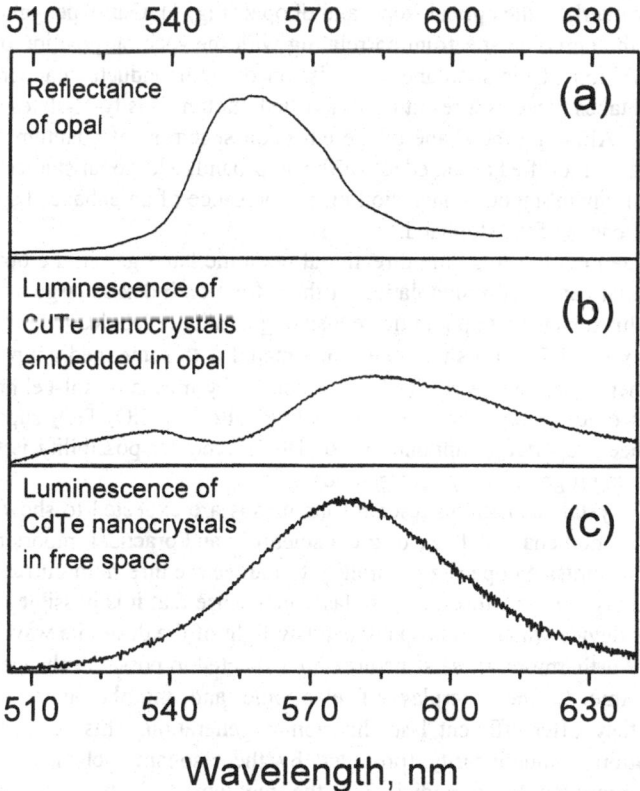

Figure 7. Modification of the spontaneous emission of CdTe nanocrystals embedded in opal[33]. Nanocrystal mean diameter is 2.4 nm. (a) Optical reflection spectrum of opal sample; (b) modified spontaneous emission spectrum of CdTe nanocrystals in opal; (c) reference emission spectrum of CdTe nanocrystals in free space.

ments on dye molecules in opals show a pronounced effect of the modified photon density of states on spontaneous emission[32]. Because of the essentially different electron and photon wavelengths, quantum dots can be embedded as isolated particles in a photonic crystal. Overlap of electron and structural resonances can be performed by means of the control of the size of semiconductor and silica colloidal particles. Preliminary experiments on the modified spontaneous emission of II-VI nanocrystals in opal-based photonic crystals have been reported by a number of authors[33,34,35,36]. Noteworthy, a photonic crystal doped with quantum dots is an example of a mesoscopic structure with simultaneous confinement of electrons and photons.

We experimentally overlapped size-dependent electron resonance of nanocrystals with the structural optical resonance of a photonic crystal by filling an opal with an aqueous colloidal solution of CdTe nanoparticles. Due to the size restriction, the absorption spectrum of CdTe nanocrystals shifts by more than 1 eV compared to the bulk (from 827 nm in bulk CdTe to 460–500 nm in nanocrystals). Quantum dots embedded in a photonic crystal exhibit a noticeable change in the luminescence spectrum when the latter overlaps the optical stop-band of opal (Fig. 7). Our exper-iments clearly show a dip in the emission spectrum correlating with the spectral position of the stop-band, i.e., inhibition of the spontaneous emission of semiconductor nanocrystals embedded in opal takes place as a result of a modified photon density-of-states within this spectral region. Although the shape of the emission spectrum of quantum dots in the photonic crystal is modified at the edges of the stop-band, additional studies are necessary to make an unambiguous conclusion on the presence of an enhanced spontaneous emission at the edges of the stopband.

Further attempts towards omnidirectional photonic band gap in the optical range are necessary that can offer a modulation of the refraction index as large as 100% and even higher. This means that opal lattice consisting of silica globules with the effective refraction index $n \approx 1.2 ... 1.3$ should be impregnated with some medium possessing $n \approx 2.5$ and higher. This can be realized, in principle, by means of sol-gel processes to get amorphous oxides. Recently a successful fabrication of SiO_2/TiO_2 superlattice in this way has been reported (Kapitonov et al. 1997). Another possibility is to develop opal replicas using high refractive compounds.

Photonic crystals containing resonant inclusions are expected to show a number of interesting phenomena which are of great scientific and practical importance. Inhibited spontaneous emission opens a possibility to reduce the threshold current in lasers. In principle it may lead to a thresholdless laser in a sense that it is possible to fabricate a thresholdless device which emits spontaneously light of the desirable wavelength and directionality. Furthermore, these structures are expected to possess enhanced resonant nonlinearities due to the interplay of electronic and morphological resonances. Additionally, they offer efficient laser harmonics generation. This becomes possible due to separation of nonlinearity (provided by the resonant inclusions) and phase synchronism (provided by periodicity of the medium), which is not possible in conventional media.

5. Conclusion

Quantum dot solids and photonic solids are examples of structural organization of matter on the mesoscopic scale. At present, colloidal crystals with a single colloidal particle on the order of electron de Broglie wavelength or photon wavelength resemble the properties of either a quantum dot solid or a photonic solid. Quantum dots can be embedded as isolated particles in a photonic solid with overlap of electron and structural resonances. Therefore colloidal structures with multi-level mesoscopic spatial organisation and interplay of electronic and morphological resonances offer fascinating ways to design and develop artificial matter with novel electronic, optical and optoelectronic properties for various applications.

Acknowledgements

The studies reviewed in the paper have been supported by Volkswagen Shiftung, ISTC, and Belarussian basic research foundation. The author acknowledge joint research and discussions with V. N. Bogomolov, U. Woggon, M. V. Artemyev, E. P. Petrov, A. N. Ponyavina, A. Eychmueller, A. L. Rogach, and I. I. Kalosha. Helpful assistance of A. M. Kapitonov, A. V. Prokofiev, and N. I. Silvanovich are also acknowledged.

References

1. Banyai, L. and S. W. Koch, S.W. (1993) *Semiconductor Quantum Dots,* World Scientific Singapore.
2. Woggon, U. (1996) *Optical Properties of Semiconductor Quantum Dots,* Springer-Verlag, Berlin.
3. Gaponenko, S.V. (1998) *Optical Properties of Semiconductor Nanocrystals,* Cambridge University Press, Cambridge.
4. Bimberg, D., M. Grundman,M., and Ledentsov, N. (1999) *Quantum Dot Heterostructures,* Wiley and Sons, London.
5. *Spectroscopy of Isolated and Assembled Nanocrystals* (1996) Eds. Brus, L., Efros, Al., T. Itoh, T. Special issue of the J. Luminescence, v. 70.
6. Bogomolov, V.N., and Pavlova, T.M. (195) Three-dimensional cluster lattices. *Semiconductors,* **29**, 428–435.
7. G. Romanov and C. Sotomayor Torres, Three-dimensional lattices of nanostructures: A template approach, in Handbook of Nanostructured Materials and Nanotechnology, Ed. H.S. Nalwa, Orlando:Academic Press 2000, pp. 231–323.
8. Pieranski, P. (1983) Colloidal crystals, *Contemp. Physics* **24**, 25-73.
9. Murray, C.B., Kagan, C.R., Bawendi, M.G. (1995) Self-organization of CdSe Nanocrystallites into three-dimensional quantum dot superlattices. *Science* **270**, 1335–1338.
10. Artemyev, M.V., Bibik, A.I, Gurinovich, L.I, Gaponenko, S.V., and Woggon, U. (1999) Evolution from individual to collective electron states in a dense quantum dot ensemble. *Phys. Rev.* B **60**, 1505-1507.
11. Artemyev, M.V., Woggon, U., and Gaponenko, S.V. (2000). Optical properties of dense quantom dot structures. Jap. J. Appl. Phys. (in press).
12. Vossmeyer, T., Katsikas, L., Giersing, M., Popovic, I.G., Diesner, K., Chemseddine, A., Eychmuller, A., and Weller, H. (1994) CdS Nanoclusters: Synthesis, Characterization, Size Dependent Oscillator Strength, Temperature Shift of the Excitonic Transition Energy, and Reversible Absorbance Shift.: *J. Amer. Chem. Societ,* **98**, 7665-7672.
13. Shklovskii, B.I., and Efros, A.L., (1982) *Electron Properties of Doped Semiconductors,* Springer-Verlag, Berlin.

14. John, S. (1987) Strong localization of photons in certain disordered dielectric superlattices, *Phys. Rev. Lett.* **58**, 2486-2487.

15. Purcell, E.M. (1946). Spontaneous emission probabilities at radio frequencies. *Phys. Rev.* **69**, 681-687.

16. Ohnesorge, B., Bayer, M., Forchel, A., Reithmaier, J.P., Gippius, N.A., and Tikhodeev, S.G. (1997) Enhancement of spontaneous emission rates by three-dimensional photon confinement in Bragg microcavities. *Phys. Rev.* B **56**, R4367 - 4370.

17. Yamanishi, M. (1995) Combined quantum effects for electron and photon systems in semiconductor microcavity light emitters. *Progress in Quant. Electron.* **19**, 1-39.

18. Pellegrini, V., Tredicucci, A., Mazzoleni, C., and Pavesi, L. (1995) Enhanced optical properties in porous silicon microcavities. *Phys. Rev.* B **52**, R14328-R14331.

19. Artemyev, M. V., and Woggon, U. (2000) Quantum dots in photonic dots. Appl. Phys. Lett. **76**, 1353-1355.

20. Bykov, V. P. (1972) Spontaneous emission in a periodic structure, *Zh. Eksp. Teor. Fiz.*, **62**, 505-513.

21. Yablonovitch, E. (1987) Inhibited spontaneous emission in solid-state physics and electronics, *Phys. Rev. Lett.*, **58**, 2059-2062.

22. John, S. (1987) Strong localization of photons in certain disordered dielectric superlattices, *Phys. Rev. Lett.*, **58**, 2486-2489.

23. Bykov, V. P. (1993) *Radiation of Atoms in a Resonant Environment.* World Scientific, Singapore.

24. Joannopoulos, J.D., Meade, R.D., and Winn J. N. (1995) *Photonic Crystals: Molding the Flow of Light*, Princeton University Press, Princeton.

25. J. Martorell, J. and N. M. Lawandy, N.M. (1990) Observation of inhibited spontaneous emission in a periodic dielectric structure, *Phys. Rev. Lett.*, **65**, 1877-1880.

26. Astratov, V. N., Bogomolov, V. N., Kaplyanskii, A. A., Samoilovich, S. M., and Vlasov, Yu. A. (1995) Optical spectroscopy of opal matrices with CdS embedded in its pores: Quantum confinement and photonic band gap effects, *Il Nuovo Cimento*, **17**, 1349-1354.

27. Bogomolov, V. N., Gaponenko, S. V., Kapitonov, A. M., Prokofiev, A. V., Ponyavina, A. N., Silvanovich, N. I., and Samoilovich, S. M. (1996) Photonic band gap in the visible range in a three-dimensional solid state lattice, *Appl. Phys. A* **63**, 613-616.

28. Gaponenko, S. V., Bogomolov, V. N., Petrov, E. P. , Kapitonov, A. M., Yarotsky, D. A., Kalosha, I. I., Eychmueller, A. A., Rogach, A. L., McGilp, J., Woggon, U., and Gindele, F. (1999) Spontaneous Emission of Dye Molecules, Semiconductor Nanocrystals, and Rare-Earth Ions in Opal-Based Photonic Crystals, *J. Lightwave Technol.* **17**, .

29. P. Pieranski, P. (1983) Colloidal crystals, *Contemp. Phys.*, **24**, 25-53.

30. Deniskina, N. D., Kalinin, D. V. , and Kazantseva, L. K. (1988) *Gem-Quality Opals, Their Synthesis and Genesis in Nature.* Novosibirsk: Nauka,

31. Bogomolov, V. N., Gaponenko, S. V., Germanenko, I. N., Kapitonov, A. M., Petrov, E. P., Gaponenko, N. V., Prokofiev, A. V., Ponyavina, A. N., Silvanovich, N. I., and Samoilovich, S. M. (1997) Photonic band gap phenomenon and optical properties of artificial opals, *Phys. Rev.E* **55**, 7619-7625.

32. Petrov, E. P., Bogomolov, V. N., Kalosha, I. I., and Gaponenko, S. V. (1998) Spontaneous emission of organic molecules in a photonic crystal, *Phys. Rev. Lett.* **81**, 77-80; (1999) *ibid.* **83**, 5401-5402.

33. Gaponenko, S. V. , Kapitonov, A. M., Bogomolov, V. N., Prokofiev, A. V. , Eychmuller, A., and Rogach, A. L. (1998) Electrons and photons in mesoscopic structures: Quantum dots in a photonic crystal, *JETP Lett.*, **68**, 142-147.

34. Romanov, S. G. , Fokin, A. V. , Alperovich, V. I., Johnson, N. P. , and De La Rue, R. M. The effect of the photonic stop-band upon the photoluminescence of CdS in Opal, *Phys. Stat. Sol. A*, **164**, 169-173.

35. Blanco, A., Lopez, C. , Mayoral, R., Migues, H. , Meseguer, F., Mifsud, A., and J. Herrero (1998) CdS photoluminescence inhibition by a photonic crystal, *Appl. Phys. Lett.*, **73**, 1781-1783.

36. Romanov, S. G., Fokin, A. V. , Tretyakov, V. V. , Butko, V. Y. , Alperovich, V. I., Johnson, N. P., Sotomayor Torres, C. M. (1996) Optical properties of ordered three-dimensional arrays of structurally confined semiconductors *J. Cryst. Growth*, **159**, 857-860.

DIELECTRIC-POLYMER NANOCOMPOSITE AND THIN FILM PHOTONIC CRYSTALS:
Towards Three-Dimensional Photonic Crystals with a Bandgap in the Visible Spectrum

C. M. SOTOMAYOR TORRES, T. MAKA, S. G. ROMANOV
Dept. of Electrical and Information Engineering

MANFRED MÜLLER, RUDOLF ZENTEL
Dept. of Chemistry

Institute of Materials Science
University of Wuppertal
42097 Wuppertal, Germany

1. Introduction

The concepts of photonic crystals and photonic bandgap (PBG) were introduced in the in late 80s [1] and in first approximation they are the optical analogous to the semiconductor and the energy gap for electrons. They are based on periodic structures formed of two or more materials, which exhibit a periodicity in real space of the dielectric function. Moreover, the refractive index contrast has to be larger than certain value, depending on the relative fill factor of the respective materials. Thus, diffraction phenomena, such as Bragg difraction, take place. There are four main aspects currently motivating research in photonic crystals: (1) The realisation of a full PBG across the electromagnetic spectrum to act, for example, as highly efficient mirrors, optical limiters and optical switches. (2) The partial realisation of the PBG for waveguiding or for "moulding" the light, which is known as photonic crystal waveguides, as opposed as refractive index waveguiding. (3) The realisation of highly efficient light sources by making use of: (i) the strong coupling in a 3-dimensional (3D) optical cavity formed by the PBG crystal and, (ii) the enhanced spontaneous emission rate of a "defect", the emission wavelength of which is highly defined spectroscopically and falls within the band-gap of the photonic crystal. (4) The use of the dispersion relations represented by the photonic band structure for ultra-refractive phenomena, super prism, spot focusing, super lens, among others, based on the changes of the group velocity in different parts of the Brillouin zone. This work attempts to address the third issue and to some degree the fourth.

However, in order to take advantage of PBG structures, light emitting devices have to be designed as a spatially extended light source integrated with a 3D photonic crystal. There are two conditions to be met for a successful design: the light source should have a well-defined emission spectrum in free space and the refractive index (RI) contrast of the photonic template should be high. In the visible part of the spectrum, photonic crystals made from colloids probably represent the most promising approach, because all-semiconductor inverted opals can, in principle, provide the

23

L. Pavesi and E. Buzaneva (eds.), Frontiers of Nano-Optoelectronic Systems, 23–39.
© 2000 *Kluwer Academic Publishers. Printed in the Netherlands.*

required refractive index contrast (RIC) [2]. However, these structures can hardly be used as PBG light sources themselves since, if inter-band transitions in the semiconductor match the stop-band frequency of the photonic crystal, the requirement of a purely real dielectric function will be broken at the upper frequency edge of the stop-band. Therefore, materials suitable to demonstrate the feasibility of PBG approach have to separate physically the PBG effect and the light emission functions into at least two independent material components.

A straightforward solution of the two-component target material is a colloidal crystal with an efficient luminescent material incorporated in its interstitial voids [3,4]. Dye solutions of different kinds represent examples of proper infill material [5,6,7]. Obviously, such composite is far from approaching a high RIC, but it offers excellent flexibility to match the stop-band with the emission band of the dye without raising dramatically the imaginary part of the dielectric function. Considering opal matrices, these can be modified prior to dye infilling in order to increase the RIC [8] or to preserve air in the opal voids as the low-RI background. However, none of these designs have been implemented so far for dye-filled opals. Therefore, all dye-opal composites studied so far possess a low RI contrast. An immediate consequence of the low contrast is a highly anisotropic PBG structure. Taking this anisotropy into account, angular resolved spectroscopy [9,10] can be used to find the PBG effect in a particular direction of the photonic crystal and to study the change of the spontaneous emission.

In this work silica and PMMA opal matrices infilled with a dye dissolved in a polymer, as well as inverse structures are used. Preliminary results using SnS_2 are also presented. By detecting the outgoing light within a small fraction of the solid angle we were able to trace the variation of the emission rate when the stop-band is probing different parts of the dye emission spectrum. The aim of this paper is to demonstrate the impact of the opal structure upon the emission of dye-polymer solutions. The fabrication of a "defect" in these quasi- 3D photonic structures in the visible is beyond the scope of this work.

2. Opals and Inverted Opals

Opal is a face-centred cubic package of identical silica balls as shown in Fig. 1 [11]. Interstitial voids are situated between adjacent balls and form an interconnected system that allows further impregnation of the opal with semiconductors [12] or a polymer solution. In what follows we distinguish between: (a) silica-based opals infilled with polymers, which themselves contain fluorescent dye molecules. (b) PMMA opals made out of the fcc arrangement of PMMA spheres, sometimes pure PMMA and sometimes PMMA spheres doped with a fluoresent dye. (c) PMMA opals infilled with another polymer which contains fluroescent dyes or an oxide semiconductor. (d) inverted opals, being the structures from which the silica or PMMA spheres have been removed leaving behind a replica of the fcc opal.

2.1. POLYMER-LOADED SILICA OPALS

2.1.1. *Sample Preparation*
Two types of silica opal were used, both with ball diameters $D = 235\text{-}240$ nm. The free volume of the total sample volume in the first type of opal was about 15% (A-opal) and 25% in the second type (B-opal). This difference was introduced by impregnating the

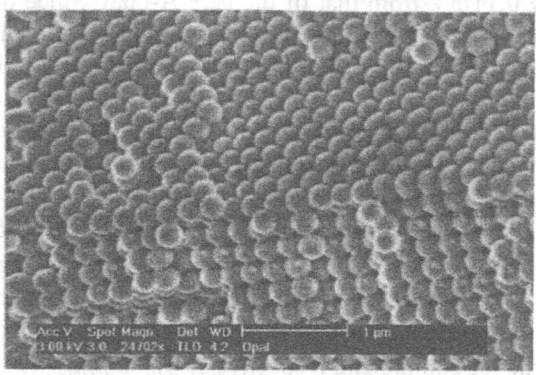

Figure 1. Scanning electron micrograph of a silica opal sample showing the fcc arrangement and relevant crystallographic facets.

voids with an extra amount of amorphous silica during the synthesis. The opal porosity was deduced from the extra weight that the opal acquires in the first five minutes after impregnation with water. Further impregnation takes up to a week to saturate and the amount of water nearly doubles the figures. This is explained by the presence of hidden porosity of the opal matrix.

A saturated solution of coumarin 6 and 1 wt% of the photoinitiator Lucirin LR44 in diethyleneglycoldimethacrylate (CP) or a perylene derivative (3,4,9,10–perylene-bis (dicyclododecylcarboximide)) in polyethylenglycol-dimethacrylate (PP) were used. Irradiation of the solution with UV light transformed it into a solid polymer network. The dye was dissolved in a solid polymer matrix with the result that, in contrast to a pure dye solution, the fluorescence is not quenched in this diluted state. Since no evaporation of the solvent occurs the filled opals can be stored for long periods. SEM inspection shows that the dye-polymer occupies the opal matrix quite homogeneously. Examination of cross-sections of trial run samples confirms that the polymerisation of the dye-polymer occurred in the whole volume of the sample. The actual amount of the polymer solution embedded in the opal voids can be deduced from Bragg diffraction.

2.1.2. *Bragg Diffraction*
The transmission spectrum of the CPA-opal (Fig.2) shows a deep minimum centred at 2.21 eV.

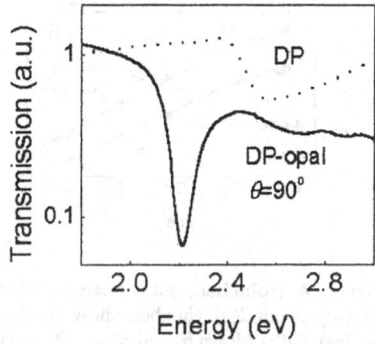

Figure 2. Transmission spectra of CPA-opal (solid line) and CP reference sample (dotted line).

It is "red" shifted by 0.11 eV from that of the bare A-opal. Since the maximum in the reflectance spectrum was found at the same energy for light scattered in the near-normal direction to the sample surface (Fig.3 top), this feature is ascribed to the Bragg resonance.

It is seen that the impregnation of opal with the dye-polymer solution causes a red shift of the resonance peak along the [111] direction of the crystal. A comparison of Bragg peak wavelengths λ_{opal} in A- and CPA-opals provides a self-consistent estimate of the ball diameter and f_{ball} in the sample. Starting from the Bragg law, the relationship $\lambda_{CPA-opal}/n_{CPA-opal} = \lambda_{A-opal}/n_{A-opal}$ holds, where n_{opal} is the average RI of the opal composite. On the other hand, using the effective medium approximation [13], the average RI of opal composite is given by $n = \Sigma n_i f_i$, where n_i and f_i are the RI and the volume fraction of the ith-component, respectively. Using $n_{ball} = 1.45$ and $n_{polymer} = 1.5$ [14], one finds $D = 236$ nm and $f_{ball} = 0.87$ for A-opal. These numbers are in excellent agreement with the results of SEM inspection. Similarly, $D = 239$ nm and $f_{ball} = 0.77$ can be deduced from the Bragg diffraction for B-opal. Since the RI of the guest exceeds that of the host, the actual ensemble of scatterers corresponds to the case of an inverted opal structure, where the RI of the connected polymer grains stands over that of the silica background. In spite of the greatly reduced RIC = $(n_{ball}/n_{void})-1$, which is about 0.45 for bare opal and only 0.05 for dye-polymer-opal composite, the relative width of the Bragg peak does not collapsed in the composite. Moreover, it becomes even better resolved due to the reduction of the background scattering (Fig.3a).

Changing the angle of the light incidence from the normal to (111) plane leads to "blue" shifts of the Bragg resonance in both bare- and infilled opals (Fig.3a). To quantify the angular dispersion of the stop-band, the Bragg law in the form

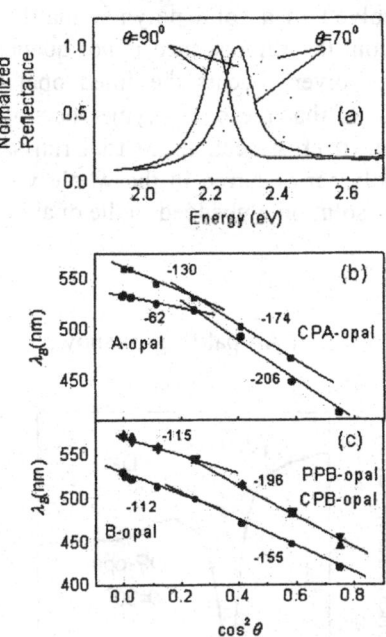

Figure 3. (a) Bragg reflectance from CPA- (solid line) and A-opals (dots) at $\theta = 90°$ and 70°. (b) Angular dispersion for CPA- (squares) and A-opals (circles). Numbers show the slopes. (c) Angular dispersion for PPB- (up triangles), CPB- (down triangles) and B-opals (circles).

$\lambda_B = 2d\sqrt{n_{eff}^2 - \cos^2\theta}$ was used. The linearity of the resonance central wavelength λ_B dependence upon $\cos^2\theta$ was proved for a variety of opal-based compounds [15]. Fig.3b,c illustrate different angular dispersions for A-opal and B-opal caused by the difference in the void volume. After infilling opal with the polymer, a "red" shift of the dispersion due to the increase of the average RI and a change of in the slope of this dependence were observed. The $\lambda(\cos^2\theta)$ dependence for the A-opal shows two distinct parts characterised by different slopes. The tangent to the steeper part of the curves changes from -206 for A-opal to -174 $nm/\cos^2\theta$ for CPA-opal, whereas the tangent to the less steep part of the curves changes from -62 to -130 $nm/\cos^2\theta$ (Fig.3b). A similar behaviour was observed in PPB-opal, where the polymer was embedded in a more open opal structure (Fig.3c). The "red" shift of the dispersion curve of PPB-opal from that of original B-opal appears more pronounced reflecting the higher volume fraction of polymer in the composite. However, the change of the dispersion curve after infilling looks different for CPA-opal and PPB-opals, being more pronounced for PPB-opal, whereas dispersions for, e.g., CPA- and PPB-opals are almost indistinguishable (Fig.3c). However, this effect may arise from inhomogeneities.

The above shows how sensitive the stop-band dispersion is not only to the variation of the RIC but also to the volume fraction of scatterers. The obvious reason for the dispersion change in infilled opals is the reconstruction of the photonic energy band structure resulting from the inversion of the scattering ensemble. Due to the complexity of the computational task, when the simulation faces the transmitance/reflectance problem for an arbitrary direction in a 3D photonic crystal, there is so far no recipe available on how to interpret the dispersion of the stop-band in opal composites. By increasing the RI contrast, a squeezing of the angular dispersion was observed in impregnated opals [9] in agreement with the general trend anticipated from the comparison of the band structure calculations for bare [16] and inverted opals [17].

2.2. PMMA OPAL-LIKE THIN FILM CRYSTALS

2.2.1. Preparation
PMMA balls for polymer-based photonic crystals were prepared by a modified emulsion polymerisation from the monomer-solution. In some cases a small amount of fluorescent dye (Coumarin 6) was added to the solution to study the PBG effect upon emission from such photonic crystal. The diameter of the balls was controlled by varying the polymerisation time. Larger particles were separated by filtration and centrifugation. Films with an area of several cm^2 were deposited on hydrophilized microscope glass slides. Balls self-assembled in the face-centred cubic *fcc* package with the (111) plane along the substrate. These thin film polymeric crystals consist of about 30 to 50 monolayers and possess domains extending over hundreds of micrometers. The dye molecules are homogeneously distributed in the balls and their concentration can be varied in a controlled manner. The resulting crystal offers the advantage of containing light emitting molecules inside the photonic structure while preserving the RIC of the host polymer. In Fig. 4 an SEM micrograph of the initial photonic structure is shown. The monodispersity of the spheres is less than 10% but shows smaller variations on a shorter length scale.

Vapour phase chemical deposition was used to fill the polymeric opal-like film with SnS_2. Polymer balls were dissolved after completing the impregnation, leaving a semiconductor replica film on the glass substrate (Fig.5). The filling factor of

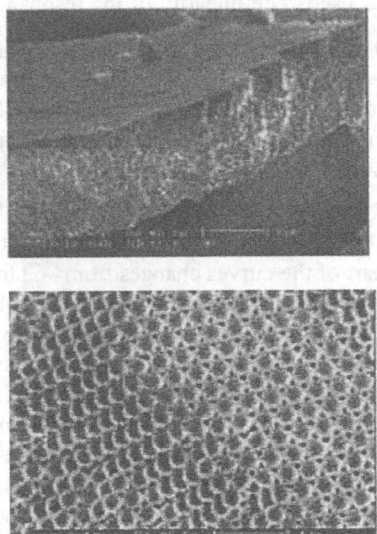

Figure 4. SEM micrographs of PMMA opaline film on a glass substrate.

Figure 5. SEM micrograph of an inverted opaline structure.

the PMMA-air structure $f_{ball} = 0.74$ is far from being optimal and the RIC is about 1.5 to 1. By contrast, in the inverted structure the fraction of high RI material $1 - f_{ball}$ is closer to the optimum value, whereas the RIC depends on the actual density of a semiconductor and varies from 1.4-1.9 to 1.

2.2.2. *Bragg Diffraction*

Transmission and reflectance spectra were measured by illuminating the sample with white light. The angular dispersion of the Bragg diffracted light was studied by measuring reflectance spectra at different angles θ between the (111) axis of the *fcc* crystal and the beam axis, the scattered light was collected within a solid angle of 2°. Changes in the light diffraction have been quantified using the Bragg law. It is known, that the angular dispersion of diffraction peaks mimics the dispersion of the stop-band in the $E - k$ space [18]. Ball diameters extracted from the Bragg diffraction are in good agreement with SEM data, if the RI for PMMA is taken as $n = 1.4893$ [14].

Both changes of the effective RI and the filling factor of the high-RI material (f_{hRI}) lead to the shift of the Bragg resonance and the increase of the width of the photonic bandgap in the semiconductor replica as compared with the polymer template. A comparison of angle-resolved reflectance spectra is shown in Fig. 6.

The reflectance of the polymeric opal shows a relative stop-bandwidth of $\Delta E/E_0 \approx 5\%$. The "blue" shift of the Bragg resonance in the replica due to the decrease of the filling factor is partly balanced by the "red" shift due to the increase of the RIC. Simultaneously, both factors serve to increase the photonic bandwidth to $\Delta\lambda/\lambda_0 = 13\%$. It is worth mentioning that the distortion of the lattice, which is bigger in the replica, also contributes to the broadening of Bragg peaks.

The rate of angular dispersion of the stop-band is governed by the effective RI, correspondingly, it is stronger for the replica, but decreases with increasing of the density of SnS_2. However, the total improvement of the PBG in the replica compared with the polymer template is the result of the much wider stop-band, that allows to overlap them more effectively for different angles. Expressing this overlap in terms of the ratio of the stop-bandwidth ΔE to the shift of the resonance frequency E_{shift} taken between 0 and 40 degrees as $\delta = \Delta E/E_{shift}$, the δ-factor improves from 0.5 to 1.3.

3. Optical Studies

3.1 REFLECTANCE AND LUMINESCENCE OF DYE-POLYMER-LOADED OPALS

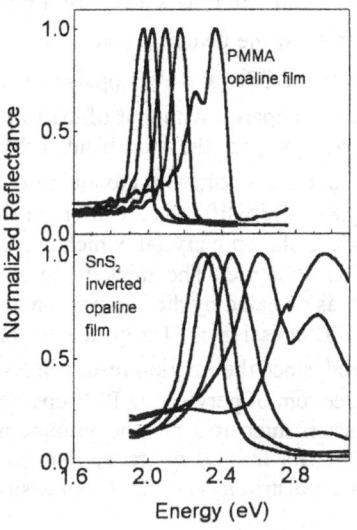

Figure 6. Angle resolved reflectance spectra of PMMA opaline film (top) and inverted SnS_2 opal film (bottom). Spectra for angles 5, 20, 30, 40 and 50 degrees from left to right.

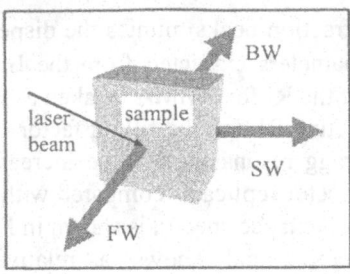

Figure 7. Schematic view of the front, back and side window configurations.

The samples under study were thin platelets of silica opal nanocomposites with the (111) plane of the ball package facing the wider side. Platelets of bulk dye-polymer compound were used as reference samples. Transmission and reflectance spectra were measured by illuminating the sample with monochromatic light and detecting the scattered light within a 0.0012π sr fraction (or about 2^0) of a solid angle. The angular dispersion of the Bragg diffracted light was studied by measuring reflectance spectra at different angles θ between the [111] axis of opal and the beam axis. The angle was varied between $\theta = 90^0$ and 40^0. The symmetric Bragg configuration was used. The photoluminescence (PL) was excited with 457.9 nm radiation using front, back and side window configurations, denoted as FW, BW and SW, respectively (Fig.7).

The FW PL was excited and collected from the same surface, whereas the BW PL was excited on the front sample surface and collected from the back surface after travelling inside the material. The SW PL was excited through the front surface and collected from the cleaved edge of the sample. The PL was collected within 0.0003π sr and 0.0075π sr solid angle formed by a collimating diaphragm placed between the sample and the spectrometer, which permits angle resolved measurements over the same range of angles θ as in reflectance measurements.

The optical properties of the samples are similar and therefore, in what follows, we will consider the most representative example of the CPA-opal (CP in A-opal, with a ball volume fraction $f_{ball} = 0.81$) and refer to spectra of PPB-opal (PP in B-opal, $f_{ball} = 0.75$), where necessary, to show the general trend.

Fig.8 shows the PL spectra of CPA-and PPB-opals at $\theta = 90^0$ and 70^0 and their corresponding reference samples. Compared with that of the reference sample, the PL intensity decreases where it overlaps with the stop-band. FW and BW PL spectra recorded at the same θ -angles appear similar, with the main exception of the depth of the PBG-related minimum (Fig.8a). This difference can be attributed to the emission from the near-surface volume of the photonic crystal, which is only slightly affected by the low RI contrast of the PBG structure. The thicker the sample, the deeper the minimum of PL spectrum, as it was revealed by the comparison of 0.3 and 0.9 mm thick samples. The anisotropy of the PL emission is clearly demonstrated if PL spectra for different θ – angles are compared, since the rotation results in a shift of the stop-band.

In the more complex spectrum of perylene in PPB-opal (Fig.4b), the stop-band partly suppresses the dye emission intensity, but the minimum cannot be resolved straightforwardly by comparison with the PP reference. Nevertheless, the stop-band effect can be clearly identifed as a redistribution of the PL intensity between the two

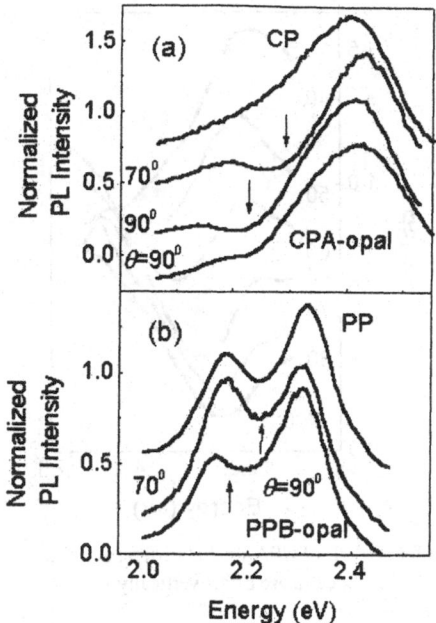

Figure 8. (a) Normalised PL spectra of CPA-opal. Bottom curve is for FW. Middle curves for BW PL at $\theta = 90^0$ and 70^0. Top curve for CP reference sample. (b) PL spectra of PPB-opal at $\theta = 90^0$ and 70^0. Top curve is that of the PP reference. Curves are offset vertically.

main emission bands of the perylene when the stop-band is shifted by the θ rotation of the crystal.

Assuming that the electronic transitions between vibronic states of dye molecules dissolved in the polymer are unaffected by the presence of silica-polymer interfaces, the difference between the reference and opal composite spectra can be taken as resulting from the effect of the PBG environment upon the dye emission. The PL intensity in CPA-opal relative to the PL intensity of the CP-reference at the same frequency, corresponds to the difference in the density of available optical modes. Therefore, the ratio of CPA-opal and CP-reference spectra represents a reasonable approximation to the spectrum of the density of photon states (DOSP).

The purpose of studying the anisotropic stop-band by angular resolution, is to fit the solid angle of the light detection within the angular width of the stop-band. This happens when the appearance of the DOSP spectrum no longer changes if finer collimation is used (Fig.5). For example, for $\theta = 90^0$ the DOSP along the [111] direction is a good choice. The depth of the dip the DOSP approaches its limit here. This limit provides evidence of intrinsic disorder in the crystal, which results in leaky modes detected on top of the eigenmodes of the photonic crystal within the stop-band. If poor collimation is used, the emission carrying information on stop-band modes, will be mixed up with the continuum of the free-space optical modes at the detector entrance. This admixture will depend on the extent to which the solid angle of the beam divergence compares to the angular (or k-space) width of the stop-band [10].

Changing the angle, the dip in the DOSP spectrum moves to higher energies in agreement with the angular dispersion of the stop-band (Fig.5). If $\theta \le 50^0$, the DOSP dip is no longer resolved in the PL spectrum of CPA-opal because the dye-polymer solution has an appreciable absorption in the range $2.55 - 2.65$ eV. This absorption suppresses the coherent scattering of electromagnetic waves and the PBG effect. By contrast, the

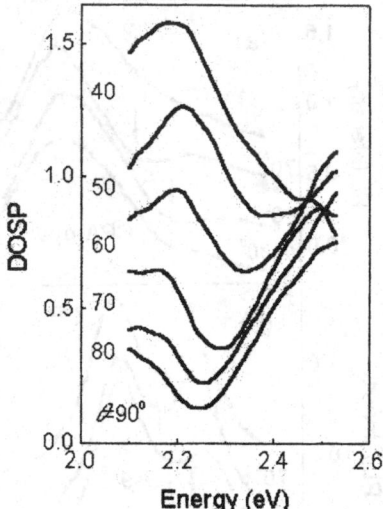

Figure 9. Local DOSP spectra for BW PL of CPA-opal at different angles θ. Sample thickness 0.3mm. Spectra are offset vertically.

peak in the reflectance spectrum, centred at 2.64 eV when $\theta = 40^0$, can be interpreted as the stop-band because the Bragg diffraction does not involve all modes of the photonic crystal and is therefore less sensitive to the dephasing of scattered light. In PPB-opal (Fig.6a, b) the perylene absorption gives rise to a large number of leaky modes at energies above 2.2 eV. This is why the dip in the local DOSP can hardly be resolved when $\theta \leq 70^0$.

The anisotropy of the PBG structure of the CP-opal leads to a frequency-selective directionality diagram of the emission (Fig.7a). For example, the emission intensity at 2.4 eV in the [111] direction is twice as large as that at $\theta = 50^0$. The directionality of the emission depends on the frequency, which provides further evidence for the PBG origin of this effect. An additional proof comes from the

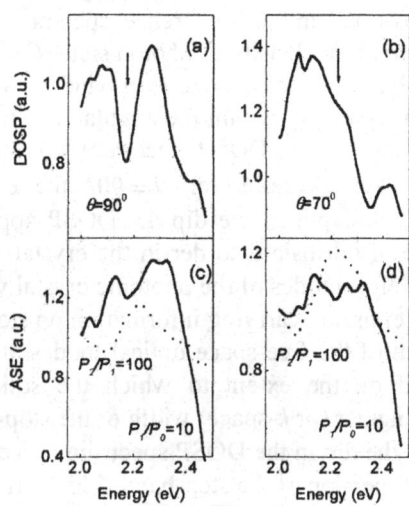

Figure 10. (a, c): Local DOSP spectrum and the normalised spectra of ASE-P (solid) and ASE-L (dot) of PPB-opal at $\theta = 90^0$. Pumping level ratios are indicated at curves. (b, d): the same but for $\theta = 70^0$.

comparison of angular dependencies of the CPA-opal PL intensity recorded under FW and BW configurations (Fig.11b,c). The FW PL intensity hardly shows an angular dependence (Fig.10c), because corresponding PL spectra possess a shallow stop-band-related minimum. By contrast, the BW PL emission shows a strong directionality, especially at 2.4 eV (Fig.10b). Moreover, the increase of the CPA-opal thickness from 0.3 to 0.9 mm leads to a factor of five enhancement of the directionality at that energy.

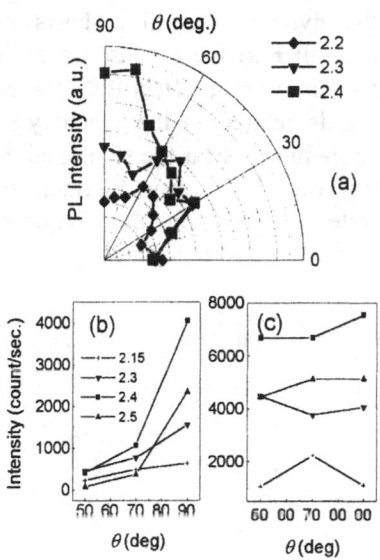

Figure 11. (a): Directionality diagrams for BW PL intensities of CPA-opal with thickness 0.3mm at different energies. (b): BW PL intensity of 0.9mm thick CPA-opal at different angles. (c): the same for FW PL.

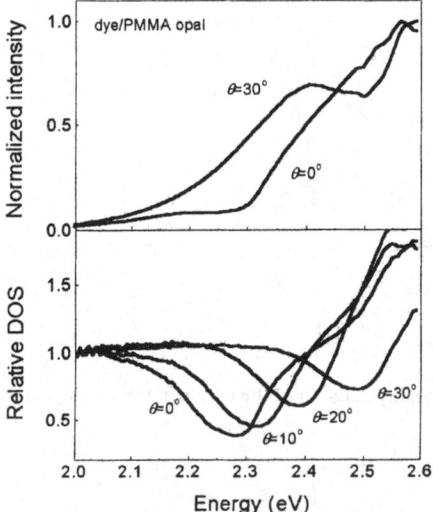

Figure 12. Relative dip in the density of photonic states obtained from PL measurements.

3.2 PHOTOLUMINESCENE OF PMMA-BASED OPALS

The Bragg configuration was used to study the photoluminescence (PL) as the angle was varied between $\theta = 0°$ and $\theta = 50°$ and the PL was collected within a 6° solid angle. To demonstrate the relative changes in the density of photonic states in the PBG region PL spectra are shown normalised to the PL spectrum on unstructured Coumarin-PMMA film (Fig. 12).

The anisotropy of the photonic band structure leads to the self-focusing of the emission (Fig. 13). The fingerprints of the PBG are reproduced at different angles for different frequencies as a dip in the directionality diagram. By contrast, the corresponding diagram of the dye-polymer film shows no obvious wavelength dependence (Fig. 14). This is similar to the results concerning dye-polymer loaded opals [19]. By analogy with the super prism effect [20], the emission self-focusing in the incomplete photonic crystal is related to the topology of dispersion planes for electromagnetic waves with the frequency near the stop-band, because the vector of the group velocity of propagating wave is always normal to the dispersion plane. Accordingly, the focusing rate depends on the frequency. Modelling of this phenomenon for our 3D photonic crystals is in progress.

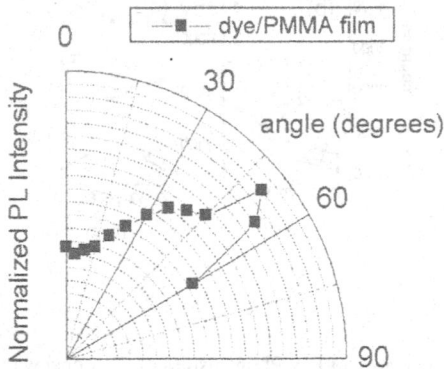

Figure 13. Directionality diagram of the emission from the polymer-dye reference sample.

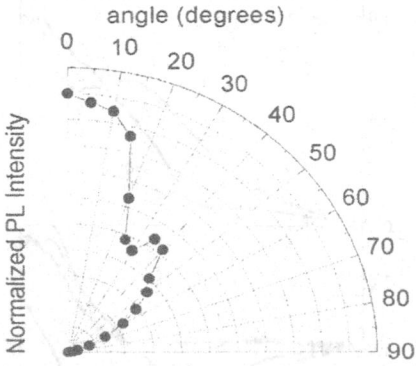

Figure 14. Directionality diagram of the emission from the dye-polymer-opal film.

3.3. AMPLIFIED SPONTANEOUS EMISSION IN POLYMER-LOADED SILICA-OPALS

The dependence of PL spectra upon excitation power provides a ground to examine the stop-band effect upon the spontaneous emission rate. To reveal this dependence we plot the ratio of PL spectra measured at different excitation power. A flat and uniform ratio spectrum corresponds to the linear regime of the luminescence, whereas some inhomogeniety may indicate the interplay between the radiation and electronic population of emitter energy levels. The optical pumping levels were set to about two orders of magnitude below those needed to saturate electronic transitions and degrade the dye emission. Therefore, we assume that the power-induced change of the PL spectrum is the result of the feedback from the photonic band structure to the dye energy levels. In other words, these ratio spectra indicate changes in the spontaneous emission rate.

The increase of the spontaneous emission rate above the mean level, which follows the increase of the number of photons in the photonic crystal, is an indication of the enhancement of the spontaneous emission. In order to compare ratio spectra at different pumping levels, they were normalised to their magnitude at an energy just below the range swept by the stop-band, when the angle θ is changed.

Two different sorts of responses with pumping power were observed in CPA-opal. Fig. 15 shows data for ratio spectra at $\theta=90^0$ as compared with a similar set taken at $\theta=70^0$. At low pumping power, the ratio spectra show two peaks at the stop-band edges, separated by a minimum centred at the stop-band frequency. By changing the angle of the light detection, this minimum follows the angular dispersion of the stop-band. At higher pumping level, the ratio spectra show a θ-independent peak near the centre of the energy range, which is swept by the stop-band. Enhanced or amplified spontaneous emission (ASE) bands, bound to the stop-band and detected in the low-pumping regime (ASE-P band) are detected first. Under higher pumping levels, another band is observed (ASE-L) centered at 2.3 eV, which shows no relationship to the stop-band. The angular dependence of the stop-band centre, dip and maximum of ratio spectra are summarised in Fig.16. The ratio spectra for the CP reference sample are clearly different from those of CPA-opal: with increasing pumping the PL intensity of the reference sample increases in the range above 2.5 eV, where the emission and absorption bands of Coumarin 6 overlap.

The difference behaviour of ASE-P and ASE-L bands is probably related to the difference in feedback mechanisms. The interplay of electronic levels and the photonic band structure is most likely the reason for the emergence of the ASE-P band. The probability of photons to be released from rovibronic states in dye molecules with k-vector, which match the stop-band direction, depends on the DOSP which, in turn, is reduced at the energy of the stop-band and increased at the stop-band edge energies. Therefore, the spectrum of excitation-relaxation equilibrium is distorted by the presence of the stop-band, compared to the free space, and a fraction of emitters accumulates energy in dressed states. In the energy range of the stop-band variation, the energy relaxation goes through non-radiative transitions using the closely spaced rovibronic states. As soon as the radiative transition becomes allowed at the stop-band edge, the accumulated energy is released. For a periodic ensemble of scatterers the frequency selection for ASE-P band is provided by the stop-band.

The CP-opal being an incomplete photonic crystal does not show an omnidirectional bandgap and the frequency selection varies with direction. This

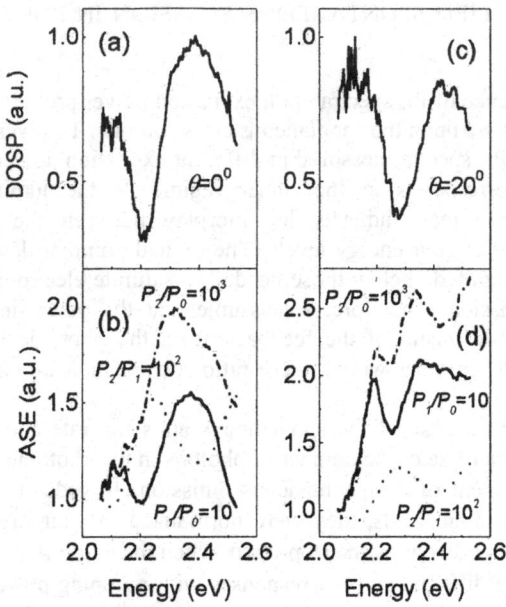

Figure 15. (a, c) Local DOSP and corresponding ASE spectra for $\theta = 90^0$. Solid line -ASE-P bands; dashed line– ASE-L band, dotted line– resultant ASE at high power excitation. Power increment is shown at curves. (c, d) The same for $\theta = 70^0$.

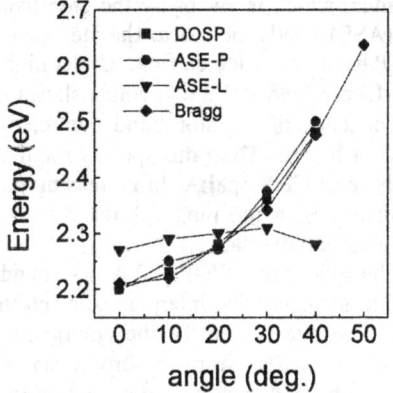

Figure 16. Angular dispersions of the Bragg peak (diamonds), local DOSP (squares), ASE-P (circles) and ASE-L (triangles) bands.

Anisotropy dramatically reduces the effectiveness of the stop-band-related feedback because the fraction of directions, along which photons emitted with a given energy form the ASE band, depends on the solid angle of the stop-band overlap. The remaining photons with this frequency undergo incoherent multiple scattering and, become partially localised. This feedback process, which leads to the formation of ASE-L band, is analogous to lasing in disordered multiple-scattering media [21]. The dye molecules become excited after absorbing energy from the pumping beam. Then emitted photons travel through the gain medium and some of them can be amplified before leaving the crystal. Initially, the intensity of the emission by such light source is proportional to the density of excited dye molecules and the cross-section of spontaneous radiative emission. Localisation increases the light-matter interaction time, determining the self-organisation process for ASE at high pumping levels. The self-selected frequency corresponds to the condition of the strongest localisation. Random scattering is less effective than the coherent one in its impact upon the emission. Therefore, the ASE-L band requires a higher number of photons for it to be detected. This is the justification for separating the self-organised and controlled mechanisms of the ASE frequency selection by the excitation power.

The side window arrangement was used to quantify the difference in the scattering strength between CP reference and CPA-opal. The light source inside the sample was simulated by the exciting the PL in the volume of the sample through the side window. By changing the distance between the PL light source and the edge of the sample the attenuation length was extracted. For the CP-reference sample, an approximately linear intensity drop with distance from the edge was observed over the emission band energy range, whereas at 2.5 eV the intensity decays exponentially. The exponential drop is caused by self-absorption due to overlapping emission and absorption bands. In the case of CPA-opal, the measured PL intensity decreases exponentially as $I \sim exp(x/l)$ within the examined energy range. This strong attenuation of the emission intensity, compared to the reference sample, clearly shows light localisation due to the presence of scatterers. The characteristic attenuation length decreases almost linearly from 0.54 mm at 2.2 eV to 0.41 mm at 2.4 eV and drops to 0.29 mm at 2.5 eV. The change in the attenuation length is caused by self-absorption.

For a homogeneous gain medium the SW technique can be used to study the ASE band [22]. However, as the scatterers are embedded in such a medium, the emission spectrum is not qualitatively different from the FW PL [23]. This situation is altered again if the scatterers are arranged in a periodic manner. Fig.17a compares BW and SW PL spectra of CPA-opal. The BW PL shows the dip at the energy where the emission is filtered by the stack of (111) planes. The SW PL spectrum is quite similar to that of the clear CP solution. Apparently, more than one set of opal planes contributes to the spectrum, which is why it has lost the directionality. However, the SW PL spectrum depends on the distance between the laser spot and the sample edge. In the same way as

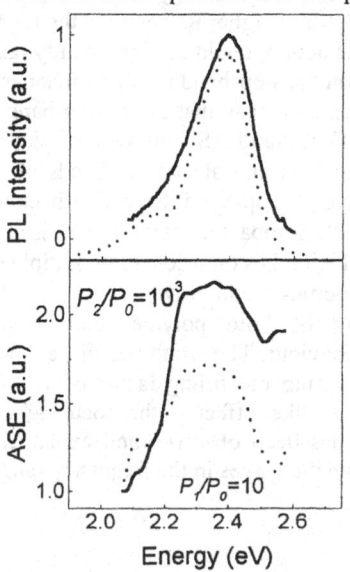

Figure 17. (a) Normalised SW (solid line) and BW PL spectra of CPA-opal. Curves are offset vertically, (b) SW ASE band at different increments of the excitation power increase.

the dependence of the spectrum on sample thickness, self-absorption progressively cuts the high energy part of the emission band with increasing distance. It is worth noting that emission band squeezing occurs much faster with increasing distance in CPA-opal than in the clear CP solution, which is tentatively explained in terms of photon localisation.

The SW PL ratio spectra of CPA-opal contain a broad ASE band, which spans the whole energy range of the stop-band variation from 2.2 to 2.4 eV (Fig.17b). It shows the gain increase with increasing pumping intensity. Apparently, this is an indication of the frequency range over which photon localisation takes place. The ratio peak is spread over a wide spectral range because for the SW PL all crystal planes take part in localisation.

4. Conclusions and Outlook

The study of different dyes and different opals was used to demonstrate the general character of the PBG effect on emission properties of dye-impregnated opal. The optical properties of dye-polymer solutions embedded in opals have shown an incomplete PBG-related anisotropy as the main feature. This anisotropy is comparable to that of the bare opal, whereas the RI contrast is about three times lower for dye-polymer loaded opals. The local DOSP spectrum was extracted by comparing PL spectra of dye-polymer opals and clear dye-polymer solutions. The angular dispersion of the DOSP minimum was found to be similar to the angular dispersion of the Bragg resonance, but the DOSP appears much more fragile with regard to the loss of coherence due to the self-absorption. We conclude that PL spectroscopy provides a more realistic estimate of the stop-band behaviour compared to Bragg diffraction because in the latter the externally generated light cannot be coupled fully to the eigenmodes of the photonic crystal.

The anisotropy of the stop-band structure has different implications for the emission from dye-polymer-opal composites. The redistribution of the emission intensity over the stop-band spectral range together with the angular dependence of the stop-band energy lead to the directionality of emission without losses. Matching the emission band maximum to the stop-band energy can be used to squeeze the angular width of the emission diagram. With a higher RI contrast, the thickness of the photonic crystal necessary for antenna-like action, could be significantly reduced.

The ASE, which relies on the stop-band feedback mechanism, exhibits a strong anisotropy of the gain distribution similar to that of the stop-band. For a broadband dye emission, the anisotropy of the DOSP and ASE in low RIC photonic crystal is limited by the Stokes shift between emission and absorption bands of the dye. With further increase of the contrast, the angular stop-band dispersion becomes squeezed and one may expect enhancement of the PBG impact upon the dye emission. Besides, the role of another feedback mechanism, which relies on incoherent multiple scattering, is expected to be diminished if the PBG properties are improved.

Concerning PMMA opals, both polymer and semiconductor structures demonstrate similar photonic behaviour. The width and dispersion of the stop-band has been greatly improved by decreasing the filling factor of a high RI component and increasing the RIC. An antenna like effect – the focusing of the emission from incomplete photonic crystals - has been observed and explained as the result of the specific dispersion of electromagnetic waves in the frequency range of PBG.

Acknowledgements
We are grateful to D. Chigrin, David Cassagne, Jesus Manzanares and Christian Jounin
for helpful discussions. Work partly funded the the EU IST project PHOBOS

5. References

1 E. Yablonovitch, *Phys. Rev. Lett*, **58** 2059 (1987); S. John, *Phys. Rev. Lett.* , **58**, 2486 (1987)
2 JEGJ Wijnhoven, W. L. Vos, *Science*, **281**, no.5738, 802 (1998); A.Zakhidov, R.H. Baughman, Z. Iqbal, C.Cui, I.Khayrullin, AS.O. Dantas, J.Marti, V.G.Ralchenko, *Science*, **282**, 897 (1998)
3 S.G. Romanov, A.V. Fokin, V.I. Alperovich, N.P. Johnson, R.M. De La Rue, *Physica Status Solidi* **163**, 169 (1997)
4 Yu. A. Vlasov, K. Literova, I. Pelant, B. Honerlalage, V.N. Astratov , *Appl. Phys. Lett.* **71**, 1616 (1997)
5 K.Yoshino, K.Tada, M.Ozaki, A.A. Zakhidov, R.H.Baughman, *Jpn.J.Appl.Phys.*, **36**, L714 (1997); K.Yoshino, S.B.Lee, S.Tatsuhara, Y. Kawagishi, M.Ozaki, A.A.Zakhidov, *Appl.Phys.Lett.*, **73**, 3506 (1998)
6 T. Yamasaki, T. Tsutsui, *Appl.Phys. Lett.*, **72**, 1957 (1998)
7 E. P. Petrov, V. N. Bogomolov, I. I. Kalosha, S. V. Gaponenko. *Phys.Rev.Lett.*, **81**, 77 (1998)
8 S. G. Romanov, A. V. Fokin, V. Y. Butko, C.M. Sotomayor Torres, *Physics Solid State*, **38**, 1825 (1996)
9 S.G. Romanov, *J.Nonlinear Optical Physics & Materials*, **7**, 181 (1998)
10 S.G. Romanov, A.V. Fokin, R.M. De La Rue, *Appl. Phys. Lett.***74** (1999)
11 V. G. Balakirev, V. N. Bogomolov, V. V.Zhuravlev, Y. A. Kumzerov, V. P. Petranovsky, S. G. Romanov and L. A. Samoilovich, *Crystallogr. Rep.,* **38**, 348 (1993)
12. Romanov S. G. and Sotomayor Torres, C. M. (2000) 3D Lattices of Nanostructures: The template approach, in H S Nalwa (Ed), *Handbook of Nanostructured Materials and Nanotechnology*, Volume 4, Academic Press, USA, pp232-323 (1999)
13 W.L.Vos, R.Sprik, A.von Blaadren, A.Imhof, A.Lagendijk and G.H.Wegdam, *Phys.Rev.B*, **53**, 16231 (1996); R.D.Pradhan, I.I.Tarhan and G.H.Watson, *Phys.Rev B*, **54**, 13721 (1996); V.N. Bogomolov, N.V.Gaponenko, A.V. Prokofiev, A.N.Ponyanina, N.I. Silvanovich, and S.M. Samoilovich, *Phys.Rev. E*, **55**, 7619 (1997); V. Yannopapas, N. Stefanou and A. Modinos, *J. Phys.: Cond. Matter*, **9**, 10261 (1998); C. López, H. Míguez, L. Vázquez, F. Meseguer, R. Mayoral and M. Ocaña, *Superlat. and Microstruct,*. **22**, 399, 1997
14 J. Brandrup, E.H. Immergut, *Polymer Handbook*, 3d ed., New York, 1989, p.VI-461
15 C. Lopez, L. Vázquez, F. Meseguer, R. Mayoral, M. Ocana, H. Miguez, *Superlattices And Microstructures*, **22**, 399 (1997); H.Miguez, A.Blanco, F.Mersequer, C. Lopez, *Phys. Rev. B*, 1999, **59**, 1563
16 D. Cassagne, *Annales de Physique*, **24**, 1 (1998)
17 K. Busch, S. John, *,Phys. Rev. E*, **58**, 3896 (1998)
18 V. Yannopapas, N. Stefanou, A. Modinos, *J. Phys.: Cond. Matter*, **9**, 10261, 1998
19 S.G. Romanov, T. Maka, C. M. Sotomayor Torres, M. Mueller, D. Allard and R. Zentel, *Appl. Phys. Lett.* **75**, 1057 (1999)
20 H. Kosaka, T. Kawashima, A. Tomita, M. Notomi, T. Tamamura, T. Sato and S. Kawakami, *Phys. Rev. B*, **58**, R10096, 1998
21 S.John and G. Pang, *Phys. Rev. A*, **54**, 3642 (1996)
22 D.S. Wiersma, M.P. van Albada, A. Lagendijk, *Nature*, **373**, 203 (1995)
23 N.M. Lawandy, R.M. Balachandran, A.S.L. Gomes, E. Sauvain, *Nature*, **368**, 436 (1994); N.M. Lawandy, R.M. Balachandran, *ibid.* **373**, 204 (1995)

QUANTUM WIRES AND QUANTUM DOTS FOR OPTOELECTRONICS: RECENT ADVANCES WITH EPITAXIAL GROWTH ON NONPLANAR SUBSTRATES

E. KAPON
Department of Physics
Swiss Federal Institute of Technology Lausanne (EPFL)
1015 Lausanne, SWITZERLAND

1. Introduction

The lateral quantum confinement imposed on electrons and holes in semiconductor quantum wires (QWRs) and quantum dots (QDs) has been predicted to bring about significant advantages for optoelectronic device applications. Early on, the increasingly sharp density of states (DOS) achieved with more degrees of confinement was expected to dramatically enhance optical absorption and emission due to the spectral confinement of the reduced DOS. Furthermore, it was anticipated that the lateral confinement would increase the exciton binding energy beyond what is obtained with two-dimensional (2D) quantum well (QW) structures, giving rise to enhanced linear and nonlinear optical excitonic effects. In addition, the extremely small number of charge carriers residing in short wire segments or small arrays of dots makes these nano-optoelectronic structures suitable for integration with novel electronic devices inherently designed for low power consumption, such as conducting nanowires and single electron transistors.

In the late 1980's and early 1990's, a myriad of fabrication techniques compatible with optoelectronic applications have been proposed and demonstrated with some success. A few of these approaches have reached some level of maturity by now and have been fruitful in unraveling physical mechanisms that are relevant to low-dimensional optoelectronic devices and in the demonstration of novel optoelectronic devices. Among these techniques one should mention cleaved-edge overgrowth (CEO) for producing T-shaped QWRs and related QD structures [1], Stranski-Krastanow (SK) growth of strained QDs [2], and epitaxial growth over nonplanar substrates for producing V-groove QWRs and other, related QWR and QD structures [3].

Figure 1 depicts schematic cross sections of these three types of low-dimensional structures. In all cases, the lateral quantum confinement is achieved due to what can be viewed as a perturbation in an otherwise 2D QW heterostructure. In the T-shaped QWRs, the stem-QW part lowers the potential of the top-QW part in a narrow section whose width is equal to the thickness of the stem-QW. In the SK dot structures the strain-driven growth results in thicker islands positioned on top of a thin, QW wetting layer; the lateral thickness variation provides the lateral confinement via the increase in

41

L. Pavesi and E. Buzaneva (eds.), Frontiers of Nano-Optoelectronic Systems, 41–64.
© 2000 *Kluwer Academic Publishers. Printed in the Netherlands.*

confinement energy with reducing QW thickness. Similar lateral thickness variations in QW layers grown on nonplanar substrates introduce the lateral quantum confinement [3]. In all these structures, large effective variations in both the conduction and valence band edges can yield lateral confinement of both electrons and holes, which is important for optoelectronic applications. Furthermore, in all these approaches the effective barriers of the low-dimensional structure consist themselves of various types of low-dimensional structures. For example, the T-shaped wires are bound by QW regions, the SK dots grow on QW wetting layers, and the V-shaped wires are connected to sidewall- and vertical- QWs (see section 2). Furthermore, QDs made using an extension of the CEO method grow connected to QWR and QW sections [4], and pyramidal QDs are self-formed with QWR and QW barriers [5]. These features provide new opportunities for controlling carrier injection into these structures, as will be demonstrated in section 5. Finally, another common feature of all these approaches is that in all cases the interfaces are formed *in situ*, which yields structures with high optical quality. In particular, the SK growth mode and epitaxy over nonplanar substrates involve self-ordering of the wires and dots, which is important for surmounting limitations associated with traditional lithography techniques, e.g., limitations due to resolution and coverage areas on the wafer.

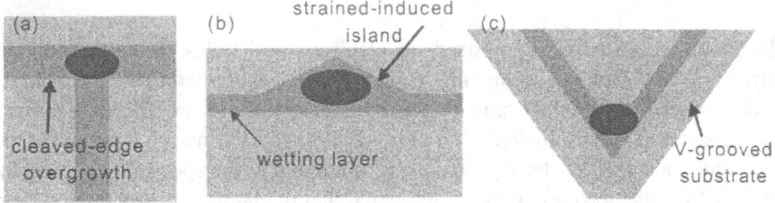

Figure 1. Schematic cross sections of low-dimensional nanostructures prepared by (a) cleaved edge overgrowth; (b) Stranski-Krastanow strain-driven growth; and (c) growth over nonplanar substrates. Darker regions correspond to lower bandgap parts. Ellipses describe the confined wavefunctions

Epitaxial growth over nonplanar substrates has been proposed as a technique for producing QWR and QD heterostructures with *in situ* formed interfaces [3]. In this approach, growth over a patterned, nonplanar substrate results in faceting at particular crystal planes for which the surface energy is minimized. Upon further growth of a QW heterostructure capable of *transverse* (i.e., in the growth direction) quantum confinement of charge carriers, e.g., a GaAs/AlGaAs QW heterostructure, the QW layers develop *lateral* (i.e., in the substrate plane) thickness variations. This, in turn, produces lateral variations in the effective bandgap via the transverse quantum confinement effect. The result is the formation of lateral potential wells for electrons and holes. To achieve lateral *quantum* confinement, the thickened QW sections need to be sufficiently narrow (typically less than a few tens of nm). An important element in achieving such narrow structuring without relying on lithography is the self-limiting growth of appropriate nanofacets during the nonplanar epitaxy (see section 2). In this sense, this approach is utilizing a *seeded self-ordering* strategy, in which the position of the QWR or QD is determined by lithography, whereas the details of the structure are defined by the growth dynamics themselves. It thus has the potential of reaping the structural advantages of each of these elements. Such self-ordered growth of QWRs and QDs over nonplanar substrates has been investigated extensively with both molecular beam epitaxy (MBE) and with organometallic chemical vapor deposition (OMCVD). Different configurations of QWRs and QDs result with these two methods due to the different nature of growth rate anisotropy in the two cases [6].

In this paper, we review recent progress achieved with QWRs and QDs made using nonplanar OMCVD growth with particular attention to the implications on future optoelectronics technology. The structure and optical properties of lattice-matched GaAs/AlGaAs and strained InGaAs/AlGaAs QWRs are discussed in section 2, the impact of exciton localization due to wire disorder are summarized in section 3, and the formation and optical properties of pyramidal QDs are the subject of section 4. Section 5 describes several optoelectronic device applications of V-groove QWRs and pyramidal QDs, and section 6 brings the conclusions.

2. V-Groove Quantum Wires

Lateral patterning of a semiconductor heterostructure can be accomplished *during* the epitaxial growth process provided there is a mechanism for directing the adatoms to the desired nucleation sites. This is contrary to the growth of 2D QW structures, in which the nucleation sites are usually selected randomly. Such control of lateral adatom fluxes J_s can be achieved by introducing controlled lateral variations in the surface chemical potential μ_s, which is related to the flux via $J_s = -\nabla\mu_s$. Lateral patterning of μ_s can be obtained using a variety of mechanisms, including strain, capilarity and entropy of mixing effects. More quantitatively, the surface chemical potential of the component i of an alloy can be written

$$\mu_s = \mu_{s0} + \Omega_0[\sigma_\tau(\xi)]^2 / 2E + \Omega_0[\gamma(\theta) + \gamma''(\theta)]\kappa(\xi) + k_B T \ln x_i(\xi) \quad (1)$$

where Ω_0 is the atomic volume, x_i the mole fraction, σ_τ the tangential surface stress, E the elastic modulus, κ the surface curvature, and $\gamma(\theta)$ the surface free energy that is a function of the inclination angle θ. Lateral patterning that makes use of the strain term has been extensively investigated in the context of Starnski-Krastanow growth of strained QDs on planar substrates [7]. The structures discussed here rely heavily on the capilarity and entropy of mixing effects.

2.1 SELF-ORDERING MECHANISMS

Lateral patterning of the surface chemical potential via the capilarity term in equation (1) can be accomplished using growth on a nonplanar substrate. Consider the epitaxial growth of thin films in an etched channel. The initial phase of the growth consists of the establishment of facets that reshape the profile of the etched groove. These facets are formed in specific crystallographic planes, typical of the particular system in question, such that the surface energy of the structure be minimized. The growth rates of these facets then determine the evolution of the surface profile. For example, for channels etched in the [01-1] direction on a (100) GaAs substrate, OMCVD growth of GaAs/AlGaAs layers proceeds such that (100) facets form at the bottom of the groove, quasi-{111}A facets develop on the sidewalls, and short sections of {311}A facets evolve at the boundaries between these two planes. Since the growth rate at the quasi-{111}A facets is faster than that at the (100) one, a sharp V-groove structure with vanishing width would develop at the bottom. However, when a sufficiently narrow width of the (100) bottom facet is reached, the capilarity effect produces lateral fluxes that are large enough to increase the growth rate of the (100) facet so as to counteract the initial growth rate anisotropy between the (100) and quasi-{111}A facets. This gives rise to a self-limiting, steady state width of the bottom (100) facet (as well as the {311}A) facets), given by [8]

$$l_s = \left(\frac{2\Omega_0 D_s \tau r_s}{k_B T \Delta r} \right)^{1/3} \qquad (2)$$

where τ is the lifetime for adatom incorporation, D_s the diffusion coefficient on the sidewalls (which has an Arrhenius temperature dependence), r_s the growth rate on the sidewalls, and Δr is the difference between the latter and the growth rate at the bottom facet. It can be seen that the self-limiting facet width depends strongly on the surface diffusion of the adatoms involved. Thus, the groove profile (and in particular its width) can be controlled by fixing this diffusion length, which can be accomplished by selecting, e.g., the growth temperature and the type of adatoms.

This latter method is especially useful for constructing QWR and QD structures. First, extremely narrow V-grooves are prepared using self-limiting growth of layers characterized by a small diffusion length of the adatoms, e.g., AlGaAs layers with sufficiently high Al content. Groove widths as narrow as a few nm can be achieved in this way with typical OMCVD growth temperatures [8]. Subsequently, a thin QW layer of lower Al content (and thus lower bandgap) is deposited on the self-limiting groove. The longer effective diffusion length now results in a wider groove profile, which yields a crescent-shaped, QW film at the bottom of the groove. A higher bandgap AlGaAs layer finally caps the structure. In this way, a QWR structure is self-formed due to the characteristic growth dynamics at the highly curved surface of the nonplanar substrate.

An example of a vertical stack of GaAs/AlGaAs QWRs grown on a V-grooved substrate is shown in the dark-field transmission electron microscope (TEM) cross section of Fig. 2. Notice that each GaAs QWR is bounded by GaAs QW layers forming on the sidewalls of the grooves. In addition, note the dark stripe running through the center of the groove. This Ga-rich AlGaAs region is formed during growth of the AlGaAs barrier as a result of the longer surface diffusion length of the Ga adatoms. The Ga-rich region is typically less than 20nm wide, and hence has been termed *vertical quantum well* (VQW); in fact, quantum confinement effects have been observed in the optical spectra of such VQWs [9]. It is worth noting that the lower Al content at the center of the self-limiting groove introduces a significant entropy of mixing term (see equation (1)) into the growth dynamics and modifies the self-limiting shape of the V-groove in an important way [8]. One should also mention that the self-limiting profile of the AlGaAs groove, which is perturbed during the growth of the GaAs QWR layer, is quickly retained upon deposition of the AlGaAs barrier layer above the wire. This makes possible the vertical stacking of nominally identical wires in such a groove, as is obvious in Fig. 2. Similar arrays of InGaAs QWRs can be formed on self-limiting GaAs V-grooves[10].

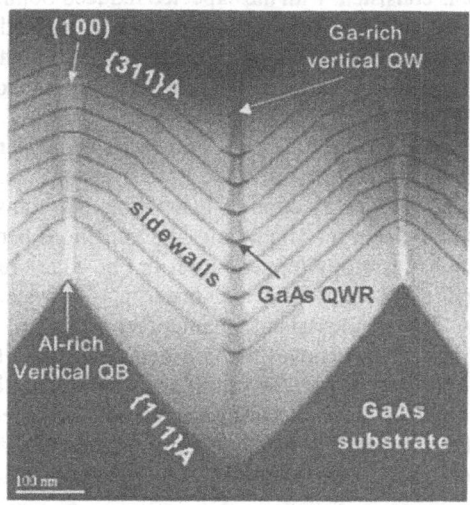

Figure 2. TEM cross section of a vertically-stacked, multiple QWR V-groove structure.

2.2   LUMINESCENCE AND ABSORPTION SPECTRA

The luminescence and absorption characteristics of low-dimensional structures are of prime importance concerning their assessment for optoelectronic device applications. The *in situ* formation of the V-groove QWRs results in high luminescence efficiency and negligible interface recombination effects, as in high quality QW structures.   One difficulty that may arise with these QWR heterostructures is that, for sufficiently large wire-to-wire separation (typically more than ~1μm), the luminescence emanating from the QW sections in between the wires dominates the optical spectra as a large fraction of the carriers recombine in these well regions. However, for wire arrays with sub-μm spacing and at high enough temperature, the carrier diffusion length becomes comparable to or larger than the wire separation and the wire luminescence becomes dominant.

Photoluminescence excitation (PLE) spectra of the QWRs contain important information about the 2D quantum confinement features and the importance of Coulomb correlation effects in these 1D systems.   Figure 3(a) shows a typical low temperature PL and PLE spectra of a GaAs/AlGaAs V-groove QWR array grown by low-pressure (20mb) OMCVD. The PL spectra of such wires exhibit a peak due to the recombination at the wires, at photon energies controlled by the wire thickness and with linewidths depending on the interface roughness (typically 5-10meV). The PLE spectra are shown for excitation with linearly polarized light, with polarization direction either parallel or perpendicular to the wire axis. One important feature of the PLE spectra is the series of peaks associated with transitions between 1D electron and hole subbands; the calculated transition energies are also given for comparison. The fact that only a single set of peaks is observed, rather than one corresponding to excitons and another to free carrier transitions, is consistent with the expected reduced Sommerfeld factor in 1D system [11,12]. Thus, the enhanced Coulomb correlation in 1D leads to the suppression of the singularity associated with free carrier recombination. Similar spectra have been obtained using absorption spectroscopy of V-groove QWRs embedded in a V-shaped optical waveguide [13].

Another feature related to the 1D nature of the sample is the marked polarization anisotropy of the PLE signal. Essentially, the optical absorption for light polarized along the wires is associated with electron to heavy-hole-like transitions, whereas for perpendicular polarization electron to light-hole-like transitions become allowed. The details of the polarization anisotropy (upper curves in Fig. 3) are related to the mixing between heavy hole and light hole states, which occur in a 1D system even at the center of the Brillouin zone due to the lateral quantum confinement [14]. In strained InGaAs/GaAs QWRs, the strain removes the degeneracy between the heavy and light hole bands at the Brillouin zone center. This eliminates the valence band mixing effects and hence quenches the polarization anisotropy at the relevant spectral range [15]. It should be noted that the strained QWRs do exhibit marked polarization anisotropy in their PL spectra, which makes them useful for applications in polarization-controlled light-emitters [15].

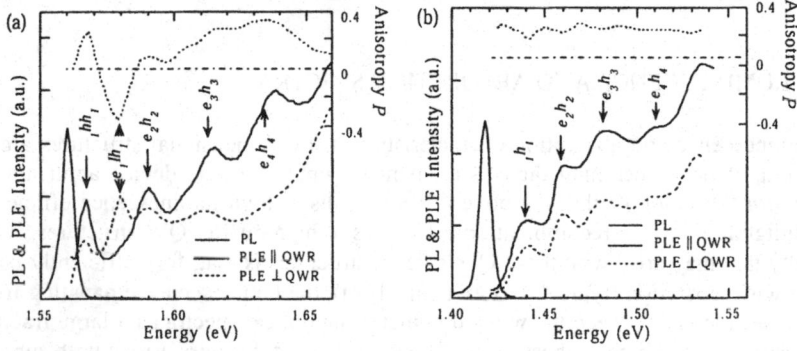

Figure 3. PLE spectra of (a) lattice-matched GaAs/AlGaAs and (b) strained InGaAs/AlGaAs V-groove QWR structures.

2.3   TWO-DIMENSIONAL QUANTUM CONFINED STARK EFFECT

The modification in the optical spectra brought about by the application of an electric field play an important role in different optoelectronic devices such as optical waveguide modulators and switches. We have investigated theoretically and experimentally the effect of an electric field on the electronic states and optical spectra of V-groove QWRs.

A unique effect of the electric field in V-groove QWRs is illustrated in Fig. 4, which shows the calculated wavefunctions of the ground electron state in such a wire for different electric fields [16]. For electric fields applied in the growth direction, the electron wavefunction is squeezed into the lower, sharper corner of the crescent as a result of the electric force applied in that direction. This is very similar to the effect in a 2D QW layer, except that in this case an effective increase in the lateral confinement is obtained as well. For the opposite orientation of the electric field, however, a dramatically different effect takes place, namely, the ground state wavefunctions splits into two *in-phase* parts due to the influence of the sharp edge of the heterostructure potential. This splitting of the wavefunction also takes place in the excited, odd wavefunctions whose central lobes split as well due to the application of an electric field [16]. The splitting decreases the overlap with the corresponding hole wavefunctions, which are squeezed in the opposite corner and are not split. At the same time, it increases the overlap with other hole wavefunctions, thereby increasing the strengths of otherwise weak (or nominally forbidden) transitions. This 2D quantum confined Stark effect (QCSE) is therefore significantly different than what is obtained in conventional 2D QWs.

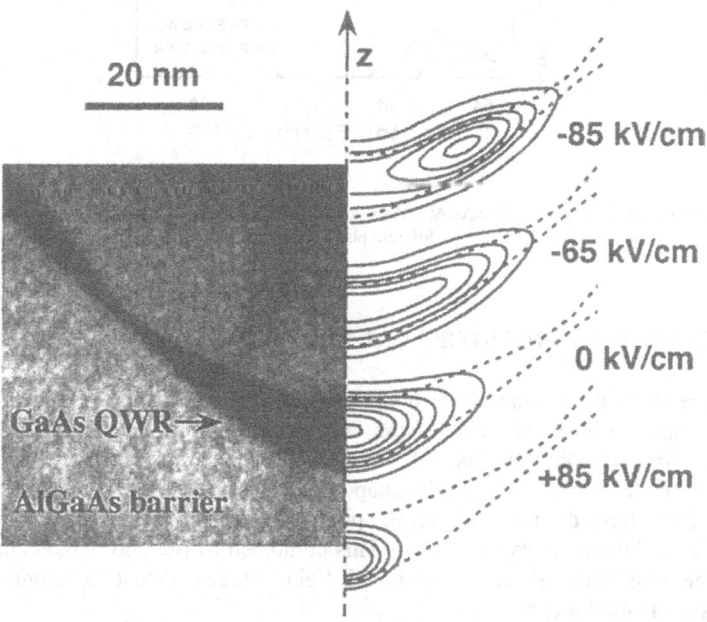

Figure 4. Left: TEM cross section of a GaAs/AlGaAs V-groove wire. Right: calculated wavefunctions of the ground electron state in the wire shown on left for different electric fields (positive fields correspond to electric fields in the growth direction).

This 2D QCSE was investigated experimentally by measuring the PL and PLE spectra of GaAs/AlGaAs V-groove wires embedded in a pn junction at different diode biases (see Fig. 5) [17]. The observed Stark shifts are in agreement with the calculated ones. Moreover, the characteristic polarization anisotropy of the PLE spectra evolves with the application of the electric field as a result of the level shifting and the modified mixing taking place in the valence band.

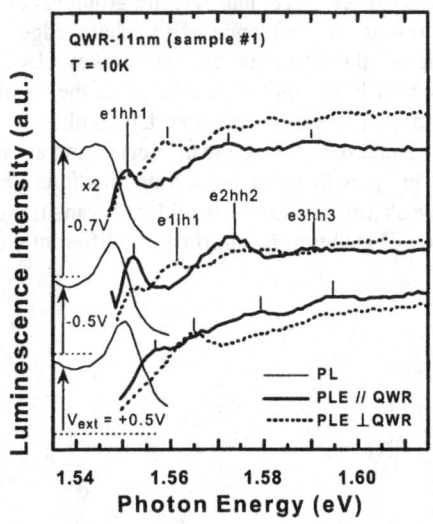

Figure 5. PL and PLE spectra of GaAs/AlGaAs V-groove QWR embedded in a pn junction, measured for different bias potentials.

2.4 QUANTUM WIRES IN OPTICAL MICROCAVITIES

The nature of the interaction between electron-hole pairs and the electromagnetic field, e.g., the characteristics of the emission and absorption spectra, depend both on the details of the electronic states as well as on the features of the electromagnetic modes. Whereas the electronic states can be shaped using quantum confinement effects as in the QWR structures discussed above, the photon modes can be controlled using optical waveguide and microcavity structures. This combined carrier and photon confinement offers the possibility of designing novel light sources, detectors, modulators and switches with new functionality.

As an example, we consider here the integration of V-groove QWR structures with planar Bragg microcavities. Such structures were produced by first preparing a GaAs/AlGaAs multilayer Bragg mirror of high reflectivity using conventional OMCVD growth. The surface of this Bragg mirror wafer was then pattern to produce a periodic array of V-grooves. Finally, the patterned wafer was regrown, again using OMCVD, with an InGaAs/GaAs single-QWR structure and a top GaAs/AlGaAs Bragg mirror. The layer thicknesses were selected such that the resulting strained QWRs are embedded at the center of a high finesse planar microcavity, with the wires overlapping with the peak of the confined photon mode. A TEM cross section of the resulting QWR microcavity structure is shown in Fig. 6 [18]. Note that the second growth step can be readily carried out because it involves growth of GaAs on GaAs, as opposed to growth on AlGaAs that usually suffers from surface oxidation and contamination. The absence of any traces of the regrown boundary in the TEM image attests to the high quality of this interface. Measurements of the finesse of the regrown cavity (>1000) confirm the high optical quality of the structure.

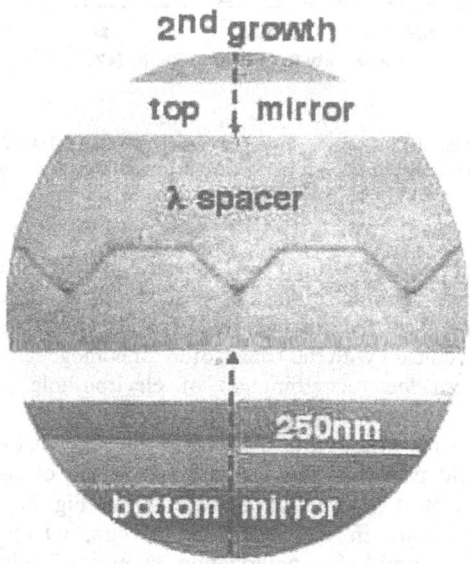

Figure 6. TEM cross section of a strained InGaAs/GaAs V-groove QWR structure embedded in a Bragg optical microcavity.

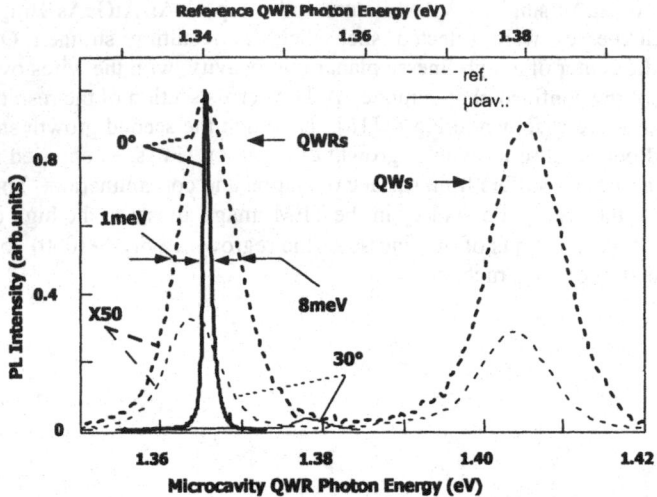

Figure 7. Comparison of the photoluminescence spectra of strained InGaAs/GaAs V-groove QWRs with and without an optical microcavity structure. Spectra for two different emission angles (with respect to the growth direction) are shown.

Figure 7 shows the measured low-temperature PL spectrum of the QWR microcavity structure, superimposed on the PL spectrum of a reference InGaAs/GaAs QWR sample grown without the microcavity [18]. The cavity photon mode, which is tuned in this case to coincide with the center of the inhomogeneously braodened QWR emission line, induces the recombination of electron-hole pairs with matching recombination energy. Narrow emission lines (1 meV) are obtained with these QWR microcavity structures for emission in the growth direction that couples well into the optical mode far field pattern. The spectral line shape of the emission in other directions resembles that of QWRs without a cavity (see Fig. 7). The QWRs preserve the polarization anisotropy in their emission spectra, which makes such QWR microcavity structures useful for polarization control of microcavity-based light emitting diodes (LEDs) and vertical cavity surface emitting lasers (VCSELs).

3. Localized Excitons in V-Groove QWRs

Interface roughness in QW heterostructures is known to yield exciton localization at potential minima generated by this roughness [19]. This results in excitons confined in all three directions, with important modifications in static and dynamic recombination features. A similar effect is expected in QWRs that exhibit potential fluctuations along the axis of the wire. In this case, however, the confining potential perpendicular to the wire is well defined, whereas only the potential distribution along the wire has a random character. Signatures of such 3D exciton confinement have been observed in disordered

QWRs of different types [20,21]. Here, the features of exciton localization in V-groove wires are summarized.

3.1 DISORDER IN V-GROOVE WIRES

The potential fluctuations along V-groove wires have their origin partly in lithography imperfections taking place during preparation of the channels. Slight width variations in the etching mask translate into undulations in the height of the sidewalls of the grooves. After epitaxial growth, these undulations are retained to a large extent, and are superimposed by surface details related to the evolution of the different crystalline facets.

One useful way of visualizing this surface morphology is using atomic force microscope (AFM) images measured inside the V-groove on the upper grown layer. Figure 8 shows an example of such a top-view (flattened) AFM image. It can be seen that the quasi-{111} sidewalls undulate with a periodicity of ~1μm, whereas the {311}A facets show a quasi-periodic structure on a finer scale. The (100) facet at the bottom of the V-groove is considerably flatter, showing monolayer-step height variations. It can be concluded that potential variations along these wires should have typical length scale in the 10-100 nm range. Similar morphology is also expected at the interface between the core and the barrier layers of the wire.

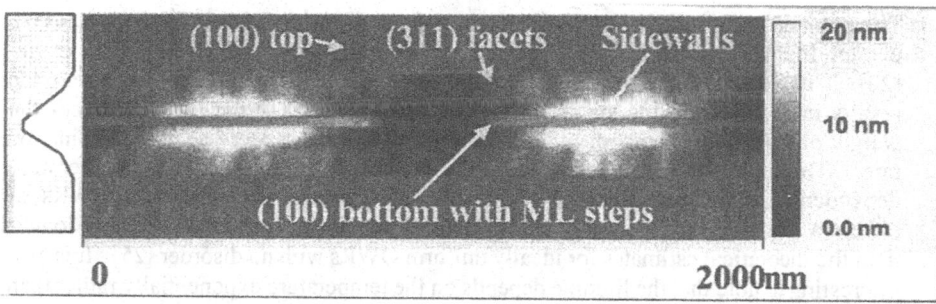

Figure 8. AFM top-view image of a V-groove QWR structure. The left part shows a schematic cross-section of the groove, aligned with the AFM image.

3.2 EVIDENCE FOR EXCITON LOCALIZATION

Several features in the optical spectra of the V-groove QWRs show evidence for exciton localization due to the interface roughness discussed above. One feature is the finite Stokes shift, i.e., the difference in energy between the peak of the PL spectrum and the lowest energy peak in the PLE spectra. This Stokes shift arises due to the fact that the PL peak is associated with localized excitons that relax to the bottom of the QWR potential well at the low energy tail of the DOS distribution. On the other hand, the peak of the PLE (or absorption) spectrum occurs near the peak of the DOS distribution at higher energies. The Stokes shift generally increases with decreasing QWR thickness (typically 3-20meV). Moreover, it decreases with increasing temperature and finally vanishes at temperatures for which the excitons become delocalized due to thermal activation [22].

A more direct evidence for exciton localization is provided by spatially resolved PL spectra of the QWRs. High spatial resolution can be obtained by fabricating apertures in opaque masks deposited on top of the sample, allowing the observation of PL spectra from wire segments much shorter than the optical wavelength. A series of such spatially resolved PL spectra is shown in Fig. 9 [21]. The apertures were fabricated in this case using electron beam lithography in Al thin films deposited on a GaAs/AlGaAs V-groove QWR array with a pitch of 0.5µm. It can be seen that as the aperture width approaches the typical length scale of the potential fluctuations, extremely sharp (~100µeV) fetaures appear in the PL spectra, representing the emission spectra of excitons trapped in the potential minima. For apertures containing several such minima, several peaks can be obtained, each corresponding to a different QD-like structure; the peak energy depends on the size of the QD structure (i.e., its extension along the axis of the wire), which determines the quantum confinement energy. For wide apertures, the spectral line shape retains its inhomogeneously broadened profile, which corresponds to a dense superposition of a multitude of localized exciton spectra originating from QD-like structures of different sizes.

3.3   EFFECT ON EXCITON DYNAMICS

The main effect of localization on the exciton dynamics is manifested by the increase in exciton lifetime with decreasing extension of the confining potential well along the wire [23]. This is because the confinement of the center of mass wavefunction of the exciton results in a corresponding spread in the wavefunction in k-space, which reduces the weight of the radiative components at k=0, thus resulting in a decrease in recombination rate. This effect is illustrated in Fig. 10, which shows the measured temperature dependence of the radiative carrier lifetime in V-groove GaAs/AlGaAs QWRs of different thicknesses [24]. At T=0, the lifetimes are about 400ps, considerably longer than the theoretical estimates for ideally uniform QWRs with no disorder [25]. It is also interesting to note that the lifetime depends on the temperature exponentially rather than as $T^{1/2}$, as predicted for ideal wires [25]. More direct measurements of the localized exciton lifetime as a function of the size of the confining QD have shown a clear linear dependence between the recombination rate and the dot size, as expected [26].

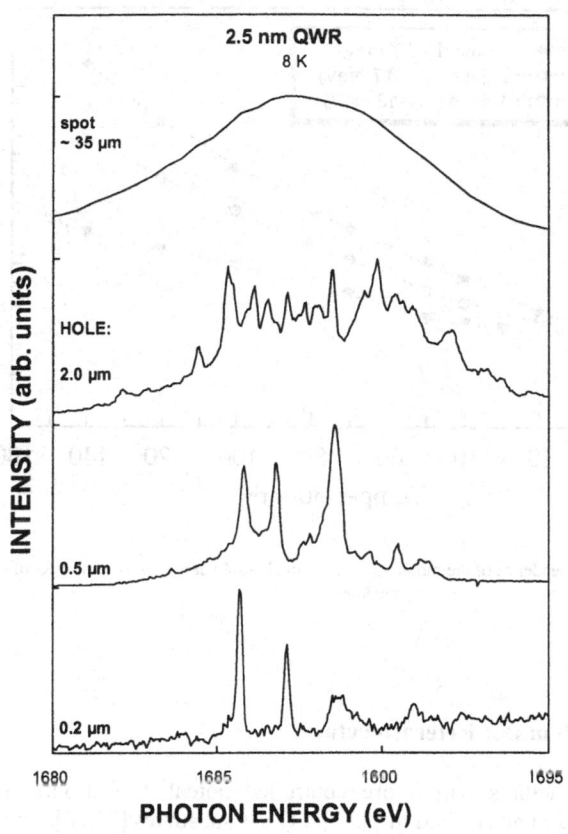

Figure 9. Low-temperature PL spectra of a GaAs/AlGaAs V-groove QWR structure measured through apertures of different diameters.

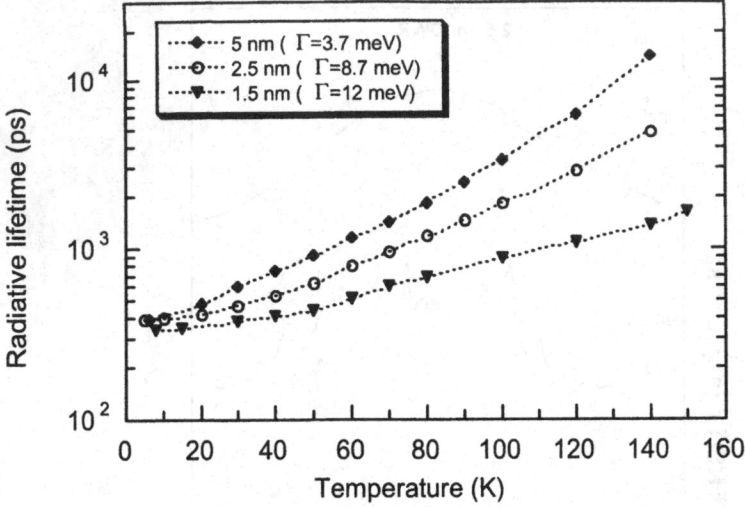

Figure 10. Temperature dependence of the radiative lifetime in GaAs/AlGaAs V-groove QWRs of different thickness.

4. Pyramidal Quantum Dot Heterostructures

Quantum dot heterostructures with more controlled potential distributions can be obtained using OMCVD in pits etched in {111}B GaAs substrates [27,28]. In this case, faceting and capilarity effects similar to those observed in the V-groove QWR structures give rise to an inverted-pyramid, self-limiting surface morphology in which QDs with well defined positions and shapes can be deposited. Again, this combination of lithography for determining the site for nucleation of the dots and the self-limiting growth yields seeded self-ordering of QDs with structural properties that are difficult to obtain with completely spontaneous self-ordering methods.

4.1 STRUCTURE OF PYRAMIDAL QUANTUM DOTS

Figure 11(a) shows a top-view scanning electron microscope (SEM) image of GaAs/AlGaAs pyramidal QDs after OMCVD growth. The upper surface of each inverted pyramid is defined by three intersecting near-{111}A surfaces. Between two of these facets, GaAs V-groove QWRs are formed, just as in the case of V-groove structures made on (100) GaAs substrates. However, in the pyramidal structures, a lens-shaped GaAs QD is formed as well at the point where these three wires meet. In addition, GaAs QW layers grow on the near-{111} facets. In the AlGaAs barrier layers, a Ga-rich VQW is produced above and below each GaAs QWR. A new feature of the pyramidal structure is the formation of a vertical QWR (VQWR) running through the center of the pyramid. Multiple QD structures can be formed simply by vertically stacking GaAs dots, as for wires in the case of V-groove structures. Much of this structural information has been obtained using cross-sectional AFM imaging techniques [29].

For optical characterization as well as applications requiring dots placed next to a convex surface, it is useful to produce similar QD heterostructures, but with up-right positioned (rather than inverted) pyramids. This has been accomplished by etching the GaAs substrate and supporting the concave part of the pyramids using a Au film [29]. An SEM image of the up-right pyramids is shown in Fig. 11(b). The lens effect produced at the convex, sharp tip of the pyramid helps increasing the PL luminescence efficiency of the dots, permitting measurements of PL spectra at extremely low excitation levels (see below).

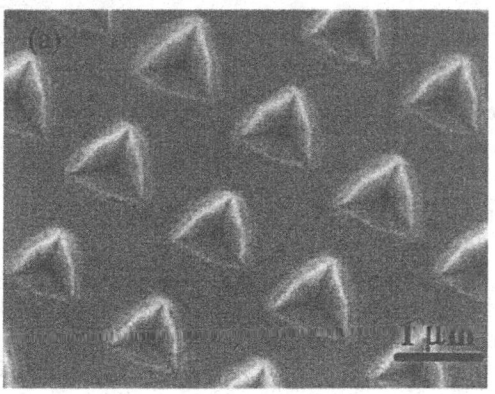

Figure 11. SEM top view of (a) as-grown and (b) etched pyramidal GaAs/AlGaAs QD structures.

4.2   LUMINESCENCE SPECTRA

The low-temperature PL spectrum of a single QD-pyramid and the corresponding PL image are shown in Fig. 11. Three groups of lines are evident in the PL spectrum. Their origin has been established using micro-PL and cathodoluminescence (CL) spectroscopy [29]. The lowst-energy lines arise from recombination in the GaAs QDs, the intermediate-energy ones from the GaAs QWRs and the highest energy features are due to recombination in the GaAs QWs. Notice the large energy difference between the QD and the QWR features, which indicate deep (more than 100meV) QD potential wells. The sub-structure in the QWR and QW parts of the spectrum is due to exciton localization induced by interface roughness. The origin of the multiple lines in the single-QD spectrum is discussed below. Such measurements, repeated over a large number of QD heterostructures grown over mm^2 areas have shown variations in the peak energy of the QD feature of only ~10meV. This uniformity is significantly better than what is achieved with completely spontaneous self-ordering techniques.

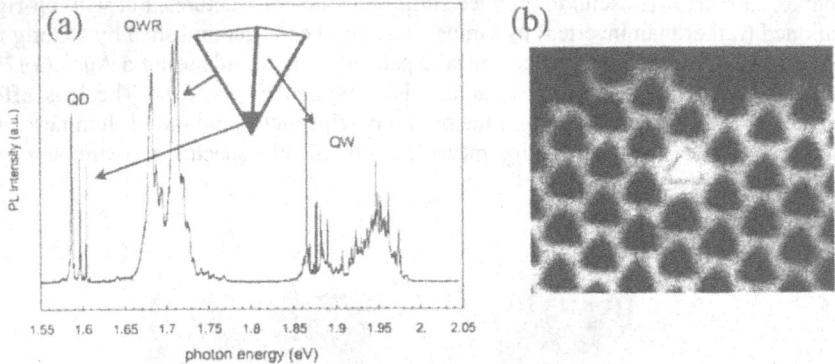

Figure 12. Low temperature PL spectrum (a) and PL image (b) of a single GaAs/AlGaAs QD pyramid.

4.3   MULTI-CHARGED EXCITONS IN QUANTUM DOTS

The high PL efficiency of the pyramidal dots permits studies of their optical properties under extremely low excitation conditions. A particularly interesting regime corresponds to the excitation of a single electron-hole pair at a time within the dot. The behavior of such a system is of special importance for applications in single-photon emitter applications.

A typical evolution of the PL spectrum of a single pyramidal QD structure as a function of excitation power is shown in Fig. 13 [30]. Only the spectral part corresponding to the ground state of the dot is shown for simplicity. Two excitation regimes are indicated. The one up to powers of 580pW corresponds to a single electron-hole pair excitation, whereas more than one pair may exist simultaneously for higher excitation levels. Multiple lines are observed even in the single-pair regime, attributed to few-particle, charged exciton states. The extra charge in the QDs has its origin in the impurity background of the AlGaAs barrier layers. Under the particular OMCVD growth conditions selected for these structures, carbon impurities are incorporated into the lattice as donors, releasing electrons that transfer to the nearby dots in the absence of photoexcitation. Under optical excitation, electron-hole pairs are generated. Initially, the electrons neutralize the impurities and the holes find their way into the dots, where they recombine in the presence of the other electrons. Since the hopping of the electrons from the neutralized donors to the dots is relatively slow, the steady-state number of electrons in the dots can be tuned by varying the excitation power [30]. As a result, the spectrum evolves with the excitation power even in the single-pair regime, showing emission lines due to single pairs in the presence of one, two, three,... electrons. At excitation levels beyond the single pair regime, multiple-exciton states are also possible. This picture is consistent with the power dependence of the integrated PL intensity (see inset in Fig. 13) and has been confirmed by model calculations [30].

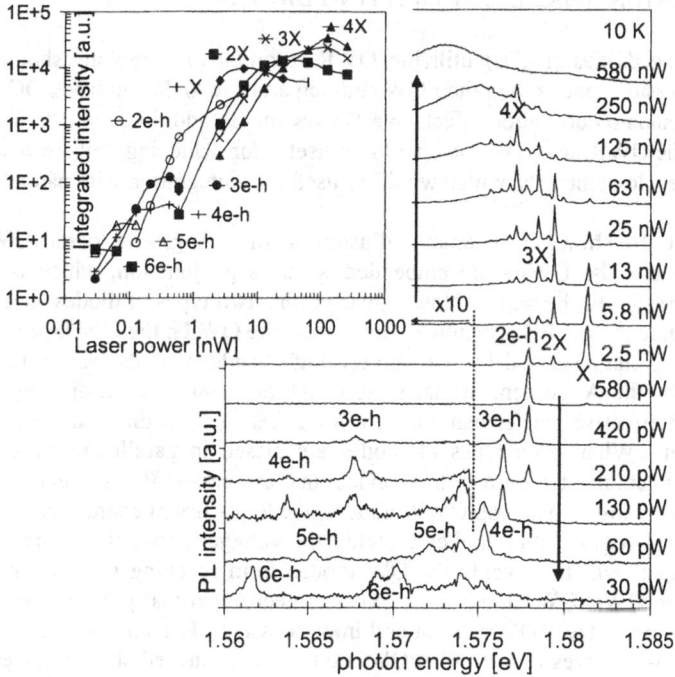

Figure 13. Power dependence of PL spectra (main panel) and integrated intensity (inset) of a single, etched GaAs/AlGaAs QD pyramid.

These results show that the single-particle model for QD luminescence spectra, suggesting extremely sharp, single lines for single-QD structures, is inadequate. This implies that optoelectronic devices relying on arrays of QDs should not only be perfectly uniform in terms of the potential distribution of the dots. Rather it should also have the same impurity environment and the same number of excited electron-hole pairs in order to reap the benefits of a δ-function like DOS distribution.

5. Applications in Optoelectronic Devices

The nonplanar quantum structures exhibit high optical quality, resulting from the fact that their interfaces are formed *in situ*, without the intervention of potentially damaging post-growth processing. Furthermore, they are capable of laterally confining both electrons and holes. This makes them attractive for applications in novel optoelectronic devices. Some examples of such applications are described in what follows, with an emphasis on prospects for achieving device performance that is beyond the capabilities of conventional, QW-based optoelectronics technology.

5.1   QUANTUM WIRE LIGHT EMITTING DIODES

Light emitting diodes (LEDs) utilizing QWR recombination regions should exhibit narrower emission spectra than their QW counterparts due to the modified DOS and the enhanced Coulomb correlation effects in a 1D system. In addition, the extremely small volume of the QWR active region should be useful for achieving devices that operate with extremely low currents, which would be useful for integration with nanoelectronics devices.

Figure 14 shows a schematic illustration of a V-groove based QWR light emitting diode. The QWRs are embedded within a pn junction, which is used for injecting carriers into the wires. Note, however, that two types of diodes, connected in parallel, form in these LED structures. In between the QWRs, the QW layers form one type of diode, characterized by a relatively high bandgap of its barrier material as determined by the Al content in that region. At the QWR sites, a different diode is formed, with effective barrier material of a lower bandgap, as dictated by the Ga-rich VQW region. When both types of diodes are biased in parallel using a common contact, as is actually the case in this device, one can expect that current would flow first through the lower-bandgap QWR diode as the light-current characteristic switches on first at that region. For higher currents the voltage across the entire structures remains pinned, which prevents the QW diodes from reaching the turn-on voltage. Thus, in these QWR/QW diodes there exists a mechanism that promotes *preferential carrier injection* at the VQW portion and into the wires. Evidence for this process is given by the I-V curves of the QWR/QW diodes, which indeed show a lower turn-on voltage as compared with that in a QW diode with the same AlGaAs barrier material as in its QW diode sections [31].

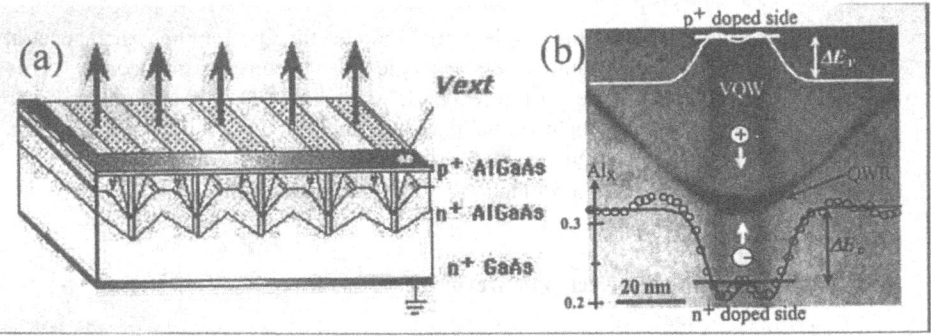

Figure 14. (a) Schematic illustration of a V-groove QWR light emitting diode. (b) TEM cross section of a V-groove QWR embedded in a light emitting diode structure. Carrier injection paths are schematically indicated.

Evidence for this preferential injection process is provided in Fig. 15 [31]. Whereas the PL spectra show emission lines from both the QWR and the QW regions, the electroluminescence (EL) spectra show uniquely QWR emission up to room temperature. Note that population of excited 1D subbands can be identified in the EL spectra at elevated temperatures.

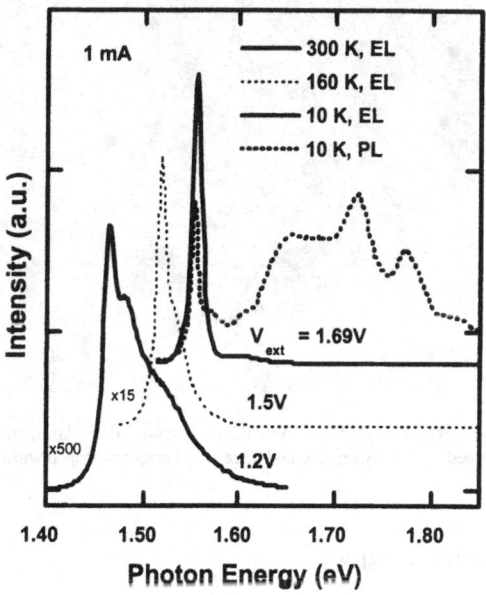

Figure 15. Comparison of photoluminescence (PL) and electroluminescence (EL) spectra of V-groove QWR light emitting diodes.

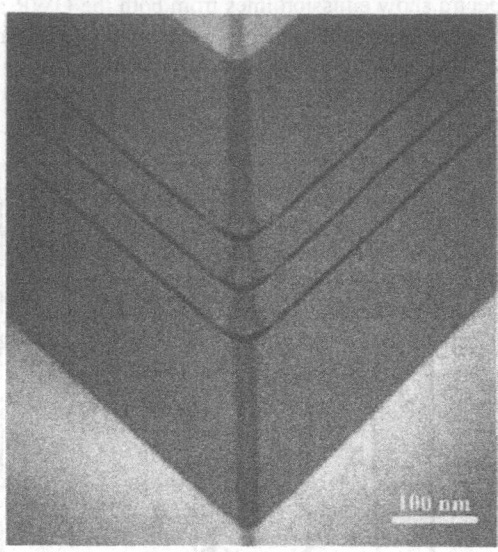

Figure 16. TEM cross section of a 3-QWR GaAs/AlGaAs V-groove diode laser, showing the AlGaAs V-shaped optical waveguide and three GaAs crescent-shaped wires.

5.2   QUANTUM WIRE LASERS

Different configurations of V-groove QWR diode lasers have been demonstrated and studied [32, 33]. Figure 16 shows a TEM cross section of a 3-QWR laser with a longitudinal cavity configuration, in which the wires run along the optical waveguide. The V-shaped optical waveguide self-formed around the wires provides tight, 2D waveguiding, with optical mode widths as small as ~0.5μm. This tight waveguiding was confirmed in scanning near field optical microscopy (SNOM) measurements of such devices under operation [34].

The increased importance of Coulomb correlation in a 1D system might lead to excitonic lasing of 1D lasers, with direct benefits to their performance related to the possibly higher excitonic gain. Such excitonic lasing mechanism has been suggested by the lasing spectra of T-shaped QWR lasers [1]. To investigate more directly the possible excitonic nature of the lasing process in 1D lasers, the PLE, PL and lasing spectra of optically pumped, multi-QWR V-groove lasers similar to that of Fig. 16 were compared [35]. The measured spectra of these devices, operated at low temperature, are shown in Fig. 17. Comparison of the lowest energy peaks in these spectra demonstrates that lasing takes place due to transitions between conduction and valence band states associated with the ground 1D subband. Furthermore, an estimation of the carrier density in the wires at threshold (1-2 10^5 cm^{-1}) indicates that lasing occurs well below the expected Mott transition, and therefore should be of excitonic nature. The fact that lasing takes place at the low energy part of the PL line may suggest that localized excitons contribute to the lasing process.

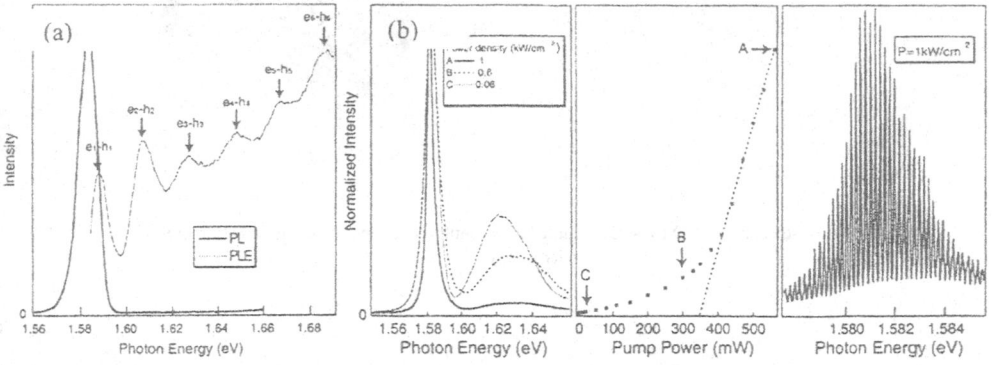

Figure 17. (a) PL and PLE spectra of a multi-QWR V-groove laser structure. (b) left to right: lasing spectra, L-I curve and a high resolution lasing specrum of the V-groove QWR laser structure of (a).

5.3 QUANTUM DOT SCANNING TIPS

One of the most exciting applications of nano-photonics devices involves the use of single-QDs for illumination and light detection, e.g., for use in single molecule probing and manipulation. The single-QD pyramidal structures offer an additional advantage, namely, the fact that the QD in these structures is situated in close proximity (~50-100nm) to a very sharp tip (~10-20nm radius of curvature). This makes these single-QD structures potentially useful in SNOM applications.

One possible implementation of such a QD-based scanning probe device is illustrated in Fig. 18. The pyramid, containing the QD near its tip, is mounted on a cantilever that is employed for scanning the tip across the surface of interest. In an illumination mode of operation, the QD can be excited either optically, e.g., by a laser beam, or electrically. In the latter case, the QD can be embedded in a pn junction similar to the structure used in the QWR LEDs described in section 5.1. Variations in the detected intensity of the QD emission as the probe is scanned near a surface can then be used to obtain a SNOM image of the surface. It is interesting to note that in the electrically pumped version, it should be possible to utilize the VQWR structure running across the center of the pyramid in order to achieve preferential injection into the QD, similar to the preferential injection into the QWRs discussed in section 5.1. Furthermore, the same QD diode can then be used to detect light during the surface scanning when the pn junction is reversed biased, yielding a QD photodiode SNOM device. If such devices can be realized with efficient light emission and/or detection, they might offer a good alternative to traditional, optical fiber based SNOM tips that suffer from high light transmission losses.

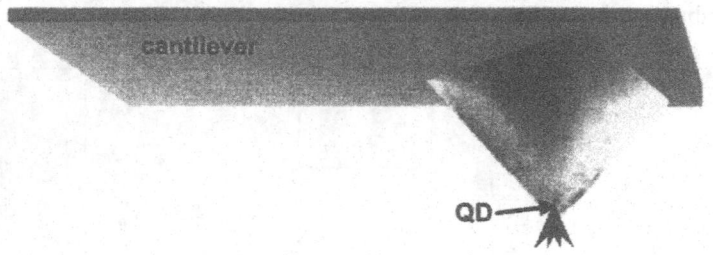

Figure 18. Illustration of a QD pyramid mounted on a cantilever for application in scanning probe microscopy.

6. Conclusions and Future Directions

Considerable progress has been achieved in recent years in developing high quality semiconductor QWR and QD structures using OMCVD growth on nonplanar substrates. These structures are particularly useful for applications in optoelectronic devices because of the *in situ* formation of all interfaces and the significant degree of control over the possible configurations allowed by the fabrication process. Several applications of these structures have already been demonstrated, particularly good quality QWR LEDs and diode lasers, and new exciting applications, particularly those relying on single-QDs, appear feasible. The challenge for future work in this area remains in the further perfection of this technology, particularly in reducing localization effects in V-groove QWRs and in controlling the environment of pyramidal QDs, as well as in the demonstration of optoelectronic devices with a clear advantage over traditional QW-based optoelectronics technology.

7. Acknowledgements

I would like to thank my collaborators at the Laboratory of Nanostructures for their contributions to the work reviewed in this manuscript, and Yann Ducommun for his help in preparing the manuscript.

8. References

1. Wegscheider, W. Pfeiffer, L. N. Dignam, M. M. Pinczuk, A. West, K. W. McCall, S. L. and Hull, R. (1993) Lasing from Excitons in Quantum Wires. *Physical Review Letters* **71**, 4071-4074
2. Goldstein, L Glas, F Marzin, J. Y. Charasse, M.N. and Le-Roux, G. (1985) Growth by Molecular Beam Epitaxy and Characterization of InAs/GaAs Strained-Layer Superlattices. *Applied Physics Letters* **47**, 1099-1101
3. Kapon, E. Tamargo, M. C. and Hwang, D. M (1987) Molecular Beam Epitaxy of GaAs/AlGaAs Superlattice Heterostructures on Nonplanar Substrates. *Applied Physics Letters* **50**, 347-349

4. Wegscheider, W. Schedelbeck, G. Abstreiter, G. Rother, M and Bichler, M. (1997) Atomically precise GaAs/AlGaAs Quantum Dots fabricated by Twofold Cleaved Edge Overgrowth. *Physical Review Letters* **79**, 1917-1920

5. Hartmann, A. Ducommun, Y. Loubies, L. Leifer, K. and Kapon, E. (1998) Structure and Photoluminescence of Single AlGaAs/GaAs Quantum Dots Grown in Inverted Tetrahedral Pyramids. *Applied Physics Letters* **73**, 2322-2324

6. Biasiol, G. and Kapon, E. (1999) Mechanism of Self-Limiting Epitaxial Growth on Nonplanar Substrates. *Journal of Crystal Growth* **201/202**, 62-66

7. Xie, Q. Madhukar, A. Chen, P. and Kobayashi, N. P. (1995) Vertically Self-Organized InAs Quantum Box Islands on GaAs(100). *Physical Review Letters* **75**, 2542-2545

8. Biasiol, G. and Kapon, E. (1998) Mechanisms of Self-Ordering of Quantum Nanostructures Grown on Nonplanar Surfaces. *Physical Review Letters* **81**, 2962-2965

9. Martinet, E. Gustafsson, A. Biasiol, G. Reinhardt, F. Kapon, E. and Leifer, K. (1997) Carrier Quantum Confinement in Self-Ordered AlGaAs V-Groove Quantum Wells. *Physical Review B* **56**, R7096-R7099

10. Constantin, C. Martinet, E. Rudra, A. Leifer, K. Lelarge, F. Biasiol G. and Kapon, E. (1999) Organometallic Chemical Vapor Deposition of V-Groove InGaAs/GaAs Quantum Wires Incorporated in Planar Bragg Microcavities. *Journal of Crystal Growth* **207**, 161-173

11. Ogawa, T. Takagahara, T. (1991) Optical Absorption and Sommerfeld Factors of One-dimensional Semiconductors. Physical Review B **44**, 8138-8156

12. Rossi, F. and Molinari, E. (1996) Coulomb-induced Suppression of Band-Edge Singularities in the Optical Spectra of Realistic Quantum-Wire Structures. *Physical Review Letters* **76**, 3642-3645

13. Martinet, E. Dupertuis, M.-A. Sirigu, L. Oberli, D. Y. Rudra, A. Leifer, K. and Kapon, E. (2000) Direct Observation of New Transitions in the Absorption Spectra of a V-Groove Quantum Wire Waveguide. *Physica Status Solidi* (in print)

14. Vouilloz, F. Oberli, D. Y. Dupertuis, M.-A. Gustafsson, A. Reinhardt, F. and Kapon, E. (1997) Polarization Anisotropy and Valence Band Mixing in Semiconductor Quantum Wires. *Physical Review Letters* **78**, 1580-1583; Vouilloz, F. Oberli D. Y., Dupertuis, M.-A. Gustafsson, A. Reinhardt, F. and Kapon, E. (1998) Effect of Lateral Confinement on Valence Band Mixing and Polarization Anisotropy in Quantum Wires. *Physical Review B* **57**, 12378-12387

15. Martinet, E. Dupertuis, M.-A. Reinhardt, F. Biasiol, G. Kapon, E. Stier, O. Grundmann, M. and Bimberg, D. (2000) Separation of Strain and Quantum Confinement Effects in the Optical Spectra of Quantum Wires. *Physical Review B* (in print)

16. Dupertuis, M.-A. Martinet, E. Weman, H. and Kapon, E. (1998) Quantum Confined Stark Effect in Quantum Wires: Wavefunction Splitting and Cascading. *Europhysics Letters* **44**, 759-765

17. Weman, H. Martinet, E. Dupertuis, M.-A. Rudra, A. Leifer, K. and Kapon, E. (1999) Two-Dimensional Quantum-Confined Stark Effect in V-Groove Quantum Wires: Excited State Spectroscopy and Theory. *Applied Physics Letters* **74**, 2334-2336

18. Constantin, C. Martinet, E. Rudra, A. and Kapon, E. (1999) Observation of Combined Electron and Photon Confinement in Planar Microcavities Incorporating Quantum Wires. Physical Review B **59**, R7809-R7812

19. Gammon, D. Snow, E. S. Shanabrook, B. V. Katzer, D. S. Park, D. (1996) Homogeneous Linewidths in the Optical Spectrum of a Single Gallium Arsenide Quantum Dot. *Science* **273**, 87-90

20. Hasen, J. Pfeiffer, L. N. Pinczuk, A. Song-He, West, K. W. and Dennis, B. S. (1997) Metamorphosis of a Quantum Wire into Quantum Dots. *Nature* **390**, 54-57

21. Vouilloz, F. Oberli, D. Y. Dwir, B. Reinhardt, F. and Kapon, E. (1998) Observation of Many-Body Effects in the Excitonic Spectra of Semiconductor Quantum Wires. *Solid State Communications* **108**, 945-948

22. Oberli, D. Y. Dupertuis, M.-A. Reinhardt, F. and Kapon, E. (1999) Effect of Disorder on the Temperature Dependence of Radiative Lifetimes in V-Groove Quantum Wires. *Physical Review B* **59**, 2910-2914

23. Citrin, D. S. (1992) Radiative Lifetimes of Excitons in Quantum Wells: Localization and Phase-Coherence Effects. *Physical Review B* **47**, 3832-3841

24. Oberli, D. Y. Vouilloz, F. Dupertuis, M.-A. Fall, C. J. and Kapon, E. (1995) Optical Spectroscopy of Semiconductor Quantum Wires. *Il Nuovo Cimento* **17D**, 1641-1650

25. Citrin, D. S. (1992) Long Intrinsic Radiative Lifetimes of Excitons in Quantum Wires. *Physical Review Letters* **69**, 3393-3396

26. Bellessa, J. Voliotis, V. Grousson, R. Wang, X. L. Ogura, M. Matsuhata, H. (1998) Quantum-Size Effects on Radiative Lifetimes and Relaxation of Excitons in Semiconductor Nanostructures. *Physical Review B* **58**, 9933-9940

27. Sugyiama, Y. Sakuma, Y. Muto, S. and Yokoyama, N. (1995) *Japanese Journal of Applied Physics* **34**, 4384-4386

28. Hartmann, A. Loubies, L. Reinhardt, F. and Kapon, E. (1997) Self-Limiting Growth of Quantum Dot Heterostructures on Nonplanar {111}B Substrates. *Applied Physics Letters* **71**, 1314-1316

29. Hartmann, A. Ducommun, Y. Leifer, K. Kapon, E. (1999) Structure and Optical Properties of Semiconductor Quantum Nanostructures Self-Formed in Inverted Tetrahedral Pyramids. *Journal of Physics: Condensed Matter* **11**, 5901-5915

30. Hartmann, A. Ducommun, Y. Kapon, E. Hohenester, U. and Molinari, E. (2000) Few-Particle Effects in Semiconductor Quantum Dots: Observation of Multi-Charged Excitons. *Physical Review Letters* **84**, 5648-5651

31. Weman, H. Martinet, E. Rudra, A. and Kapon, E. (1998) Selective Carrier Injection into V-Groove Quantum Wires. *Applied Physics Letters* **63**, 2959-2961

32. Kapon, E. Hwang, D. M. and Bhat, R. (1989) Stimulated Emission in Quantum Wire Semiconductor Heterostructures. *Physical Review Letters* **63**, 430-433

33. Kapon, E. (1993) Quantum Wire Lasers Grown by OMCVD on Nonplanar Substrates. *Optoelectronics Devices and Technologies* **8**, 429-460

34. Ben-Ami, U. Nagar, R. Ben-Ami, N. Scheuer, J. Orenstein, M. Eisenstein, G. Lewis, A. Kapon, E. Reinhardt, F. Ils, P. and Gustafsson, A. (1998) Near-Field Scanning Optical Microscopy Studies of V-Grooved Quantum Wire Lasers. *Applied Physics Letters* **73**, 1619-1621

35. Sirigu, L. *et al.* (2000) *Physical Review B Rapid Communications* (in print).

QUANTUM DOT LASERS

H. SCHWEIZER, J. WANG, U. GRIESINGER, M. BURKARD, J.
PORSCHE, M. GEIGER, AND F. SCHOLZ
4. Phys. Inst. Universitat Stuttgart,
Pfaffenwaldring 57, 70550 Stuttgart, Germany.
T. RIEDL, AND A. HANGLEITER
Institut fur Technische Physik Universitat Braunschweig,
Mendelsohnstr. 2, 38106 Braunschweig, Germany.

Abstract

Downsizing of the active region of semiconductor lasers to zero dimension has different effects on static and dynamic laser parameters. Due to the thermodynamics of low dimensional electron hole plasmas the most prominent effects are the threshold reduction and the improved temperature stability. Based on the changed density of states also effects on the modulation response and beam quality of the laser are theoretically expected but not yet demonstrated. This obvious discrepancy between theoretical expectations and experimental results must be related to additional effects as carrier transport and relaxation. After a brief review of nano fabrication aspects of dot lasers as self assembled methods as well as lithography based implantation and etching methods, laser device properties based on thermodynamics in low dimensional systems will be discussed. Strong emphasis will be put on carrier dynamics (transport, recombination, and relaxation). On the footing of a rate equation approach we discuss the static and dynamic properties of quantum dot lasers as a function of the dot array filling factor. Also some applications and device properties will be discussed. Based on periodic dot arrays gain coupled dot distributed feedback lasers can be realized which result in an improved side mode suppression ratio of the laser emission. From modulation experiments extremely low dynamic chirp of dot lasers can be observed making quantum dot lasers possibly promising candidates for high speed communication in the long wavelength range if modulation response limitations can be solved.

1. Introduction

Dot lasers are nanometer optical devices which combine length scales defining the confinement of photons (100 nm) and length scales defining the confinement of electrons and holes (10 nm) in the solid. Both scales are strongly different but must be controlled better than 0.2 nm to realize macroscopic filter structures. In Fig. 1 a typical resonator structure is shown with an active region consisting of quantum wires and dots and passive

65

L. Pavesi and E. Buzaneva (eds.), Frontiers of Nano-Optoelectronic Systems, 65–84.
© 2000 *Kluwer Academic Publishers. Printed in the Netherlands.*

66

regions serving as waveguide and cladding regions. For the fabrication of low dimensional lasers a number of techniques have been developed. Below we give a list of fabrication techniques to realize wire and dot structures.

1. Fabrication: Growth techniques
   1.1. growth on vicinal substrates (FLS fractional layer super-lattices [1] by MOVPE
   1.2. growth on misoriented substrates (tilted super-lattices (TSL)[2]) by MBE
   1.3. SSL- (serpentine superlattice [3,4]) technique
2. Fabrication: Lithographic techniques
   2.1. over-growth of patterned strained layers (buried stressors [5])
   2.2. over-growth of etched V-grooves (an extension of the buried crescent laser technique to nanometer sizes) [6,7]
   2.3. V-groove and migration (AlGaAs/GaAs material system) growth along $(111)_A$-planes for V-grooves in [01-1]-direction [8,9]
   2.4. lithography based patterning (dry etching and overgrowth) techniques [10,15]
   2.5. control of surface recombination- and dead layer- effects [11,15-17].
   2.6. local implantation enhanced intermixing (IEI) techniques [18-26]
   2.7. control of ion straggling, defects beneath local masks [22].
   2.8. focused ion beam (FIB) technique (wire and dot structures) [27]
3. Fabrication: Self assembled growth techniques (self assembled dots=SAD)
   3.1. growth of dots [28-32]

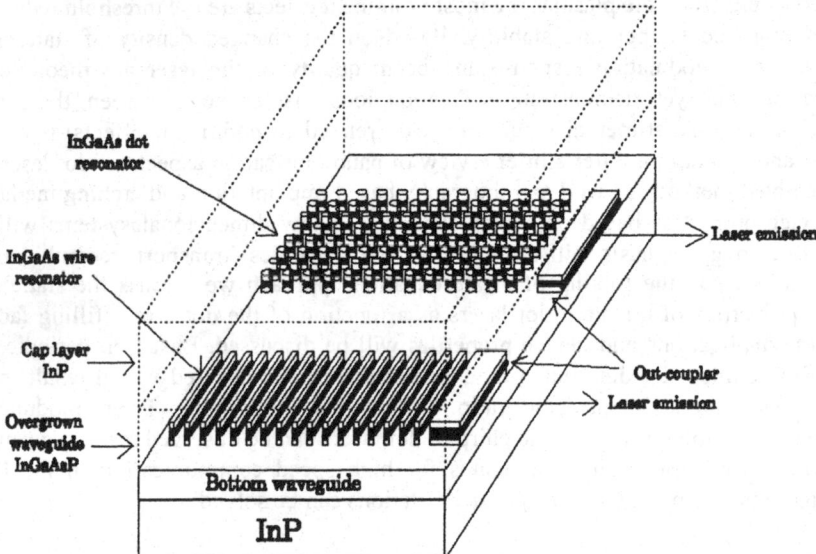

Figure 1. Principle of a low dimensional laser structure of a InGaAs wires and dots (active region) embedded in a InGaAsP waveguide and a InP cladding layer. The active wire and dot region can be realized by different fabrication techniques (see text).

TABLE 1 Device data of quantum dot lasers with self assembled dots (SAD) in the active region.

Group	Date Ref	Structure Fabrication	Stacks	λ (RT) nm	i_{th} (RT) A/cm^2
A.F.Ioffe Inst. TU.Berlin	4/97 [33,34]	GaInAs/AlGaAs MBE vertic. coupl	10	1000	60
U. Austin	5/97 [35,36]	InAlGaAs/ GaAs Surface Emitter		970	1100
TU.Berlin	7/97 [37]	InAs/GaAs+(GaInAs) MOCVD	3 1	1060 1000	220 181
Fujitsu Lab Atsugi.	7/96 [30,38]	GaInAs/GaAs MBE	3	1105	5000
U. Michigan	7/96 [39]	GaInAs/GaAs MBE	1	1028	600
U. Cal.	8/96 [40]	GaInAs/GaAs MBE	1	1020	---
Inst. Microstr.Sci Canada.	11/96 [41,42]	InAlAs/ AlGaAs MBE	1	723	4800
MRC, U. Texas ,	11/98 [43]	InGaAs/GaAs MBE	-	1300	270
MPI. Stuttg U. Stuttg U. Stuttg.	4/98 [44,45] 12/99 [46]	InP/GaInP MBE MOVPE	3 3	705 705	265@150 k 5000
U. New Mexico	6/99 [47]	InAs/InGaAs MBE	1	1250	26
NEC	9/99 [48]	InAs/InGaAs MBE	3	1250	76

A list of some dot laser data is given in Tab. 1. From the list we can individuate the trend that dot lasers with larg filling factors of the active region show improved threshold current densities. This threshold behavior is remarkably different to the behavior of 2D lasers where single well lasers for suitably adjusted resonator lengths nearly always show the lowest threshold current densities. Furthermore, besides the thermodynamically expected high characteristic temperature T_0 for the laser threshold of a dot laser (see below) nearly all dot lasers show a strong reduction of T_0 at higher temperatures combined with a strong reduction of the quantum efficiency. See for instance the InGaAs/GaAs material system [43,49]. These effects are attributed to the strongly modified carrier transport and carrier relaxation in low dimensional lasers and will be discussed in the next section.

2. Electronic effects/optical transitions/carrier dynamics

2.1 DENSITY OF STATES

The discussion of the advantages of low dimensional systems with respect to device operation parameters are usually based on the changed density of states in these systems [50]. The effects for various dimensions d can be immediately seen by studying the density of states D(E). Assuming spherical dispersion relation E(k) the number of states in k-space can be simply derived from the volume of a d-dimensional sphere with constant energy E (Eq. 32). The density of states follows from the derivative of the volume with respect to energy (Eq. 33). In Eq. 34, D(E) for different dimensions are given. Based on this density of states, the chemical potential μ of these systems can be calculated by inversion of the carrier density n(d) dependence on the temperature, where:

$$n(d) = \frac{4\pi}{h^d} \sum_{j=1}^{N} \left(2m_j KT\right)^{d/2} F_k(\eta),$$

(1)

$$F_k(\eta) = 1 + \int_0^\infty \frac{\varepsilon^k}{1 + e^{\varepsilon - \eta}} d\varepsilon,$$

(2)

$$\eta = \frac{\mu^{id}}{KT} \quad \text{with} \quad k = \frac{d}{2} - 1, \quad d = 3,2,1.$$

(3)

The result is the well known dimension dependence of the chemical potential on the density:

$$\mu \propto n^{2/d}.$$

(4)

By inverting Eq. 1, we end up with the well known dependency of the transparency density of on temperature,

$$n \propto T^{2/d}.$$

(5)

If we assume in a first approach no further density and temperature dependence of the carrier lifetime in a laser, Eq. 5 predicts an extremely week dependence of the laser threshold current on temperature [50] for low dimensional lasers [51].

The strong dependence of the chemical potential on the dimensionality shown by Eq. 4 can be nicely demonstrated in GaAs/AlGaAs quasi wire structures with different wire widths (see Fig. 2 and [52,53]). The main result for the chemical potential is a steeper increase of the potential value for smaller dimensions according to the decrease of the average number of states. This is demonstrated by the calculation depicted in Fig. 2, left, where chemical potential changes from $\mu^{(2)} = n^l$ (2D-limit) to $\mu^{(1)} = n^2$ (1D-limit) according to Eq. 4. This behavior is the reason for the laser threshold density reduction in low dimensional systems. Furthermore, the dimension dependence of the k-space occupation determines the temperature dependence of the density of transparency, which gives us the characteristic temperature T_0, if we neglect in a first approach the temperature dependence of the carrier lifetime in the LD-EHP:

$$T_0^{(d)} = \left[\frac{\partial \ln\left(n_{tr}^{(d)}\right)}{\partial T} \right]^{-1} \approx \frac{2}{d}$$

(6)

This increase of the characteristic temperature can be demonstrated again by calculating the density of transparency of GaAs/AlGaAs wires as function of temperature (Fig. 3) based on the experimental data of Fig. 2. Going to lower dimensions this temperature dependence develops also continuously becoming an infinitely large value of T_0 in the 0D-limit [50].

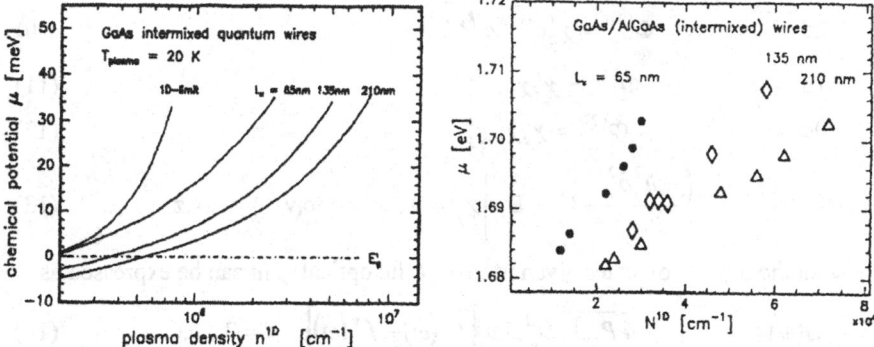

Figure 2 Left: Calculated chemical potential of quasi one dimensional GaAs wires v.s. carrier density including excact 1D-limit. Right: measured chemical potentials of GaAs/AlGaAs wires as function of carrier density for different wire widths.

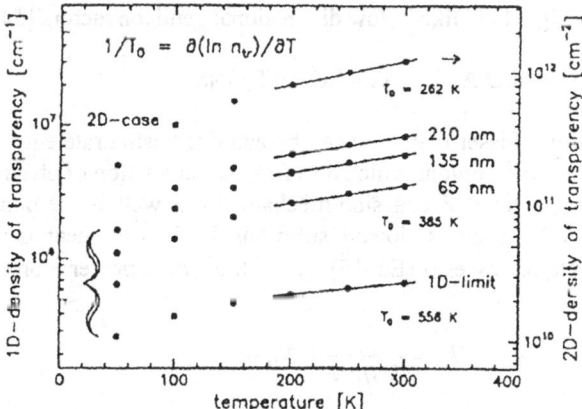

Figure 3. Calculated densities of transparency v.s. temperature for various one dimensional GaAs wires including 1D-limit.

2.2 OPTICAL TRANSITIONS

The optical transitions in a low dimensional gain medium can be described by a wave function (Eq. 7) consisting of a central cell function (Eq. 8) which describes the polarization of the optical transition and an envelope function (Eq. 9-12) which describes the subband selection rules of the optical transitions between different subbands. The envelope function is calculated from the Schrödinger equation 13.

Wave functions:

$$\Psi^{(d)i} = u_i \Phi_x^{(d)i} \quad d = 3,\ldots,0 \quad i = c, v \tag{7}$$

$$u_i = \text{central cell function e (S-type), h (P-type)} \tag{8}$$

Envelope functions [54,56]:

3d: $$\Phi^{(3)i} = e^{i\vec{k}\vec{r}} \tag{9}$$

$$2d: \qquad \Phi^{(2)i} = \chi_z^i e^{i\overrightarrow{k_x}\,x} e^{i\overrightarrow{k_y}\,y} \tag{10}$$

$$1d: \qquad \Phi^{(2)i} = \chi_z^i \chi_x^i e^{i\overrightarrow{k_y}\,y} \tag{11}$$

$$0d: \qquad \Phi^{(2)i} = \chi_z^i \chi_x^i \chi_y^i \tag{12}$$

$$\left(\frac{-\hbar^2 \partial_j^2}{2m_j} + V_j - E_j \right) \chi_j = 0, \qquad i=c,v, \quad j=x,y,z. \tag{13}$$

Together with the density of states given in Eq. 34 the optical gain can be expressed as

$$G(\omega) = \frac{4\pi^2 e^2}{ncm_o^2} \sum_{c,v} \left| \overrightarrow{eP_{c,v}} \right|^2 D_{c,v}^d(\omega) \left[f^c(\omega) - f^v(\omega) \right], \quad d = 3,\ldots,0. \tag{14}$$

Results on gain calculations can be found in [57]. The matrix elements are given in Eq. 35 in section 7.2. Due to the strong kp-interaction of the valence bands in 1D and 0D structures strong mixing occurs between heavy and light hole states. Therefore hole states appear as heavy or light hole like in low dimensional semiconductors [54,55].

2.3 LASER THRESHOLD AND RATE EQUATIONS

A simple description of laser operation can be achieved using rate equations for carriers (electrons and holes) and photons which describe the interaction of electrons and photons by stimulated and spontaneous emission mechanisms as well as the pump process (Eqs. 43,44 in section 7.3). Their stationary solutions lead to a linear dependence of the threshold density N_{th} on losses α (Eq. 15) and of the output power S on the pump current I (Eq. 18):

$$N_{th} = \frac{\alpha_g}{\partial G/\partial N} + N_{tr} \tag{15}$$

$$\text{for } G_r = \frac{I}{qV} \quad \Rightarrow \quad I_{th} = qV \frac{N_{th}}{\tau_S} \tag{16}$$

$$\text{for } \frac{1}{\tau} = A + BN_{th} \quad \Rightarrow \quad I_{th} = qVN_{th}\left(A + BN_{th}\right) \tag{17}$$

$$S = \frac{1}{Gv_{gr}qV}(I - I_{th}) \tag{18}$$

with α_g the optical resonator loss, G the optical gain which is equal to α_g, G_r the pump rate radiative contribution, N_{tr} the density of transparency, v_{gr} the group velocity of light, V the volume of the active area, A the linear recombination coefficient, and B the bimolecular recombination coefficient.

In a low dimensional laser the stationary solutions (Eqs. 18) are modified (modification of G_r) due to carrier transport processes and carrier relaxation processes [58-61]. A main result is the modification of the effective capture time τ_{cap}^{eff} in low dimensional lasers

resulting in a diffusion contribution τ_{diff} and in a scaled up capture time contribution which arises from the conservation law of particles in the active region-barrier scattering processes:

$$\tau_{cap}^{eff} = \tau_{diff} + \frac{V_b}{V_{ac}}\tau_{cap}^{Q} \tag{19}$$

Relation 19 can be nicely verified by measuring the modulation response of wire and dot lasers with various ratios of V_b/V_{ac} (see Fig.4 left as well as Fig. 11).

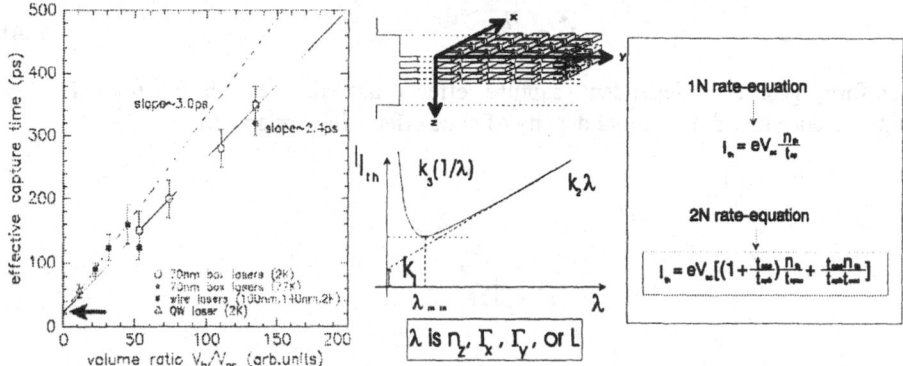

Figure 4 Left: measured effective capture time of wire and dots lasers v.s. volume ratio V_b/V_{ac}. Center: Geometry of a dot laser (top) and threshold current as function of universal structural factor λ Right: Expressions for the threshold current I_{th} for 1N (only active region)- and 2N-(barrier and active region) rate equation models.

The description of the dot or wire DFB laser operation can be obtained by a 2-level (active and passive regions) rate equation approach shown in Fig. 4, center and right (see also section 7.3). Similar to the example of QW lasers the laser threshold current can be expressed in terms of a structural factor $\lambda = n_z\Gamma_x\Gamma_yL$ (Fig. 4, center), with n_z=number of QWs or of dot stacks in growth direction, Γ_x=optical filling factor in x-direction, Γ_y=optical filling factor in y-direction, L=resonator length. Usually laser thresholds are represented as functions of resonator lengths L and the other structural factors are parametrically used. From a plot of the threshold current as shown in Fig.4 center, the threshold current minimum can be obtained in terms of the coefficients k_1, k_2, and k_3 (with $k_1 \approx \alpha_c$ and α_{wg}, $k_2 \approx V_{ac}$ and α_{wg}, $k_3 \approx \alpha_{wg}$, where α_c =1/L ln(1/R) optical out coupling, α_{wg} =optical loss in the waveguide, and V_{ac}=active volume). The modification of the barrier transport and carrier relaxation in low dimensional systems modifies the effective capture time (Eq. 19) and therefore the recombination coefficients in the laser threshold equation (Eq. 17) will be modified by a factor F_B which describes the carrier scattering between barrier and active region:

$$F_B = 1 + \frac{\tau_{cap}^{eff}}{\tau_{s,B}^{eff}}, \tag{20}$$

$$B^* = F_B B, \tag{21}$$

$$A^* = AF_B + \frac{\tau_{cap}^{eff}}{\tau_{s,B}^{eff}\tau_{esc}}. \tag{22}$$

For large filling factors as in QW and bulk lasers F_B is 1. For small filling factors especially in wire and dot lasers F_B can become much larger than 1. This modifies strongly the threshold current and the quantum efficiency by scattering effects. Compared to lasers with $F_B \approx 1$ the threshold current increases and the quantum efficiency η decreases:

$$I_{th} = qVN_{th}\left(A^* + B^* N_{th}\right) \tag{23}$$

$$\eta^* = \frac{\eta}{F_B} \tag{24}$$

Therefore, geometry dependent capture effects described by the factor F_B can counterbalance low dimensional density of state effects in wire and dot lasers.

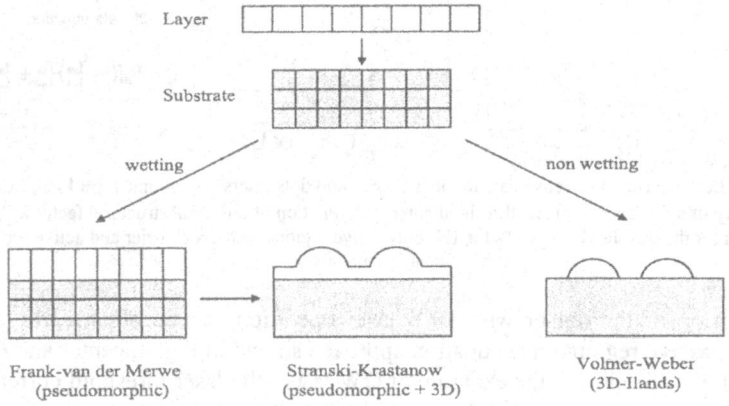

Figure 5 Different epitaxial growth modes for strained layer growth. Left bottom Fig. shows 2D growth of a strained layer. Center bottom shows formation of 3D island arising after from wetting layer growth. Right bottom Fig. shows 3D island growth without wetting layer growth.

3. Fabrication

This section gives a brief and of course not complete overview on nano fabrication techniques for wire and dot lasers. The realization of dot and wire lasers requires a flexible definition and fabrication of large scale structures as optical resonators and filters as well as definition of the device periphery for carrier injection. On the other hand on a small length scale, mainly two techniques can be distinguished:

- Direct epitaxial growth   Self assembled growth
                   Selective growth on non-planar substrates [62-64]
  $\Rightarrow$ reduced flexibility with improved material quality
- Lithography based   Implantation
                       Dry etching
  $\Rightarrow$ enhanced flexibility, but for some cases reduced material quality

3.1 SELF ASSEMBLED TECHNIQUES

Self assembly of quantum dots has revitalized standard epitaxial growth for growth of quantum dots in the active region of a laser. The balance of the free enthalpy of layer, interface and substrate, $\Delta\sigma =\sigma_l+\sigma_i-\sigma_s$, determines the way the growth proceed. The growth of epitaxial layers can be classified according to the principle depicted in Fig. 5. In the left side the Frank-Van der Merwe [65] is shown with dominating substrate term (dominant $\sigma_s \rightarrow \Delta\sigma \leq 0$), which results in a pseudomorphic layer growth (wetting). In the center the Stranski-Krastanow mode is shown [66], which is a strain driven self-assembled growth mode ($\sigma_i=\sigma_i^0+ \sigma_l(strain)$) and results in wetting layer growth and subsequent island formation. The right side shows the Vollmer-Weber growth modus [67], where σ_l is large resulting in a 3D island growth without wetting layer. Examples for Stranski-Krastanow growth mode of dots are GaInP/InP dots [46] that are emitting in the interesting 650 nm range which is useful for plastic fiber data communication. First laser results at room temperature were achieved in this system recently (Tab. 1). In Fig. 6, gain spectra and the electrical laser characteristic are shown.

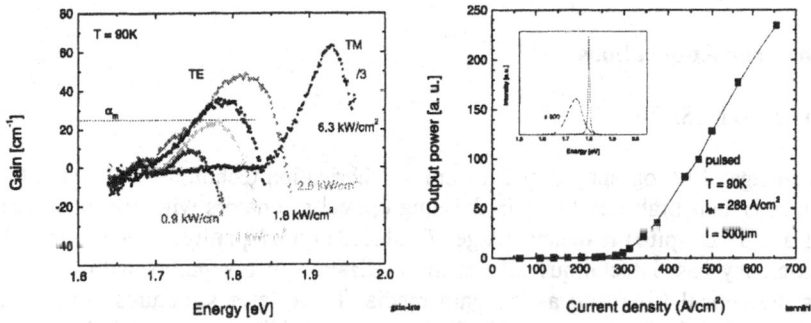

Figure 6 Left: Gain spectra of InP/GaInP quantum dots for different optical pump power. According to quantization conditions the transitions at lowest energies show TE characteristics (heavy hole like) and higher transitions are light hole like. Right: Pulsed electrical output chararcteristic of an InP- dot laser. Inset shows emission around 1.72 eV (x500) for 180 A/cm² and emission around 1.8 eV for 370 A/cm².

3.2 LITHOGRAPHY BASED METHODS

A very flexible method for wire and dot realization is depicted in Fig 7. Ion implantation using nanometer structured gold masks (top of Fig. 7, left) allows the definition of very small non-implanted islands. After a rapid thermal annealing process which cause local intermixing of the GaAs and AlGaAs layers (mid of Fig. 7, left), dots or wires form according to the implantation mask geometry. Instead of intermixing, by using the disordering of a crystal (e.g. disordering of ordered GaInP), potential wells with a depth of the order kT and dot- and wire sizes down to 55 nm and 35 nm, respectively, can be obtained [18,68]. A further attractive possibility is the combination of e-beam lithography and epitaxial overgrowth (Fig. 7, right). This technique allows a most flexible wire and dot definition with the additional advantage to use the wire and dot array directly as optical filter. Many results reported in this contribution are based on this fabrication method.

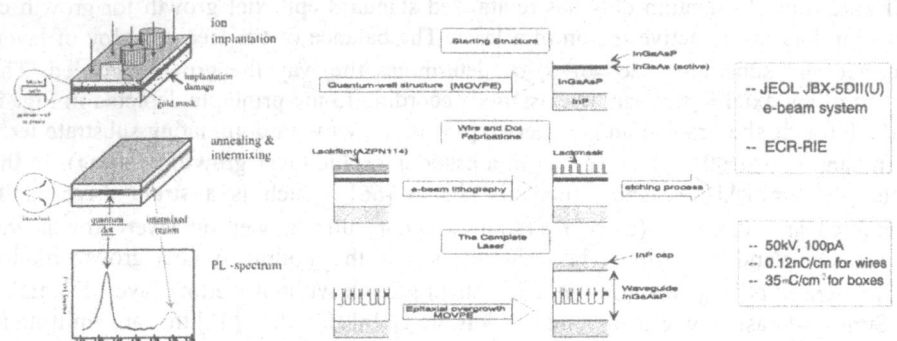

Figure 7 Left: Principal process steps of the lithography based dot laser production using implantation methods. Right: Principal process steps of the lithography based dot laser production using dry etching methods for dot array patterning.

4. Devices and Applications

4.1 DOT DFB LASERS

The advantage of lithography based nanometer fabrication techniques compared to SAD techniques is their high flexibility in defining optical resonators with arbitrary geometry (Figs. 1 and 8). Despite the disadvantage of reduced optical quality of the fabricated dots, the lithography based technique allows the realization of resonators with dots serving both as the optical filter and as the gain media. These laser structures are also called complex coupled DFB lasers [12,13,69] because in addition to a periodic real index function the resonator possesses also a periodic imaginary complex function. By fine-tuning the filling factor (Fig. 8, left and centre) and keeping the DFB period constant, detailed and systematic dynamic investigations of carrier transport and carrier relaxation can be accomplished which is hardly achieved in SAD structures. An impression of the quality of the optical resonators can be also obtained by studying sectional views of wire and dot DFB resonators by AFM. In Fig. 8, right we show the sectional view of a wire DFB laser for the long wavelength emission range around 1550 nm. An example of a dot DFB laser emission at room temperature is given in Fig. 9. From the measured T_0 one obtains the expected increase of T_0 from 42 K (2D-laser) to 57 K (0D-lasers) for lasers on the same wafer. The increase of T_0 is not as high as expected from the pure phase space discussion (section 2.1). This must be attributed to the barrier-active region interaction mainly described by the factor F_B (Eq. 20). From the below threshold spectra we have also obtained the optical gain in dot DFB lasers as function of pump current (Fig. 9,right). As expected the dot laser shows higher differential gain than the 2D laser. On the other hand gain compression also occurs reducing the maximum gain and the maximum modulation bandwidth (see below). The strong barrier-active region interaction of dot lasers can be analysed by measuring the threshold current density as function of the lateral geometrical resonator filling factor $\Gamma_{geo}=\Gamma_x\Gamma_y$ shown in Fig. 10. As can be seen from the different contributions of barrier and active region currents, even at the threshold current minimum (Γ_{geo} ≈0.25 for InGaAs/InGaAsP-dot lasers) the barrier current has roughly the same magnitude as the active region current and it increases steeply for non-optimal filling factors due to high barrier carrier density (Fig. 10). For higher filling factors the dot laser approaches the QW-laser case and the current is mainly determined by the active region current.

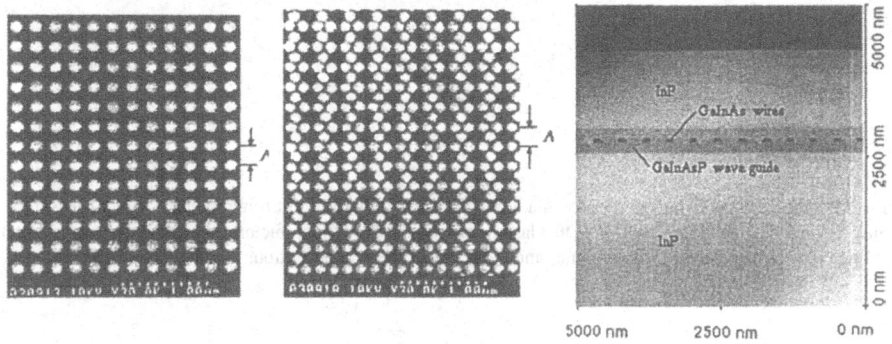

Figure 8 Left: AFM picture of a dot array with a simple DFB period. Center: Analogous to left but with filling factor modified DFB period. Right: AFM picture of a sectional view of a GaInAs/GaInAsP/InP wire DFB laser.

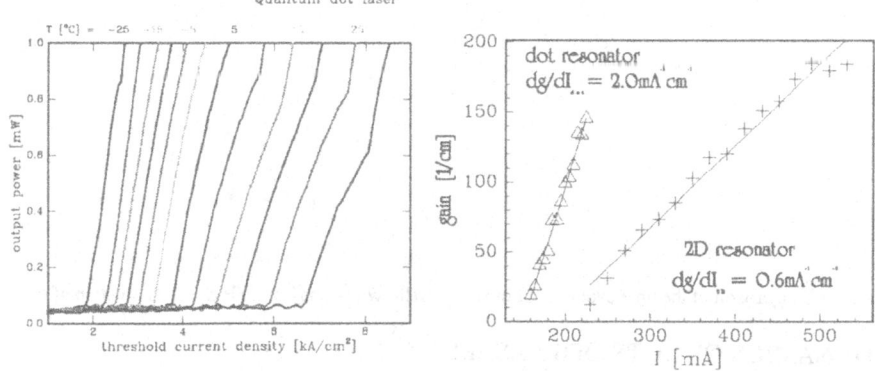

Figure 9 Left: P(I) characteristic of a dot DFB laser for various temperatures from -25°C to +25°C. Right: Comparison of differential gain values dg/dI for QW- and dot-laser.

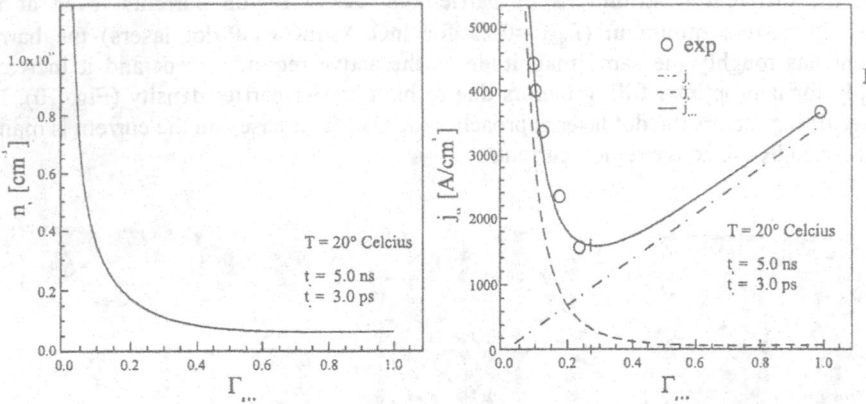

Figure 10 Left: Calculated barrier density of a box DFB laser for various dot filling factors. Right: Measured total current density (open circles) of box DFB lasers for various array filling factors and calculated total current (solid line). The barrier current (dashed line) and the active current contributions (dashed dotted line) are also shown.

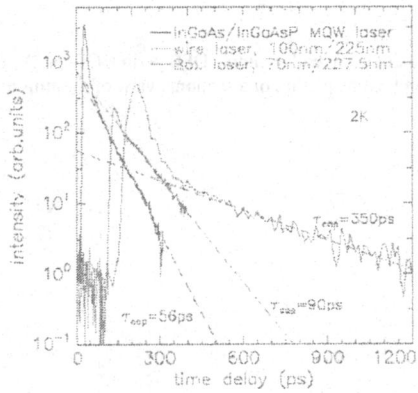

Figure 11 Comparison of the time evolution of quantum well-, wire-, and dot DFB lasers as function of time.

4.2 DYNAMICS OF DOTS DFB LASERS

From the analysis of the rate equations 51-53 in section 7.3 one finds the well known modulation band width given by:

$$f_{3dB} = \frac{2\sqrt{2}\pi}{K}, \tag{25}$$

$$K = 4\pi^2 \left[\tau_{cap}^{eff} + \frac{\varepsilon}{v_{gr} \, \partial g / \partial n} \right], \tag{26}$$

$$\tau_{ph} = \frac{1}{\Gamma \alpha_g v_{gr}}. \tag{27}$$

An attempt to increase the modulation band width results simultaneously in an increase of the gain compression via the effective capture time (Eq. 19. Therefore in a low dimensional laser structure only two time constants, the photon life time and the effective capture time determine the maximum modulation band width f_{3dB} of the laser, no matter what the differential gain value is:

$$K = 4\pi^2 \left[\tau_{ph} + \tau_{cap}^{eff}\right].$$

(28)

The modulation band width (Eq. 25) depends on one side on the differential gain $\partial g / \partial n$ which determines the relaxation oscillation frequency of a laser, and on the other side on the damping of the relaxation oscillations which are mainly determined by the gain compression factor ε. These dependencies can be summarized by the so called K-factor given in Eq. 26. A further analysis of the 2-level model based on the rate equations 51-53 shows that the gain compression factor ε and the differential gain $\partial g / \partial n$ are related by the strongly geometry dependent effective capture time (Eq. 19) as in:

$$\varepsilon = v_{gr}\tau_{cap}^{eff}\frac{\partial g}{\partial n}.$$

(29)

This relation represents a strong restriction of the maximum achievable modulation band width and leads to strong constrains for the optimisation of the modulation band width of a dot / wire laser.

This strong dependence of the laser response can be seen directly in Fig. 11, where we compare QW wire and dot DFB lasers with respect to their modulation response. The lower the dimension the stronger the capture time effects on delay-on time and pulse width of the lasers.

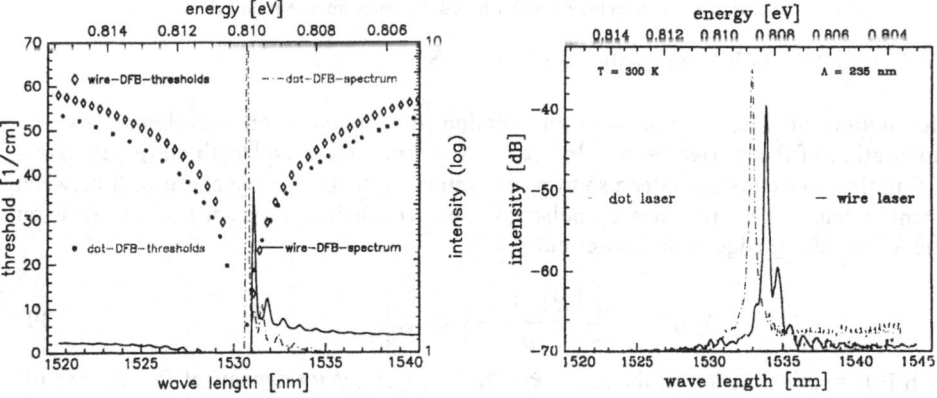

Figure 12 Left: Calculated laser threshold and below threshold spectra of wire (solid black line, black diamonds) and dot (red dashed line and red solid dots) DFB lasers. Wire: κ_i=33 cm⁻¹, κ_g=80 cm⁻¹ Dot: κ_i=33 cm⁻¹, κ_g=130 cm⁻¹. Right: Experimental wire (solid black line) and dot (dashed red line) DFB spectra at room temperature.

4.3 SPECTRA OF DOTS LASERS

Dot and wire arrays with periodic distances between wires or dots are suitable for optical filters with complex coupling between light field and grating. In Fig. 12, left the calculated mode threshold (dot filter: solid red circles, wire filter: open diamonds) of a dot and a wire DFB laser is depicted. Due to the higher imaginary contribution to the whole coupling factor, the dot DFB laser shows a stronger asymmetry of the threshold values and therefore also a stronger side mode suppression ratio. This theoretical expectation is demonstrated in Fig. 12, right. Compared to a conventional QW DFB laser or even to a wire DFB laser the dot DFB laser shows strongly improved side mode suppression ratio of the order of 40 dB despite the very wide transversal resonator cavity of 64 μm. This makes dot DFB lasers suitable candidates for improved DFB laser emission and for low cost optical single mode lasers devices as fabrication tolerances to achieve transversal single mode in these lasers are strongly relaxed due to the 2-dimensional DFB-grating.

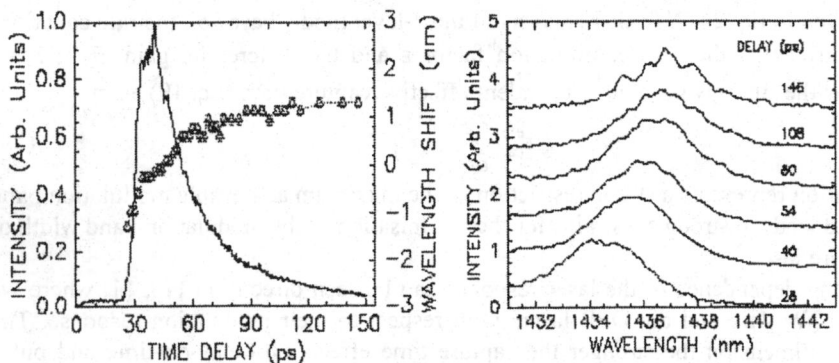

Figure 13 Dynamical wavelength chirping of a QW DFB laser. Left: Intensity and wavelength shift as function of time. Right: DFB emission for various delay times from 28 ps to 146 ps.

4.4 CHIRP IN LOW DIMENSIONAL LASERS

Modulation of lasers enforces a modulation of the emission wavelength by the modulation of the carrier density. In Fig. 13 we depict the wavelength chirp (triangles) during the laser emission after a short ps excitation pulse. Due to the phase shift between intensity pulse and carrier density pulse one observes mainly a red shift of the emission due to the falling edge of the carrier pulse:

$$\Delta v(t) = -\frac{\alpha}{2\pi} \left[\frac{d \ln P(t)}{dt} + \kappa P(t) \right] \tag{30}$$

with P(t) =optical power of the laser, $\kappa = 2\Gamma \varepsilon / V_{ac} \eta h v$, Δv =dynamical frequency shift and $\alpha = -4\pi/\lambda \left(\frac{\partial n}{\partial N} \Big/ \frac{\partial g}{\partial N} \right)$. The overall wavelength shift given in Eq. 30 consists of a structure independent shift according to a change in photon density (and therefore carrier density) and a structure dependent shift expressed by the factor κ.

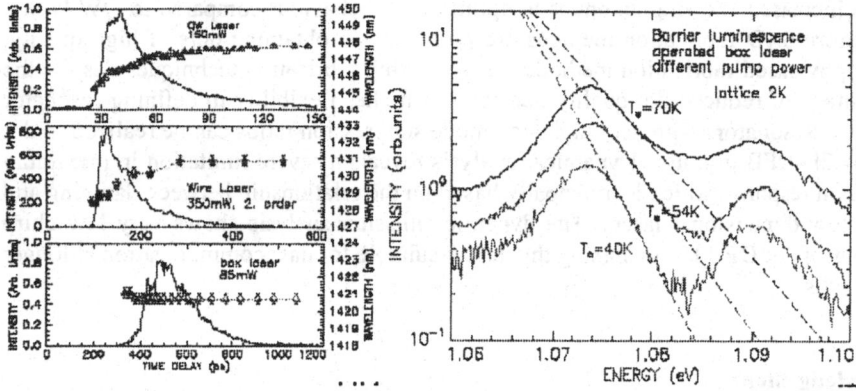

Figure 14 Left: Comparison of emission intensity and wavelength chirp of QW-, wire-, and box-lasers as function of delay time. Right: Comparison of the barrier emission of a box laser for different pump powers.

The overall wavelength shift in low dimensional lasers depends not only on the carrier density change in the active region. Also the carrier density change in the passive waveguide region contributes to a refractive index change and therefore to a wavelength change. In Fig. 14, left we compare the wavelength shift of QW- wire- and dot DFB lasers. The QW and the wire DFB lasers show comparable shifts as expected from the carrier density changes in the active region and in the barrier. The wavelength change of the dot DFB laser, however, is surprisingly low because it shows in the range of the measurement error no wavelength shift. The finding of strong chirp reduction in dot DFB lasers remains even if we change the pump power by a factor of 4. As the barrier filling factor is larger in dot DFB lasers than in wire or QW DFB lasers, also in dot DFB lasers a chirp effect should occur. However, a more detailed analysis reveals also that the barrier carriers experience an increase in carrier temperature with increased pump level

$$\Delta n_{eff} = \frac{\partial n}{\partial N} \Delta N + \frac{\partial n}{\partial T} \Delta T , \qquad (31)$$

(Fig. 14, right). This we attribute to Auger heating [70]. Taking into account the measured increase in carrier temperature in the total refractive index change (Eq. 31) we find nearly a balance between carrier induced refractive index change and temperature induced refractive index change leading to the low observed chirp values in dot DFB lasers.

5. Conclusions

In this contribution we discussed the density of states related phenomena and the transport related phenomena influencing dot laser operation. From this analysis we find that the filling factor of a dot laser most prominently characterized by the effective carrier capture time of dots determines strongly laser threshold, quantum efficiency, and modulation band width in a counterbalancing scheme with respect to density of states. For obtaining high power or high modulation band width, the filling factor in dot lasers must be increased showing an opposite optimisation behaviour compared to QW lasers. Furthermore SAD fabrication methods are promising in obtaining dots of high quality. Lithography based fabrication methods are interesting fabrication techniques despite the disadvantage of reduced dot quality due to their higher flexibility in defining resonator structures. Resonators with very high side mode suppression ratios can be realized in dot arrays as 2D-DFB gratings. Dynamical analysis reveals a severe limitation in maximum modulation response which is principally based on the relationship between damping and gain in low dimensional lasers. The dynamic emission analysis shows very low chirp behaviour of dot DFB lasers making this lasers suitable for data communication at longer wavelengths.

Acknowledgement

The authors would like to thank M.H. Pilkuhn for steady support of the project. For technical assistance we would like to thank E. Kohler for assistance in MOVPE growth, P. Burkard and H. Gräbeldinger for their support in lithography and dry etching and M. Jetter for computational support. This work was financially supported by the Deutsche Forschungsgemeinschaft and the European Community Esprit Project NANOPT .

6. Appendix

6.1 DENSITY OF STATES

Volume of a sphere in d-dimensional k-space:

$$N^{(d)} = \frac{2}{(2\pi)^d} \frac{1}{L^{3-d}} \int_{E=const} d^d k = \frac{2}{(2\pi)^d} \frac{1}{L^{3-d}} \frac{\pi^{d/2}}{(d/2)!} \left(\frac{2m}{\hbar^2}\right)^{d/2} E^{d/2} \quad [\text{cm}^{-3}] \quad (32)$$

Density of states in d dimensions:

$$D^{(d)}(E) = \frac{dN^d}{dE} = \frac{2}{(2\pi)^d} \frac{1}{L^{3-d}} \frac{\pi^{d/2}}{(d/2-1)!} \left(\frac{2m}{\hbar^2}\right)^{d/2} E^{d/2-1} \quad [\text{cm}^{-3}\text{meV}^{-1}] \quad (33)$$

Some examples of densities of states for different dimensions

$$D^{(3)}(E) = 2\frac{2\pi}{(2\pi)^3} \left(\frac{2m}{\hbar^2}\right)^{3/2} \sqrt{E - E_g} \quad [\text{cm}^{-3}\text{meV}^{-1}] \quad (34)$$

$$D^{(2)}(E) = 2\sum_n \frac{\pi}{(2\pi)^2} \frac{1}{L_z} \left(\frac{2m}{\hbar^2}\right) H(E - E_n) \quad [\text{cm}^{-3}\text{meV}^{-1}]$$

$$D^{(1)}(E) = 2\sum_{n,l} \frac{1}{(2\pi)} \frac{1}{L_x L_y} \left(\frac{2m}{\hbar^2}\right)^{1/2} \frac{1}{\sqrt{E - E_x(n) - E_y(l)}} \quad [\text{cm}^{-3}\text{meV}^{-1}]$$

$$D^{(0)}(E) = 2\sum_{n,l,k} \frac{1}{L_x L_y L_z} \delta\big(E - E_x(n) - E_y(l) - E_z(k)\big) \quad [\text{cm}^{-3}\text{meV}^{-1}]$$

6.2 MATRIX ELEMENTS

$$\left|\vec{e}\vec{P_{cv}}\right|^2 = \sum_{m_s,n}\left|\sum_{m_j,l}\left\langle u^c_{m_s}\left|\vec{e}\vec{p}\right|u^v_{m_j}\right\rangle\left(J^{(D)c,v}_{m,n,l}\right)\right|^2 \tag{35}$$

$$\left\langle u^c_{m_s}\left|\vec{e}\vec{p}\right|u^v_{m_j}\right\rangle \qquad \text{(polarization selection rule)} \tag{36}$$

$$J^{(D)c,v}_{m,n,l} = \left\langle \Phi^{(D)c}_n(r)\left|\Phi^{(D)v}_{m_j,l}(r)\right.\right\rangle \qquad \text{(subband selection rule)} \tag{37}$$

CB:
$$\left|u^c_{m_s}\right\rangle = \left|u^c_{1/2}\right\rangle = i\left|S\uparrow\right\rangle \tag{38}$$

$$\left|u^c_{m_s}\right\rangle = \left|u-^c_{1/2}\right\rangle = i\left|S\downarrow\right\rangle \tag{39}$$

VB:
$$\left|u^v_{m_j}\right\rangle = \left|v^j_{m_j}\left(P_{x,y,z}\uparrow,\downarrow\right)\right\rangle \tag{40}$$

HH: $\qquad\qquad m_j=\pm 3/2$ (41)
LH: $\qquad\qquad m_j=\pm 1/2$ (42)

Non zero elements identical for all dimensions d=0 … 3:

$$\left\langle S\left|p_x\right|P_x\right\rangle = \left\langle S\left|p_y\right|P_y\right\rangle = \left\langle S\left|p_z\right|P_z\right\rangle = P_{cv} = \sqrt{\frac{mE_p}{3}}\,,$$

E_p=20.17 interband matrix element for InGaAs [71]. E_p=22.71 interband matrix element for GaAs [55]. The symbols in the above equations have the usual meanings.

6.3 RATE EQUATIONS

1-N (1 level) rate equation for carrier density N and photon density S:

carriers:
$$\dot{N} = G_r - \frac{N}{\tau_s} - v_{gr}GS \qquad [\text{cm}^{-3}\text{s}^{-1}] \tag{43}$$

$$S = \frac{\left|E_x(z)\right|^2}{\hbar\omega} \qquad [\text{cm}^{-3}]$$

$$\tau_s = A + BN + CN^2 \qquad \text{(spontaneous emission time)}$$

photons:
$$\dot{S} = \Gamma v_{gr}GS - \frac{S}{\tau_{ph}} + \Gamma\beta BN^2 \qquad [\text{cm}^{-3}\text{s}^{-1}] \tag{44}$$

$$\beta = \frac{v_{gr}\left(\partial G/\partial N\right)\tau_s}{V} \qquad \text{(spontaneous emission factor)}$$

$$\frac{1}{\tau_{ph}} = \Gamma v_{gr}\alpha_g \qquad \text{(photon lifetime)}$$

$$\Gamma G = \Gamma\alpha_g = \Gamma\alpha_{ac} + (1-\Gamma)\alpha_{wg} + \frac{1}{2L}\ln\left(\frac{1}{R_1 R_2}\right) \qquad \text{(resonator loss balance)} \tag{45}$$

where, $G_r = I/(qV_{ac})$ is the pump rate, v_{gr} the light group velocity, $G = \left(\frac{\partial G}{\partial N}\right)\left(N - N_{tr}\right)\left(1 - \varepsilon S\right)$, $\left(\frac{\partial G}{\partial N}\right)$ the differential gain coefficient, A,B,C the monomolecular, the bimolecular and the Auger recombination coefficients, Γ the optical filling factor, α_g the resonator loss, α_{ac} the resonator loss in the active region, α_{wg} the loss in the waveguide region.

2-N (2 levels) rate equation for carrier density N, photon density S and carrier density n_B in the barrier region

$$\frac{\partial n_B}{\partial t} = D\frac{\partial^2 n_B}{\partial x^2} - \frac{n_B}{\tau_s^B} - f(x)\frac{n_B}{\tau_{cap}^Q} \tag{46}$$

$$\frac{\partial N}{\partial t} = R_{cap} - \frac{N}{\tau_s} - v_{gr}\left(\frac{\partial G}{\partial N}\right)\left(N - N_{tr}\right)\left(1 - \varepsilon S\right)S \tag{47}$$

$$\frac{\partial S}{\partial t} = \Gamma v_{gr}\left(\frac{\partial G}{\partial N}\right)\left(N - N_{tr}\right)\left(1 - \varepsilon S\right)S - \frac{S}{\tau_{ph}} - \Gamma\beta BN^2 \tag{48}$$

When the diffusion term in (48) is approximated by an effective lifetime

$$\tau_{cap}^{eff} = \tau_{diff} + \frac{V_B}{V}\tau_{cap}^Q \tag{49}$$

equations (48)-(50) become

$$\frac{\partial n_B}{\partial t} = G(t) - \frac{n_B}{\tau_{cap}^{eff}} - \frac{n_B}{\tau_s^B} + \frac{NV}{\tau_{cap}^{eff}V_B} \tag{50}$$

$$\frac{\partial N}{\partial t} = \frac{n_B V_B}{\tau_{cap}^{eff}V} - \frac{N}{\tau_s} - \frac{N}{\tau_{esc}^{eff}} - v_{gr}GS \tag{51}$$

$$\frac{\partial S}{\partial t} = \Gamma v_{gr}GS - \frac{S}{\tau_{ph}} - \Gamma\beta BN^2 \tag{52}$$

where τ_{cap}^Q is the LO-phonon capture time, V the active are volume, V_B the barrier volume, R_{cap} the carrier capture rate from barrier to active region.

References

[1] T. Fukui, H. Saito, and Y. Tokura, Appl. Phys. Lett. 55, 1958 (1989).
[2] J. Gaines et al., J.Vac. Sci. Technol. B 6, 1378 (1988).
[3] M. Miller et al., J. Cryst. Growth 111, 323 (1991).
[4] M. Miller et al., Phys. Rev. Lett. 68, 3464 (1992).
[5] Z. Xu, M. Wassermeier, Y. Li, and P. Petro_, Appl. Phys. Lett. 60, 586 (1992).
[6] E. Kapon et al., Appl. Phys. Lett. 60, 477 (1992).
[7] M. Walther et al., Phys. Rev. B 45, 6333 (1992).
[8] IIIA-planes are formed by group III atoms and IIIB-planes are formed by group V atoms. Therefore growth rates of III-V crystals are higher on IIIA-planes than on IIIB-planes.
[9] E. Kapon, Quantum Well Lasers - Quantum Wire Semiconductor Lasers, peter s. zory, jr. ed. (Academic Press, INC., Boston San Diego
New York London Sydney Tokyo Toronto, 1993).
[10] B. Maile et al., Appl. Phys. Lett. 57, 807 (1990).
[11] G. Lehr et al., Appl. Phys. Lett. 61, 517 (1992).
[12] U. A. Griesinger et al., Photon. Technol. Lett. 7, 953 (1996).

[13] U. A. Griesinger et al., IEEE Phot. Technol. Letters 8, 587 (1996).
[14] U. A. Griesinger et al., J.Vac. Sci. Technol. B 14 (6), 4058 (1996).
[15] A. Forchel et al., Microelectronic Engineering 32, 317 (1996).
[16] B. Maile, Dissertation, Universit. at Stuttgart, 4. Phys. Inst. ,Pfa_enwaldring 57, 1990.
[17] R. Germann, Dissertation, Universit. at Stuttgart, 4. Phys. Inst., Pfa_enwaldring 57, 1990.
[18] M. Burkard et al., Appl. Phys. Lett. 70, 1290 (1997).
[19] J. Cibert et al., Appl. Phys. Lett. 49, 1275 (1986).
[20] H. Leier et al., Microelectronic Engineering 11, 43 (1990).
[21] H. Leier et al., J. Appl. Phys. 67, 1805 (1990).
[22] H. Leier et al., Appl. Phys. Lett. 56, 48 (1990).
[23] C. Vieu et al., J. Appl. Phys. 71, 5012 (1992).
[24] F. E. Prins, G. Lehr, H. Schweizer, and G. W. Smith, Appl. Phys. Lett. 63, 1402 (1993).
[25] F. D. Prins et al., Jap. Journ. Appl. Phys. 32, 6228(1993).
[26] M. Burkard et al., J.Vac. Sci. Technol. B 12, 3677(1994).
[27] W. Beinstingl et al., J.Vac. Sci. Technol. B 9, 3479 (1991).
[28] N. Kirstaedter et al., Europhys. Lett. 30, 1416 (1994).
[29] N. Kirstdter et al., APL 69, 1226 (1996).
[30] H. Shoji et al., Jap. Journ. Appl. Phys. 35, L903 (1996).
[31] H. Shoji et al., IEEE Journal OF Selected Topics IN Quantum Electronics 3 (2), 188 (1997).
[32] D. Bimberg et al., IEEE Journal OF Selected Topics IN Quantum Electronics 3 (2), 196 (1997).
[33] S. Zaiztsev et al., Superlattices and Microstructures 21, 559 (1997).
[34] V. Ustinov et al., J. Crystal Growth 175, 689 (1997).
[35] N. Chand et al., Appl. Phys. Lett. 58, 1704 (1991).
[36] D. Hu_aker et al., Appl. Phys. Lett. 70, 2356 (1997)29
[37] F. Heinrichsdor_ et al., Appl. Phys. Lett. 71, 22 (1997).
[38] H. Shoji et al., Appl. Phys. Lett. 71, 193 (1997).
[39] K. Kamath et al., Electron. Lett. 32, 1374 (1996).
[40] R. Mirin, A. Gossard, and J. Bowers, Electron. Lett. 32, 1732 (1996).
[41] S. Fafard et al., Science 274, 1350 (1996).
[42] K. Hinzler et al., Journal Physika EMSS8, 68 (1997).
[43] D. Hu_aker et al., Appl. Phys. Lett. 73, 2564 (1998).
[44] M 7. K Eberl et al., Appl. Phys. Lett. submitted, (1998).
[45] T. Riedel, E. Fehrenbacher, A. Hangleiter, and M. Z. K. Eberl, Appl. Phys. Lett. submitted, (1998).
[46] P. Porsche, A. Ruf, M. Geiger, and F. Scholz, J. Crystal Growth 195, 591 (1998).
[47] G. Liu et al., Electron. Lett. 35, 1163 (1999).
[48] H. Saito, K. Nishi, Y. Sugimoto, and S. Sugou, Electron. Lett. 35, 1561 (1999).
[49] D. Bimberg, M. Grundmann, and N. Ledentsov, Quantum dot heterostructures, p. 328 (John Wiley, & Sons,, Chichester, 1999).
[50] Y. Arakawa and H. Sakaki, Appl. Phys. Lett. 40, 939 (1982).
[51] The chemical potential of dot systems can not be approximated by theintegral over states it must be summed up over discrete energy levels.
[52] H. Schweizer et al., phys. stat. sol. (b) 173, 331 (1992).
[53] H. Schweizer et al., Superlattices and Microstructures 12, 419 (1992).
[54] M. Willatzen, T. Tanaka, and Y. Arakawa, IEEE J. Quantum Electron. 30, 640 (1994).
[55] U. Bockelmann and G. Bastard, Phys. Rev. B 45, 1688 (1992).
[56] E. Kane, The kp-method in: Semiconductor and Semimetals (R.K. Willardson and A.C. Beer Vol. 1 Physics of III-V Compounds Academic Press, New York, 1966).
[57] M. Asada, Y. Miyamoto, and Y. Suematsu, IEEE J. Quantum Electron. QE 22, 1915 (1986).
[58] J. Wang et al., Appl. Phys. Lett. 69, 287 (1996).
[59] J. Wang, U. Griesinger, and H. Schweizer, Appl. Phys. Lett. 69, 1585 (1996).
[60] J. Wang, U. Griesinger, and H. Schweizer, Appl. Phys. Lett. 70, 1152 (1997).
[61] J. Wang et al., IEEE Journal of Selected Topics in Quantum Electronics 3, 223 (1997).
[62] G. Biasiol, E. Kapon, Y. Ducammun, and A. Gustafsson, Phys. Rev. B 57, R9416 (1998-II).
[63] S. Ishida, Y. Arakawa, and K. Wada, APL 72, 800 (1998).
[64] A. Hartmann, L. Loubies, F. Reinhaedt, and E. Kapon, APL 71, 1314 (1998).
[65] F. Frank and J. van der Merwe, Proc. Roy. Soc. LondonA198, 205 (1949).
[66] I. Stranski and L. Krastanov, Akad. Wiss. Wien Kl. IIb 146, 797 (1938).
[67] M. Volmer and A. Weber, Z. Phys. Chem. 119, 277 (1926).

84

[68] M. Burkard et al., J. Appl. Phys. 82, 1 (1997).
[69] U. A. Griesinger et al., IEEE Phot. Technol. Letters 7, 953 (1995).
[70] A. U. and, Appl. Phys. Lett. {, (1998).
[71] E. Zielinski et al., J. Appl. Phys. 59, 2196 (1986).

NANOSTRUCTURED SILICON AS AN ACTIVE OPTOELECTRONIC MATERIAL

L.T. CANHAM

Sensors & Electronics Division, DERA, St Andrews Rd, Malvern, Worcs, WR14 3PS, UK

Abstract

This review paper charts the progress made over the last decade in realising a variety of optoelectronic devices using nanostructured porous silicon. Although waveguides, modulators and detectors are covered, emphasis is given to visible light-emitting diodes whose efficiency is now 1%. Although their performance is still inadequate for optical interconnect applications, two promising approaches for further major advances are concluded to be the fabrication of more monodisperse nanostructures and the exploitation of optical microcavities.

1.   Introduction

A decade has now passed since the demonstration that silicon can emit visible light efficiently at room temperature (1). This review summarises progress made towards an all-Si optoelectronic technology using nanostructured porous Si. It addresses briefly the motivation with regard to "optical interconnect", the mechanisms of light emission, the best performance to date of LED's, waveguides, modulators, detectors, the prospects for Si nanostructure lasers, and finally the wide range of alternative fabrication routes being developed for nanostructuring. It does not cover the use of its optoelectronic properties in Si solar cells, nor in Si microsensor systems, and thus provides an update of two specific earlier reviews (2,3) and complements more recent reviews on nanoscale Si light emitters (4,5).

2.   Silicon-Based Optoelectronics

CMOS circuitry dominates the current $150 billion semiconductor market due to the astonishing power of silicon electronic integration technology. By the year 2010, state-of-the-art 3cm wide chips are forecast to have 1000 million transistors with minimum feature sizes of 100nm, and running at 4GHz clock speeds (6). The

85

L. Pavesi and E. Buzaneva (eds.), Frontiers of Nano-Optoelectronic Systems, 85–97.
© 2000 *British Crown. Printed in the Netherlands.*

combination of ever-increasing chip size and decreasing feature size, however, has already raised major chip-interconnection issues. In the past 30 years electrical interconnects have matured from single layers of evaporated aluminium to high density, multilevel structures, but still fall short of the demand from high performance circuits (7). Optical interconnections have thus been of interest for some time, and at all levels in digital computers; between mainframes, modules, boards, chips and even points within a large chip (8). The implementation of 3D free-space optical interconnects between chips, for example, is predicted to provide in excess of terabit/second bandwidth (9). In contrast to the dominance of silicon in electronics, the field of optoelectronics utilises a diversity of materials for emitting, guiding, modulating and detecting light (10). For some the future of Si-based optoelectronics lies in "hybrid" solutions (11), for others the utilisation of more photonic functions by silicon itself (12). The diverse application areas that all-Si based solutions offer cover not only optical interconnect, but cheap integrated displays, fibre-optic systems, optoelectronic microsensors and imaging systems (13).

3.   Visible Light from Porous Si (pSi)

High porosity silicon films were first reported to yield efficient red luminescence under photoexcitation (PL) in 1990 (1), but within a year, orange, yellow and green output had already been attained (3). Indeed, a characteristic feature of PL from pSi is the wide spectral tunability achievable. More recently, the peak of the broad emission band has been tuned all the way from the bulk Si band gap at 1150nm in the near IR through the whole visible spectrum to 400nm in the violet (14,15).

External quantum efficiencies (EQE) for photoexcited thin films lie in the range 1-10% (3) but recent confocal analysis has shown that this arises from small regions of the skeleton with much higher PL efficiency (16). The associated radiative decay times are strongly wavelength dependent (ns in the blue to nearly 1ms in the near IR) (3), as shown in figure 1. The involvement of non-radiative Auger processes in the photoexcited nanostructure is evident from PL saturation and other non-linear optical effects (17). Both bandstructure calculations (18) and spectroscopic characterisation (3,14) have established that nanocrystalline Si (undulating quantum wires and dots) retains the phonon-assisted features of radiative recombination in bulk silicon, at least over the 2-10nm size regime. For extremely small Si nanocrystals (< 1nm) however, theory predicts a further substantial increase in oscillator strength for luminescence, with values comparable to those of direct gap semiconductors (19, 20).

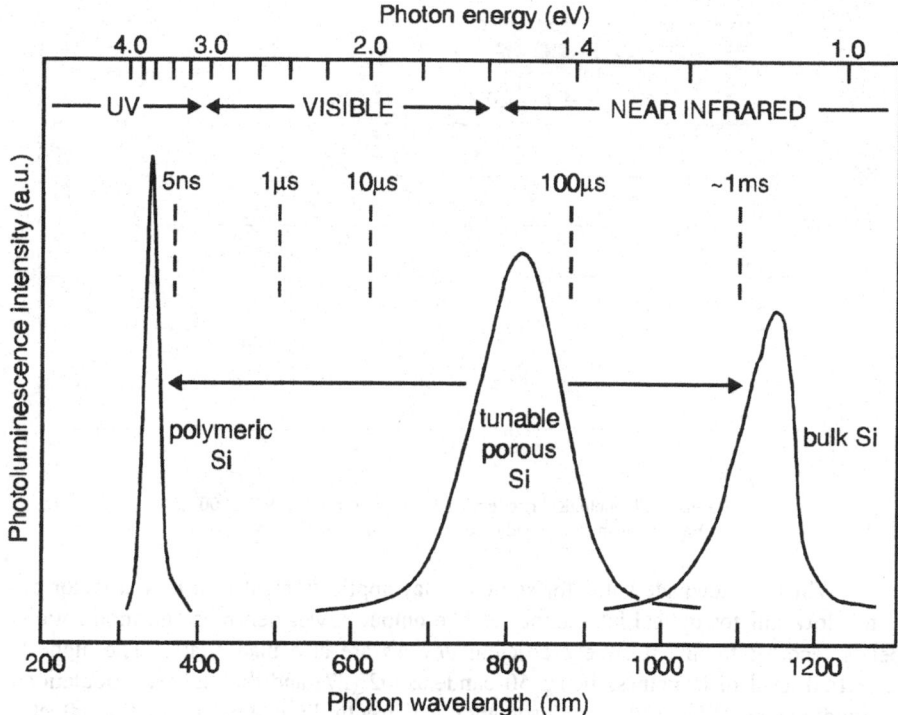

Figure 1. Tunability of pSi nanocrystallite luminescence and its decay time. Typical photoluminescence spectra of polysilane, porous silicon and bulk silicon are compared. The PL lifetimes for porous silicon emission at selected wavelengths are taken from references 14, 15 and vary somewhat with layer microstructure and the type of skeleton passivation.

4. pSi Light Emitting Devices.

Bulk silicon LED's (21) generate near infrared light inefficiently (EQE ~0.01%) and visible light extremely inefficiently (EQE~0.000001%). Nonetheless, the image-intensified visible light emitted from a range of VLSI circuitry is increasingly being used as a diagnostic tool for assessing hot carrier effects, thin oxide reliability, latch-up and so on (22). The first porous silicon LED, of quantified efficiency, was reported by Koyama and Koshida in 1992 (23). Device efficiency has subsequently continuously improved (see Figure 2), and is now at 1%, some 6 orders of magnitude higher than that of reverse-biased bulk Si diodes (24-29).

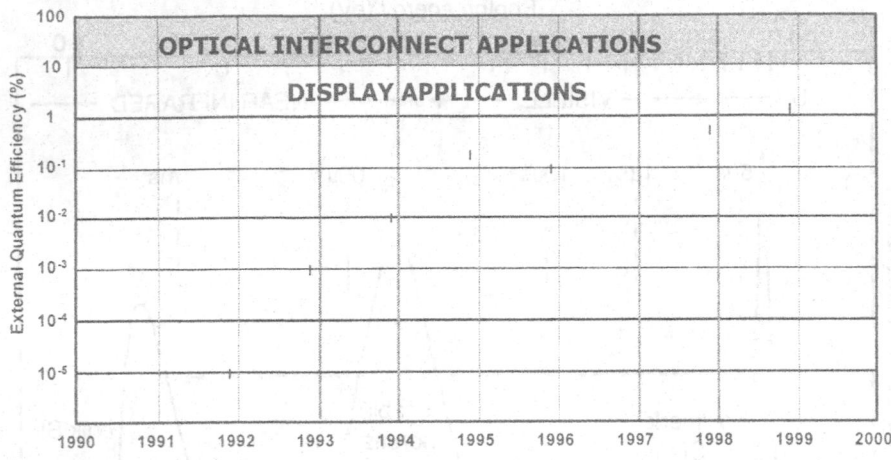

Figure 2. Chronological rise in pSi LED efficiency from 1990-2000
Values shown correspond to references 23-26, 28-29, and 34.

Whilst an adequate value for some display applications, it is at least a factor of 10 too low still for optical interconnect. The output power densities and modulation speeds reported for the relatively efficient devices are also inadequate. The highest quantified level of brightness being 50 candelas/m2 (27) and the highest modulation bandwidth about 1MHz (30). The efficiency is currently limited by percolation effects in the skeleton's inhomogeneous size distribution (31); output power by the Auger limit of one photon per nanocrystal per exciton lifetime (14); and modulation by carrier mobility and radiative decay time (30). Although most effort to-date on improving LED stability has utilised oxidation treatments (4,5), I believe a much more attractive approach is that of chemical derivitisation (32,33) of the internal surface of the porous diode. The latter can impart stability without degrading the interconnectivity of the Si skeleton. Nonetheless, oxidised pSi LED's have been successfully integrated with bipolar circuitry (34), an important step on the road to assessing the CMOS compatibility of specific device designs.

Finally, I draw attention to an alternative pSi LED application area; that of cold cathode arrays emitting electrons in vacuum (35). Such structures are under development for pumping phosphor layers in flat panel displays. Here, a high emission current to diode current efficiency ratio of 12% has been achieved to date (35).

5. pSi Waveguides, Modulators and Detectors

Bondarenko and co-workers first reported waveguiding in pSi in 1993 (36). The refractive index is tunable via porosity, microstructure, and degree of oxidation. As-etched structures have transparency at the fibre optic wavelengths of 1.3 and 1.5 μm, but guiding of visible light requires conversion to silica. A variety of approaches and designs (eg. planar, ridge and multilayer guides) have been investigated (37), but optical losses have been consistently high (> 4dB/cm). This has been attributed to interface roughness, rather than problems with Rayleigh scattering within the guiding layer itself. Very recent work, however, has achieved values of around 1dB/cm (38), which is the common initial target, regarded as a prerequisite for technological viability. Research has also been initiated in rendering such waveguides optically-active via impregnation (39).

Evidence that PSi can exhibit non-linear optical effects has been steadily accumulating since 1992 (40-44). The reported values for the third order non-linear optical susceptibility, χ^3 are typically 10^{-8} to 10^{-9} esu for picosecond timescales and 400-550 nm excitation; but as high as 10^{-1} esu for millisecond response times. The former phenomena are attributed to electronic effects, the latter to thermo-optic effects (2). Indeed, given the incredibly low values of thermal conductivity achievable with pSi (45), very large changes in skeleton temperature can occur at moderate excitation levels (46). Both the non-linear optical properties (47) and electro-optic properties of bulk silicon (48) have been considered for many years as on-chip modulators of near infrared light. Thermo-optic devices in both bulk silicon (49) and silica-on-silicon (50) for both near infrared and visible wavelengths, are well documented, but suffer from high power consumption. Porous silicon appears to have significant potential here (2), as either a passive thermal isolation support, or in the active region of a modulating or switching waveguide structure. No such devices have appeared to date however.

Photodetection of visible light is the one area of optoelectronics where bulk Si technology is mature and offers high performance devices (51). Depending on the application, detector responsivity is often traded for speed. Thus planar inter-digitated MSM Si photodiode designs might be chosen for ultrafast (75GHz) operation (52), whilst vertical pn junction structures optimised when detector sensitivity is the key need. The incorporation of pSi in photodetectors for OEIC use was first proposed in 1992 (53,54) but there has not been much activity in this area since then (55,56). Certainly, respectable response times (eg.2ns (53)) and responsivity under gain have been demonstrated for visible light (eg. 6.4A/W at 810nm (55)), but no major benefits are evident (2). A more interesting target for pSi photodetectors might be the fibre-optic wavelengths, where hybrid material integration (eg Ge,InGaAs on Si) is currently being pursued.

6.   pSi Filters, Mirrors and Microcavities

The ability to modulate porosity with depth was first demonstrated in 1994 (57,58) as a means to improve the mechanical robustness of luminescent structures and realise optical filters. It is Pavesi and co-workers (59-61), however, who have pioneered the use of pSi microcavities to improve light emission characteristics via photon

90

confinement effects. The quality of these planar pSi cavities has undergone gradual improvement over the last 5 years, with regard to minimising their multilayer interface roughness (see Figure 3).

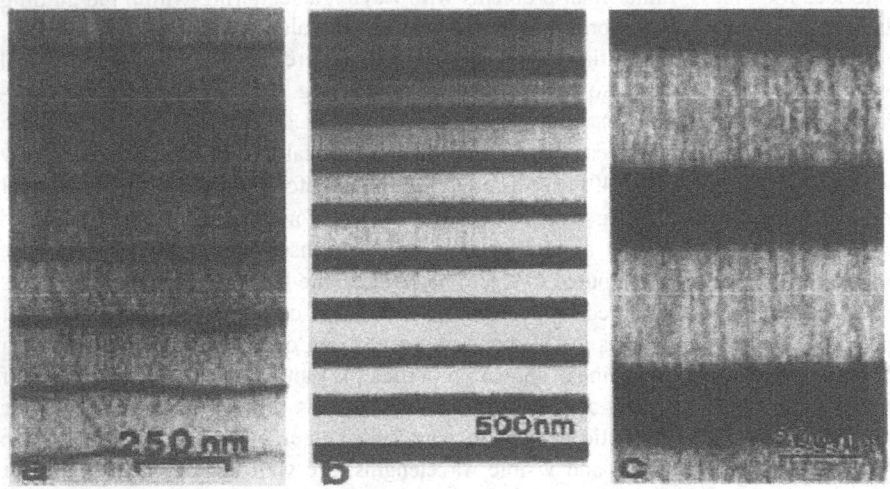

Figure 3. XTEM images of part of a pSi multilayer (a)1994 (b),(c) 2000
Image in (a) is from reference 58. Dark layers(I) correspond to lower porosity; lighter layers (II) to higher porosity. The two multilayer structures shown were both fabricated via current density modulation in p+ wafers.

This has led to dramatic reductions in PL line widths (the current record being 2nm FWHM) but not to date in radiative lifetime (61). I propose that in this regard, the fabrication of the hemispherical cavity shown in figure 4(c) would lead to much reduced lifetimes, compared with the planar designs (Figure 4(a)) studied to date or even a "post" microcavity design (Figure 4(b)). An array of such structures would result from modulated anodisation through lithographically defined micron-size areas. Under such conditions the pores spread radially, undercutting the mask, and creating a stack of hemispheres (Figure 4(c)).

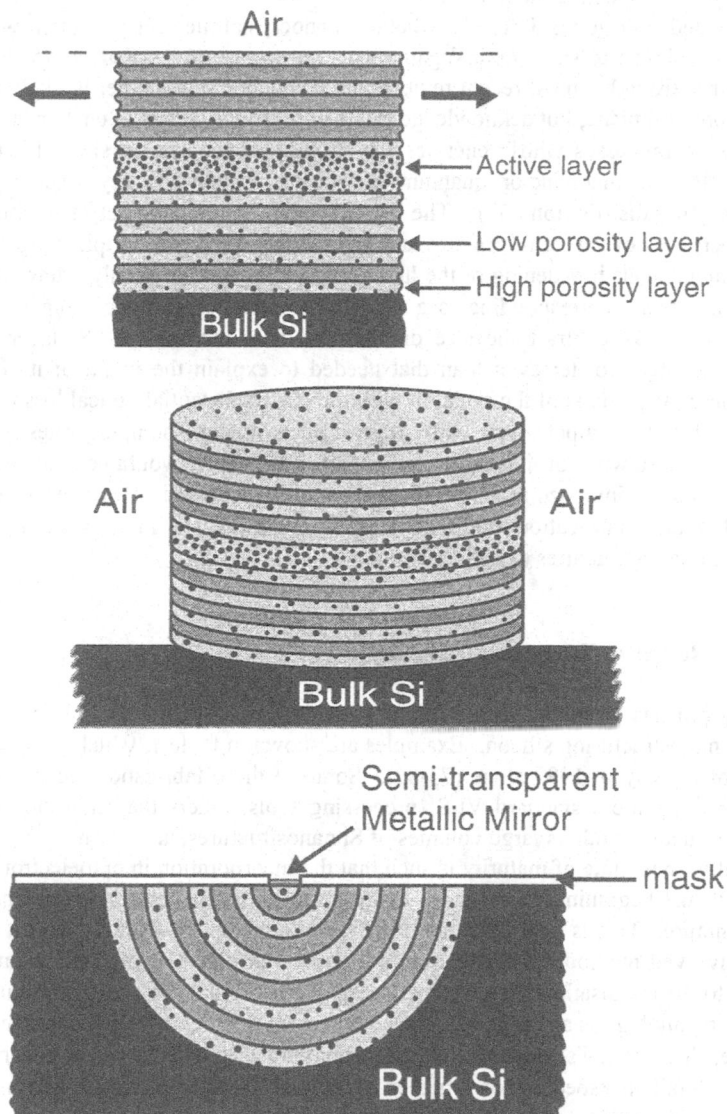

Figure 4. Schematic of (a) current planar microcavities (b) a common "post" microcavity design (c) the proposed plano-concentric microcavities

Theory predicts that the use of such curved mirrors will lower radiative lifetimes substantially more than planar or post cavities (62). Current planar microcavity pSi LED's (63) also utilise anodised p+ wafers, which provide the best quality multilayer stacks, but not the highest radiative efficiencies (1). Another important next step will thus be to realise better LED cavities in p- wafers (64), or utilise p+/p-/p+ epitaxial structures, and create the mirrors in the p+ regions. The quality of thermally oxidised cavities might already be sufficient however, to achieve photopumped lasing using impregnated dyes as the active medium (65).

One related intriguing issue is whether nanocrystalline silicon itself will eventually be developed to show optical gain, and even to lase (66). Fauchet (5) has discussed this briefly in his most recent review, and concluded that neither is likely to occur. I am more optimistic, but acknowledge it is a tremendous technical challenge, at least for radiative processes whose energies are strongly dependent on size. If one looks at the III-V semiconductor quantum dot lasing systems, a key issue was minimising the size distribution (67). The breakthrough in realising self-assembled stacks of nanocrystals of very uniform size and high density then led to rapid progress (68,69). Inhomogeneous broadening of the luminescence line width of only a few 10's of meV is now typical, corresponding to a fabrication tolerance of under 1nm for a 10nm nanocrystal. By contrast the size distribution in a luminescent pSi layer is currently significantly broader even than that needed to explain the width of its PL band (70). These larger parts of the skeleton will introduce substantial optical loss to a microcavity. Also, to compete well with Auger effects at high pumping rates, the virtually monodisperse wires or dots in the active part of the cavity would need to be in the 1nm rather than 2-5nm size regime commonly attained with pSi. However, if we turn now to alternative fabrication routes, some of these are just starting to yield more monodisperse Si nano-structures (71,72).

7.   Other Routes to Nanostructured Silicon

Light-emitting pSi has prompted research into a whole range of different fabrication approaches to nanostructuring silicon. Examples are shown in table 1, which provides a "snapshot" of activity in 1999 alone (73-87). Some of these fabrication techniques are clearly targeting more standard VLSI processing tools; others the realisation of monodisperse structures; others large volumes of Si nanostructures, and so on.

Generally, their state of maturity is such that the incorporation in optoelectronic devices is only just beginning to occur, so comparison with pSi device performance would be premature. This is well illustrated by Figure 5, which shows how interest in other, more involved methods of fabrication, are rising significantly, but with a time lag compared to the anodisation route. Nonetheless, the continual shrinkage of feature size in CMOS technology is already pushing us into an era of silicon "nanoelectronics" (88-90). Here, the material's behaviour on the nanometre length scale has to become understood in detail, irrespective of whether or not all-Si optoelectronics becomes viable. Interest in nanostructured silicon thus seems assured to continue to grow unabated, and for the foreseeable future.

TABLE 1. Alternative fabrication techniques for nanostructuring silicon (1999)

Fabrication Route	Nanostructure	Ref
Self-assembly & reactive ion etching	10-100 nm pillars	73
Ion implantation & anneal	2-5 nm nanocrystals	74
Sputtering, oxidation & anneal	3-7 nm nanocrystals	75
Laser ablation	3-43 nm wires	76
Vapour deposition, wafer bonding & anneal	2-8 nm nanocrystals	77
E-beam lithography, dry etching & oxidation of SOI wafers	10-60 nm nanocrystals	78
Scanning probe lithography & wet etching	55 nm grating	79
Cluster beam deposition	2-10 nm nanocrystals	80
Vapour deposition within host matrix	1 nm clusters	81
Solution growth within inverse micelles	1.8-10 nm nanocrystals	82
Gas evaporation, matrix isolation & suspension	1.5-7 nm nanocrystals	83
MBE Si/CaF2 multilayer deposition	1.4-1.6 nm wells	84
Micromachining of SOI wafers	45-200 nm wires	85
Mechanical ball-milling	5-40 nm nanocrystals	86
MBE a-Si deposition & anneal	2-3 nm nanocrystals	87

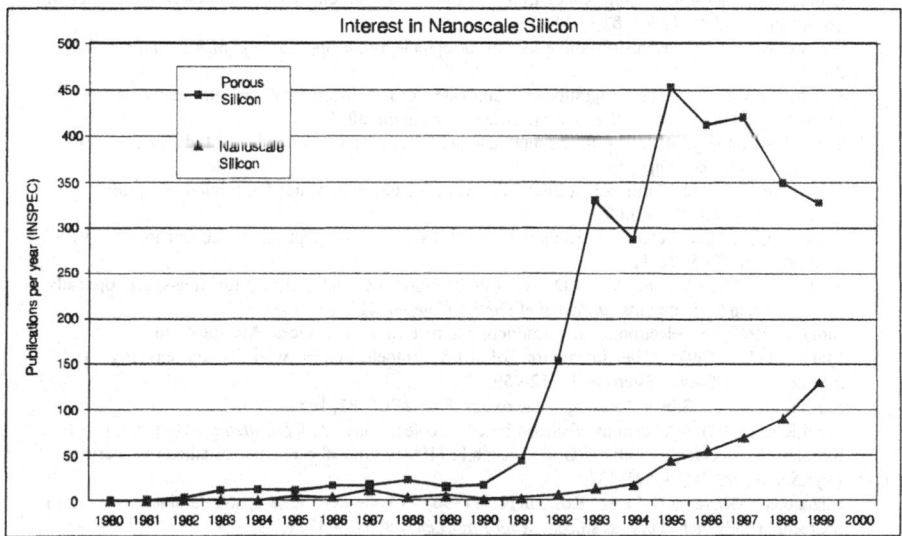

Figure 5. Rise in interest in nanoscale silicon (1980-2000) The strings used to search the INSPEC database were: (a) porous, (w) silicon not (oxide or dioxide or carbide).(b) silicon (w) quantum (w) (well or wire or dot or pillar) or silicon (w) (nanowire or nanodot or nanostruct* or nanocryst* or superlattice) or silicon (w) nanosize.

94

8. Concluding Remarks

After much worldwide study it is now established that the dominant semiconductor, silicon, is capable of emitting visible light efficiently via quantum size effects. The efficiency of porous silicon LED's has risen by 5 orders of magnitude over the last decade to 1% and is approaching commercial viability for some integrated display applications, but remains quite inadequate for optical interconnect, the main initial "driver". Some exciting areas for further pSi technological development are the realisation of more monodisperse Si nanostructures in the 1-2nm size regime; the further refinement of Si micro-cavities; the use of surface derivitisation; and the realisation of thermo-optic devices for fibre-optic communication systems. Interest in nanoscale Si has seen tremendous growth over the last decade, and continued CMOS device scaling will further support that trend.

References

1.   Canham,L.T. (1990) Silicon quantum wire array fabrication by electrochemical and chemical dissolution of wafers. *Appl.Phys.Lett.* **57**, 1046-1049.

2.   Canham,L.T., Cox,T.I., Loni,A. and Simons,A.J. (1996). Progress towards silicon optoelectronics using porous silicon technology. *Appl.Surf.Science.* **102**, 436-441.

3.   Cullis,A.G., Canham,L.T. and Calcott,P.D.J. (1997). The structural and luminescent properties of porous silicon. *J.Appl.Phys.* **82**,909-965

4.   Tsybeskov L. (1998). Nanocrystalline silicon for optoelectronic applications. *M.R.S. Bulletin.* **23** (4) 33-38.

5.   Fauchet,P.M. (1999). The integration of nanoscale porous silicon light emitters: materials science, properties and integration with electronic circuitry. *J.Lumin.* **80**, 53-64.

6.   The US Technology Roadmap for Semiconductors (1997). The Semiconductor Industry Association web page: www.semichips.org.

7.   Lu,R., Pai C.S. and Martinez,E. (1999). Interconnect technology trend for micro-electronics. *Sol. State Electron.* **43**, 1003-1009.

8.   Goodman,J.W., Leonberger,F.J., Kung,S.Y. and Athale,R. (1984) Optical interconnections for VLSI systems. *Proc.IEEE.* **72**, 850-866.

9.   Krishnamoorthy,A.V. and Miller,D.A.B. (1997). Firehose architectures for free-space optically interconnected VLSI circuits. *J.Parrallel Distrib.Comput.* **41**, 109-114.

10.   Singh,J. (1996) Optoelectronics: an introduction to materials and devices. McGraw-Hill.

11.   Mathine,D.L. (1997). The integration of III-V optoelectronics with silicon circuitry. *IEEE J.Select.Topics Quant. Electron.* **3**, 952-959.

12.   Soref,R.A. (1993) Silicon-based optoelectronics. *Proc.IEEE*, **81**, 1687-1706.

13.   Soref,R.A. (1998) Applications of silicon-based optoelectronics. *M.R.S.Bulletin*, **23** (4) 20-24.

14.   Kovalev,D., Heckler,H., Polisski,G. and Koch,F. (1999). Optical properties of silicon nanocrystals. *Phys.Stat.Solidi.* **B215**, 871-932.

15.   Mizuno,H., Koyama,H. and Koshida,N. (1996). Oxide-free blue photo-luminescence from photochemically etched porous silicon. *Appl.Phys.Lett.* **69**, 3779-3781.

16.   Credo,G.M., Mason,M.D. and Buratto.S.K. (1999). External quantum efficiency of single porous silicon nanoparticles. *Appl.Phys.Lett.* **74**, 1978-1980.

17.   Efros.Al.L., Rosen,M., Averboukh,B., Kovalev,D., Ben-Chorin,M. and Koch,F. (1997). Nonlinear optical effects in porous silicon: PL saturation and optically induced polarisation anisotropy. *Phys.Rev.* **B56**, 3875-3884.

18.   Hybertson,M. (1994). Absorption and emission of light in nanoscale silicon structures. *Phys.Rev.Lett.* **72**, 1514-1517.

19.   Sanders,G.D. and Chang,Y.C. (1992). Theory of optical properties of quantum wires in porous silicon. *Phys.Rev.* **B45**, 9202-9213.

20. Ohno,T., Shiraishi,K. and Ogawa,T. Intrinsic origin of visible light emission from silicon quantum wires: electronic structure and geometrically restricted exciton. *Phys.Rev.Lett.* **69**, 2400-2403.

21. Kramer,J., Seitz,P., Steigmeier,E.F., Ardeset,H., Delley,B. and Balten,H. (1993) Light-emitting devices in industrial CMOS technology. *Sens.Actuat.* **A37/38**, 527-533.

22. Leroux,C. and Blachier,D. (1999) Light emission microscopy for reliability studies. *Microelectron.Eng.* **49**, 169-180.

23. Koshida,N. and Koyama,H. (1992). Visible electroluminescence from porous silicon. *Appl.Phys.Lett.* **60**, 347-349.

24. Kesan,V.P., Bassous,E., Munguia,P., Pesarcik,S.F., Freema,M., Iyer,S.S. and Halbout,J.M. (1993). Electroluminescence and photoluminescence from microporous p-n junctions. *J.Vac.Sci.Technol.* **A11**, 1736-1738.

25. Steiner,P., Kozlowski,F. and Lang,W. (1994) Light-emitting porous silicon diode with an increased electroluminescence quantum efficiency. *Appl.Phys.Lett.* **62**, 2700-2702.

26. Loni,A., Simons,A.J., Cox,T.I., Calcott,P.D.J. and Canham,L.T (1995). Electroluminescent porous silicon device with an external quantum efficiency greater than 0.1% under CW operation. *Electron.Lett.* **31**, 1288-1289.

27. Chen,Y.A., Chen,B.F., Tsay,W.C., Laih,L.H., Chang,M.N., Chyi,J.I., Hong,J.W. and Chang,C.Y. (1997). Porous silicon light-emitting diode with tunable colour. *Solid State Electron.* **41**, 757-759.

28. Nishimura,K., Nagao,Y. and Ikeda,N. (1998) High external quantum efficiency of electroluminescence from photoanodised porous silicon. *Jpn.J.Appl.Phys.* **37**, L303-L305.

29. Gelloz,B., Nakagawa,T. and Koshida,N. (1999). Enhancing the external quantum efficiency of porous silicon LED's beyond 1% by a post-anodisation electrochemical oxidation. *MRS.Symp.Proc.* **536**, 15-17.

30. Cox,T.I., Simons,A.J., Loni,A., Calcott,P.D.J., Canham,L.T., Uren,M.J. and Nash,K.J. (1999). Modulation speed of an efficient porous silicon light-emitting diode. *J.Appl.Phys.* **86**, 2764-2773.

31. Hamilton,B., Jacobs,J., Hill,D.A., Pettifer,R.F., Teehan,D. and Canham,L.T. (1998) Size-controlled percolation pathways for electrical conduction in porous silicon. *Nature*, **393**, 443-445.

32. Buriak,J.M. and Allen,M.J. (1998) Lewis acid mediated functionalization of porous silicon with substituted alkenes and alkynes. *J.Amer.Chem.Soc.* **120**, 1339-1342.

33. Song,J.H. and Sailor,M.J. (1999). Chemical modification of crystal porous silicon surfaces. *Comments Inorg.Chem.* **21**, 69-84.

34. Hirschmann,K.D., Tysbeskov,L., Duttagupta,S.P. and Fauchet,P.M. (1996) Silicon-based visible light-emitting devices integrated into microelectronic circuits. *Nature*, **384**, 338-341.

35. Sheng,X., Koyama,H. and Koshida,N. (1998) Efficient surface-emitting cold cathodes based on electroluminescent porous silicon diodes. *J.Vac.Sci.Tech.* **B16**, 793-795.

36. Bondarenko,V.P., Varicheko,V., Dorofeev,A.M., Kazyuchits,N.M., Labunov,V.A. and Stelmakh,V.F. (1993) Integrated optical waveguide fabricated with porous silicon. *Tech.Phys.Lett.* **19**, 463-464.

37. Arrand,H.F., Benson,T.M., Sewell,P., Loni,A., Bozeat,R.J., Arens-Fischer,R., Kruger,M., Thonissen,M. and Luth,H. (1998) The application of porous silicon to optical wave-guiding technology. *IEEE J.Select.Topics Quant.Electr.* **4**, 975-981.

38. Vorozov,N., Dolgyi,L., Yakovtseva,V., Bondarenko,V., Balucani,M., Lamedica,G., Ferrari,A., Vitrant,G., Broquin,J.E., Benson,T.M., Arrand,H.F. and Sewell,P. (2000). Self-aligned oxidised porous silicon optical waveguides with reduced loss. *Electron.Lett.* **36**, 722-723.

39. Bondarenko,V.P., Yakovtseva,V.A., Dolgi,N., Vorozov,N.N., Kazyuchits,N.M., Tsybeskov,L. and Foucher,F. (1999) Erbium-doped oxidised porous silicon for integrated optical waveguides. *Tech.Phys.Lett.* **25**, 705-706.

40. Wang,J., Jiang,H., Wang,W., Zheng,J., Zhang,F., Hao,P., Hou,X. and Wang,X. (1992) Efficient infrared-up-conversion luminescence in porous silicon: a quantum-confinement-induced effect. *Phys.Rev.Lett.* **69**, 3252-3255.

41. Dneprovskii,V.S., Karavanskii,V.A., Klimov,V.I. and Maslov,A.P. (1993). Quantum size effect and pronounced optical nonlinearities in porous silicon. *JETP Lett.* **57**, 406-409.

42. Matsumoto,T., Hasegawa,N., Tamaki,T., Veda,K., Futagi,T., Mimura,H. and Kanemitsu,Y. (1994). Large induced absorption change in porous silicon and its application to optical logic gates. Jpn J Appl Phys. **33**, L35-36.

43. Henari,F.Z., Morgenstern,K., Blau,W.J., Karavanski,V.A. and Dneprovskii,V.S. (1995) Third-order optical nonlinearity and all-optical switching in porous silicon. *Appl.Phys.Lett.* **67**, 323-325.

44. Lettieri,S., Fiore,O., Maddalena,P., Ninno,D., Di Francia,G. and La Ferrara,V. (1999). Nonlinear optical refraction of free-standing porous silicon layers. *Optics Commun*, **168**, 383-391.

45.   Gesele,G., Linsmeier,J., Drach,V., Fricke,J. and Arens-Fischer,R. (1997). Temperature-dependent thermal conductivity of porous silicon. *J.Phys.D*, **30**, 2911-2916.

46.   Koyama, H. and Fauchet,P.M. (2000) Laser induced thermal effects on the optical properties of free-standing porous silicon films. *J.Appl.Phys.* **87**, 1788-1794.

47.   Jager,D. and Forschmann,F. (1987) Optical, optoelectronic and electrical bistability and multistability in a silicon Schottky SEED. *Solid State Electron*. **30**, 67-71.

48.   Soref,R.A. and Bennett,B.R. (1987) Electro-optical effects in silicon. *IEEE J.Quant Electr*. **23**, 123-129.

49.   Cocorullo,G. and Iodice,M., Rendina,I. (1994) All-silicon Fabry-Perot modulator based on the thermo-optic effect. *Optics Lett*. **19**, 420-422.

50.   Kasahara,R., Yanagisawa,M., Sugita,A., Goh,T., Yasu,M., Himeno,A. and Matsui,S. (1999). Low power consumption silica-based 2X2 thermo-optic switch using trenched silicon substrate. *IEEE Photon.Techn.Lett*. **11**, 1132-1134.

51.   Woodward,T.K. and Krishnamoorthy,A.V. (1999) 1 Gb/s integrated optical detectors and receivers in commercial CMOS technologies. *IEEE J.Selected Topics in Quant.Electr*. **5**,146-156.

52.   Alexandrou,S., Wang,C.C., Hsiang,T.Y., Liu,M.Y. and Chou,S.Y. (1993). A 75 GHz silicon metal-semiconductor-metal Schottky photodiode. *Appl.Phys.Lett*. **62**, 2507-2509.

53.   Zheng,J.P., Jiao,K.L., Shen,W.P., Anderson,W.A. and Kwok,H.S. (1992).Highly sensitive photodetector using porous silicon. *Appl.Phys.Lett*. **61**, 459-461.

54.   Yu,L.Z. and Wie,C.R. (1992) Fabrication of MSM photodetector on porous silicon using micromachined mask. *Electron.Lett*. **28**, 911-913.

55.   Tsai,C., Li,K.H., Campbell,J.C. and Tasch.A. (1993). Photodetectors fabricated from rapid-thermal-oxidised porous silicon. *Appl.Phys.Lett*. **62**, 2818-2820.

56.   Lee,M.K., Wang,Y.H. and Chu,C,H. (1997). High sensitivity porous silicon photodetectors fabricated through rapid thermal oxidation and rapid thermal annealing. *IEEE J.Quant.Electron*. **33**, 2199-2202.

57.   Vincent,G. (1994) Optical properties of porous silicon superlattices. *Appl.Phys.Lett*. **64**, 2367-2369.

58.   Berger,M.G., Dieker,C., Thonissen,M., Vescan,L., Luth,H., Munder,H., Theiss,W., Wernke,M. and Grosse,P. (1994). Porosity superlattices: a new class of Si heterostructures. *J.Phys.D*. **27**, 1333-1336.

59.   Pavesi,L., Mazzoleni,C., Tredicucci,A. and Pellegrini,V. (1995). Controlled photon emission in porous silicon microcavities. *Appl.Phys.Lett*. **67**, 3280-3282.

60.   Pavesi,L. (1997) Porous silicon dielectric multilayers and microcavities. *Riv.Nuovo Cimento*, **20**, 1-76.

61.   Cazzanelli,M., Vinegoni,C. and Pavesi,L., (1999) Temperature dependence of the PL of all-porous silicon microcavities. *J.Appl.Phys*. **85**, 1760-1764.

62.   Yamamoto,Y., Bjork,G., Karlsson,A., Heitmann,H. and Matinaga,F.M. (1993) Quantum state control in semiconductor pn junctions II Controlled spontaneous emission in quantum-well microcavity lasers. *Int.J.Modern Phys*. B**7**, 1653-1695.

63.   Chan,S. and Fauchet,P.M. (1999) Tunable,narrow,and directional luminescence from porous silicon light emitting devices. *Appl.Phys.Lett*. **75**, 274-276.

64.   Setzu,S., Letant,S., Solona,P., Romestain,R. and Vial,J.C. (1999) Improvement of the luminescence in p-type as-prepared or dye impregnated porous silicon microcavities. *J.Lumin*. **80**, 129-132.

65.   Canham,L.T. (1993). Laser dye impregnation of oxidised porous silicon on silicon wafers. *Appl.Phys.Lett*. **63**, 337-339.

66.   Nayfeh,M., Akcakir,O., Therrien,J., Yamani.Z., Barry,N., Yu,W. and Gratton,E. (1999). Highly nonlinear photoluminescence threshold in porous silicon. *Appl.Phys.Lett*. **75**, 4112-1114.

67.   Vahala,K.J. (1988). Quantum box fabrication tolerance and size limits in semiconductors and their effect on optical gain. *IEEE J.Quant.Electron*. **24**, 523-530.

68.   Huffaker,D.L. and Deppe,D.G. (1998). Electroluminescence efficiency of 1.3um wave-length InGaAs/GaAs quantum dots. *Appl.Phys.Lett*. **73**, 520-522.

69.   Mukai,K., Nakata,Y., Otsubo,K., Sugawara,M., Yokoyama,N. and Ishikawa,H. (2000). 1.3um CW lasing characteristics of self-assembled InGaAs-GaAs quantum dots. *IEEE J.Quant.Electr*. **36**, 472-478.

70.   Canham,L.T. (1997). Skeleton size distribution in porous silicon. in PROPERTIES OF POROUS SILICON p106-111, IEE.

71.   Holmes,J.D., Johnston,K.P., Doty,C. and Korgel,B.A. (2000). Control of thickness and orientation of solution-grown silicon nanowires. *Science*, **287**, 1471-1473.

72.   Gole,J.L., Stout,J.D., Rauch,W.L. and Wang,Z.L. (2000) Direct synthesis of silicon nanowires,silica nanospheres, and wire-like nanosphere agglomerates. *Appl.Phys. Lett.* **76**, 2346-2348.

73.   Seeger,K. and Palmer,R.E. (1999) Fabrication of ordered arrays of silicon nanopillars. *J.Phys.D.* **32**, L129-132.

74.   Brongersma,M.L., Polman,A., Min,K.S. and Atwater,H.A. (1999) Depth distribution of luminescent Si nanocrystals in Si implanted SiO2 films on Si. *J.Appl.Phys.* **86**, 759-763.

75.   Zacharias,M., Blasing,J., Veit,P., Tsybeskov,L., Hirschman,K. and Fauchet,P.M. (1999) Thermal crystallisation of amorphous Si/SiO2 superlattices. *Appl.Phys.Lett.* **74**, 2614-2616.

76.   Tang,Y.H., Zhang,Y.F., Wang,N., Lee,C.S., Han,X.D., Bello,I. and Lee,S.T. (1999). Morphology of Si nanowires synthesised by high-temperature laser ablation. *J.Appl.Phys.* **85**, 7981-7983.

77.   Kahler,U. and Hofmeister,H. (1999). Silicon nanocrystallites in buried SiOx layers via direct wafer bonding. *Appl.Phys.Lett.* **75**, 41-643.

78.   Prins,F.E., Single,C., Zhou,F., Heidemeyer,H., Kern,D.P. and Piles,E. (1999) Thermal oxidation of silicon-on-insulator dots. *Nanotechnology*, **10**, 132-134.

79.   Chien,F.S.S., Wu,C.L., Chou,Y.C., Chen,T.T., Gwo,S. and Hsieh,W.F. (1999). Nanomachining of (110)-oriented silicon by scanning probe lithography and anisotropic wet etching. *Appl.Phys.Lett.* **75**, 2429-2431.

80.   Huisken,F., Kohn,B., Paillard,V. (1999). Structured films of light-emitting nanoparticles produced by cluster beam deposition. *Appl.Phys.Lett.* **74**, 3776-3778.

81.   Dag,M., Ozin,G.A., Yang,H., Reber,C. and Bussiere,G. (1999).Photoluminescent silicon clusters in oriented hexagonal mesoporous silica film. *Adv.Mater.* **11**, 474-480.

82.   Wilcoxon,J.P. and Samara,G.A. (1999). Tailorable,visible light emission from silicon nanocrystals. *Appl.Phys.Lett.* **74**, 3164-3166.

83.   Kimura,K. and Iwasaki,S. (1999) Vibronic fine structure found in the blue luminescence from silicon nanocolloids. *Jpn.J.Appl.Phys.* **38**, 609-612.

84.   Nassiopoulo,A.G., Tsakiri,V., Ioannou-Sougleridis,V., Photopoulos,P., Menard,S., Bassani,F. and Arnaud d'Avitaya,F. (1999). Light-emitting structures based on nanocrystalline (Si/CaF2) multiquantum wells. *J.Lumin.* **80**, 81-89.

85.   Carr,D.W., Evoy,S., Sekaric,L., Craighead,H.G. and Parpia,J.M. (1999). Measurement of mechanical resonance and losses in nanometer scale silicon wires. *Appl.Phys.Lett.* **75**, 920-922.

86.   Huang,J.Y., Yasuda,H. and Mori,H. (1999). Deformation-induced amorphisation in ball-milled silicon. *Philos.Mag.Lett.* **79**, 305-314.

87.   Fujita,S. and Sugiyama,N. (1999). Visible light-emitting devices with Schottky contacts on an ultra thin amorphous silicon layer containing silicon nanocrystals. *Appl.Phys.Lett.* **74**, 308-310.

88.   Randall,J.M., Reed,M.A. and Frazier,G.A. (1987) Nanoelectronics: fanciful physics or real devices. *J.Vac.Sci.Techn.* **B7**, 1398-1404.

89.   Fountain,T.J., Duff,M.J.B., Crawley,D.G., Tomlinson,C.D. and Moffat,C.D. (1998) The use of nanoelectronic devices in highly parallel computing systems. *IEEE Trans. VLSI Systems*, **6**, 31-38.

90.   Hu,C. (1999) Silicon nanoelectronics for the 21st century. *Nanotechnology.* **10**, 113-116.

SILICON LIGHT EMITTERS: PREPARATION, PROPERTIES, LIMITATIONS, AND INTEGRATION WITH MICROELECTRONIC CIRCUITRY

P. M. FAUCHET, S. CHAN, H. A. LOPEZ, and K. D. HIRSCHMAN
Department of Electrical and Computer Engineering
University of Rochester, Rochester NY 14627, USA

Abstract

Starting with Canham's discovery in 1990 that porous silicon (PSi) can emit bright light in the visible range of the spectrum, there has been a strong interest in silicon light emitters. PSi and other light-emitting forms of silicon contain nanostructures or crystallites in the nanometer size range. Throughout most of the 1990's, the intense visible luminescence from nanoscale silicon crystallites has been a source of numerous investigations and considerable debate. Today, most of the controversies have been put to rest. However, much less has been written about nanoscale Si light-emitting devices, in part because some of their characteristics are less than ideal and not well understood. This paper reviews the status of nanoscale silicon light emitters. It starts with a survey of the manufacturing methods used to produce nanoscale Si. Next, key physical, optical, electrical, and structural properties of nanoscale Si are examined. The fabrication of electroluminescent devices (LEDs) is then discussed. We focus on the stability, efficiency, speed, and spectral characteristics of nanoscale Si light emitters. Recent results obtained on microcavity PSi LEDs and 1.5μm LEDs produced by doping PSi with erbium are discussed. Finally, the integration of PSi LEDs with microelectronic circuitry is reported and the prospects for practical devices are briefly examined.

1. Introduction

The radiative recombination of an electron with a hole in semiconductors produces photons with an energy equal to the bandgap. If the electron and the hole are located at the same point in the Brillouin zone, as in direct gap semiconductors, the radiative recombination rate is large and the radiative lifetime τ_{rad} is short (typically a few nsec) [1]. In indirect gap semiconductors such as Si, the radiative recombination of an electron located at the bottom of the conduction band with a hole located at the top of the valence band requires the participation of a phonon. As a result, the radiative lifetime is much longer, typically in the msec regime. The quantum efficiency is given by

99

L. Pavesi and E. Buzaneva (eds.), Frontiers of Nano-Optoelectronic Systems, 99–119.

$$\eta = \tau_{non\text{-}rad} / (\tau_{non\text{-}rad} + \tau_{rad}) \qquad (1)$$

where $\tau_{non\text{-}rad}$ is the lifetime for the recombination pathways that do not involve the production of photons. In direct gap semiconductors, η can be very large, typically of the order of 10% or more, because the radiative lifetime is short enough to compete with the non-radiative lifetime. However, in silicon, this is not the case and η drops by several orders of magnitude, to be as low as 10^{-4} to 10^{-5} %.

Several approaches have been investigated to improve the quantum efficiency of silicon and silicon-based alloys. These approaches, which are summarized in Ref. 2, amount to either (1) confining the electrons and holes in small regions of space where there are few or no non-radiative recombination centers, or (2) to modifying the electron and hole wavefunctions in such a way that the radiative recombination rate is strongly enhanced. Among these approaches, those that rely on confinement, for example around isoelectronic impurities, have sometimes led to very good quantum efficiencies (up to 10% in some cases) and acceptable electroluminescence efficiencies (approaching 1%), but only at cryogenic temperatures. The physical reason for this is that the confinement energies are small. Thus, when kT becomes large enough, for example when T = 300K, the carriers acquire enough kinetic energy to overcome the confining potential and thus may leave the defect-free regions.

Since 1990, there has been a renewed interest in the preparation, properties, and device applications of porous silicon (PSi) and other silicon structures containing crystallites in the submicron to nanometer size range. This results from the discovery of the intense, visible luminescence from nanoscale silicon crystallites [3]. Quantum efficiencies greater than 10% at room temperature and 50% at cryogenic temperatures have been demonstrated [4], because the confinement potential is very large (typically, much greater than 1 eV) and the surface passivation can be excellent. This luminescence has been a source of numerous investigations and considerable debate, which are reviewed in Refs. 5-11.

2. Preparation and Materials Science

Uhlir [12] made PSi for the first time in 1956, by anodically etching a p-type silicon wafer in a solution containing HF. PSi is formed below a critical current density J_{crit} in forward bias, whereas above J_{crit} electropolishing takes place [13]. It is also possible to produce PSi starting from an n-type Si wafer by photogenerating the holes that are necessary for the dissolution of the Si wafer. PSi is classified as macroporous, mesoporous, or microporous, in order of decreasing pore sizes [14]. The term nanoporous silicon is also used when the size of the Si crystallites is in the nanometer regime. The nanocrystallite size can be measured by transmission electron microscopy (TEM), X-ray diffraction (XRD) and Raman spectroscopy. Strongly luminescent samples are found to contain nanocrystallites of sizes below 4 nm (nanoporous Si) [15]. The dependence of the luminescence intensity and peak wavelength with the average crystallite size and the porosity for PSi films made from p-type wafers is plotted in Figure 1.

The open structure of very high porosity films, which produce the brightest PL, makes them easily destroyed by the capillary stresses present during drying. After supercritical drying [16,17], a technique widely used for preparing silica aerogels, PSi films survive intact even when the porosity exceeds 95%. For porosities below 85%, normal drying is usually sufficient. The surface of PSi is naturally passivated by silicon

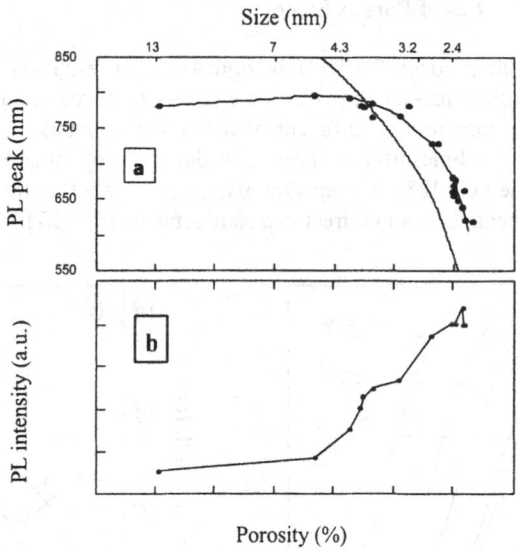

Figure 1. Photoluminescence peak wavelength a) and intensity b) vs. porosity (linear scale) and crystal size (nonlinear scale) for porous silicon films made from 6 Ω cm p-type silicon wafers. In a), the thin line that fits the data is a guide to the eye and the thick line is the excitonic bandgap.

hydride bonds. An excellent passivation is required for achieving a high luminescence efficiency since the radiative lifetime in PSi remains very long (in the μsec to msec range). The fragile Si-H bonds can be broken by UV illumination, heating, exposure to various gases and liquids, and even the presence of a large electric field. When one Si-H bond is broken, one dangling bond is formed on the nanocrystal surface. This dangling bond acts as a non-radiative recombination center and, as a result, that nanocrystal becomes "dark", which means that it does not luminesce. When enough dangling bonds are formed in the sample, the PL efficiency drops precipitously.

Silicon dioxide and its interface with crystalline silicon can be defect-free. When the Si-H bonds are replaced by Si-O bonds, the stability of the luminescence efficiency of PSi becomes excellent [18]. The high temperature treatment necessary for the formation of a high quality silicon/SiO_2 interface leads to a very rapid oxidation rate which may produce an unacceptably large oxide thickness and eventually for T > 1100°C to the full conversion of PSi into a porous glass [19]. To reduce the oxidation rate, the high temperature oxidation is performed in an atmosphere containing only a small percentage of oxygen [20]. A very thin, very high quality oxide layer (a few monolayers at most) coats the Si nanocrystallites, allowing efficient and stable PL and electroluminescence (EL) [21]. Recently, chemical stabilization [22,23] has shown promise for stabilizing the luminescence efficiency of PSi, but to date, SiO_2 remains the only passivation agent that is compatible with the microelectronic industry requirements.

3. Optical Properties of Porous Silicon

An increase in the porosity P affects the optical properties. The transmission spectra of several 20 μm-thick free-standing PSi films of various porosities are shown in Figure 2a from which the absorption coefficient of the Si nanocrystals is obtained as shown in Figure 2b [24]. The blueshift in absorption is due to the opening of the gap as a result of quantum confinement [25]. The spectra also demonstrate that Si nanocrystals as small as 2.4 nm still behave as an indirect gap semiconductor [26,27].

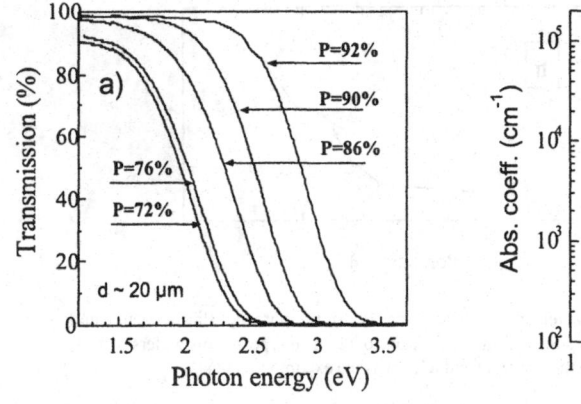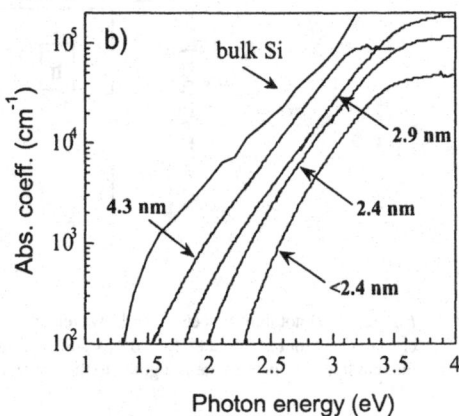

Figure 2. a) Transmission spectra of several 20 μm thick free-standing PSi films (5-7 Ω cm p-type Si substrate) of various porosities. b) Absorption spectra for Si nanocrystals with sizes from 4.3 nm to 2.4 nm deduced from the transmission spectra of a). The absorption of c-Si is shown for comparison. The saturation beyond 3eV is an artifact of the measurement procedure.

Nanoscale Si can be made to luminesce efficiently throughout the visible and in the near infrared [18], as shown in Figure 3. When the crystallites are large and well passivated by SiO_2, as is the case with mesoporous silicon after oxidation, the PL efficiency is high and the low-temperature PL spectrum is identical to that of high purity c-Si [28]. The reason for the good PL efficiency is not that Si has become a direct gap semiconductor but rather that the crystallites are defect-free and well-passivated. These conditions can be achieved more easily with smaller crystallites, in which the opening of the bandgap and the susbsequent shift of the luminescence into the visible are produced by quantum confinement. Theory [29] shows that the increase of the bandgap scales as $(size)^{-1.4}$. Thus, the PL energy also increases with decreasing sizes, provided that the radiative recombination occurs at or near the bandgap of the silicon nanocrystals (some of the

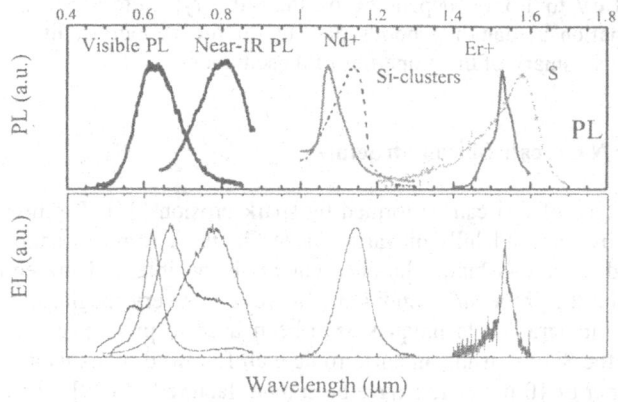

Figure 3. Room temperature PL and EL spectra for PSi passivated by SiO₂ and prepared with sub- 5 nm crystallites (visible and near IR luminescence), large (>>10nm) crystallites (Si-clusters), or doped with Nd, Er or S. PSi LEDs can be made to emit from the visible to 1.5 μm.

weak PL observed in PSi, such as the often reported broad peak near 1.6 eV, is likely due to defects [30]).

An interesting question is whether recombination involves free excitons (when the PL energy should equal the one-electron bandgap minus the binding energy of the exciton), or one of the carriers [31] or the exciton [32] involved in the luminescence is trapped on the surface of the nanocrystals (when the PL energy is expected to be below the excitonic bandgap). We have demonstrated that the measured bandgap energy is in good agreement with the calculated excitonic bandgap and that the PL peak energy is consistently below this bandgap, with the clear trend of an increasing Stokes shift between bandgap and PL energies with decreasing sizes [24], at least for nanocrystals whose surface is passivated by both hydrogen and oxygen. Recently, it has been shown that the luminescence wavelength of oxygen-free PSi can be tuned throughout the visible range, including deep in the blue [33,34]. We also have demonstrated that the luminescence red shifts immediately after exposure to oxygen, consistent with the formation of a specific Si-O bond on the surface of the nanocrystallites [34]. From these experiments, we can conclude that quantum-confined free excitons produce the luminescence in oxygen-free samples, whereas at least one carrier (the electron) is trapped at a Si-O bond for small crystallites exposed to oxygen.

Figure 3 shows several PL bands that are extrinsic. In particular, the emission near 1.5 μm is due to the incorporation of Er in PSi, either by ion implantation or electrochemically [35]. From cryogenic temperatures to room temperature, the Er-related PL broadens but its integrated efficiency remains nearly constant. Although not shown in Figure 3, PSi can also emit brightly in the blue (>2.5 eV) for strongly oxidized samples or in the infrared, between 0.8 and 1.3 eV for samples containing a large concentration of dangling bonds [9]. In both cases, the PL is not produced by radiative recombination inside the quantum dots. The blue PL, whose decay is fast (~ 1 nsec [36]), is related to recombination centers in SiO₂. The infrared PL which can be tuned from 0.8 eV to 1.3 eV depending on the nanocrystallite bandgap, is associated with recombination at dangling bonds present on the nanocrystallite surface. Refs. 5-10 provide a summary of the properties of these bands.

4. Other Nanoscale Silicon Structures

Another type of PSi can be formed by spark erosion [37]. The resulting Si surface is covered by pits and hills of various sizes in the micron to nanometer scale. Spark processed silicon is highly luminescent in the visible and the emission wavelength depends on the processing conditions. However, its very rough surface is unsuitable for devices. Lithographic techniques have been used to produce Si nanostructures. After thinning the Si core using an anisotropic etch followed by oxidation, Si nanostructures of the order of 10 nm or less have been manufactured [38,39]. Photoluminescence and even electroluminescence have been demonstrated but the density of the Si nanostructures is low, which severely limits their usefulness in devices.

Another interesting method to produce luminescent Si nanocrystals is to oxidize a microcrystalline film deposited by CVD. Oxidation decreases the diameter of the Si crystallites to the appropriate size (less than 5 nm) and passivates the surface of each nanocrystallite [40]. The advantages of this procedure are its simplicity and compatibility with Si microelectronic technology. However, the highest reported efficiency is still more than one order of magnitude below that of a medium-quality PSi film and the crystallite size distribution is not well controlled.

Si quantum dots can be prepared by implanting Si into SiO_2 and then annealing to form Si nanocrystallites. The luminescence of these structures can be tuned throughout the visible by controlling the processing steps [41]. However, the control on the size of the crystallites and the crystallite packing density are limited. Nanocrystallites can also be prepared in the gas phase, and then collected on a glass plate or in a matrix. Gas phase synthesis of Si quantum dots, followed by gas phase oxidation and size selection, has produced the highest luminescence efficiency in Si to date (quantum efficiency of 50% at low temperatures [4]). However, the production rate of these crystallites is too low to make them technologically useful.

Silicon nanocrystals can be also prepared by the controlled crystallization of amorphous silicon (a-Si) layers [42-44]. In our work, amorphous Si/SiO_2 multilayers are first grown by plasma enhanced chemical vapor deposition (PECVD) or by RF magnetron sputtering. The a-Si thickness ranges from ~ 2.5 nm to 25 nm and that of the SiO_2 from 1.5 nm to 6 nm. The crystallization in both types of samples is performed in two steps: first, rapid thermal annealing (RTA) at 800°C-900°C, then furnace annealing both done in an atmosphere of argon and nitrogen. After crystallization, the Si layers contain many nanocrystals that are randomly oriented, spherical, and all of the same size, equal to the thickness of the original amorphous Si layer. The quantum efficiency of these samples has been measured to approach 1%.

5. A Brief History of Electroluminescence From Porous Silicon

Electroluminescence (EL) was observed in PSi shortly after Canham's report of strong room-temperature PL, first in solution during anodic oxidation [45] and then in solid-state devices [46-49]. Strong (0.1-1% efficiency), voltage-tunable EL has even been reported under cathodic polarization of n-type PSi using liquid contacts [50]. Unfortunately, the use of a liquid contact makes these structures impractical for large-scale applications and these devices are intrinsically short-lived as the PSi layer is consumed. In the following, we restrict ourselves to a discussion of the solid-state light-emitting devices (LEDs) because of their technological importance. PSi LEDs of different types have been reported. A typical device consists of a transparent or semitransparent contact (metal, ITO or conducting polymers) and a <1 to 10 μm thick PSi layer on a crystalline silicon (c-Si) substrate (p or n-type) [46-49,51]. Because of the poor transport properties of PSi [52,53] and the technological difficulties in making good contacts to the porous layer, the threshold voltage and current density for EL have for a long time been very high, typically $V \sim 10$ V and $J \sim 10$ mA/cm^2. At present, the best numbers reported are threshold voltage ~2 V and current density <<1 mA/cm^2 [54]. Until 1995, PSi LEDs had a low EL external quantum efficiency (0.01%) and degraded rapidly and irreversibly. More recently, devices that are stable and have a much higher efficiency have been demonstrated [55,56]. In the remainder of this section, we will address the stability of the PSi LEDs, their efficiency, their speed, their emission wavelengths, and their compatibility and integration with accepted microelectronic procedures. In particular, we will discuss in detail two recent results we have achieved, namely the demonstration of narrow-band, tunable microcavity LEDs and of relatively efficient 1.5 μm LEDs after doping with erbium.

6. The State-of-the-art in 2000

6.1 LONG-TERM STABILITY

The most common cause of degradation is the depassivation of the Si nanocrystals. As mentioned before, the Si-H bonds that naturally passivate PSi are very fragile and can be easily broken by exposure to light, ambient air, moderate temperatures, and large electric fields. To improve the stability, the fragile Si-H bonds should be replaced by the stronger Si-O bonds and the devices should be engineered to provide better heat sinking and greater mechanical stability. When these improvements are implemented, no degradation has been observed after several weeks of operation under ambient conditions [55,56]. Simple semitransparent metal / partially oxidized PSi / c-Si LEDs were found to be stable over 100 hours of continuous operation but relatively inefficient (power efficiency < 0.01%). The quality of the oxide and the Si/SiO$_2$ interface has been improved by using elevated temperatures [55,56]. The temperature treatment takes place in a diluted oxygen atmosphere to slow down the conversion of Si into SiO$_2$ and keep the oxide thickness to a few monolayers, thus preventing the formation of thick SiO$_2$ barriers that might impede electrical transport. Furthermore, the device structures include a low porosity Si layer and a doped μc-Si layer between the PSi region and the

106

top metal contact. Figure 4 shows the results of stability tests, performed under pulsed drive conditions well above EL threshold. No degradation is seen for several weeks. The power efficiency of the best PSi LED of this type is near 0.1%. Although stability for thousands of hours under realistic conditions has not yet been demonstrated, this is not the primary reason for the low level of commercial interest. We believe that a multi-year focused research effort would likely lead to stability performances that are acceptable for commercial applications.

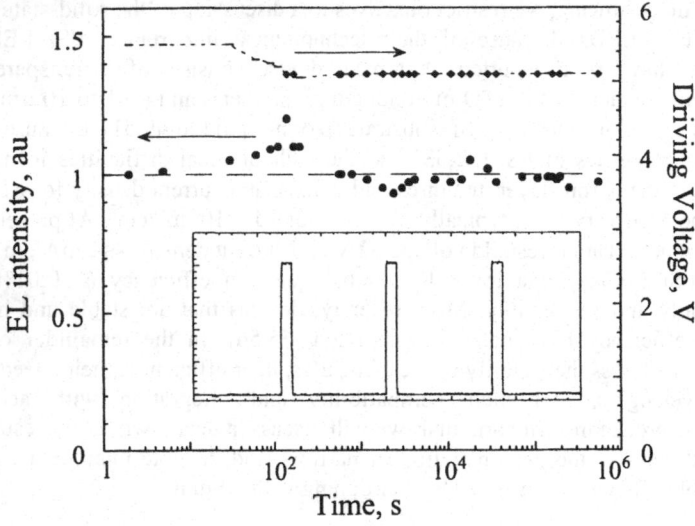

Figure 4. Stability of the electroluminescence with time.

6.2 EFFICIENCY

In the early 1990's, the efficiency of PSi LEDs quickly increased from ~0.0001% to ~0.01% mostly through better device engineering and processing. However, until 1995, the EL efficiency still remained two to three orders of magnitude below the typical PL quantum efficiency of 1% or better. Part of the problem is probably unavoidable. Indeed, to bring the PL efficiency above 1%, it is necessary to increase the porosity to values (see Fig. 1) that make this material very fragile and incompatible with processing techniques. The difficulties in injecting carriers from the contact into the PSi and the poor transport properties in PSi are the other main reasons for the large difference between PL and EL efficiency. Making a good quality electrical contact to a highly porous structure such as PSi is difficult. Evaporation is a line-of-sight technique that produces contacts only with the top nanostructures. Electropolymerization to form an intimate contact has also been demonstrated [51] but the efficiency of the LEDs was not improved.

In 1995, several groups reported a power efficiency that was increased to greater than 0.1% [54,55,57,58]. The internal quantum efficiency, defined as the ratio of the number of photons generated inside the PSi layer to the number of carriers injected inside the PSi layer, was estimated to be > 1 %, which is comparable to the PL efficiency. The first efficient LED that was operational in the cw mode of operation and had a power efficiency of 0.1% [54] also had a record low threshold voltage (2 V) and current density (1 μA/cm^2). The best of these devices was stable for hours. Our PSi LEDs also have a power efficiency ~ 0.1% but they are brighter and are stable for weeks with no detectable degradation [55,56]. Electron injection does not take place from a heavily doped μc-Si top contact directly into the PSi layer, but rather into a mesoporous buffer layer. This "buffer" layer serves at least two purposes. First, the quality of the contact with the top electrode measured by the C-V technique can be excellent, because the porosity is low. Second, we speculate that the mesoporous layer acts as a graded bandgap layer between the μc-Si layer and the PSi layer. The thickness of the PSi layer is also kept at ~ 1 μm to minimize losses during transport. As will be discussed later, these LEDs are processed in semiconductor fabrication line compatible conditions, including high temperature treatments. Improvements in the LED efficiency have stopped since 1995, with one possible exception [59]. The lack of progress on this front is the major reason for the lack of industrial interest in PSi LEDs. For both optical interconnects and optical displays, the necessary brightness and power dissipation requirements require LEDs with at least one order of magnitude larger efficiencies. Unless a breakthrough is achieved in the near future, PSi LEDs will not find widespread commercial use although niche applications are possible.

6.3 RESPONSE TIME

We have measured the frequency response of PSi LEDs by modulating the applied voltage and characterizing the modulated EL [52,60]. The device is held at a large forward bias, well above threshold for detectable EL. A small AC voltage is applied and

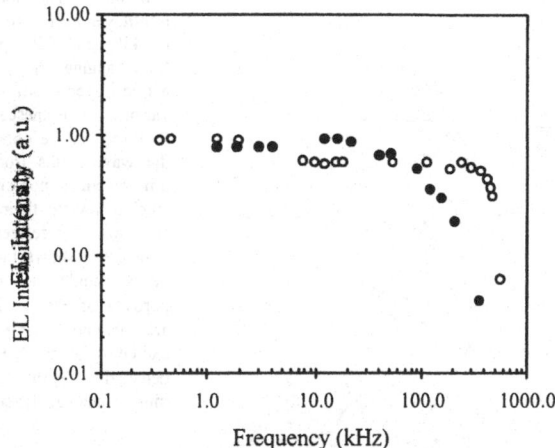

Figure 5. Frequency modulation response of the electroluminescence of porous silicon p-n junction LED's with (O) 0.2 μm junction depth and 5 μm porous silicon layer, (●) 1 μm junction depth and 10 μm porous silicon layer. The 3 dB bandwidth can exceed 100 kHz.

the amplitude of the AC EL is monitored as a function of frequency. Figure 5 shows that the magnitude of the AC EL signal remains constant up to a critical frequency beyond which it drops quickly. The 3-dB frequency, defined as the frequency at which the AC EL has dropped by a factor of 2, is the inverse of the transit time and determines the maximum frequency response of the diode in this small-signal analog test. Thus, the response time of the EL is limited by the time it takes carriers to cross the PSi layer, and not the PL lifetime, which is usually shorter. By making the PSi layer thinner than 1 μm or by using a PSi p-n junction, the 3-dB freqency can exceed 1 MHz. Another important mode of operation for LEDs is the pulsed mode of operation, in which a voltage pulse is applied to the device to produce a pulse of light. In LEDs using oxidized PSi with a measured PL lifetime of ~ 30 ns, the rise and fall times were measured to be as short as 20 nsec [61,62]. However, these numbers indicate that direct modulation of PSi LEDs will not be attractive for optical interconnects, which may require GHz modulation speeds.

6.4 INTRINSIC SPECTRUM AND MICROCAVITIES FOR NARROW, TUNABLE EMISSION

Since the PL can span the spectrum from the infrared to the blue, it is reasonable to expect that PSi LEDs can cover the same range of wavelengths. As early as 1992, LEDs

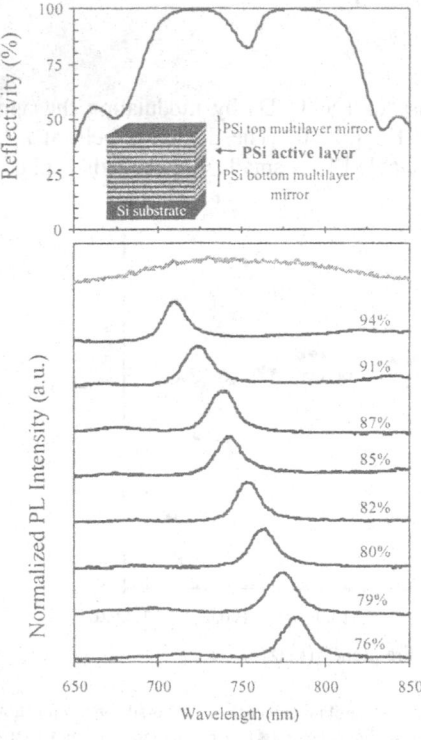

Figure 6. PSi microcavity resonator structure on a p^+ (0.008-0.012 Ω-cm) c-Si substrate (top inset). The reflectivity spectrum (top) of a microcavity resonator, with active layer porosity of 82%, is shown to have a reflectivity dip of ~80% at 750 nm. The top and bottom multilayer mirror contain 6 periods of 43% and 62% porosity layers. By changing the porosity of the active layer from 76% to 94%, various luminescence spectra (bottom) can be obtained, spanning the wavelengths from 780 to 700 nm (average porosity values are labeled next to their respective PL spectra). For reference, the broad gray curve shows the PL spectrum of a single active layer with porosity of ~80%. All PL spectra are obtained using the 514 nm excitation of an Ar+ ion laser and detected with an optical multichannel analyzer detector.

with peak wavelengths ranging from the deep red to the blue were demonstrated [63]. The spectral width of the emission was very wide, comparable to that observed in photoluminescence experiments. We have also been able to demonstrate PSi LEDs emitting at 1.1 μm using mesoporous silicon [64]. These devices exhibit a narrower spectrum but the efficiency was several orders of magnitude lower than that for red LEDs.

Because PSi consists of a matrix containing a range of silicon nanocrystal sizes, the characteristic luminescence is broadband. Since control of the size distribution is extremely difficult, an alternative method is needed to reduce the luminescence bandwidth. This can be achieved through the use of PSi interference filters and multilayer mirrors [65-67]. A typical device structure is shown in the top of Figure 6. To obtain high reflectivity mirrors, a large difference in refractive indices of the single layers is necessary. A p+ substrate is chosen because the range of porosities that can be formed is wide (30% - 95%) [68], yielding values of refractive index from 2.69 to 1.06 [69]. Anodization begins when a constant current is applied between the silicon wafer and the electrolyte. The porosity is a linear function of the current density for a specific HF concentration and current density interval [68]. Therefore a periodic pulse alternating between two different current densities is a convenient way of forming multilayer PSi films. Changing the current density does not affect the PSi layer previously formed because silicon dissolution occurs at the silicon/electrolyte interface [69]. To stabilize the PSi multilayer structures after anodization, all samples are thermally oxidized in an oxygen ambient at 900° C for ten minutes. Oxidation induces a blue-shift in peak reflectivity due to a change in the refractive index of the layers, which must be taken into account. A good quality multilayer structure requires its peak reflectivity to remain constant, even after oxidation. For our multilayer structures, the peak reflectivity does not decrease upon thermal treatment.

In the structures discussed, the multilayer mirrors have alternating porosities and thicknesses of 43%; 80 nm and 62%; 160 nm. All porosity values are estimated using a porosity dependence on current density plot [68] and values for the thickness are obtained through SEM micrographs. For fixed values of porosities or refractive indices, the reflectivity of a multilayer mirror increases as the number of periods increases. Nearly 100% reflectivity is achieved for a multilayer mirror containing 10 periods. However, in our devices, we opted for a compromise between high reflectivity and thickness (number of periods), because transport through a thick PSi film is difficult [52]. For a multilayer mirror containing 6 periods, 88% reflectivity centered at 760 nm is attainable [70]. Therefore all subsequent PSi devices are fabricated with multilayer mirrors each containing 6 periods.

The microcavity resonator is constructed by inserting an active, highly luminescent PSi layer in between two multilayer mirrors, creating a Fabry-Pérot cavity, in which significant line narrowing of the luminescence spectra is achievable [72]. The active layer porosity varies from 76% - 94%, with an approximate thickness of 150 nm. High porosity active layers can be used because the top multilayer mirror stabilizes the fragile film and prevents it from flaking off the substrate. The reflectivity spectrum for a microcavity resonator has a characteristic double peak, with a sharp dip between the two maxima. This dip in reflectivity determines the luminescence transmission bandwidth of the embedded active layer. For our devices, a reflectivity dip to 80% is obtained with an active layer thickness of 150 nm, as shown at the top of Figure 6. To further increase the magnitude of the reflectivity dip, an increase in thickness of the active layer is necessary. Again, due to transport reasons, a compromise must be made between minimum reflectivity dip and active layer thickness. By merely changing the porosity of the active layer, thereby affecting its refractive index, the peak luminescence wavelength can be tuned from 780 nm - 700 nm [72]. The bottom of Figure 6 shows the room temperature PL from several PSi microcavity resonators with varying active layer porosities. The experimental microcavity resonator PL spectra are in agreement with computer simulated plots generated with the previously reported parameters for the multilayer mirror and active layer.

LEDs can be made using these structures [72]. A first step toward this goal was taken by Araki et al. who demonstrated a narrowing of the EL linewidth in a metal/PSi/multilayer structure [73]. Our work further examines the narrowing of EL in a PSi microcavity resonator, and in addition, shows the tunability of the EL spectra and its high luminescence directionality. A thin (~ 15 nm), semi-transparent layer of Au is sputtered onto the top of the sample, with a circular area of 0.079 cm^2. At 750 nm, the Au layer is approximately 50% transparent. PL spectra from samples through the Au contact were also measured to compare with EL. The current density-voltage characteristics of these devices, is nearly symmetrical. The EL spectra for several PSi microcavity resonators with different active layer porosities (76% - 94%) are shown in Figure 7. As expected, the peak luminescence blue-shifts as the porosity of the active layer increases and the EL spectra are narrow. The full width at half maximum (FWHM) for each EL spectrum is ~ 20 nm, one order of magnitude lower than the typical FWHM of PSi. The spectral shape is independent of the bias; however, the EL intensity increases as the current increases. Continuous operation of PSi microcavity devices at 25 mA/cm^2 show that no degradation in EL is observed for up to one hour. Currently, the power efficiency for these devices is much lower (by more than 2 orders

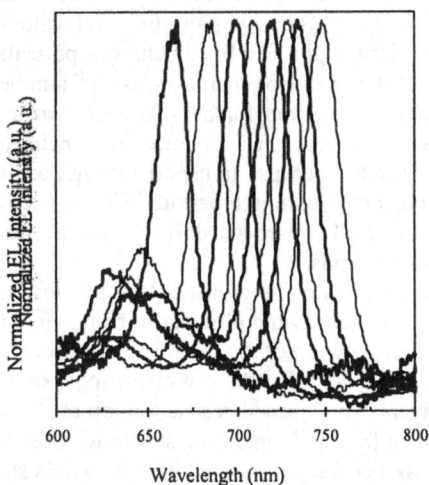

Figure 7. Room temperature EL from oxidized PSi microcavity resonators with varying active layer porosity. All spectra obtained at a reverse bias of ~100 V.

of magnitude) than that of the best PSi LEDs [54,56,59]. This can be attributed to several factors: an overall increase in thickness of the PSi microcavity resonator (~ 3 µm) which severely limits transport, the presence of a multilayer structure which may cause discontinuities across different interfaces, and high reflectivity of the top contact which prevents the light from escaping.

Control of the radiation pattern to obtain a high angular concentration of output emission maybe very important for some device applications. The room temperature PL radiation pattern for a PSi microcavity resonator has been reported to concentrate in a 30° cone around the normal axis [74]. For angles far from the normal axis, a decrease in the coupling between excitons and cavity modes exists, due to the shift in the cavity mode resonant wavelength [75]. This effect is also seen for EL [72], as shown in the top of Figure 8. The highest EL intensity for a microcavity resonator (dark circles) occurs at 0°, which is normal to the emitter. Within a cone of 30°, the intensity is drastically reduced, which coincides with PL from a microcavity resonator. As a reference, the open circles represent the PL radiation pattern for a single layer of PSi detected under the same experimental setup as EL from a microcavity resonator. The bottom of Figure 8 shows the relative EL spectra of a microcavity resonator detected at various emission angles. As the emission angle deviates from the normal axis, the resonant wavelength slightly shifts and the intensity decreases.

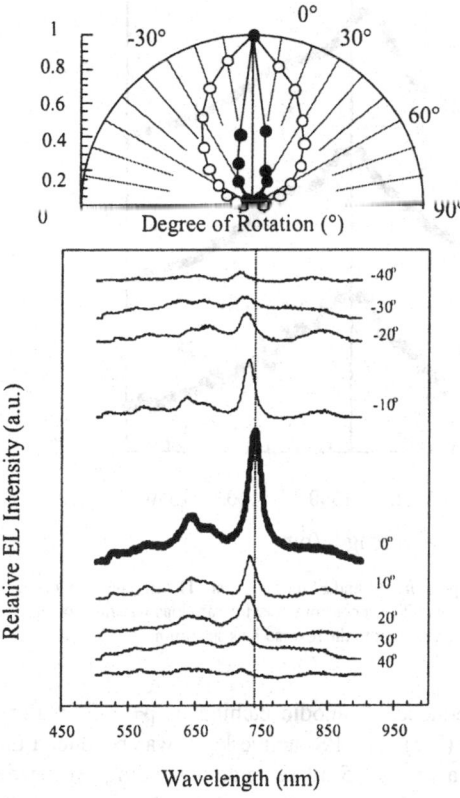

Figure 8. Luminescence radiation pattern (top) for a PSi microcavity resonator (EL,●) and PSi active layer (PL,O). The relative EL spectra of a PSi microcavity resonator at various emission angles are shown at the bottom. As the emission angle deviates from the normal axis, shifts in the resonant wavelength and decreases in intensity are evident.

6.5 EXTRINSIC SPECTRUM OF ERBIUM DOPED POROUS SILICON LED'S

After the initial report of weak 4*f*-shell luminescence from Er-doped silicon [76], a vast technology driven interest in the field developed. The radiative atomic-like transition at 1.54 μm from Er is very important because it matches the window for maximum transmission in silica based optical fibers. Numerous semiconductors from group IV [77] and group III-V [78] to various glasses [79] have been doped with Er, all of them exhibiting infrared luminescence. Out of all the Er-doped materials, silicon based materials are the most studied, because silicon dominates the microelectronic industry and Er-doped materials eventually need to be integrated with silicon to produce intra-chip and chip-to-chip optical interconnects. The very large surface area to volume ratio makes the PSi matrix very accessible for Er-doping, as well as a host for large concentrations of oxygen necessary for erbium emission. Doping of PSi has been achieved by ion implantation [80], diffusion [81,82], and electrochemical deposition [83]. Cathodic electrochemical deposition is preferred because it offers the advantages of deeper erbium penetration (10-20 μm), lower cost, and simplicity of processing.

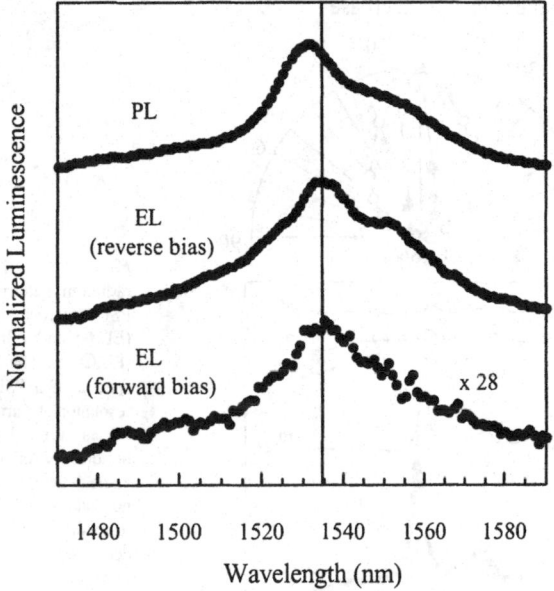

Figure 9. Normalized room-temperature PL and EL comparison. The EL spectra were taken at a constant current of ~ 6 mA. The spectrum under forward bias was multiplied by a factor of 28 with respect to the reverse-bias spectrum.

The PSi used in this study was produced by anodic etching of p+ *c*-Si wafers in a hydrofluoric acid-ethanol electrolyte (3:2). The PSi active layer was produced under a current density of 90 mA/cm^2 over a time of 5-50 seconds depending on the desired thickness. The thickness of the PSi active layer varied from 0.5 to 5 μm and its porosity was ~46 % (calculated gravimetrically). After etching, the samples were rinsed with ethanol and a pore cleaning step was performed before doping [84]. The cleaning step facilitates ion infiltration, by removing SiO_x pore blocking species, resulting in an increase in erbium concentration. The PSi surface roughness also decreases, improving the adhesion of the electrical contact. After cleaning, the samples were rinsed and immersed in a saturated $ErCl_3·6H_2O$/ethanol solution and negatively biased relative to the Pt electrode in order to introduce the erbium ions into the PSi matrix. The incorporated erbium concentration is estimated to be 10^{19} cm^{-3} by taking 10% of the total charge of the electrochemical process [83]. After doping, the samples were oxidized for 10 minutes at 950°C in a dilute oxygen environment and densified for 5 minutes at 1100°C in a nitrogen environment. For electrical contacts, a thin (~13 nm), semi-transparent gold film with a circular area of 0.044 cm^2 was sputtered on top of the sample.

Since our first report of weak EL from Er-doped PSi [85], we have improved the stability, efficiency, and structure, as well as characterized the properties of the LEDs [86]. The typical I-V characteristic of a 3 μm thick Er-doped PSi active layer device is nearly symmetrical. The high resistance observed in the I-V plot is due to the poor transport through the oxidized PSi, but as the thickness of the Er-doped active layer, the annealing time, and the annealing temperature decrease, improvements in the transport are observed. Figure 9 compares the normalized PL and EL spectra. Both EL spectra were taken at a similar current of ~6 mA, but the peak intensity is about 28 times higher for reverse bias (Figure 10a). The EL line shape and peak position in forward and reverse biases are almost identical, but compared to the PL, they are shifted ~2 nm toward lower energy. The difference in position may be a result of optically and electrically exciting Er centers in different microenvironments within the active layer. In all measurements, the very weak silicon band-edge emission and the lack of visible and/or defect luminescence show that Er centers are the most efficient radiative sites.

In reverse bias, the EL detection threshold current is lower (1.4 mA) and the EL detection threshold voltage is higher (18 V) than in forward bias (Figure 10a) [86].

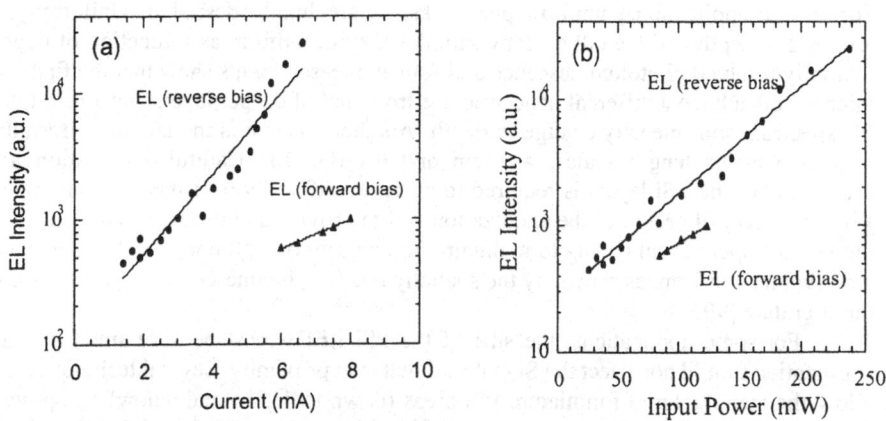

Figure 10. EL intensity as a function of driving current (a) and device input power (b) for both forward (s) and reverse (l) bias conditions.

Figure 10b shows the EL intensity dependence at 1.54 μm on the measured input electrical power. The device exhibits an exponential EL dependence in both biasing conditions as a function of input power, and a lower EL turn on input power (25 mW) for reverse bias, as well as higher output power efficiencies. The estimated external quantum efficiency is ~ 0.01 %. This value is of the same order as what has been reported from Er-doped *c*-Si devices [87]. Preliminary results suggest that decreasing the thickness of the Er-doped PSi active layer and lowering the annealing temperature improve the efficiency of the devices. The devices showed little degradation even after several hours of operation.

Figure 11 shows the temperature dependence of the EL intensity at a constant current of 1.5 mA [86]. The small temperature range is limited by the poor transport properties of the device at lower temperatures. For the same voltage, the current through the device drops by more than one order of magnitude from 300 to 200 K (larger current drops are observed in reverse bias). Within this temperature range, there is a clear difference in temperature dependence for the two biasing conditions. Under reverse bias, the EL intensity decreases by a factor of 24 and under forward bias by a factor of 2.6 when the temperature increases from 240 to 300 K. The EL temperature dependence in forward bias is very similar to the PL dependence, suggesting that similar excitation processes exist. On the other hand, the reverse bias temperature dependence is very different, suggesting a different excitation mechanism. It has been proposed that EL in reverse bias occurs by impact of hot electrons [88,89]. In these reports the temperature quenching is stronger for forward bias EL, but in our work we observe the opposite. If in our case excitation in reverse bias occurs by hot electrons, then the larger temperature quenching might be a result of the poor transport through the Er-doped PSi matrix. The poor transport (worse in reverse bias) could greatly reduce the excitation effectiveness of the hot electrons. Further studies are in progress to clarify the EL excitation mechanisms and the strong temperature dependence of the Er-doped PSi devices.

6.6 INTEGRATION WITH MICROELECTRONICS

For device applications, uniform porous layers are highly desirable. Uniformity is required in depth and laterally. Many samples are not uniform as a function of depth. Spatially resolved photoluminescence and Raman measurements show that the first few microns often have a different nanostructure from that of deeper layers. As a result, the PL spectrum and intensity change in depth. Another concern is the lateral uniformity, over two major length scales, > 1 cm and 0.1-100 μm. Careful anodization and processing of the PSi layers is required to achieve uniformity over these length scales. Without a careful design of the anodization cell to provide a uniform current density or the use of supercritical drying to maintain the integrity of high porosity PSi, the layers are often not uniform, as shown by the spatially resolved luminescence maps reported in the literature [90].

For some applications, the size of the PSi LEDs must be very small and the anodization should not affect the Si material that is in proximity. Several techniques that allow the manufacture of miniature PSi areas (down to 1 μm^2 and below) and protect the c-Si regions immediately adjacent to them have been developed [91-93]. They include localized amorphization by ion implantation to prevent anodization, localized low-energy ion bombardment to seed the pore formation, and protection of selected areas by a combination of silicon nitride and photoresists. Thus, PSi LEDs of sizes down to 1 μm appear feasible.

By using a combination of the techniques discussed above, we have built an integrated bipolar transistor/PSi LED structure [94]. The complete structure, shown in Figure 12, was fabricated using accepted silicon microelectronic fabrication procedures. It was possible to turn the LED on and off by applying a small current pulse to the base of the bipolar transistor. The PSi LED itself was fabricated using the material's passivation techniques and device design approaches described above. The devices had the following specifications at room temperature: electroluminescence peak from 1.7 to 2.0 eV; detectable light emission at an applied voltage of ~ 2V and a current density of ~ 10mA/cm^2; maximum light intensity of ~ 1mW/cm^2; highest external power efficiency ~ 0.1%; modulation bandwidth exceeding 1MHz; and stability without degradation for several weeks of continuous operation. The driver transistor was

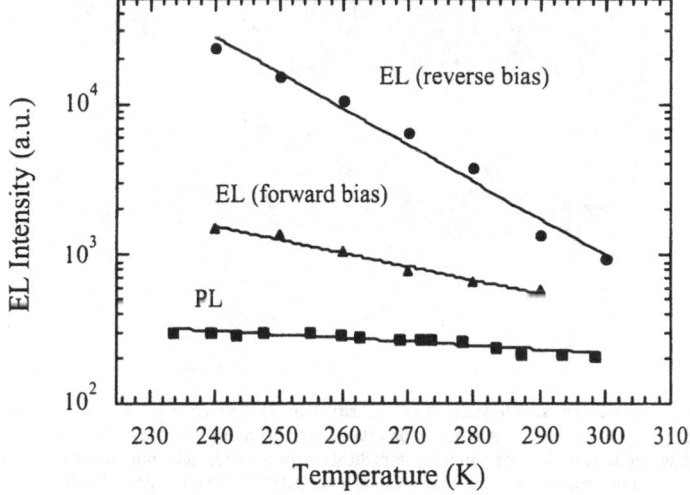

Figure 11. Temperature dependence of PL (■) and EL intensity under forward (▲) and reverse (●) bias. Data for EL were taken at a constant current of 1.5 mA.

connected in a common-emitter configuration and it modulates light emission by amplifying a small base input signal and controlling current flow through the LED. Structures of various sizes were fabricated, with the active area ranging from 0.005 to 2 mm^2.

The standard bipolar process was modified as explained in Ref. 94. In short, the transistor was fabricated first and then protected using Si$_3$N$_4$ prior to the fabrication of the LED. Figure 12 shows the resulting integrated device. Both the LED and the bipolar transistor performed as expected when tested individually, showing that our fabrication procedure was successful. We then used the base of the transistor to turn on and off the

116

electroluminescence, demonstrating for the first time full integration of a porous silicon LED with silicon microelectronic circuitry.

Since this work, we have demonstrated compatibility between PSi LEDs and CMOS technology [95]. Arrays of individually addressable elements have been constructed in an N by N grid. This represents a concrete step toward practical optical displays that goes beyond our earlier demonstration of a simple seven element display in which each element had to have its own power supply.

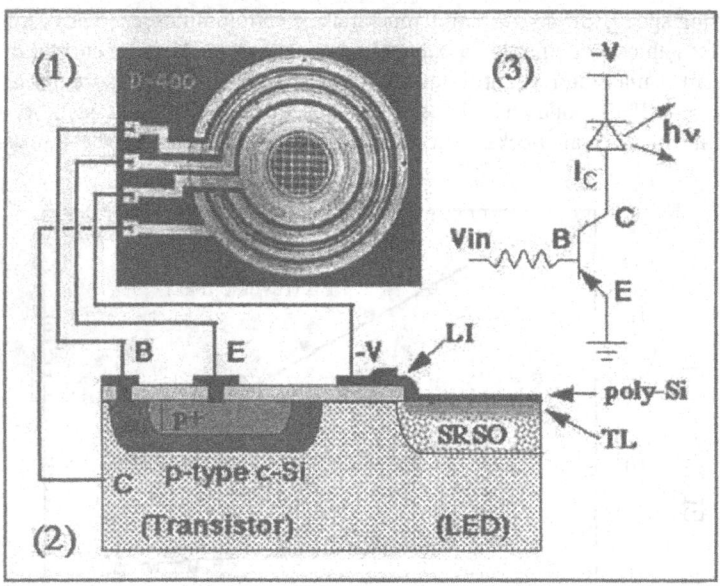

Figure 12. Integrated bipolar transistor/PSi LED structure. (1) Top view of a structure with a 400 μm active light-emitting area. (2) A cross-section starting (on the right) from the center of the LED and ending (on the left) with the *pnp* transistor (SRSO is the oxidized light-emitting PSi, TL is the mesoporous layer, LI is the local interconnect). (3) The equivalent circuit.

7. Conclusions

The performances of porous silicon light emitting diodes have been briefly reviewed. The most significant obstacles to the commercial use of these devices have been identified. Even if no large scale application of PSi LEDs ever takes place, the scientific community has learned a great deal about silicon nanostructures and about nanoscience and nanotechnology in general. This knowledge could prove important for LEDs made of other nanoscale semiconductors or for other uses of nanoscale silicon, for example in nanoelectronics or biosensing.

This work was supported in part by grants from the U.S. Army Research Office and the U.S. National Science Foundation. This review is inspired by three previous manuscripts: P. M. Fauchet, *Proceedings of the International School of Physics "Enrico Fermi," Course CXLI Silicon-Based Microphotonics: from Basics to Applications*, edited by O. Bisi, S. U. Campisano, L. Pavesi, and F. Priolo (Italian Physical Society, IOS Press, Amsterdam, Oxford, Tokyo, Washington DC, 1999), pp 163-190; S. Chan and P. M. Fauchet, Appl. Phys. Lett. **75**, 274-276 (1999); and H. A. Lopez and P. M. Fauchet, Appl. Phys. Lett. **75**, 3989-3991 (1999).

References

1. C.F. Klingshirn, Semiconductor Optics (Springer, Berlin, 1995)
2. D.J. Lockwood, *Light Emission in Silicon: From Physics to Devices*, Semiconductors and Semimetals, Vol. 49 (Academic Press, San Diego, 1998)
3. L.T. Canham, Appl. Phys. Lett. **57**, 1046 (1990)
4. K.A. Littau, P.J. Szajowski, A.J. Miller, A.R. Kortan, and L.E. Brus, J. Phys. Chem. **97**, 1224 (1993)
5. P. M. Fauchet, L. Tsybeskov, C. Peng, S. P. Duttagupta, J. von Behren, Y. Kostoulas, J. V. Vandyshev, and K. D. Hirschman, IEEE Jour. Selected Topics in Quantum Electron. **1**, 1126-1139 (1995)
6. P.M. Fauchet, J. Luminesc. **70**, 294 (1996)
7. R.T. Collins, P.M. Fauchet, and M.A. Tischler, Physics Today **50**, 24 (January 1997)
8. A.G. Cullis, L.T. Canham, and P.D.J. Calcott, J. Appl. Phys. **82**, 909 (1997)
9. P. M. Fauchet, in *Light Emission in Silicon: From Physics to Devices*, D.J. Lockwood editor, Semiconductors and Semimetals, Vol. 49 (Academic Press, San Diego, 1998), pp 206-252
10. L.T. Canham, ed., *Properties of Porous Silicon*, Electronic Materials Information Service Datareviews Series No 18 (INSPEC, The Institution of Electrical Engineers, London, UK, 1997)
11. P.M. Fauchet, J. von Behren, K.D. Hirschman, L. Tsybeskov, and S.P. Duttagupta, Phys. Stat. Sol. (a) **165**, 3 (1998).
12. A. Uhlir, Jr., Bell Syst. Tech. J. **35**, 333 (1956)
13. R.L. Smith and S.D. Collins, J. Appl. Phys. **71**, R1 (1992)
14. P. M. Fauchet, in *Pits and Pores: Formation, Properties, and Significance for Advanced Luminescent Materials*, edited by P. Schmuki, D. J. Lockwood, H. Isaacs, and A. Bsiesy (The Electrochemical Society, Pennington, NJ, 1997), pp 27-60
15. P.M. Fauchet and J. von Behren, Phys. Stat. Sol. (b) **204**, R7 (1997)
16. L.T. Canham, A.G. Cullis, C. Pickering, O.D. Dosser, T.I. Cox, and T.P. Lynch, Nature **368**, 133 (1994)
17. J. von Behren, E.H. Chimowitz, and P.M. Fauchet, Adv. Mater. **9**, 921 (1997)
18. P.M. Fauchet, L. Tsybeskov, S.P. Duttagupta, and K.D. Hirschman, Thin Solid Films **297**, 254 (1997)
19. L. Canham, MRS Bulletin **18**, 22 (July 1993)
20. L. Tsybeskov, S.P. Duttagupta, K.D. Hirschman, and P.M. Fauchet, in *Advanced Luminescent Materials*, D.J. Lockwood, P.M. Fauchet, N. Koshida, and S.R.J. Brueck, editors (The Electrochemical Society, Pennington, NJ, 1996), pp 34-47
21. L. Tsybeskov, S.P. Duttagupta, K.D. Hirschman, and P.M. Fauchet, Appl. Phys. Lett. **68**, 2058 (1996)
22. M. Warntjes, C. Vieillard, F. Ozanam, and J.N. Chazalviel, J. Electrochem. Soc. **142**, 4138 (1995)
23. J.M. Buriak and M.J. Allen, J. Am. Chem. Soc. **120**, 1339 (1998)
24. J. von Behren, T. van Buuren, M. Zacharias, E.H. Chimowitz, and P.M. Fauchet, Solid State Commun. **105**, 317 (1998)
25. V. Lehmann and U. Gosele, Appl. Phys. Lett. **58**, 856 (1991)
26. P.D.J. Calcott, K.J. Nash, L.T. Canham, M.J. Kane, and D. Brumhead, J. Phys.: Condensed Matter **5**, L91 (1993)

118

27.   J. von Behren, Y. Kostoulas, K. B. Ucer, and P. M. Fauchet, J. Non-Cryst. Solids **198-200**, 957 (1996)
28.   L. Tsybeskov, K.L. Moore, D.G. Hall, and P.M. Fauchet, Phys. Rev. B **54**, R8361 (1996)
29.   C. Delerue, G. Allan, and M. Lannoo, Phys. Rev. B **48**, 11024 (1993)
30.   S. Prokes, Appl. Phys. Lett. **62**, 3244 (1993)
31.   F. Koch, V. Petrova-Koch, and T. Muschik, J. Lumin. **57**, 271 (1993)
32.   G. Allan, C. Delerue, and M. Lannoo, Phys. Rev. Lett. **76**, 2961 (1996)
33.   H. Mizuno, H. Koyama, and N. Koshida, Appl. Phys. Lett. **69**, 3779 (1996)
34.   M. V. Wolkin, J. Jorné, P. M. Fauchet, G. Allan, and C. Delerue, Phys. Rev. Lett. **82**, 197 (1999).
35.   L. Tsybeskov, S.P. Duttagupta, K.D. Hirschman, P.M. Fauchet, K.L. Moore, and D.G. Hall, Appl. Phys. Lett. **70**, 1790 (1997)
36.   L. Tsybeskov, Ju.V. Vandyshev, and P.M. Fauchet, Phys. Rev. B **49**, 7821 (1994)
37.   R.E. Hummel, A. Morrone, M. Ludwig, and S.S. Chang, Appl. Phys. Lett. **63**, 271 (1993)
38.   A.G. Nassiopoulos, S. Grigoropoulos, D. Papadimitriou, and E. Gogolides, Phys. Stst. Sol. (b) **190**, 91 (1995)
39.   S.H. Zaidi, A.-S. Chu, and S.R.J. Brueck, Mat. Res. Soc. Symp. Proc. **358**, 957 (1995)
40.   H. Tamura, M. Ruckschloss, T. Wirschem, and S. Veprek, Appl. Phys. Lett. **65**, 1537 (1994)
41.   V. Petrova-Koch, T. Fischer, K. Sheglov, and F. Koch, in *Advanced Luminescent Materials*, D.J. Lockwood, P.M. Fauchet, N. Koshida, and S.R.J. Brueck, editors (The Electrochemical Society, Pennington, NJ, 1996), pp 382-392
42.   L. Tsybeskov, K.D. Hirschman, S.P. Duttagupta, P.M. Fauchet, M. Zacharias, P. Kohlert, J.P. McCaffrey, and D.J. Lockwood, in *Quantum Confinement: Nanoscale Materials, Devices, and Systems,* edited by M. Cahay, J.P. Leburton, D.J. Lockwood, and S. Bandyopadhyay (The Electrochemical Society, Pennington, NJ, 1997), pp 134-145
43.   L. Tsybeskov, K.D. Hirschman, S.P. Duttagupta, M. Zacharias, P.M. Fauchet, J. McCaffrey, and D.J. Lockwood, Appl. Phys. Lett. **72**, 43 (1998)
44.   L. Tsybeskov, K. D. Hirschman, S. P. Duttagupta, P. M. Fauchet, M. Zacharias, J. P. McCaffrey, and D. J. Lockwood, Phys. Stat. Sol. (a) **165**, 69 (1998)
45.   A. Halimaoui, C. Oules, G. Bomchil, A. Bsiesy, F. Gaspard, R. Herino, M. Ligeon, and F. Muller, Appl. Phys. Lett. **59**, 304 (1991)
46.   A. Richter, P. Steiner, F. Kozlowski, and W. Lang, IEEE Electron Device Lett. **12**, 691 (1991)
47.   N.M. Kalkhoran, F. Namavar, and H.P. Maruska, Mat. Res. Soc. Symp. Proc. **256**, 84 (1992)
48.   E. Bassous, M. Freeman, J.-M. Halbout, S.S. Iyer, V.P. Kesan, P. Munguia, S.F. Pesarcik, and B.L. Williams, Mat. Res. Soc. Symp. Proc. **256**, 23 (1992)
49.   N. Koshida and H. Koyama, Appl. Phys. Lett. **60**, 347 (1992)
50.   A. Bsiesy, F. Muller, M. Ligeon, F. Gaspard, R. Herino, R. Romestain, and J.-C. Vial, Phys. Rev. Lett. **71**, 637 (1993)
51.   N. Koshida, H. Koyama, Y. Yamamoto, and G.J. Collins, Appl. Phys. Lett. **63**, 2655 (1993)
52.   C. Peng, K.D. Hirschman, and P.M. Fauchet, J. Appl. Phys. **80**, 295 (1996)
53.   M. Ben-Chorin, in *Properties of Porous Silicon*, Electronic Materials Information Service Datareviews Series No 18 (INSPEC, The Institution of Electrical Engineers, London, UK, 1997), pp 165-175
54.   A. Loni, A.J. Simons, T.I. Cox, P.D.J. Calcott, and L.T. Canham, Electron. Lett. **31**, 1288 (1995)
55.   L. Tsybeskov, S.P. Duttagupta, K.D. Hirschman, and P.M. Fauchet, in *Advanced Luminescent Materials*, edited by D.J. Lockwood, P.M. Fauchet, N. Koshida, and S.R.J. Brueck (The Electrochemical Society, Pennington, NJ, 1996), pp 34-47
56.   L. Tsybeskov, S.P. Duttagupta, K.D. Hirschman, and P.M. Fauchet, Appl. Phys. Lett. **68**, 2058 (1996)
57.   J. Linnros and N. Lalic, Appl. Phys. Lett. **66**, 3048 (1995)
58.   A.J. Simons, T.I. Cox, A. Loni, L.T. Canham, M.J. Uren, C. Reeves, A.G. Cullis, P.D.J. Calcott, M.R. Houghton, and J.P. Newey, in *Advanced Luminescent Materials*, edited by D.J. Lockwood, P.M. Fauchet, N. Koshida, and S.R.J. Brueck (The Electrochemical Society, Pennington, NJ, 1996), pp 73-86.
59.   B. Gelloz, T. Nakagawa, and N. Koshida, Appl. Phys. Lett. **73**, 2021 (1998)
60.   C. Peng and P.M. Fauchet, Appl. Phys. Lett. **67**, 2515 (1995)
61.   S. Lazarouk, P. Jaguiro, S. Katsouba, G. Masini, S. La Monica, G. Maiello, and A. Ferrari, Appl. Phys. Lett. **68**, 1646 (1996)

62.   J. Wang, F.-L. Zhang, W.-C. Wang, J.-B. Zheng, X.-Y. Hou, and X. Wang, J. Appl. Phys. **75**, 1070 (1994)
63.   P. Steiner, F. Kozlowski, H. Sandmaier, and W. Lang, Mat. Res. Soc. Symp. Proc. **283**, 343 (1993)
64.   L. Tsybeskov, K.L. Moore, S.P. Duttagupta, K.D. Hirschman, D.G. Hall, and P.M. Fauchet, Appl. Phys. Lett. **69**, 3411 (1996)
65.   M. G. Berger, M. Thönissen, R. Arens-Fischer, H. Münder, H. Lüth, M. Arntzen, and W. Theiβ, Thin Solid Films **255**, 313 (1995)
66.   M. G. Berger, C. Dieker, M. Thönissen, L. Vescan, H. Lüth, H. Münder, M. Wernke, and P. Grosse, J. Phys. D: Appl. Phys. **27**, 1333 (1994)
67.   S. Frohnhoff and M. G. Berger, Adv. Mater. **6**, 963 (1994)
68.   L. Pavesi, La Rivista del Nuovo Cimento **20**, 1 (1997)
69.   W. Theiβ, Surface Science Reports **29**, 91 (1997)
70.   S. Chan, L. Tsybeskov, and P. M. Fauchet, Mat. Res. Soc. Symp. Proc. **536**, 117 (1999)
71.   C. Peng and P. M. Fauchet, Appl. Phys. Lett. **67**, 2515 (1995)
72.   S. Chan and P. M. Fauchet, in *Silicon-based Optoelectronics*, edited by D. C. Houghton and E. A. Fitzgerald, SPIE Proc. 3630, 144 (1999)
73.   M. Araki, H. Koyama, and N. Koshida, Appl. Phys. Lett. **69**, 2956 (1996)
74.   L. Pavesi, C. Mazzoleni, A. Tredicucci, and V. Pellegrini, Appl. Phys. Lett. **67**, 3280 (1995)
75.   A. Tredicucci, Y. Chen, V. Pellegrini, and C. Deparis, Appl. Phys. Lett. **66**, 2388 (1995)
76.   H. Ennen, J. Schneider, G. Pomrenke, and A. Axmann, Appl. Phys. Lett. **43**, 943 (1983)
77.   S. Coffa, F. Priolo, G. Franzo, V. Bellani, A. Carnera, and C. Spinella, Phys. Rev. B **48**, 11782 (1993)
78.   H. Ennen, U. Kaufmann, G. Pomrenke, J. Schneider, J. Windscheif, and A. Axmann, J. Cryst. Growth **64**, 165 (1983)
79.   A. Polman, J. Appl. Phys. **82**, 1 (1997)
80.   F. Namavar, F. Lu, C. H. Perry, A. Cremins, N. M. Kalkhoran, J. T. Daly, and R. A. Soref, Mat. Res. Soc. Symp. Proc. **358**, 375 (1995)
81.   A. M. Dorofeev, N. V. Gaponenko, V. P. Bondarenko, E. E. Bachilo, N. M. Kazuchits, A. A. Leshok, G. N. Tryanova, N. N. Vorosov, V. E. Borisenko, H. Gnaser, W. Bock, P. Becker, and H. Oechsner, J. Appl. Phys. **77**, 2679 (1995)
82.   H. A. Lopez, X. L. Chen, S. A. Jenekhe, and P. M. Fauchet, J. Lumin. **80**, 115 (1999)
83.   T. Kimura, A. Yokoi, H. Horiguchi, R. Saito, T. Ikoma, and A. Sato, Appl. Phys. Lett. **65**, 983 (1994)
84.   H. A. Lopez, S. Chan, L. Tsybeskov, H. Koyama, V. P. Bondarenko, and P. M. Fauchet, Mat. Res. Soc. Symp. Proc. **536**, 135 (1999)
85.   L. Tsybeskov, S. P. Duttagupta, K. D. Hirschman, P. M. Fauchet, K. L. Moore, and D. G. Hall, Appl. Phys. Lett. **70**, 1790 (1997)
86.   H. A. Lopez and P. M. Fauchet, Appl. Phys. Lett. **75**, 3989 (1999)
87.   S. Coffa, G. Franzo, and F. Priolo, MRS Bulletin **23** (4), 25 (1998)
88.   S. Coffa, G. Franzo, F. Priolo, A. Pacelli, and A. Lacaita, Appl. Phys. Lett. **73**, 93 (1998)
89.   H. Isshiki, H. Kobayashi, S. Yugo, T. Kimura, and T. Ikoma, Appl. Phys. Lett. **58**, 484 (1991)
90.   E. Ettedgui, C. Peng, L. Tsybeskov, Y. Gao, P.M. Fauchet, H.A. Mizes, and G.A. Carver, Mat. Res. Soc. Symp. Proc. **283**, 173 (1993)
91.   S.P. Duttagupta, P.M. Fauchet, C. Peng, S.K. Kurinec, K. Hirschman, and T.N. Blanton, Mat. Res. Soc. Symp. Proc. **358**, 647 (1995)
92.   S.P. Duttagupta, C. Peng, L. Tsybeskov, and P.M. Fauchet, Mat. Res. Soc. Symp. Proc. **380**, 73 (1995)
93.   P. Schmuki, L.E. Erickson, and D.J. Lockwood, Phys. Rev. Lett. **80**, 4060 (1998)
94.   K.D. Hirschman, L. Tsybeskov, S.P. Duttagupta, and P.M. Fauchet, Nature **384**, 338 (1996)
95.   K. D. Hirschman, Ph.D. Thesis, University of Rochester, April 2000

VISIBLE LIGHT EMISSION FROM A NEW MATERIAL SYSTEM: Si/SiO₂ SUPERLATTICES IN OPTICAL MICROCAVITIES

L. PAVESI, G. PUCKER, Z. GABURRO, M. CAZZANELLI
INFM and Department of Physics, University of Trento,
via Sommarive 14, 38050 Povo-Trento, Italy.
P. BELLUTTI
ITC-IRST, Microsystem Division, 38050 Povo-Trento, Italy.

Abstract

In order to add optical functionalities to silicon based microelectronics, we decided to develop Si/SiO₂ superlattices where quantum confinement effects should drive Si to become a good emitter. A further improvement it is possible when one couples the beneficial effects of low dimensional electronic systems (as in Si superlattices) with the enhancement of the spontaneous emission rate that occurs in an optical microcavity. Not to lose the fundamental goal to add optical functionality to electronic circuits we have performed the growth of this material in an industrial environment by using standard CMOS equipments. In this paper, we will present the status of our research

1. Introduction on light emitting silicon

The enormous increase of communication technology (information highway) has increased the demand for efficient and low-cost optoelectronic functions. Recent trends in technology is to look for solutions at novel materials (organics, polymers), or at the effects that reduced dimensionality causes on more conventional semiconductors (quantum wells, wires and dots) or at the modified photon statistics in photonic band-gap materials. Impressive progresses in all these fields have been achieved during the last years. On the other hand, Si has been considered as an outcomer in this field [1]. In the past, several times the death of Si has been announced due to some physical limits: e. g. sub-micron technology, high frequency operation, etc. However new functionalities have been added to Si in order to keep it in the play (see, e.g., high frequency applications of Si/Ge transistors). Indeed, due to the maturity of Si technology, which is extensively used for forming VLSI integrated circuits, we believe that starting from two key materials, silica and Si, it is possible to put Si into the optoelectronics arena. The attractive features of Si are well known. However, Si is characterised by an indirect band-gap and by a weak electro-optic effect. It is therefore not suitable for the implementation of fundamental optical functions such as light emission and modulation. At the moment the only viable solution to the integration of optical and electronic functionalities in the

121

L. Pavesi and E. Buzaneva (eds.), Frontiers of Nano-Optoelectronic Systems, 121–136.
© 2000 *Kluwer Academic Publishers. Printed in the Netherlands.*

electronic and telecommunication industries is the hybrid approach where Si microelectronics is coupled with III-V optoelectronics.

1.1 STATE OF THE ART ON VISIBLE LIGHT EMISSION FROM SILICON

Silicon is an indirect band-gap material and momentum conservation requires phonon assistance in the photon emission process. Even though well passivated and dislocation free Si shows internal quantum efficiency values larger than 1 % at low temperature, the luminescence lifetimes are very long (ms at room temperature) exposing the radiative process in strong competition with fast non-radiative recombinations. Si is a poor emitter indeed. Recently, room temperature light emission has been demonstrated to be possible in Si when it is turned into a low dimensional system [2-7] or when selected active impurities [8] and/or new phases [9] are inserted in the Si lattice. Many claims of a future role of Si in photonic applications have since then appeared. [10-12] Porous Si (PS) was the first material that showed room temperature luminescence with high quantum efficiencies.[3] Its structure is formed by a disordered array of nanocrystalline Si with sizes in the nm range. The room temperature luminescence was due to the sum of three facts:

i) quantum confinement which causes an opening of the Si band-gap and an increased radiative emission rate.

ii) carrier localisation where non-radiative recombination centres are isolated into "dark" silicon nanocrystals (NS).

iii) stable passivation of the interface through hydrogen or, better, oxygen bonding.

Electroluminescence (EL) devices were also demonstrated. Since the first devices with low quantum efficiency 10^{-5} %, the today systems have external quantum efficiency (EQE) in the 1% range and show long term stability. Integration with electronic driving circuits has been also demonstrated [13]. The five orders of magnitude improvement in the EQE of these devices were mostly due to the good passivation of the NS surfaces by oxidation treatments, the improved transport properties caused by better injection contacts as provided by a p/n junction and the removal of leakage current paths by selective oxidation.

Despite these interesting properties the use of PS in commercial device is severely questioned by its large internal surface which is highly reactive and hence determines ambient dependent properties. In addition microelectronic engineers do not well accept the wet processing used to form PS. For these reasons alternative approaches which relay on the formation of low dimensional Si passivated by SiO_2 layers have since appeared. [5,6] Two main approaches have been followed: Si quantum dots (referred as Si nanocrystals, [5] and Si superlattices (SL) [6]. Both fulfil the above cited items i)-iii): low dimension of the Si structures in the 1-5 nm range, good passivation of the surfaces or interfaces by SiO_2 and elimination of non-radiative recombinations by spatial localisation of carriers. A further interesting point is the fact that most of the techniques used in the literature to grow these structures are VLSI compatible which leads to an easy integration of electronic and optoelectronic functions. Mostly all the produced materials show room temperature luminescence with EQE of the order of some percent. EL in these structures was weak (EQE$\approx 10^{-5}$) due mainly to the injection problem caused by the poor transport properties of the oxide. Si superlattice formed by cyclic deposition of amorphous Si and oxidation, have shown a relation between the thickness of the Si layers and the peak and intensity of the luminescence. Quantum confinement has been thought as the cause of the emission shift. Since the first report of photoluminescence at room temperature from (SiO₂/Si) superlattices [6] different growth techniques for the preparation of efficient light emitting superlattices were investigated, e. g. MBE [6], sputtering [14] and LP-CVD [15]. While these studies confirm that the interesting properties of SL are independent on the growth techniques, they have not yet approached the possibility of their preparation in a production line.

1.2 MICROCAVITY

The spontaneous emission properties of a material depend on the coupling with the photon modes. In fact when a semiconductor layer is placed in between an optical microcavity a modification of the spontaneous emission occurs.[16] An amplitude build-up due to multiple in-phase reflections of optical waves is responsible of the variation of the emission: the unique feature of the microcavity is to concentrate the field intensity into the resonant cavity mode by as much as it suppresses it in non-resonant modes. These effects allow the control of the spontaneous emission as the spontaneous lifetime will almost be unchanged. In the weak coupling case (when the lifetime of the photon mode in the cavity is shorter than the one of the electronic excitation) multiple reflection leads to build-up of light emission in a given direction thus leading to directionality. Another useful effect is line narrowing due to the spectral width of allowed photon modes. These two effects have been used both to increase efficiency of light extraction from LED and displays and in reducing chromatic dispersion in optical fibre systems. The physics of PS based microcavities is described in [17].

2. Room temperature luminescence from (Si/SiO₂) multilayers grown in an industrial low pressure-chemical vapour deposition reactor

Single and multiple Quantum Wells (QW) of Si separated by wide band gap barriers (e.g. SiO₂, CaF₂ or Al₂O₃) have been prepared by different techniques such as oxygen implantation (SIMOX) [18,19], molecular beam epitaxy [20-24] sputtering [25,26] and chemical vapour deposition (CVD) [27-29]. Among the published results, photoluminescence (PL) spectra from SiO₂/Si/SiO₂ QW or Si/SiO₂ superlattices can be classified according to two categories: a) spectra dominated by a peak at 1.65 eV which increases in intensity with decreasing Si layer thickness [18,19], b) spectra where a thickness dependent shift of the PL is accompanied by a concurrent intensity increases [20,28]. While spectra of type B can be interpreted within the quantum confinement model, spectra of type A are explained by recombinations at oxygen defects formed at the Si/SiO₂ interface. Both experimental and theoretical studies make clear that the observation of PL from Si quantum wells depends not only on the Si-thickness but also on the properties of the Si/SiO₂ interfaces. These studies confirm that different growth techniques can be used to obtain Si/SiO₂ superlattices showing PL, but none has demonstrated the possibility of their preparation in a microelectronic production line yet. It is the aim of this work to close this gap by means of the development of a simple and

124

CMOS compatible process sequence to grow photoluminescent SiO₂/Si superlattices by using an LP-CVD industrial reactor on 4" Si wafers. The choice to prepare these structures in a CMOS fabrication line results in some severe limitations. In particular, the parameters in LP-CVD (e.g. temperature and deposition pressure) can only be changed in a narrow range, to exclude interference with other processes in the fabrication line. On the other hand, if feasible, it would turn out in an easy and compatible transfer of the process in an already defined Si CMOS process.

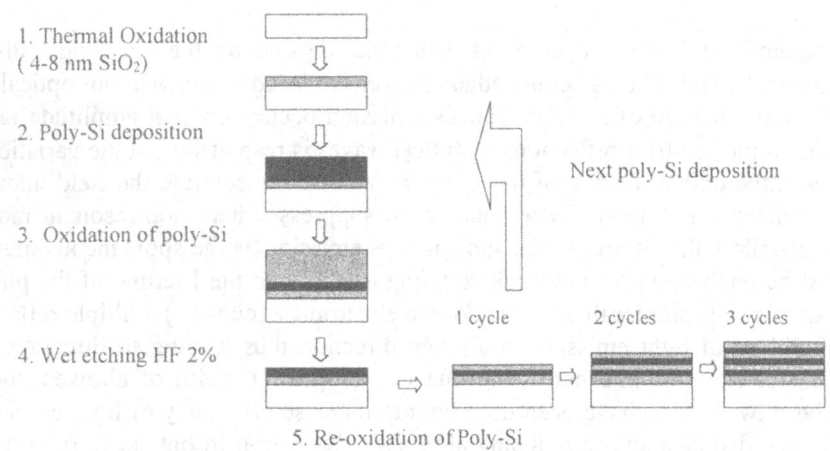

Figure 1. Schematic representation of the process used to obtain sub 3 nm poly-Si layers. The steps 2-5 are repeated 1, 2 or 3 times to obtain 1, 2 or 3 periods of the Si/SiO₂ -layers.

2.1 GROWTH OF Si/SiO₂ MULTILAYERS

The growth of Si/SiO₂ nanometric multilayers requires control of periodic deposition of both Si and SiO₂. In standard CMOS processes the polycrystalline Si (poly-Si) layers are in the range from 50 nm to some hundreds of nanometer and a roughness of the poly-Si layer of some nm is not critical. Since in Si/SiO₂ multilayers both the light emission intensity and the light emission energy depend on the Si-layer thickness, the Si-layer thickness and roughness are fundamental parameters which have to be controlled and their repeatability has to be assured. These stringent requirements make quite challenging the use of a direct deposition of very thin Si layer with conventional LP-CVD furnace. On the other hand, the preparation of thin SiO₂ layers (down to some nm) by thermal oxidation of Si is a common process in CMOS technology, for the preparation of thin gate-oxides. Hence the control of the SiO₂ layers is much easier to achieve, being based on already standard and well-experimented processes. Our processing approach is based on the following considerations: i) using poly-Si deposition conditions which give thin reproducible and uniform layers, and ii) reduction of these Si layers to the thickness needed by means of an oxidation process. By sequentially repeating the poly-Si deposition and oxidation, multilayered structures can be obtained. Fig. 1 gives a schematic outline of the process. Details of the process are given in [29].

TABLE 1. Structure of the samples investigated in this study. Layer thickness refers to estimations from the oxidation rates and control measurements using TEM and single-line ellipsometry (HeNe laser 633nm). DBR stands for dielectric Bragg mirror (see section 3).

Sample name	Substrate Resistivity [Ω/cm]	1st SiO$_2$ layer Thickness [nm]	Multi-layer		
			Si layer Thickness [nm]	SiO$_2$ layer Thickness [nm]	Number of periods
A1	8-15	4	2.5	4	1
A2	8-15	4	2.2	4.6	1
A3	8-15	4	1.8	5.4	1
A4	8-15	4	1.5	6	1
A5	8-15	4	1.1	3.8	1
B1	0.01-0.02	5.2	1.8	4.4	1
B2	8-15	8.4	1.8	4.4	2
B3	0.01-0.02	5.2	1.8	4.4	3
C1	8-15	8	1.1	6	1
C2	8-15	8	1.1	6	2
C3	8-15	8.4	1.1	6	3
D1	Fused silica		20	native oxide	-
D2	Fused silica		5	30	1
D3	Fused silica		1.5	6	1
E1	3x[Si/SiO$_2$] DBR on c-Si		1.1	6	3

Table 1 reports the main parameters of the samples whose PL we will discuss in the following. Here we detail their rationale. We prepared a series of samples (named A1-A5) where the thickness of a single poly-Si layer was changed by increasing re-oxidation times from 10 min for sample A1 to 50 min for sample A5. Series A1-A5 demonstrates the effect of decreasing Si thickness. We prepared two series of samples (named B and C) to test the effects of a periodic multilayer sequence. They were formed by 1, 2 and 3 periods of Si/SiO$_2$ layers grown by using a poly-Si deposition duration of 180 s and a first 100 min long oxidation. Series B had a 10 min re-oxidation time whilst series C had a longer re-oxidation time of 30 min. Hence, samples of series C are expected to have thinner poly-Si layers and thicker capping oxide layer with respect to series B. In one case we deposited the nanometric Si structures on a mirror formed by a dielectric Bragg reflector (sample E1). The Bragg reflector was constituted by three periods of a 120 nm thick SiO$_2$ and a 45 nm thick Si layers followed by a top SiO$_2$ layer of 260 nm. A fourth series of samples, named D, was grown on 4" polished fused quartz substrates. In PL measurements no difference between the front and the back sides of these samples was noticed. Samples of series D offer some advantages for optical characterisation and allow measurements of transmission spectra.

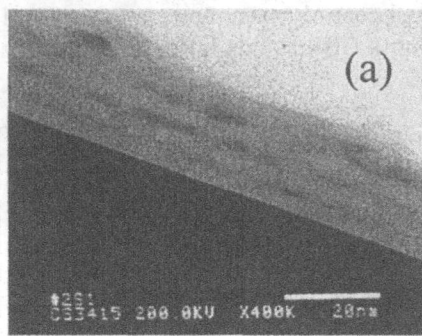 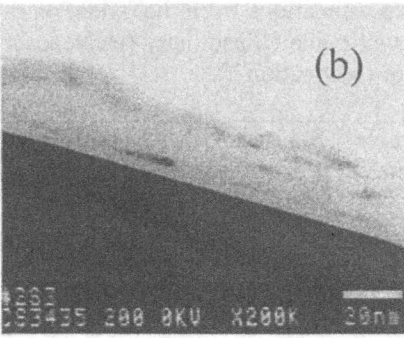

Figure 2. TEM images of Si/SiO$_2$ multilayered samples (Si dark bands, SiO$_2$ light bands): (a) sample B3 and (b) sample C3.

2.2 RESULTS

2.2.1 *Microscopy imaging of the Si/SiO$_2$ multilayers*

Fig. 2 shows typical TEM images for samples B3 (Fig. 2a) and C3 (Fig. 2b). The SiO$_2$ layers have a thickness of 5 nm and are very uniform in thickness. The thickness of the Si-layers of sample B3 as measured in the TEM image are 2.5 nm, 2.5 nm and 5 nm, respectively, with a roughness of about ± 0.5 nm.

The TEM image of sample C3 (Fig. 2b) is representative of samples with thin Si layers. It is worth noticing that the contrast between the poly-Si layers and the SiO$_2$ depends on the orientation of the poly-Si layer. When the thickness of Si is at least 1 nm and the orientation of the crystal plane gives a high contrast between Si and SiO$_2$, the Si layers can be identified for all three layers (see, e. g., at the right hand side of the image). Some evidences that the Si layers might be interrupted in some points due to complete oxidation of the poly-Si can be observed. There is also some evidence of the existence of isolated Si nanocrystals in all three layers.

2.2.2 *Optical absorption*

To overcome the problems due to the unknown reflectance of the structure, we plotted in Fig. 3 the trasmittance data as $\ln(I_0/I)/d_{Si}$ where I_0 is the reference beam intensity measured as the transmitted intensity trough the bare fused silica substrate, I is the transmitted intensity through the sample and d_{Si} the overall Si thickness expressed in cm. Roughly this quantity is related to the absorbance of the Si multilayer.

The absorbance line-shape for sample D2 resembles qualitatively the absorption coefficient for bulk Si in the same energy range [20] with maxima due to direct transitions at 3.5 and 4.2 eV. Also the value at 4.2 eV is very close (27×10^6 cm^{-1} versus 22.5×10^6 cm^{-1}) to the one measured in Ref. [20]. The major difference is a smooth transition from the edge of the direct transitions around 3.5 eV versus lower energies, which can be attributed to the spectral dependence of the refractive index. For this reason, a reliable estimation of the indirect band gap from the spectra in Fig. 3 is hopeless. The absorption spectra obtained for sample D3 differs remarkably from the one of sample D2. The maximum in the curve is strongly blue shifted to 4.5 eV and a second maximum appears at about 4.9 eV. Moreover, a broadening of the absorption spectra is evident. This absorption line-shape is very similar to the one of PS [31]. Similar results were also obtained from calculations of the dielectric function ε_2 of $(Si/SiO_2)_n$ superlattices [32].

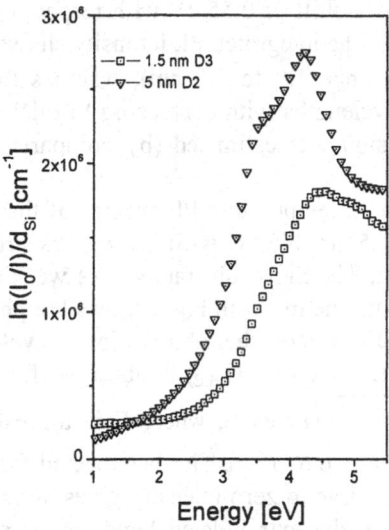

Figure 3. Spectral dependence of the transmission (at 300 K) of samples D2 (down triangles) and D3 (squares) plotted in the form $\ln(I_0/I)/d_{si}$. (d_{si} = thickness of the Si layer in cm).

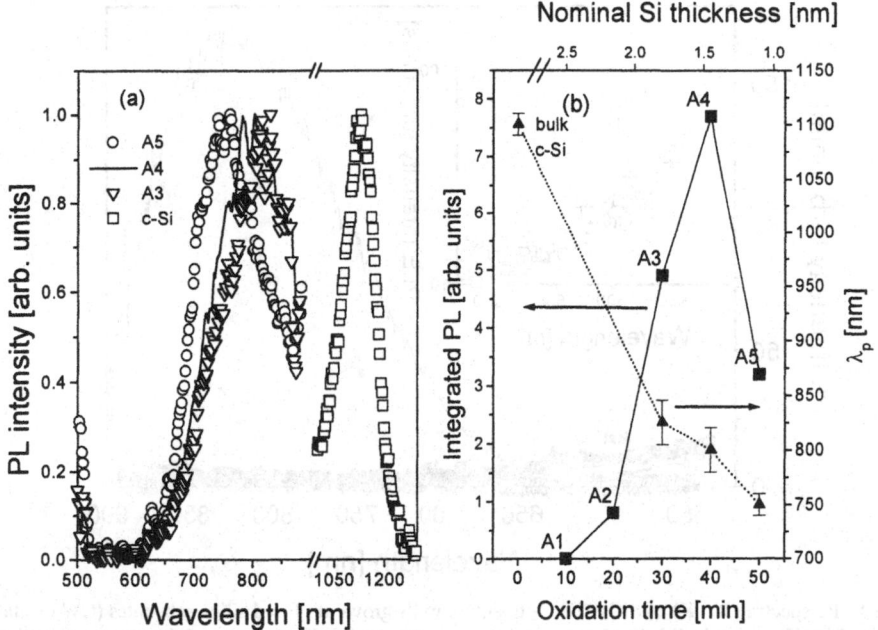

Figure 4. Room temperature PL spectra from the $SiO_2/Si/SiO_2$ quantum-wells with different Si-thickness: (a) shift in the peak maximum, (emission maxima were normalised to 1), (b) integrated PL (range 600 nm - 900 nm) and peak maximum λ_p as function of Si thickness.

128

2.2.3 *PL measurements*

Figure 4 shows the room temperature PL spectra from the A series. Samples with thick Si-layers A1 and A2 show no or very weak emission in the range from 500 to 900 nm, whereas the samples A3, A4, and A5 show a wide PL band. The longer the re-oxidation time, the thinner the Si layer thickness is and the higher the energy of the PL maximum is. A maximum blue-shift of 0.55 eV with respect to the energy gap of bulk Si is observed for sample A5. The integrated PL intensity shows a distinct maximum for samples A3, A4 (thickness range 1.8 to 1.5 nm), whereas the peak maximum λ_p is strongly shifted to lower wavelengths with decreasing Si thickness. The external quantum efficiency of this emission was estimated (by comparison with known standard samples) to be around 1%.

Figure 5 shows the room temperature PL spectra of the samples grown on SiO_2 substrates. Sample D3 with 1.5 nm thick poly-Si layer gives strong PL at 825 nm. Sample D2 has no observable PL. The SiO_2 substrate shows weak and broad PL extending from 530 nm to about 700 nm. The insert in Fig 5 shows the photoluminescence excitation spectrum of sample D3, where the observation wavelength was 825 nm. A significant increase in the PLE intensity (I_{PLE}) is observed for wavelengths lower than 475 nm. A graph of $\sqrt{I_{PLE} \cdot E}$ versus E, where E is the excitation energy, gives a straight line in the range 475 nm to 625 nm. This is expected for an indirect gap material (see [33]). Extrapolating this curve to zero intensity gives an indirect band-gap at 1.6 ± 0.05 eV (or 765 nm). Hence, the extrapolated band-gap is about 0.1 eV above the energy of the PL maximum, i. e. 825 nm.

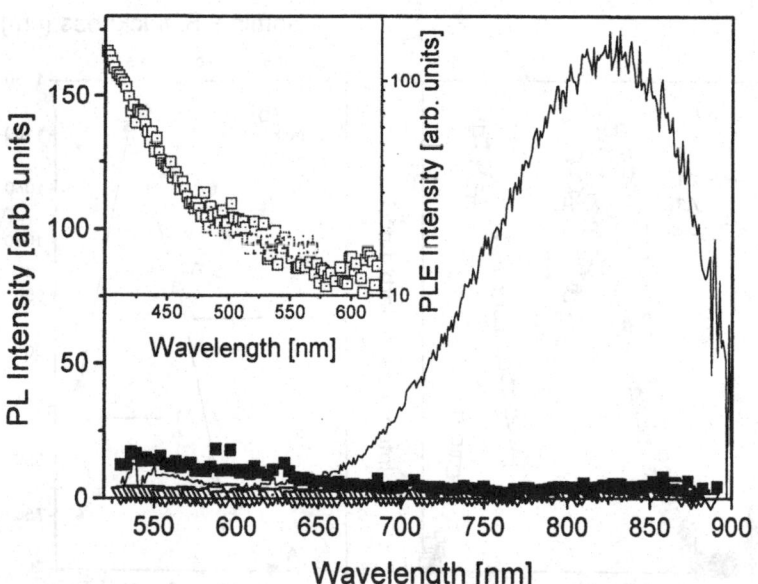

Figure 5. PL spectra (300 K) of SiO_2/Si/SiO_2 quantum wells grown on fused silica substrates (CW excitation 514 nm, 50 mW): sample D3 (Si 1.5nm) (solid line), sample D2 (Si 5 nm) (solid squares), fused silica substrate (down triangles). Insert: PLE spectrum (300 K) of sample D3 (observation wavelength 825 nm) using the ligth of a 1000 W Xe-arc lamp dispersed by a 25 cm single monochromator for excitation.

The temperature dependence of the PL for samples E, C3 and C2 is summarised in Fig. 6. The PL integrated in the range 650-880 nm (displayed by open symbols) is going through a flat maximum at about 150 K, whereas the PL integrated in the range 550-650 nm (displayed by solid symbols) decreases monotonically by increasing the temperature.

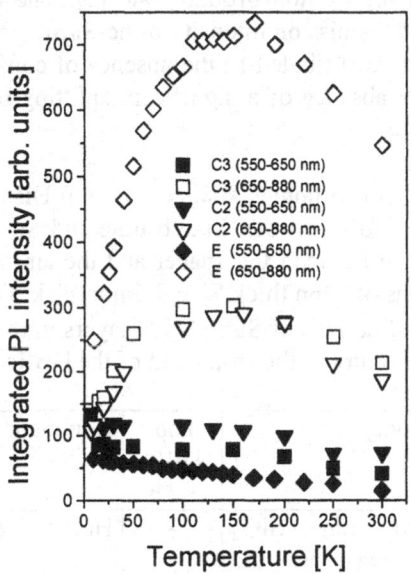

Figure 6. Temperature dependence of integrated PL for the samples C2 (triangles), C3 (squares and E (diamonds) for the integration limits 550 nm -650 nm (solid symbols) and 650 nm - 880 nm (open symbols).

2.3 DISCUSSION

Let us compare our results with theoretical and experimental expectations for excitons quantum-confined in low dimensional Si (see [2] and references therein). The quantum confinement model was used to explain many experimental data for such different systems as PS, nanocrystalline Si, and Si superlattices [1].

The first experimental evidence comes from transmission spectra (Fig. 3) where a blue shift of the direct transitions is observed when d_{Si} decreases from 5 nm to 1.5 nm. From PLE spectra (Fig. 5) the indirect band-gap of a 1.5 nm thick Si layer (sample D3) was estimated to be about 1. 6 eV resulting in a quantum-confinement energy of 0.5 eV, a value similar to the one observed for amorphous-Si/SiO$_2$ superlattices [20]. Larger (3-20 nm) Si layers show within the accuracy of our data no shift of the direct transitions to higher energies and hence the quantum-confinement effects are negligible in this thickness range.

The second experimental evidence is the luminescence. Studies indicate that i) the ideal Si thickness to obtain the maximum emission is about 1.5 nm for amorphous-Si/SiO$_2$ superlattices [20] and 2.2 nm for SiO$_2$/Si/SiO$_2$ quantum wells [18] and ii) the emission is observed for a Si thickness range of ± 0.5 nm in this area. The temperature dependence of the emission shows a trend similar to the one observed for PS [2] and different from the one observed for Si/SiO$_2$ quantum wells [18] and nc-Si/SiO$_2$ superlattices with d_{si} in the 3.5 to 20 nm thickness region [26]. The justification of the maximum in the PL intensity in the range 100 K to 150 K is usually based on a simple two levels model. Two states with different radiative transition probabilities and separated by an energy gap of some meV are formed in the Si nano-crystals. For the origin of the states, the quantum confinement (QC) model claims that the two levels are due to the singlet and triplet states of the exciton ground state [2]. The difference between the temperature dependence of the emission intensity of nc-Si/SiO$_2$ superlattices reported in [26] and of our samples can be attributed to the absence of confinement in the thick Si layers of [26] and hence the absence of a significant splitting between the singlet and triplet state.

TABLE 2. Description of samples studied: Bottom Distributed Bragg Reflectors (DBR) is the DBR between Si-substrate and SiO$_2$ spacer, whereas the top DBR is the one between the spacer and the air. The superlattice is formed by 3 repetitions of 2 nm thick Si and 5 nm thick SiO$_2$ barrier layer. a refers to the number of periods of Si/SiO$_2$ λ/4 layers in the DBR, while b to the wavelength of the centre of the stop band of the DBRs in nm.

sample	Bottom DBR a / b	Cavity	Top DBR a / b	Comment
E1	3 / 750	270 nm SiO$_2$ followed by the superlattice	-	Half active cavity
E2	3 / 750	270 nm SiO$_2$	-	The same as E1 without the superlattice
E3	3 / 750	Superlattice in the center of the SiO$_2$ spacer of 600 nm	2 / 750	Active λ-microcavity
F1	2 / 690	425 nm SiO$_2$	2 / 690	Passive λ-microcavity

3. Visible luminescence from a Si superlattice embedded in high-quality Si/SiO$_2$ optical microcavities

Microcavities consisting of two Distributed Bragg Reflectors (DBR, periodic repetition of λ/4 thick Si or SiO$_2$ layers) separated by a SiO$_2$ λ-thick spacer layer, where λ is the central cavity wavelength, were realised on 4' p-type <100> Si wafers. All the oxide layers of the DBRs were approximately 115 nm thick and they were obtained by a wet-oxidation process at 975° C. The first oxide layer was grown on the Si substrate then a 115 nm thick poly-Si layer was deposited in a LPCVD furnace from SiH$_4$ (at 280 mTorr and 620°C). Part of the poly-Si layer was then oxidised to obtain a final poly-Si layer of about 45 nm and a corresponding 115 nm thick SiO$_2$ layer. This process sequence of deposition and oxidation steps was repeated to form the desired number of periods. The central SiO$_2$ layer (named spacer) was deposited from tetra-ethyl ortho-silicate (TEOS). For microcavities realisation, only the first half of the spacer was deposited. Then, the nanometer thick SiO₂/Si superlattice was formed. Finally, the second half of the spacer and the top DBR were grown by cyclic deposition and oxidation processes. The different samples studied are described in Table 2.

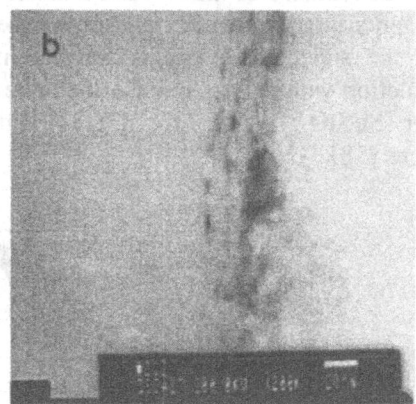

Figure 7. TEM images of the microcavity sample E3. The darker regions correspond to the Si layers. A magnified image of the central region of the microcavity is shown in Fig. 7b. The three superlattice periods are embedded within a thick SiO₂ layer.

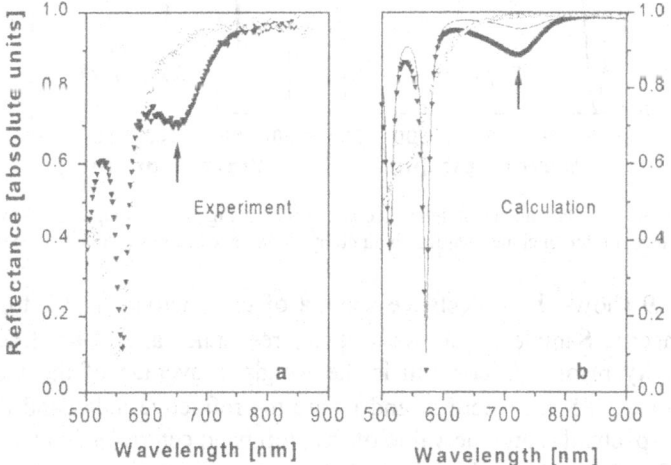

Figure 8. Measured (a) and calculated reflectance spectra (b) of the half-cavity E1 (triangles) and E2 (circles) samples. The calculations are performed by using the transfer matrix approach [17] with the Si parameters of Ref. [39] but for the superlattice where the same imaginary part of the refractive index k as for Si (line) or a five-fold larger one (triangles) has been used. The circles refer to the calculation for sample E2.

Figure 7 shows the TEM images of the microcavity E3. The microcavity structure is clearly observed. In the centre of the SiO₂ spacer, a thin grey line reveals the presence of the nanometer thick SiO₂/Si superlattice. A blow-up of this region is reported in Fig. 7b. The superlattice is formed by three Si layers which are only partly seen as dark lines.

132

Figure 8 reports the measured (a) and calculated (b) reflectance spectra of the half-cavity E1 (with superlattice) and E2 (without superlattice) samples. No difference are observed apart for the presence of a quite pronounced minimum in the reflectance of sample E1 at about 640 nm. As shown in Fig. 8b, the presence of the superlattice causes the appearance of the 640 nm feature. In addition, this reflectance minimum is deeper when a larger imaginary part of the dielectric function is used to describe the Si layers in the superlattice. In this way an increase in the imaginary part of the dielectric function with respect to bulk Si can be inferred, which was both theoretically predicted for $(Si/SiO_2)_n$ superlattices [36,37] and experimentally observed in a-Si/SiO_2 superlattices [38].

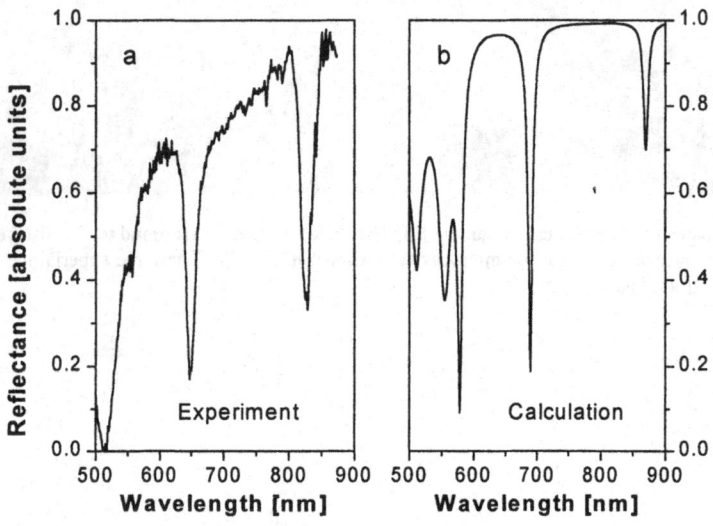

Figure 9. Measured (Fig. 9a) and calculated reflectance spectra (Fig. 9b) of sample E3. A wavelength dependent refractive index and the thickness values obtained by TEM have been used in the calculations.

Fig. 9 shows the reflectance spectra of the microcavity E3 together with the computed spectra. Sample E3 shows a cavity resonance at 830 nm ($\Delta\lambda$= 13 nm, Q = 65). This cavity resonance position is the weighted average of the value due to the spacer (λ_c=$n_c L_c$= 885 nm where n_c and L_c are the refractive index and the thickness of the spacer, respectively) and the value of the stop band centre (estimated to be 750 nm). Beside the 830 nm cavity mode, a second resonance is observed at 650 nm. For an ideal planar microcavity (constant reflectivity), this second resonance is expected to be at $2/3 \cdot \lambda_c$=590 nm. The discrepancy is explained by the following reasons: 1) the DBRs are centred at 750 nm thus shifting the cavity mode to larger wavelengths; 2) both the imaginary and the real parts of the refractive index of Si are non-zero below 650 nm, thus resulting in a further shift to larger wavelengths. The calculation reported in Fig. 9b, where these effects are considered, gives the two cavity modes at 870 nm and 689 nm, values which are in reasonable agreement with the experimental ones. The observed difference in the lineshape between calculated and experimental spectra is probably due to scattering effects (roughness of the SiO_2/Si interfaces), which are not included in the calculation.

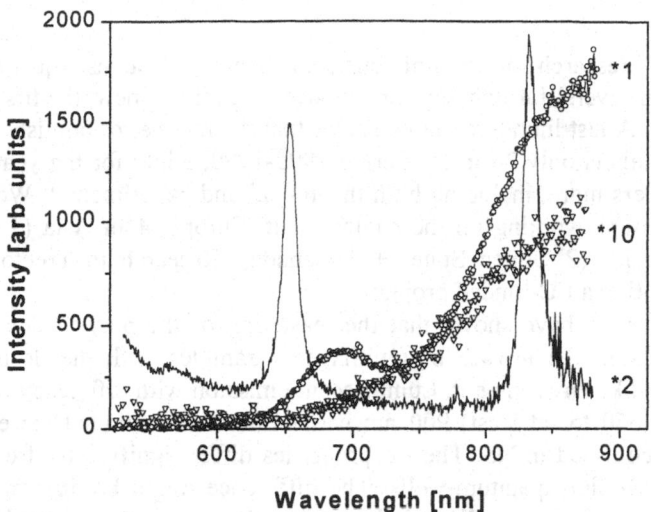

Figure 10. Room temperature photoluminescence spectra of the half-cavity E1 sample (circles), the microcavity E3 sample (straight line) and the reference superlattice B3 sample (triangles). The excitation wavelength was 488 nm. Spectra of the E3 sample and the B3 sample have been multiplied by 2 and 10, respectively.

Fig. 10 shows the photoluminescence (PL) of the samples E1, E3 and B3. Samples F1 and E2 showed no appreciable luminescence. The reference B3 sample shows a broad emission band peaked at 870 nm (1.4 eV). A similar emission feature is observed for the half-cavity E1 sample, but it is considerable stronger. The 20-fold luminescence enhancement can be related to two contributions: 1) the high-reflecting mirror, reflecting both the exciting laser and the emitted light, which explains a luminescent enhancement by factor of $\approx$ 4, 2) the larger roughness of the superlattice layers (see Fig. 7b), with respect to the roughness of the superlattice of sample B3 [9]. This roughness increases the carrier localization and hence, enhances the luminescence [20].

The photoluminescence of sample E3 shows two cavity modes at 830 nm and 650 nm, respectively, and weak emission at higher energies due to leaky cavity modes. The intensity of the 830 nm emission equals the one of the half cavity E1 sample in the same spectral region. However, excitation of the superlattice is less effective for sample E3, since the exciting laser light is partly reflected and absorbed by the top DBR. Moreover, sample E3 shows intense emission at the 650 nm resonance and at even higher energies. At such high energies the emission from sample E1 was very low. This high energy emission is assigned to oxygen related defects, which are induced by the presence of the Si/SiO$_2$ superlattice. Microcavities without a superlattice inside the spacer do not show any PL. The reason for which the high energy emission in sample E3 is significantly larger than in sample E1 is not entirely clear. It may be possible that the exciton-photon coupling in the microcavity modifies the ratio between non-radiative and radiative recombinations at 650 nm. Instead of relaxing to low energy states, excitons couple with the resonance optical mode at 650 nm and recombine radiatively. This idea needs further deepening in the future.

4. Conclusions

World wide, the research on two-dimensional Si/SiO$_2$ systems (quantum wells, superlattices, multi-layers) for light emitting diodes is relatively new, the first paper was published in 1995. A fast literature search shows that the number of published papers in international journals is only 26 in the period 1995-1999, while for the year 2000 one can expect 15 papers more, including both theoretical and experimental. We are aware of 14 different groups working on the subject, 8 in Europe, 4 in Asia (1 China + 3 Japan), 3 in America (2 United States + 1 Canada). Research in Trento started in november 1998 within a EC-funded project.

In this paper we have shown that the mastering of the poly-Si deposition and oxidation has allowed the growth of multilayered samples with the desired optical properties in a fab-like environment. Luminescent emission with efficiency in excess of 1% ranging from 650 to (at least) 900 nm was found. We attributed the emission to exciton quantum confined in NS. The PL properties differ significantly from the ones reported for SiO$_2$/Si/SiO$_2$ quantum-wells. The difference might be due to the crucial role of the roughness of the poly-Si/oxide interfaces. We demonstrated also high quality Si/SiO$_2$ microcavities formed by LPCVD. The optical properties of Si superlattices embedded in the microcavity differ considerably from those of bulk Si and of bare Si/SiO$_2$ superlattices. The emission from the cavity is narrowed, spatially concentrated, and enhanced with respect to the emission from SiO$_2$/Si superlattices in the absence of the cavity. The microcavities have been grown with the same furnaces as the Si multilayers.

The main advantages of the proposed deposition process are: i) the process sequence does not require any modifications of the furnaces, which would beneficially impact on its implementation in an already existing microfabrication plant; ii) the process can easily be upgraded to implement the growth of both very thick and very thin Si layers which can be used to embed nanometric Si multilayers into other complex optical structures, e. g. waveguides or microcavities. In comparison with other approaches reported in the literature, our process requires three steps to grow one Si/SiO$_2$ double layer (1 deposition and 2 oxidation processes), whereas more sophisticated technologies relay on 2 deposition or 1 deposition and 1 oxidation process. To our opinion, the additional oxidation process in our case is an acceptable compromise to meet the CMOS compatibility requirements.

The main difficulties we faced in the processing were the characterisations of both the furnace and the processes to obtain the desired thin poly-Si layer as well as its "on line" control, i.e. a reliable, fast and non destructive technique to measure the deposited layer thickness. Here one can clearly expect future improvements resulting in a better control of the Si layer thickness. When in-line CMOS compatibility, cost and throughput are an issue, the proposed deposition technique has many advantages with respect to other techniques such as MBE or PECVD where better control on the nanometric scale can be reached.

Based on this deposition method, we developed simple light emitting devices [40]. EL in the range 500 to 900 nm, EQE at 650 nm of 5×10^{-5}, highest power output 670 pW, stability over long period, time decay of EL in the 10^{-6} s range were observed.

The EL mechanism is ascribed only to a minor extent to recombination of e-h pairs in NS. It is attributed to hot-electron recombination in the Si wafer. Si/SiO_2 multilayers act as majority carrier electrical injectors. The EQE results are similar to the state-of-the-art worldwide. We believe that the main problem in the low efficiency EL is carrier injection. A solution might be a lateral injection scheme which we are currently investigating. Looking at the EQE in PL one could expect an overall upper-limit of EQE in the 1-5 % range. One can foresee a severe limitation for the modulation speed (GHz never achievable without external modulator). Improved LED can be realized by introducing microcavities in the active regions. For an improved microcavity-LED the main problems are the carrier injection and to a less extent the design of the LED.

5. Acknowledgement

This work has been supported by EC through the SMILE project (ESPRIT MEL-ARI 28741), by MURST through COFIN99 (S. Modesti) and by INFM through PRA2000-RAMSES. We acknowledge C. Spinella for the TEM measurements and K. Gatterer for the absorbance measurements

6. References

[1] Silicon based microphotonics: from basics to applications, eds. O. Bisi, S. U. Campisano, L. Pavesi, F. Priolo (IOS press, Amsterdam 1999).
[2] O. Bisi, S. Ossicini and L. Pavesi, Surface Science Reports 264, 1 (2000).
[3] L.T Canham, Appl. Phys. Lett. 57, 1045 (1990).
[4] A.G. Cullis and L.T. Canham, Nature 353, 335 (1991)
[5] W.L. Wilson, P.F. Szajowski and L.E. Brus, Science 262, 1242 (1993)
[6] Z.H. Lu, D.J. Lockwood, and J.-M., Baribeau, Nature 378, 258 (1995)
[7] K.D. Hirschman, L. Tsybeskov, S.P. Duttagupta, P.M. and Fauchet, Nature 384, 338 (1996)
[8] G. Franzò, F. Priolo, S. Coffa, A. Polman, and A. Carnera, Appl. Phys. Lett. 64, 2235 (1994)
[9] D. Leong, M. Harry, K.J. Reeson, and K.P., Homewood, Nature 387, 686 (1997)
[10] D.A. Miller, Nature 378, 238 (1995)
[11] S.S. Iyer, and Y.-H. Xie, Science 260, 40 (1993)
[12] Silicon based Optoelectronics. MRS Bulletin (April 1998, volume 23, No. 4).
[13] P. M. Fauchet, J. Lumin. 80, 53 (1999).
[14] M. Zacharias, L. Tsybeskov, K.D. Hirschman, P.M. Fauchet, J. Bälsing, P. Kohlert, P. Veit, J. Non-Cryst. Solids 227-230, 1132 (1998)
[15] L. Heikkil, T. Kuusela, H.-P. Hedman, H. Ihantola, Appl. Surface Science 133 84 (1998)
[16] Confined electrons and photons, eds. C. Weisbuch and E. Burstein (Plenum Press, Boston 1995).
[17] L. Pavesi, La Rivista del Nuovo Cimento 20, 1 (1997).
[18] Y. Takahashi, T. Furuta, Y. Ono, T. Ishiyama and M. Tabe, Jpn. J. Appl. Phys. 34, 950 (1995).
[19] Y. Kanemitsu, S. Okamoto, Physical Review B 56, 15561 (1997).
[20] Z.H. Lu, D.J. Lockwood, J. M. Baribeau, Nature 378, 258 (1995).
[21] B. T. Sullivan, H. J. Labbè J. Lumin. 80, 75 (1999).
[22] J. Keränen, T. Lepistö, L. Ryen, S. V. Novikov, E. Olsson, J. Appl. Phys. 84, 6827 (1998).
[23] M. Zacharias, L. Tsybeskov, K.D. Hirschman, P.M. Fauchet, J. Bälsing, P. Kohlert, P. Veit, J. Non-Cryst. Solids 227-230, 1132 (1998).
[24] L. Heikkil, T. Kuusela, H.-P. Hedman, H. Ihantola, Appl. Surface Science 133, 84 (1998).
[25] J.M. Baribeau, D.J. Lockwood, Z.H. Lu, H.J. Labbè, S.J. Rolfe, G.I. Sproule, J. Lumin 80, 417 (1999).
[26] L. Tsybeskov, K.D. Hirschman, S. P. Duttagupta, M. Zacharias, P. M. Fauchet, J. P. McCaffrey, and D.J. Lockwood, phys.stat.sol (a) 165, 69 (1998).

136

[27] L. Tsybeskov, K.D. Hirschman, S. P. Duttagupta, M. Zacharias, P. M. Fauchet, J. P. McCaffrey, and D.J. Lockwood Appl. Phys. Lett. **72**, 43 (1998).
[28] L. Khriachtchev, M. Räsänen, S. Novikov, O. Kilpelä, J. Sinkkonen, J. Appl. Phys. **86**, 5601 (1999).
[29] V. Mulloni, R. Chierchia, C. Mazzoleni, G. Pucker, L. Pavesi, Phil. Mag. B **80**, 705 (2000).
[30] H.R. Philipp, E.A. Taft, Phys. Rev. **120**, 37 (1960).
[31] N. Koshida, H. Koyama, Y. Suda, Y. Yamamoto, M. Araki, T. Saito, K. Sato, Appl. Phys. Lett. **63**, 2774 (1993).
[32] M. Nishida, Phys. Rev. B **59**, 15789 (1999).
[33] A. Kux, M. Ben Chorin, Phys. Rev. B **51**, 17535 (1995).
[34] Y. Sakurai, K. Nagasawa, H. Nishikawa, Y. Ohki, J. Appl. Phys. **86**, 370 (1999).
[35] F. Iacona, G. Franzo, C. Spinella, J. Appl. Phys. **87**, 1295 (2000).
[36] N. Tit, M.W.C. Dharma-Wardana, J. Appl. Phys. **86** 387 (1999).
[37] E. Degoli, S. Ossicini, (submitted to Surface Science).
[38] S.V. Novikov, J. Sinkkonen, O. Kilpelä S.V. Gastev, J. Vac. Sci. Technol. B **15** 1471 (1997).
[39] Data in Science and Technology, Semiconductors Group IV Elements and III-V Compounds, Springer Verlag 1991, p. 20.
[39] M. Luppi, E. Degoli, S. Ossicini, phys. stat. solidi a (accepted for publication).
[40] Z. Gaburro, G. Pucker, P. Bellutti, L. Pavesi, Solid State Communications **114**, 33 (2000).

SILICON NANOSTRUCTURES IN Si/SiO$_2$ SUPERLATTICES FOR LIGHT EMISSION APPLICATIONS : POSSIBILITIES AND LIMITS

A.G. NASSIOPOULOU, T. OUISSE and P. PHOTOPOULOS

IMEL/NCSR "Demokritos", P.O.Box 60228, 153 10 Aghia Paraskevi Attikis, Athens Greece

1.   Introduction

Silicon is by far the dominant material in today's microelectronics, its low dimensional structure constitutes also one of the most fascinating fields of research in materials science and technology. Quantum size effects, which occur when crystallite sizes go below $\cong$ 5nm and become comparable to the exciton Bohr radius, are at the origin of material properties, which find interesting applications in new types of devices. Single electron transistors and memories constitute one of such families, based on the well known Coulomb blockade effect. Light emitting devices (LEDs) and displays, based on quantum confinement, constitute another area of primary importance in technological applications. The main reason being the inefficiency of bulk silicon to emit light and the lack of C-MOS compatible efficient light emitters for integrated optoelectronics.

Although the first and mostly investigated low dimensional silicon material is porous silicon (for reviews see papers 1-3), an increasing interest in other forms which are fabricated by C-MOS compatible processing, and often offer improved material stability, is currently observed. Different deposition and annealing techniques, as well as lithography and etching, are used to produce either quantum dots in different host materials [4-8], or quantum wires and pillars [9-10], or quantum wells and quantum walls [9].

A key issue in all cases is to control the size of the structures and their size distribution with the maximum accuracy. In the nanometer scale, lithography and etching techniques are at their limits. So deposition and growth techniques seem to be more promising. An interesting approach is nanoscale silicon layers intercalated with a wide band gap material, in which size control is made by controlling layer thicknesses. As insulating layers, two different materials were studied in the literature, namely silicon dioxide (SiO$_2$) with a band gap of $\cong$ 8.5 eV [11] and calcium fluoride (CaF$_2$) [6] with a much larger band gap of $\cong$ 12 eV. SiO$_2$ offers important advantages, since it is the most common insulator used in microelectronics, with excellent properties even in very thin films. Calcium fluoride on the other hand is grown epitaxially on silicon (111), but the main drawback in applications is the incompatibility with silicon processing.

Si/SiO$_2$ superlattices were first fabricated by Lockwood et al. [11] in 1996 by using molecular beam epitaxy (MBE) of silicon at room temperature and UV ozone exposure. Excellent control of both Si and SiO$_2$ thicknesses was possible. Both Si and

L. Pavesi and E. Buzaneva (eds.), Frontiers of Nano-Optoelectronic Systems, 137–146.
© 2000 *Kluwer Academic Publishers. Printed in the Netherlands.*

SiO$_2$ in their layers were amorphous. Similar a-Si/SiO$_2$ superlattices were also grown by Novikov et al. [13] using a slightly different oxidation process.

Later, Tsybeskov et al. [14] fabricated nanocrystalline silicon (nc-Si)/SiO$_2$ superlattices by controlled recrystallization of a-Si/SiO$_2$ superlattices. The obtained silicon layers were, however, relatively thick (from 2.5 to 25 nm). Kanemitsu and Okamoto [15] fabricated a Si/SiO$_2$ single quantum well by using thermal oxidation of a SIMOX wafer. In all the above cases of low dimensional silicon in Si/SiO$_2$, one of the expected properties due to quantum confinement was light emission in the visible range. Novikov [13] and Lockwood [11] reported a weak photoluminescence (PL) from their amorphous structures. Tsybeskov et al. [14] did not report any PL from their recrystallized layers, since they were relatively thick. From the single quantum well in reference [15], a weak asymmetric PL spectrum was observed, one component of which was attributed to quantum confined states.

Our recent results on Si/SiO$_2$ superlattices [16], fabricated by successive cycles of low pressure chemical vapour deposition (LPCVD) and subsequent oxidation, show enhanced PL emission, tuned from 900 nm to 650-700 nm by reducing the silicon thickness from 4-5 nm down to 1-1.5 nm. EL at the same wavelenghts is also observed [17]. In this paper we report on tunable PL and EL from Si/SiO$_2$ superlattices, with one to 5 periods of Si/SiO$_2$. Saturation effects in electroluminescence due to non-radiative Auger recombination, Coulomb charging and/or the quantum confined Stark effect will also be discussed.

2. Fabrication of Si/SiO$_2$ superlattices

The structures were fabricated in silicon processing facilities. Low pressure chemical vapor deposition (LPCVD) was used to deposit each silicon layer, part of which was oxidized to form Si/SiO$_2$. The same procedure was repeated several times to form a superlattice. The substrates were n-or p-type (100) silicon wafers with a resistivity of 5-10 Ω cm and an initial pad oxide was grown on the substrate, which constituted the initial barrier to bulk silicon.

The oxidation time in the fabrication process was critical, since it determined the silicon layer thickness in each bilayer. The initial LPCVD silicon layers, deposited at 580 °C, 300mTorr, were initially amorphous and they were crystallized during the temperature rise for oxidation. The oxidation processing was done at 900 °C and the oxidation time ranged between 30 and 120 min. For an initial silicon layer of 12 nm, the final silicon thickness in each bilayer ranged between 6nm (30 minutes oxidation) and below 1.5 nm (90 minutes oxidation). Superlattices were fabricated by repeating the process used to form one bilayer. LPCVD silicon films, 12 nm thick, were used, oxidized at 900°C for 50 minutes. Up to 6 periods were added.

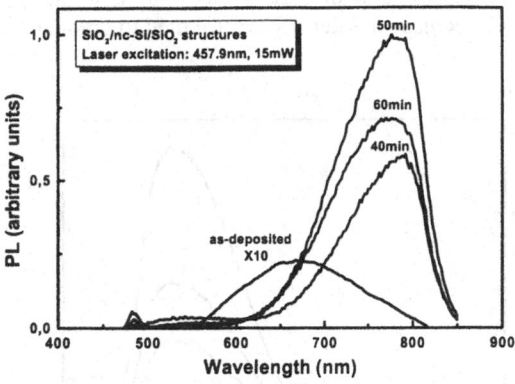

Fig.ure 1. PL spectra from silicon films, 12 nm thick, deposited by LPCVD on an oxidized silicon substrate and subjected to a high temperature thermal processing at 900°C at oxygen ambient for different durations (zero for as-deposited, 40, 50 and 60 minutes). The intensity of the spectrum from the as-deposited film was very low and it was

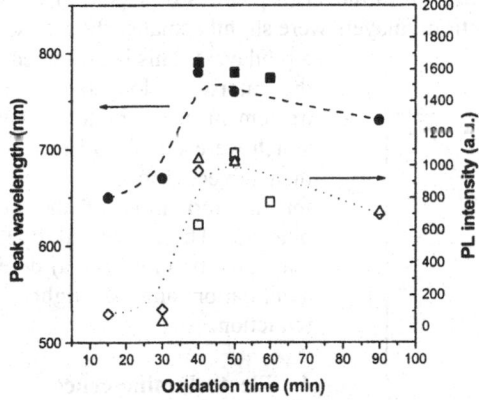

Figure 2. Variation of PL peak wavelength (left vertical axis, solid symbols) and PL intensity (right axis, arbitrary units, open symbols) as a function of the oxidation time for three series of samples (represented by circles, squares or triangles). The silicon substrate was p-type and the pad oxide was 4nm thick. The as-deposited silicon layer was 12 nm thick.

3. Photoluminescence

For PL measurements, an argon laser line at 457.9 nm was used. As-deposited silicon layers exhibited weak broadband lumminescence extended from 500 to 800 nm with two maxima, one at 530-550 nm and the other at $\cong 650$ nm.

The first peak was in general unstable and it disappeared during the measurement [18]. Its origin is attributed to defects in SiO_2. The second peak is attributed to quantum confinement.

In samples which were thermally treated and oxidized to produce the SiO_2 bilayers a drastic increase in PL emission, compared with the emission from the as-deposited layer, was observed.

The peak position depends on the oxidation time (see fig 1) For an oxidation time from 15 to 35 minutes, the peak wavelength red-shifts from 650 to 800 nm and for more prolonged oxidation it gradually blue-shifts down to $\cong 700$ nm (fig. 2). PL totally disappeared with an oxidation time longer than 120 minutes.

The above behavior supports the mechanism of light emission from silicon crystallites in a SiO_2 matrix. The enhanced PL compared with the as-deposited layer is due to crystallization of the layer from the amorphous phase. The PL red-shift coincides with the gradual formation of silicon crystallites. When all the layer is transformed into silicon crystallites surrounded by SiO_2, further oxidation results in size reduction of the internal silicon core. This explains the blue-shift of the

140

luminescence in films oxidized for a longer time than that corresponding to the maximum in wavelength and intensity. The decrease in intensity is attributed to gradual disappearance of some crystallites by full oxidation.

PL from superlattices had similar characteristics as that from one bilayer, while PL intensity increased by increasing the number of bilayers. In samples with two periods, PL intensity was roughly the double of that obtained from samples with one period. For more than 2 periods, a superlinear increase in intensity was observed, while peak position slightly blue shifted. These results are illustrated in figs.3 and 4. In order to explain this behavior,

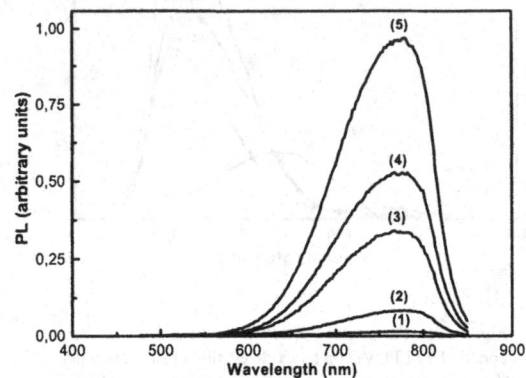

Figure 3. PL spectra from superlattices with a different number of periods of Si/SiO$_2$ bilayers.

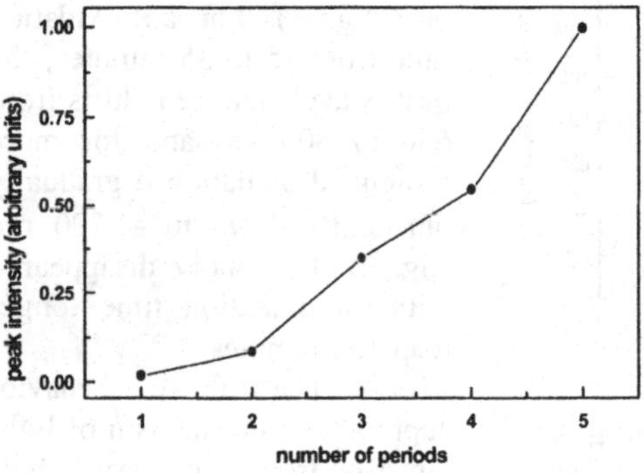

the samples were subjected to extensive characterization by cross sectional transmission electron microscopy (TEM). From the TEM pictures of superlattices, we have seen that crystallite sizes of the bottom bilayers were slightly smaller than those of top bilayers. This is explained by the more prolonged thermal treatement of the bottom layers, which were initially formed and then subjected to thermal cycles for the formation of the next bilayers. These thermal cycles may have two results : a) defect annihilation and b) slight size reduction.

4. Electroluminescence

Exciton luminescence from silicon nanocrystallites is limited by the difficulty to inject carriers into confined structures. When nanocrystallite size is reduced, quantum confinement becomes more effective, but at the same time the difficulty to inject carriers is increased. For exciton luminescence, both electron and hole injection is needed and this is an additional difficulty due to the asymmetry of barriers for electrons and holes in silicon dioxide. Hole injection is less probable and much thinner SiO$_2$ layers are needed for tunneling in a vertical structure. EL from a single layer of nanocrystallites sandwiched between two

Figure 4. Variation of PL peak intensity from superlattices as a function of the number of periods.

SiO$_2$ layers was first studied. The device structure is shown in fig. 5. The substrates were p-type silicon wafers with resistivity in the range of 5-10 Ù.cm. A pad oxide, 5-6 nm thick, was used and an LPCVD film, 10-12 nm thick, was then deposited on top of it at 580°C, 300mTorr and oxidized at 900°C. The oxidation time was 50 minutes. The onset of EL from these devices was about 5-6 V.

EL spectra (see fig. 6) were similar to PL ones at relatively low voltages, up to 8.5V. For higher voltages, EL intensity was increased and at the same time the peak was shifted to lower wavelengths. A saturation in intensity and a peak broadening was observed at higher voltages [17]. In devices with different oxidation times, corresponding to a difference in crystallite sizes, EL was blue shifted by increasing the oxidation time, following the same behavior as PL [17].

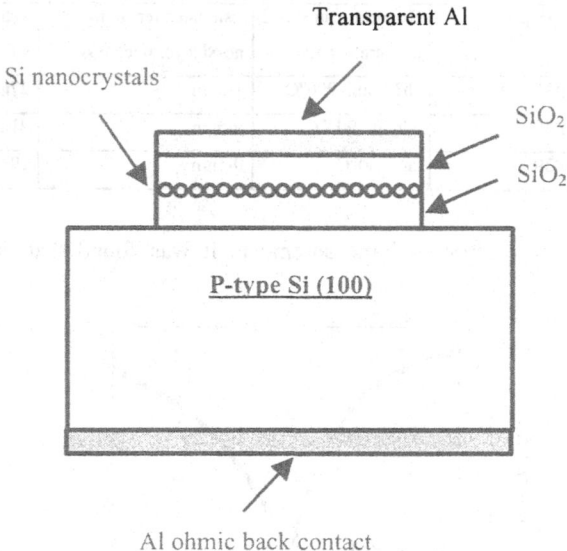

Figure 5. Structure of a vertical device of a silicon layer of quantum dots sandwiched between two thin SiO$_2$ layers.

5. Electrical characterization of light emitting structures

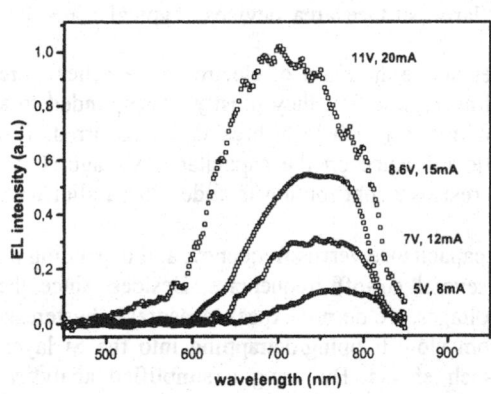

Figure. 6 EL from a device as that illustrated in fig. 5

A series of samples with different oxidation times of the silicon layer was prepared as in table I. The oxidation time used and the expected thicknesses of the silicon layer and the top oxide are indicated. The samples were named P37, P38 and P39 and the gate surface area of the devices was in the range of 2.1×10^{-4} cm^2 to 1mm^2. Their electrical characteristics were investigated by means of complex impedance measurements as a function of frequency, with the aim to extract same relevant physical parameters and to link them to the electroluminescence data.

TABLE I: Oxidation conditions and characteristics of samples described in §5.

Sample	Oxidation time and temperature	Estimated remaining nc-Si layer thickness	Estimated t_{OX2} value	Nominal t_{OX1} value
P37	1h30min, 900°C	1-2nm	27nm	4nm
P38	50min, 900°C	4-5nm	21nm	4nm
P39	2h, 900°C	0-1nm	29nm	4nm

From I-V measurements it was found that the vertical current through the structures was always much higher than that expected from structures with a high quality oxide. With a thickness equal to the pad oxide thickness, in the range of 5-20 nm, and with the gate surface area used, the current should be harshly measurable at low voltages. We checked the quality of the pad oxide by fabricating devices with only the bottom thermal oxide (thickness around 3-5 nm), and the obtained current was much reduced. From fully processed devices as in fig. 5, the current was large, with a dispersion of approximately one order of magnitude around the average for large surface area devices. Typical static I-V characteristics are indicated in fig. 7.

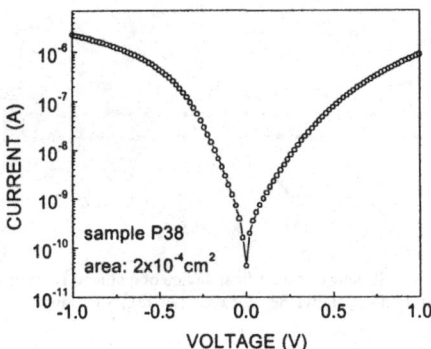

Figure 8. Typical I-V characteristics of sample P38 before stress.)

The obtained currents at voltages above the onset of electroluminescence were too high to account for a direct tunneling regime, but they mostly corresponded to a relatively leaky insulator, or to an insulator in which local breakdown occurred. The oxide leakage has, in general, a drastic influence on the capacitance-voltage (C-V) characteristics of the devices, since a resistive contribution is added in parallel with each of the insulating layers.

This induces a decrease of the capacitance versus frequency, and depending on the specific device structure one or several cut-off frequencies. Besides, since the injection current is high, even at low voltages, we do not expect measurable hysteresis effects in the CV curves, resulting from slow trapping/detrapping into the Si layer. Indeed we did not observe any of such effects. By using a simplified analytical modeling, we extracted some significant parameters of the structures.

We first considered P38, in which the oxidation time of the poly-Si layer was the shortest one (50min), and was thus expected to have the thicker residual Si layer. The capacitance always decreased with frequency (see inset of Fig.8 for a typical example). However, depending on the tested device, the capacitance exhibited either one or two cut-off frequencies (see Fig.8 for the latter case). We *never* found a device exhibiting more than two cut-off frequencies, and the lowest one greatly varied from one device to another. Qualitatively and *quantitatively*, these features can be explained by a very simple model, assuming that the top and the bottom oxides exhibit different leakage properties, and that conduction inside the polysilicon layer, even bad, is much easier to achieve than through the oxides, so that the corresponding equivalent resistance is negligible compared to that of the oxides. In the accumulation regime, the total structure can then be described by a double RC cell circuit, as depicted in the inset of Fig.9. The top and bottom oxide thicknesses t_{OX1} and t_{OX2} having the same order of magnitude, the equivalent parallel capacitance will exhibit one or two cut-offs in the

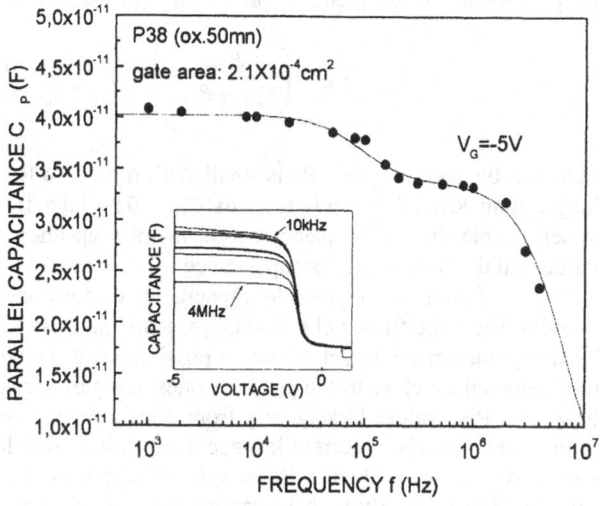

Figure 8. Parallel capacitance versus frequency of a device from sample P38 with a 50min oxidation time, in the accumulation regime. The inset shows CV curves of the same device with the frequency as a parameter. The solid line is a fit of the experimental points with the same parameters as those given in Figure 9.

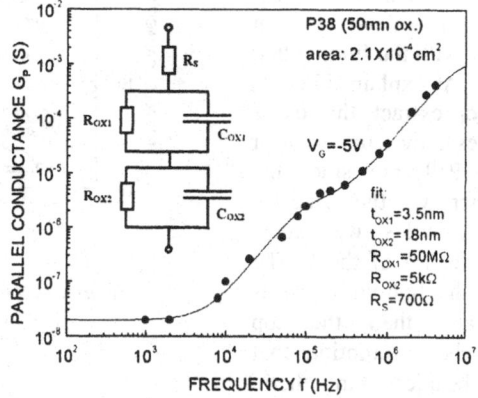

Figure 9. Parallel conductance of the same device as in Fig.8. The inset shows the equivalent circuit used for fitting the data.

144

investigated frequency range, mainly depending on the ratio between R_{OX1} and R_{OX2}. We shall not write here the (cumbersome) analytical expression for the impedance, but only quote the low frequency approximation of the parallel capacitance:

$$C_{LF} \cong \left(\frac{R_{OX1}}{R_{OX1}+R_{OX2}}\right)^2 C_{OX1} + \left(\frac{R_{OX2}}{R_{OX1}+R_{OX2}}\right)^2 C_{OX2}$$

obtained by assuming that R_S is small with respect to R_{OX1} and R_{OX2}. If R_{OX1} is much larger than R_{OX2}, C_{LF} tends towards C_{OX1}. The high frequency cut-off is due to R_S. When visible, the low frequency cut-off mainly depended on the lowest oxide resistance value and the corresponding capacitance.

Typical fits of the parallel capacitance C_P and conductance G_P are given in Figs. 8 and 9. From the fit, we obtained t_{OX1}=18nm and t_{OX2}=3.5nm. It is worth noticing that by using another device, it is always possible to fit the new data by keeping the oxide thickness values close to the previous ones, and just varying the resistance values. The R_{OX1} and R_{OX2} values largely vary from one device to another. This is quite consistent with the hypothesis of current leakage through localised defective paths, whose density varies from device to device. Besides, t_{OX2} lies in the range expected from the oxidation process. The relatively high R_S value can be ascribed to the thin semi-transparent top metal layer. The total conductance value saturates at low frequency and corresponds to the reciprocal dynamic resistance extracted from the static I-V curve. When this curve is superlinear with V, we checked that the plateau value at low f increases with voltage, as expected from the static characteristic. In general, the bottom oxide thickness is more leaky than the top (i.e. $R_{OX1} \gg R_{OX2}$). This is in agreement with the finding that $t_{OX1} < t_{OX2}$. As explained below, we cannot extract the oxide thicknesses in the same way for P37 and P39, but since the etching step was the same for all three samples, we expect $t_{OX1} < t_{OX2}$ for all of them. The fact that the bottom oxide is more leaky than the top supports the assumption that holes can be injected into the Si

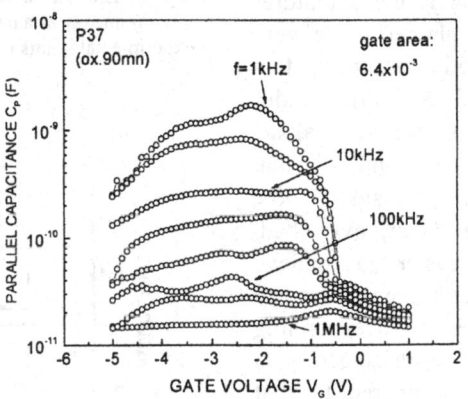

Figure 10. CV characteristics of a typical device from sample P39 (90min oxidation).

layer, and thus induce electroluminescence by recombining with electrons. It would be impossible to inject holes with a perfect oxide. More generally, we think that sample P38 unambiguously exhibits the specific electrical signature (two cut-off frequencies) resulting from the combination of a conductive polysilicon layer sandwiched between two different, leaky insulators.

In the case of P37 and P39 (respectively 90 and 120min long oxidation), the polycristalline Si layer is not continuous, as it has been verified by TEM observations [13]. The dependence of the C-V curve on voltage and frequency is indeed more intricate than for sample P38. As for P38, the curves vary from device to device, and a typical example is given in Fig.10. It can be seen from this figure that depending on f, the CV curve may exhibit several peaks, before decreasing at a high accumulation voltage. For such high voltage, the current has an exponential shape and thus the equivalent oxide resistance decreases with voltage, thus reducing the measured parallel capacitance. The leakage is generally higher than with sample P38, and the conductance is almost constant with frequency. It is thus not possible to extract reliable, quantitative parameters from the impedance measurements. But the overall results favour the hypothesis of non-homogeneous Si and SiO$_2$ layers.

6. Discussion

In vertical electroluminescent devices presented above, carrier injection is the most important limiting factor for effective light emission. Current flow through a leaky silicon dioxide offers the possibility to inject both electrons and holes into the nanocrystalline layer and thus electroluminescence with the same characteristics as photoluminescence was obtained. However, in terms of efficiency, a great difference was observed between PL and EL, indicating that optical pumping was much more effective than the electrical one. At the same detection wavelength, PL efficiency, expressed as the ratio of PL intensity to the input laser power, was approximately two orders of magnitude greater than the corresponding EL efficiency (expressed as the ratio of EL intensity per input electrical power). Another factor, which limits emission from nanocrystallites, is non-radiative Auger recombination. This process becomes predominant over radiative recombination when more than one electron-hole pair, or an additional carrier (electron or hole) is added to a single nanocrystal.

At high current flow, the probability to inject a third carrier into a crystallite becomes higher and in a normal distribution of sizes, it favorizes the larger nanocrystallites . EL quenching and saturation of the emission is expected to occur, together with a blue shift in EL wavelength. This is in agreement with the obtained experimental results (see fig. 6) and with similar effects observed in porous silicon [19].

Another effect, however, which could lead to the same result of EL saturation and quenching, is the quantum confined Stark effect. It expresses the influence of the electric field on the radiative recombination of an electron-hole pair in a semiconductor quantum dot. In Si/SiO$_2$ superlattices, the use of SiO$_2$ as a barrier material makes possible the application of very high electric fields, which should have a considerable impact on the radiation lifetime. It has been demonstrated recently by means of a variational calculation [20], that the application of an electric field such as, e.g., the one required for electron or hole injection from the gate and substrate electrodes in Si/SiO$_2$ structures as those used here, induces a significant effect on the radiative recombination lifetime. This effect is size-dependent, being more substantial in larger crystallites, and it should induce the same effect as the Auger recombination.

7. Conclusion

PL and EL from silicon nanocrystals in Si/SiO$_2$ superlattices have been demonstrated. EL exhibited tunability from the near IR down to 650 – 700 nm, by changing the silicon nanocrystallite size and /or the applied voltage.

From the obtained results, it seems that photo- and electroluminescence from nanocrystalline silicon might be promising for future applications. The main issue for improving the EL efficiency seems to be an effective carrier injection. However, some limits are imposed to the expected improvement, which are due to three different effects: a) Auger non-radiative recombination, which starts to compete with radiative recombination when more than one electron-hole pair is present in the crystallite b) The influence of the electric field on the radiative lifetime of an electron-hole pair. This effect is expected to be negligible in the case of very small crystallites and c) Coulomb charging of silicon nanocrystals, which induces Coulomb blockade effects.

This work has been done within the ESPRIT MEL-ARI project "SMILE".

References

[1] A.G. Cullis, L.T. Canham and P.D.J. Calcott, *J. Appl.Phys.* **82** (3) 909 (1997)

[2] P.M. Fauchet, *IEEE Journ,. of selected Topics in Quant. Electr.* **4**, No 6, 1020 (1998)

[3] D. Kovalev, H. Heckler, G. Polisski and F. Koch, *Phys. Stat. Sol. (b)* **215**, 871 (1999)

[4] N. Lalic and J. Linnros, in *Proceedings of the E-MRS, J. Lumin.* **80**, 263 (1999)

[5] Q.Zang, S.C. Bayliss and R.A. Hult, *Appl. Phys. Lett.* **66**, 1977 (1995)

[6] F.Bassani, L.Vervoort, I.Mihalcescu, J.C.Vial, F.Arnaud d'Avitaya, *J. Appl. Phys.* **79**, 4066 (1996)

[7] A.G.Nassiopoulou, V.Tsakiri, V.Ioannou-Sougleridis, P.Photopoulos, S.Menard, F.Bassani, F. Arnaud d' Avitaya, *J. Lum.* **80**,81 (1999)

[8] P.Normand, D.Tsoukalas, E.Kapetanakis, J.A.Van Den Berg, D.G.Arnour, J.Stoemenos, C.Veu, *Electroch. Sol. Stat. Lett.* **1**(2), 88 (1998)

[9] A.G.Nassiopoulou, S.Grigoropoulos, E.Gogolides and D.Papadimitriou, *Appl. Phys. Lett.* **66**, 1114 (1995)

[10] A.G. Nassiopoulou, S. Grigoropoulos and D. Papadimitriou, *Appl. Phys. Lett.* **69**, 2267 (1996)

[11] D.J. Lockwood, Z.H. Lu and J.-M. Baribeau, *Phys. Rev. Lett.* **76** (3), 539 (1996)

[12] V.Ioannou-Sougleridis, V. Tsakiri, A.G. Nassiopoulou, F. Bassani, S. Menard and F. Arnaud d' Avitaya, *Mat. Sci. Eng. B* **69-70**, 309 (2000)

[13] S.V. Novikov, J. Sinkkonen, O. Kilpela and S.V. Gastev, *J. Vac. Sci. Techn. B* **15**(4), 1471 (1997)

[14] L.Tsybeskov, K.D.Hirschman, S.D.Duttagupta , P.M.Fauchet, M.Zacharias, J.P.Mc Caffrey and D.J.Lockwood, *Phys. St. Sol. Vol. (a)* **165**, 69 (1998) and *Appl. Phys. Lett.* **72** (1), 43 (1998)

[15] Y.Kanemitsu and S. Okamoto, *Phys. Rev. B* **56** (24), R15561 (1997)

[16] P.Photopoulos, A.G. Nassiopoulou, D.N. Kouvatsos and A.Travlos, *Appl. Phys. Lett.***76** (24), june 2000 (in press)

[17] P.Photopoulos and A.G. Nassiopoulou, *Appl. Phys. Lett.* (submitted)

[18] P.Photopoulos and A.G. Nassiopoulou, D.N. Kouvatsos, and A. Travlos, *Mater. Sci. Eng. B* **69-70**, 345 (2000)

[19] I.Mihalcescu, J.C. Vial, A. Bsiesy, F. Muller, R. Romenstein, E. Martin, C. Delerue, M. Lannoo and G. Allan, *Phys. Rev. B* **51**, 17605 (1995)

[20] T.Ouisse and A.G.Nassiopoulou, Europhysics Lett. (submitted)

FIRST PRINCIPLES OPTICAL PROPERTIES OF LOW DIMENSIONAL SILICON STRUCTURES

STEFANO OSSICINI AND ELENA DEGOLI
*Istituto Nazionale per la Fisica della Materia (INFM) and
Dipartimento di Fisica, Università di Modena e Reggio Emilia
Via Campi 213/A, 41100 Modena, Italy*

Abstract

The remarkable changes that the optoelectronic structure of silicon shows when the material is reduced to the nanometric scale have enticed many researchers to study from computational point of view the electronic and optical properties of confined Si systems, e.g Si quantum slabs, quantum wires and quantum dots. After a brief review of the actual situation, we present ab initio calculations of the optical properties of Si confined structures for two case studies: Si undulating quantum wires and Si/SiO2 quantum wells.

1. Introduction

Bulk crystalline silicon (c-Si), by far the most important semiconductor material, does not exhibit good optoelectronic properties because of the indirect gap, in the infrared region, and of the low probability of radiative electron-hole recombination. Nevertheless the desire for the integration of optoelectronic devices with silicon microelectronics has stimulated an intense effort to search Si-based materials that emit light efficiently. The surprising discovery by Canham [1] of bright, visible light emission from porous silicon (po-Si) has enticed many researchers to reactivate studies of the optical properties of silicon based nanostructures. In fact the most likely explanation for the photoluminescence (PL) in po-Si is the quantum confinement (QC) effect induced by the formation of Si nanocrystallites. Moreover after the discovery of PL in po-Si efficient visible light emission has been predicted [2] and observed [3] in Si-CaF$_2$ quantum wells, Si-SiO$_2$ superlattices [4], Si nanopillar structures produced by nanolitography [5], and in nanometer-size Si crystallites [6]. It appears, therefore, that one of the key ingredients necessary to induce PL in silicon is the presence of nanostructures of reduced dimensionality.

Low-dimensional semiconductor systems exist in three flavors: two-dimensional (2D) quantum wells or quantum slabs, one-dimensional (1D) quantum wires and zero dimensional (OD) quantum dots. Thus the calculations have mainly focused on the investigation of the electronic and optical properties of confined Si structures: quantum well, quantum wire, and quantum dot [7]. The used computational techniques range from model calculation to effective mass approximation (EMA), from tight binding (TB) to empirical pseudopotential (EPS), from Hartree-Fock to Density Functional Theory (DFT). All the calculations performed on confined Si structures give a similar picture concerning the quantum confinement effect: a widening of the band gap from the near infrared wavelength region to, and beyond, the visible range. Furthermore an enhancement of the dipole matrix element responsible of the radiative transitions is found.

Fig. 1 shows the predicted band gaps versus size as obtained from DFT-LDA calculations compiled in Ref. [8] for hydrogen terminated slabs, wires and dots. They are compared to the results obtained by empirical TB and EPS. We note that the energy gap blueshift increases on going from slabs to wires and dots, which is clearly due to the increasing

147

L. Pavesi and E. Buzaneva (eds.), Frontiers of Nano-Optoelectronic Systems, 147–160.

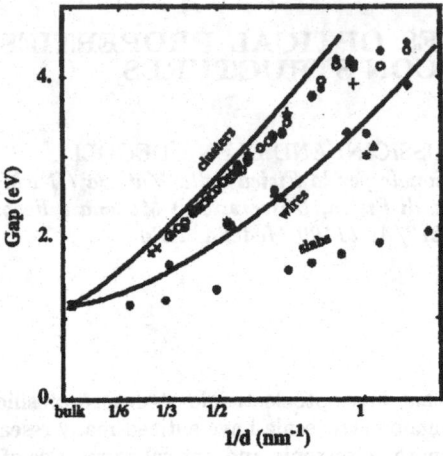

Figure 1. Energy gap versus confinement parameter 1/d for H-terminated Si dots, wires and slabs. LDA: (filled and empty dots) [8], (+) [9], (△) [10], (*) [11]. EPS: (×) [12]. ETB: [13].

number of dimensions affected by confinement effects. It is worthwhile to note that in the case of LDA a correction of 0.6 eV has been added to the energy gap values in order to overcome the well-known LDA underestimation of the Si bulk band gap.

In the case of Si quantum wells it is important to understand not only the role played by the dimension of the crystalline core, but also the role of the agents passivating the dangling bonds at the surface and of the symmetry of the Si surface. Considering Si/H(111), Si/H(001), Si/CaF$_2$(111) and Si/SiO$_2$ QW we have found that the symmetry of the lattice changes the nature of the band gap that is indirect in the Si/H(111) saturated wells and become direct in the Si/H(001) saturated wells [14]. Moreover the different saturating species play an important role in the formation of interface states that can occupy or leave free the band gap so improving or making worse the optical properties of the material.

Concerning Si quantum wires [15, 16] usually the structure investigated consists of a single, infinite wire of rectangular cross section with its axis along the [001] direction and its surfaces corresponding to the (110) and (1-10) planes of bulk Si. The two sides of the cross section are delimited by zig-zag chains of M and N Si first neighbors, this structure is usually referred as MxN. The Si atoms at the wire surfaces are usually passivated by H. The results of the band structure calculations [17] show that a direct band gap appears for $k=0$, this behavior, characteristic of the quantum wire structure, is due to the confinement effect on the six equivalent ellipsoidal conduction band minima (with different masses in the confinement plane) of crystalline Si. Table 1 collects results for the calculated band gaps of Si quantum wires against wire orientation and width [16]. In the Table LDF means ab-initio Local Density Functional calculations, whereas LDF* indicates calculations where also structural relaxation has been taken into account. Si quantum wires, whose dimensions are of the order of ∼ 20 - 30 Å, give energy gaps in the range of the experimental PL energies. These dimensions compare well with the experimental ones. Excitonic corrections to the band gaps are as large as ∼ 200 (100) meV for quantum wires of ∼ 10 (20) Å width [22, 23], they do not alter the previous conclusion.

Table 1. Results for calculated band gaps of Si quantum wires with respect to wire orientation and width.

Wire orientation and mean width (Å)	Energy gap (eV)	Method	Wire orientation and mean width (Å)	Energy gap (eV)	Method
[100] - 4.80	4.07	LDF	[100] - 6.72	3.40	LDF*
[100] - 6.72	3.38	[17]	[100] - 8.64	3.03	[18]
[100] - 11.4	2.64	LDF*	[111] - 5.51	3.63	LDF*
[100] - 15.6	2.23	[11]	[100] - 7.6	3.13	[19]
[100] - 7.68	3.29	LDF*	[100] - 7.68	3.35	LDF
[100] - 8.64	3.08	[20]	[100] - 11.52	2.73	[21]
[100] - 6.72	3.53	LDF*	[100] - 7.68	3.33	LDF
[100] - 10.56	2.88	[10]	[100] - 11.52	2.69	[22]
[100] - 14.40	2.50		[100] - 15.36	2.28	

The computed band gap reduces with increasing size. In a simple particle in a box argument one expects a d^{-2} dependence of the gap on the wire dimensions. Ossicini's calculations [18] show a slightly complicated dependence. The values can be fitted within the formula:

$$E_g(d) = E_g(\infty) + \frac{c_1}{d} + \frac{c_2}{d^2} \ (eV) \tag{1}$$

where c_1 and c_2 are constants. It is interesting to note that also the calculations performed for Si wires with axis along the (111) direction give similar results [19], showing that in Si quantum wires the band gap widening is governed only by size effects.
An important point under discussion is the nature of the Si confined band gap: it is really a direct one or it is pseudodirect? In order to solve this problem Yeh et al. [24] have performed a careful study of the character of the wire near gap states. The results demonstrate that the valence band maximum (VBM) originates mainly from coupling of the two highest bulk valence bands at an off-Γ k point. This large projection shows that this is not a surface states even if the wave function is localized near the surface. The conduction band minimum (CBM) comes, however, mainly from the lowest two bulk conduction bands at a different off-Γ k point. The fact that the VBM and CBM wire states project into bulk states of different k wave vectors proves that the Si quantum wire band gap is pseudodirect and not direct.
Another interesting output of the calculations are the radiative lifetimes that are directly related to the inverse of the oscillator strengths. Table 2 shows some calculated values for radiative lifetimes for the lowest energy transition. The experimental values for the slow transition in po-Si are of the order of ~ 30 μsec [25]; thus Si quantum wires, whose band gap agrees with photoluminescence energies, show computed lifetimes for the lowest energy transitions that are larger by a factor 5-10 than the experimental ones. It is interesting to note that the calculations (both empirical [26] and first-principle [20]) show that, within a given wire, several low energy transitions (only 0.1-0.2 eV above the lowest ones) exist with widely different radiative lifetimes that depend also on the symmetry of the states involved in the transition. Moreover for smaller wires (width less than 10 Å) these lifetimes can be in the order of nanoseconds. These behaviors compare well with the results of a recent experiment [27] on po-Si from oxide-free porous Si samples. This experiment shows a nice correlation between photoluminescence emission energies and photoluminescence decay times. Porous Si samples with red, green or blue emission give lifetime that follow a sort of universal curve from microseconds (red emission) to

Table 2. Calculated radiative recombination lifetimes for the lowest energy transition of Si quantum wires

Wire width (Å)	Radiative lifetime $\tau(\mu sec)$	Ref.
7.68	9.24	
15.36	69.4	[25]
23.04	533	
30.72	167	
7.68	8	[26]
8.64	20	
14.40	380	[24]
13.44	900	[19]

nanoseconds (blue emission).

Concerning the Si dots here we will summarize the discussion regarding the dependence of the energy band gap on the size and shape of the Si crystallites and the different possible channels for the recombination processes [28]. As far as for the first point the calculated

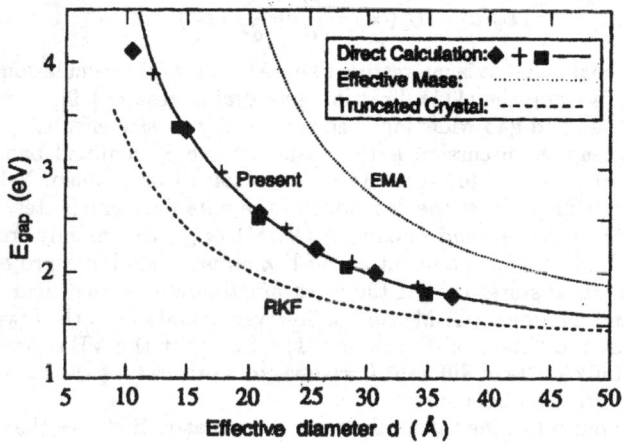

Figure 2. Band gap versus the effective diameter d (Å) for three prototype quantum dot shapes. (Diamonds); spheres, (+): rectangular boxes, (squares): cubic boxes. Also shown are EMA results [29] and model calculation results (RKF) [30]. After Ref.[31]

band gaps relative to three different shapes of Si crystallites covered by hydrogen (see fig. 2) [31] show that all the data corresponding to different shapes collapse into a single unified curve, whereas the EMA and model calculation results are far away from this curve in opposite directions. This is another strong indications that in order to give reliable predictions for the energy gap of Si crystallites one needs to use a method able to provide a good description of the Si bulk band structure. The curve of the energy gap dependence on the thickness can be fitted expressing the effective diameter of the dots in Å by the equation:

$$E_g(d) = 1.167 + \frac{88.34}{d^{1.37}} \ (eV) \tag{2}$$

Once more the exponent of 1/d dependence is an intermediated value between 1 and 2. The discussion about the possible channels for the radiative recombination starts from the observed energy difference between absorption and luminescence in PS. Such a behavior might be consistent with the existence of localized states.

The existence of such localized states that reduce the luminescence energy with respect to the absorption one has been demonstrated by Allan et al. [32] and by Baierle et al [33, 34]. In the first case from total energy calculations, the existence of self-trapped excitons at some surface bonds of silicon crystallites has been shown. These give a luminescence energy almost independent of size and originate an huge Stokes shift. The stabilization of the self-trapped excitons is obtained by dimer bonds passivations by hydrogen atoms at the surface of the Si nanocrystallites. In the second calculation it has been demonstrated that Si dots in their excited states relax to highly distorted equilibrium configurations, giving rise to new transitions involving localized states that lower the emission threshold with respect to the absorption. Both mechanisms point to the fact that the Si small structures are intrinsically subject to significant distortions when excited.

In the next chapter we will present our theoretical calculations for some special Si confined structures.

2. The Theoretical Method

The self-consistent electronic structures of the quantum confined wells, wires and dots are calculated by means of the Linear Muffin-Tin Orbital method in the Atomic- Sphere Approximation (LMTO-ASA) within the supercell technique. Exchange and correlation effects are included within the DFT-LDA. Once the self-consistent electronic properties have been calculated, the optical properties have been computed by evaluating the imaginary part of the dielectric function in the optical limit:

$$\epsilon_2^\alpha(\omega) = \frac{4\pi^2 e^2}{m^2 \omega^2} \sum_{v,c} \frac{2}{V} \sum_k |<\psi_{c,k}|p_\alpha|\psi_{v,k}>|^2 \delta[E_c(k) - E_v(k) - \hbar\omega] \tag{3}$$

where $\alpha = (x,y,z)$, E_v and E_c denote the energies of the valence $\psi_{v,k}$ and conduction $\psi_{c,k}$ band states at a k point, and V is the supercell volume[18, 35]. The ϵ_2 is calculated for photon energies up to 20 eV. To perform the summation over k we have used the tetrahedron method and a suitable number of k points. The optical absorption coefficient

$$\alpha(\omega) = \frac{\omega}{nc}\epsilon_2(\omega) \tag{4}$$

is directly related to ϵ_2, thus ϵ_2 contains the necessary information about the absorption properties of the material. Nevertheless, the ϵ_2, that contains the various inter-band transitions weighted by the momentum optical matrix element, can also indirectly give information about the photoluminescence processes through its features.

3. From undulating Si quantum wires to Si quantum dots

3.1. THE MODEL

Freshly etched porous silicon shows the structure of a crystalline silicon skeleton with a connected undulating-wire morphology while in aged porous silicon samples the presence of Si dots is predominant. In order to simulate the undulating nature of po-Si we use wires, where the rectangular cross section varies along the growth axis of the wires, instead of the usual constant cross section wire model. Since we make use of the supercell method, the wires present periodic variation of the cross-sectional size, i.e. they show the presence of narrowings and bulges along the [001] axis. This effect is obtained by superimposing elementary cells with different rectangular cross section, for example a NxM wire on a JxK one; in particular we have considered 3x3-3x4, 2x3-3x4 and 2x2-3x4 structures. The smaller section corresponds to narrowings, the larger one to bulges. To avoid Si dangling bonds, all the Si atoms at the wire and dot surfaces are passivated by H. Si atoms with no H bond are thus present in the core of wires or dots, i. e. they are in a "pure" Si crystalline environment. In addition, there are surface Si atoms with a single Si-H bond and surface Si atoms with two or three Si-H bonds. Figure 3 shows the

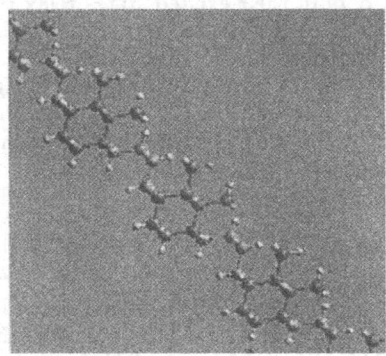

Figure 3.   Ball and stick model of the 2x2-3x4 undulating wire cell repeated along the (001) direction. White balls: H atoms, pale grey balls:Si atoms of the 2x2 cell and dark grey balls: Si atoms of the 3x4 cell.

balls and sticks model of the 2x2-3x4 undulated wire where white balls represent H, pale grey balls are the Si atoms of the 2x2 elementary cell (the narrowing) and dark grey balls are the Si atoms of the 3x4 elementary cell (the bulge). If we remove the white atoms saturating the Si dangling bonds with H we have a collection of dots that can interact each other or not reducing or enhancing the inter-grain distances.

3.2. THE ELECTRONIC AND OPTICAL PROPERTIES

In the three different undulating wires considered the bulges have even the same dimensions of 3.84×5.76 Å^2, whereas the narrowings show reducing dimensions of 3.84×3.84, 1.92×3.84 and 1.92×1.92 Å^2. In this way we can study the role of quantum confinement but also the effect of aging in the evolution from freshly etched (thicker undulating wires) to lightly oxidized (thinner undulating wires) down to heavily oxidized (dots) po-Si samples.

Figure 4 shows our calculated band structure for the three undulating wires. We note an enlargement of the gap from 3.14 eV to 3.15 eV up to 3.40 eV with reducing dimension of the restriction; these values are now intermediate between the values (3.24 eV for the 3x3 wire and 2.96 eV for the 3x4 wire) for the corresponding quantum wires with constant cross section [18]. Moreover due to the progressive reduction of the smaller cross section size, the band dispersion along the growth direction (z) is progressively reduced, in particular the band edge states become more localized and separated in energy like

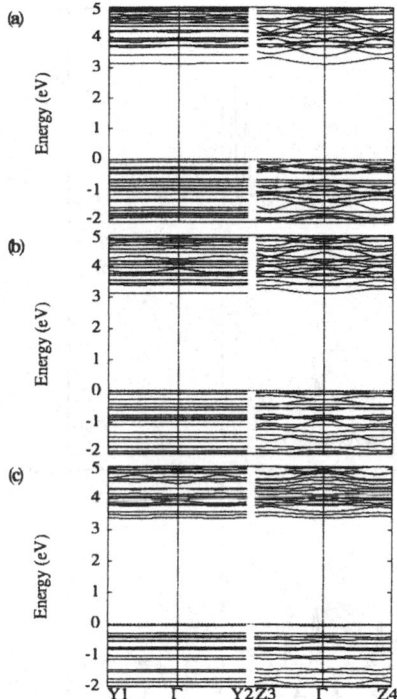

Figure 4. Calculated band structure for the (a) 3x3-3x4, (b) 2x3-3x4, and (c) 2x2-3x4 undulating quantum wires projected along two symmetry directions of the two dimensional Brillouin zone of the (001) surface. Energies (in eV) are referred to the valence band maximum.

dot states. In order to better understand the role played by the narrowings and bulges in the wire, we have considered (Fig. 5) the density of states of the 2x3-3x4 wire, projected on different sites: a silicon atom at the center of the bulges in a crystalline environment, a silicon atom bond with an H atom in the narrowing and a Si atom bonded with three H atoms at the discontinuity between bigger and smaller cross section. It is evident that the main contribution to the near band gap states of the wire originates from Si atoms located in and near the restriction, exactly the opposite situation happened for wires with constant cross section. The same occurs for the three different undulating wires considered here and for the dots too. Actually, on going from wires to dots the electronic properties show a smooth transition.

The importance of the near undulation states is also evident in the imaginary part of the calculated dielectric function where their presence produce an enhancement of the peak intensity in the low energy region. Regarding the optical properties of dots and the role of interaction, Fig. 6 shows our calculated ϵ_2 for isolated and interacting Si nanocrystallites. First of all the ϵ_2 behavior is quite similar to the experimental one determined for po-Si by Koshida et al. [36], moreover as we can see the interaction between the crystallites strongly reduces the energy gap from 3.50 eV to 1.69 eV; this is due to the presence of electronic states mostly localized at the surfaces of the dots. The tails of these states extend in the region between the dots and overlap the neighboring dots.

154

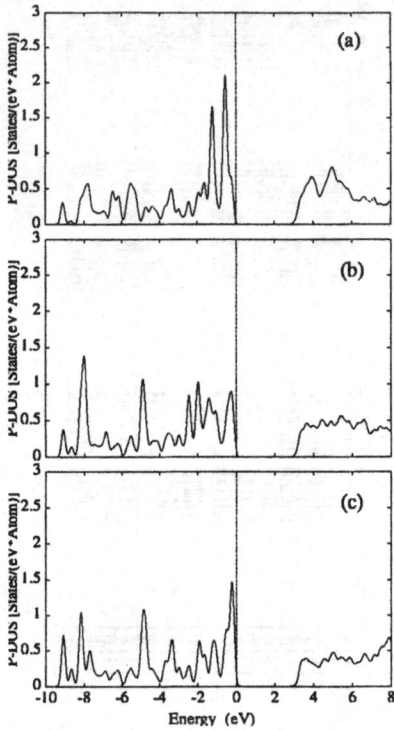

Figure 5. Density of states of the 2x3-3x4 wire projected (a) on a silicon atom at the center of the bulge, (b) on a silicon atom bonded with an H atom in the narrowing and (c) on a silicon atom bonded with three H atoms at the discontinuity between bigger and smaller cross section.

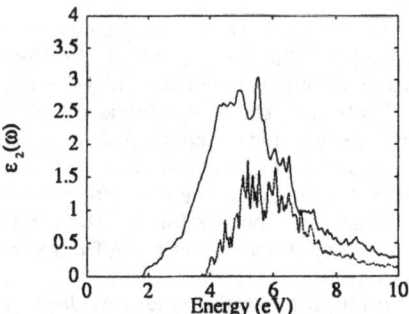

Figure 6. The imaginary part of the dielectric function, ϵ_2, for interacting (solid line) and isolated (dashed line) Si quantum dots.

They are inter-dot states and thus are less affected by the quantum confinement.

4. The quantum wells

We have considered several different Si quantum wells. Our tasks were to elucidate: i) the role of dimensions, using different thickness of the silicon embedded layers, ii) the role of symmetry, considering both (111) and (001) silicon surfaces passivated with hydrogen, iii) the role of different passivating agents like CaF_2 or SiO_2 at the Si interface and in the last case also to see iv) the role of oxygen related defects inside the structure. Table 3 shows the values of the energy gap *vs* the Si layer thickness for each studied structure.

Table 3. Calculated energy gaps of Si quantum wells with respect to the Si layer thickness.

	Size (nm)	Gap (eV)
	1.95	0.86
	1.01	1.22
Si/H(111)	0.70	1.78
	0.39	2.38
	1.28	0.86
Si/H(001)	0.73	1.14
	0.46	1.57
	2.44	0.56
	1.49	0.63
Si/CaF$_2$(111)	0.87	1.24
	0.55	0.76
Si/SiO$_2$(001)	0.543	0.82

4.1. Si/H QUANTUM WELLS

Studying the Si/H quantum wells we were interested in three aspects: the role of quantum confinement, of passivation and of symmetry. For the QC we have considered wells of different thickness, between 0.5 and 2.5 nm, choosing a variable number of Si double layers (DL's) in the well. A blue shift of the band gap due to the QC is evident in each system, as we can see from the gap values for the different structures collected in Table 3. Concerning the role of passivation we observed that the H-Si bond, being mostly covalent, gives rise to a large bonding-anti-bonding energy separation and pushes the interface states well inside the Si valence and conduction bands leaving the energy gap free of states and enhancing the gap opening. The role of symmetry has been investigated considering both (111) and (001) silicon surfaces passivated with hydrogen. The indirect gap observed in the first case [12] become direct in the second direction of growth through the folding that shifts the conduction band minima at the Γ point. Figure 7 shows this direct band gap at the Γ point, for the Si(001)-H QW with 5 layers of Si, through its band structure projected along 2 symmetry directions of the two-dimensional Brillouin zone of the (001) surface. The direct band gap of the (001) H-saturated case produces its effects also on the optical properties originating more intense transitions, in the visible range than the indirect one observed in the (111) structures. This result supports the idea that symmetry considerations can play an important role for the optical response of confined Si wells.

4.2. Si/CaF$_2$ QUANTUM WELLS

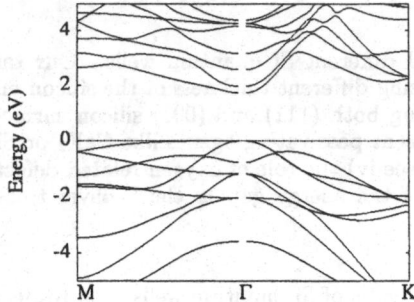

Figure 7. Calculated band structure of the Si/H(001) lattice with a Si layer thickness of 0.73 nm projected along two symmetry directions of the two dimensional Brillouin zone of the (001) surface. Energies (in eV) are referred to the valence band maximum.

In the case of Si-CaF2(111) QW's the calculated electronic properties show that the opening of the gap due to the confinement is less effective than for the H-Si systems [2, 37, 38, 39] (see Table 3). It should be noted that the Si-Ca bond at the interface is somewhat intermediate between the covalent Si-Si bond and the ionic Ca-F bond, therefore the bonding-antibonding interface states are not completely removed from the gap as in the case of the H-saturated structures. These electronic properties are reflected in the optical results: in ϵ_2 new features appear in the low energy region (between 1 eV and 2.5 eV), which are completely absent in the case of bulk Si. If we take into account the LDA underestimation of the Si energy gap, this is the optical region of interest. In the optical region there is a clear predominant role played by the band edges and in particular by the interface states whose character affects, for very thin Si slabs, the last occupied and the first unoccupied states. It is worthwhile to note that the oscillator strengths for the matrix elements between these states increase very rapidly as the thickness of the Si slabs decreases. For layers with thickness less than 2.5 nm the oscillator strengths are of the same order of magnitude as for the direct transition at Γ in bulk Si and only one order of magnitude smaller than those for GaAs. We remember that the experimentally grown Si-CaF2 QW's show photoluminescence only for Si thickness less than 2.5 nm [3,30]. Figure 8 shows the comparison between our calculated absorption spectra and the experimental data [30]. A striking resemblance between the two results is evident. In both cases we observe a blue shift of the onset and a decrease of the optical absorption with decreasing Si thickness.

4.3. Si/SiO₂ QUANTUM WELLS

We have also studied the Si-SiO₂ quantum wells, not only to elucidate the behavior of this new passivating agent but also: i) to evidence the role of oxygen and oxygen related defect in the optoelectronic properties of this structures and ii) to compare our theoretical results with some experimental data, in particular with the PL spectra of Kanemitsu and co-workers [42]. For the Si-SiO₂ confined quantum wells we use the β−crystobalite model for the SiO₂ with 2 O atoms added at the two interfaces in order to saturate the dangling bonds of the Si atoms at the center of the interfaces. To study the role of the oxygen related defects we introduce a defect in the Si/SiO₂ lattice, leaving one of the O-double bonded Si interface atoms unsaturated. To study the role of dimensionality we change the thickness of the Si layer from 0.543 nm to 1.086 nm up to 1.629 nm. For the thinner fully saturated QW we find that the system is still a

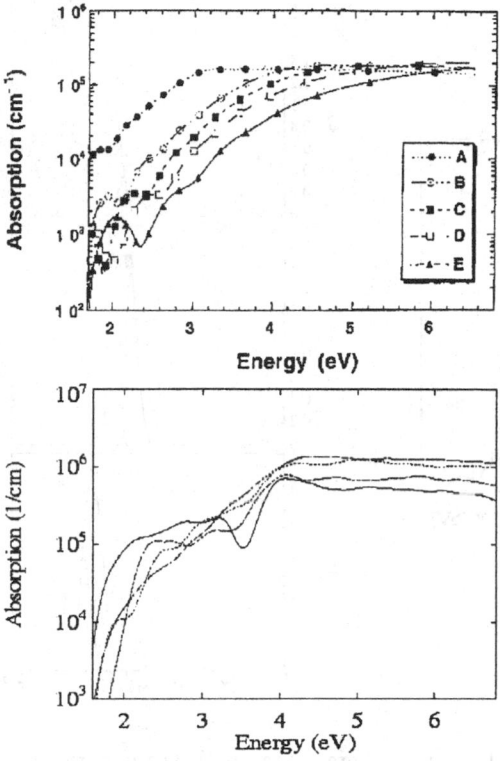

Figure 8. The absorption coefficient for the Si/CaF$_2$(111) systems. Top panel: experimental results from Ref.[40]; Si thickness (A) 5 nm, (B) 2 nm, (C) 1.6 nm, (D) 1.5 nm, (E) 1 nm. Bottom panel: theoretical results; large dashed line : 2.44 nm, small dashed line : 1.49 nm, dotted line: 0.87 nm, dash-dotted line: 0.55 nm.

semiconductor with a band gap larger than in the case of bulk silicon as a consequence of the confinement. A localized state, partially due to O-Si interaction, is found near the top of the valence band. Including the oxygen vacancies we observe the appearance of a new state in the band structure at the top of the valence band, that was absent in the fully saturated lattice. This defect state produces a closing of the gap (from 0.82 eV to 0.70 eV) that is still indirect but with the top of the valence band at a different k-point in the reciprocal space. Nevertheless, the material is still a semiconductor; this situation is quite different from that of H-covered Si quantum dots and wires, where the presence of only one dangling bond destroys the optical properties of the material in the visible region [14,25,32]. The analysis of ϵ_2 of the three partially saturated lattices with different Si thickness shows that in each case the first peak is quite asymmetric and can be fitted by two gaussian bands as Kanemitsu [42] did with his PL spectra. In fig. 9 the experimental PL spectra as a function of the Si layer thickness obtained by Kanemitsu is reported (left panel) and compared with our gaussian fit of the calculated ϵ_2 first peak (right panel) as a function of the Si layer thickness, too. Kanemitsu observed efficient PL in the visible spectral region only in very thin well sample ($\leq$2 nm) and the gaussian

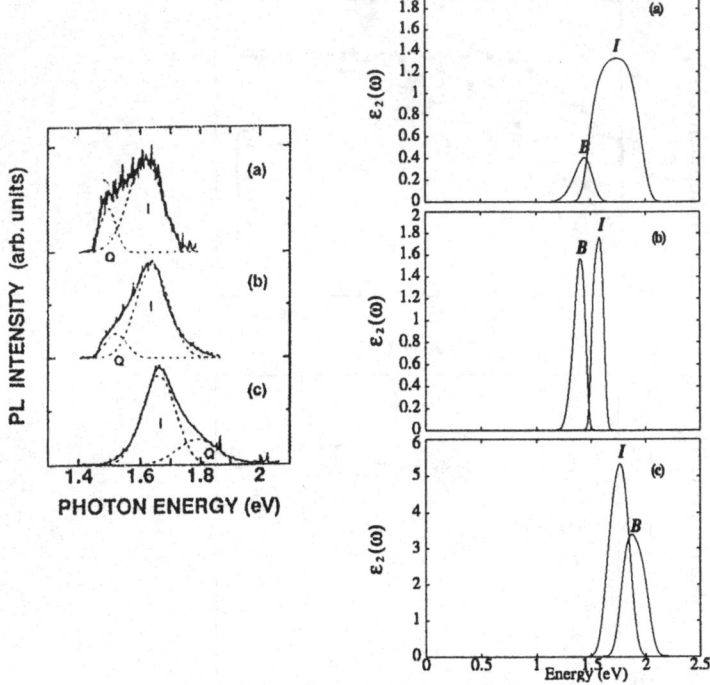

Figure 9.   Si/SiO$_2$ quantum wells: comparison with the experiment. left panel: experimental PL spectra of c-Si/SiO$_2$ single quantum wells from Ref.[42]; Si thickness (a) 1.7 nm, (b) 1.3 nm and (c) 0.6 nm. The asymmetric PL spectra can be fitted by two gaussian bands, the weak **Q** band and the strong **I** band. Right panel: theoretical results; Gaussian fit of the ϵ_2 first asymmetric peak for the (a) Si$_{[3]}$-SiO$_2$, (b) Si$_{[2]}$-SiO$_2$ and (c) Si$_{[1]}$-SiO$_2$ superlattices. The letter "I" indicates the interface gaussian band while the letter "B" indicates the bulklike gaussian band.

fit of these asymmetric spectra shows a weak band that shifts to higher energy (from ∼ 1.5 eV to ∼ 1.9 eV) with decreasing Si layer thickness (from 1.7 nm to 0.6 nm), and a strong band, at ≈ 1.65 eV, almost independent on the well thickness. The weak band (Q) is attributed to quantum confinement effects and the strong one (I) to radiative recombination in the interface region. In the same way we observe in our fitting two bands: a bulklike band (B) that shifts to higher energies (from ∼ 1.4 eV to ∼ 1.9 eV) when the thickness of the Si layer decreases, and an interface band (I), positioned at around 1.7 eV, almost unaffected by the dimensionality of the Si slab. While B is due to the radiative recombination in the 2D well (QC effect), I is due to the radiative recombination in the interface region through the defect states. The comparison between our outcomes (Fig. 9 right panel) and Kanemitsu [42] results (Fig. 9 left panel) confirms the double nature of the PL in these materials: quantum confinement and oxygen related defects.

Conclusions

We have presented theoretical results for the electronic and optical properties of 2D, 1D and 0D Si nanostructures. The obtained results have been compared with the available experimental data on similar systems. A final general comment is due about the reliability of ab initio calculations for a quantitative description of the optical properties of real systems. The recent major advances in the calculation of ground state properties of solids are essentially related to the development of the density functional theory (DFT) in the local density approximation (LDA). For the spectroscopic properties, however, LDA suffers of the drawback of the underestimation of the energy gap. Nevertheless in the last years the calculation of self-energy effects has given the possibility to overcome this problem. Moreover the inclusion of the electron-hole interactions and local field effects produces results which are in quantitative agreement with the experiments for the absorption spectra [43]. Due to the complexity of advanced calculations, it is not possible to take into account all the complicated processes that affect the PL spectra, like relaxation, impurity scattering, etc.. Nevertheless, the calculations have not only strongly helped in the interpretation of the experimental results, but also have suggested and can suggest new Si-based materials with the necessary optical properties.

*References

[1] Canham L. T. (1990), Appl. Phys. Lett. **57**, 1046 ;
[2] Ossicini S., Fasolino A., and Bernardini F. (1994) Phys. Rev. Lett. **72**, 1044;
[3] D'Avitaya F. Arnaud, Vervoort L., Ossicini S., Fasolino A., Bernardini F. (1995) Europhys. Lett. **31**, 25;
[4] Lockwood D. J., Lu Z. H., Baribeau J.-M. (1996) Phys. Rev. Lett. **76**, 539;
[5] Nassiopoulos A. G., Grigoropoulos S., Papadimitriu D., and Gogolides E. (1995) phys. stat. sol. (b) **190**, 91;
[6] Kanemitsu Y.,Uto H. , Masumoto Y., Matsumoto T., Futano T., Mimura H (1993) Phys. Rev. B **48**, 2827;
[7] Bisi O., Ossicini S., Pavesi L. (2000), Surf. Sci Rep. 38, 1-126;
[8] Delley B. and Steigmeier E. F. (1995) Appl. Phys. Lett. **67**, 2370;
[9] Hirao M.,Uda T., Murayama Y. (1993), Mater. Res. Soc.Symp. Proc. 283, 425;
[10] Read A. J., Needs R. J., Nash K. J., Canham L. T., Calcott P. D. J. and Qteish A. (1992) Phys. Rev. Lett. **69**, 1232;
[11] Buda F., Kohanoff J. and Parinello M. (1992), Phis. Rev. Lett. **69**, 1272;
[12] Wang L. -W., Zunger A. (1994), J. Phys. Chem. 98, 2158;
[13] Lannoo M., Allen G., Delerue C., in: Amato G., Delerue C., von Bardeleden H. -J. (Eds) (1997), *Structural and Optical Properties of Porous Silicon Nanosturctures*, Gordon and Breach, Amsterdam, p. 187;
[14] Degoli E., Ossicini S., Barbato D., Luppi M., Pettenati E. (2000), Mater. Sci. and Engin. B, 444;
[15] Ossicini S., Bisi O. in Amato G., Delerue C., von Bardeleden H. -J. (Eds) (1997), *Structural and Optical Properties of Porous Silicon Nanosturctures*, Gordon and Breach, Amsterdam, p. 191;
[16] Ossicini S. (1997) in *Properties of Porous Silicon*, ed. by L. T. Canham (IEE INSPEC, The Institution of Electrical Engineers, London) p.207;
[17] Dorigoni L., Bisi O., Bernardini F. and Ossicini S. (1993) Phys. Rev. B **53**, 4557;
[18] Ossicini S., Bertoni C. M., Biagini M., Lugli A., Roma G., and Bisi O. (1997) Thin Solid Films **297**, 154;
[19] Saitta A. M., Buda F., Fiumara G. and Giaquinta P. V. (1993) Phys. Rev. B **53**, 1446;
[20] Hybertsen M. S. and Needels M. (1993) Phys Rev. B **48**, 4608;

160

[21] Lee S. -G., Cheong B. -H., Lo K. -H. and Chang K. J. (1995), Phys. Rev. B **51**, 1762;

[22] Ohno T., Shiraishi K. and Ogawa T. (1992) Phys. Rev. Lett. **69**, 2400;

[23] Yeh C.-Y. , Zhang S. B. and Zunger A. (1994) Phys. Rev. B **50**, 14405;

[24] Yeh C.-Y., Zhang S. B. and Zunger A. (1993) Appl. Phys. Lett. **63**, 3455;

[25] Calcott P. D. J., Nash K. J., Canham L. T., Kane M. J. and Brumhead D. J. (1993) Phys. Condens. Matter **5**, L91;

[26] Sanders G. D. and Chang Y. C. (1992) Phys Rev. B **45**, 9202;

[27] Mizuno H., Koyama H. and Koshida N. (1996) Appl. Phys. Lett. **69**,3779;

[28] For a recent review on Si dots calculations see Delerue C. (1997) in *Properties of Porous Silicon*, ed. by L. T. Canham (IEE INSPEC, The Institution of Electrical Engineers, London) p.212;

[29] Takagahara T., Takeda K. (1992), Phys. Rev B **46**, 15578;

[30] Rama Krishna M. V., Friesner R. A. (1991), Phys. Rev. Lett. **67**, 629;

[31] Zunger A., Zhang S. B. (1996), Appl. Surf. Sci. 102, 350;

[32] Allan G., Delerue C., Lannoo M. (1997) Phys. Rev. Lett. **78**, 3161;

[33] Baierle R. J., Caldas M. J., Molinari E., Ossicini S. (1997) Solid State Comm. **102**, 545;

[34] Caldas M. J., Baierle R. J., Molinari E., Ossicini S. (1997) Mater. Sci. Forum **258,263**, 11;

[35] Ossicini S. (1998) phys. stat. solidi (a) **170**, 377;

[36] Koshida N., Koyama H., Suda Y., Yamamoto Y., Araki M., Saito T., Sato K., Sata N., and Shin S. (1993) Appl. Phys. Lett. **63**, 2774;

[37] Degoli E. and Ossicini S. (1998) Phys. Rev. B 57, 14776;

[38] Degoli E. and Ossicini S. (1999) J. of Luminescence 80/1-4, 411-415;

[39] Ossicini S., Fasolino A. and Bernardini F. (1995) Phys. Stat. Sol. (b) 190, 117;

[40] Bassani F., Mihalcescu I., Vial J. C., D'Avitaya, F. Arnaud (1997) Appl. Surf. Sci. 117/118, 670;

[41] Delerue C., Allen G. and Lannoo M. (1993) Phys. Rev. B 48, 11024;

[42] Kanemitsu Y., Okamoto S. (1997) Phys. Rev. B **56**, R15561;

[43] for a recent review see Del Sole R. (1998) phys. stat. sol. (a) **170**, 183.

SILICON NANOSTRUCTURES AND THEIR INTERACTIONS WITH ERBIUM IONS

F. PRIOLO and G. FRANZÒ
INFM, Physics Department, University of Catania
Corso Italia 57, 95129 Catania, Italy

F. IACONA
CNR-IMETEM
Stradale Primosole 50, 95121 Catania, Italy

1. Introduction

Silicon, the leading semiconductor in microelectronics industry, has for a long time been considered unsuitable for optoelectronic applications which remained the domain of III-V semiconductors and glass fibers. This is mainly due to the silicon indirect bandgap, which makes it a poor emitter, and to the absence of linear electro-optic effects. The enormous progress in communication technologies in the last years resulted in an increased demand for optoelectronic functions integrated with electronic circuits. This would allow to couple the information processing capabilities of microelectronics with the efficient interconnection properties of optoelectronics. In principle, silicon would be the material of choice, due to its mature processing technology and to its unrivaled domain in microelectronics, the main limiting step being the absence of efficient Si-based light sources.

Recently a strong effort has been hence devoted to study all of those processes able to circumvent the physical inability of silicon to emit light [1]. Since the discovery of light emission from porous silicon made in 1990 by Canham [2] a lot of work has been performed in studying silicon nanostructures. These comprehend not only porous silicon but also nanocrystals produced by several techniques [3-10], as well as silicon-insulator multilayers [11-12]. The initial problems related to the instability of the luminescence yield have finally been solved and today reliable, stable structures, compatible with the silicon technology have been fabricated. In particular, the group at the Rochester University has now produced silicon-rich silicon oxide electroluminescent devices integrated with silicon microelectronic circuitry [13]. Alternative approaches comprehend the doping of silicon with rare earths [14-22]. In this case the luminescence is due to an internal 4f shell transition of the rare earth ion excited through electron-hole recombinations within the silicon matrix. Among rare earths Er ions have been most widely studied since they emit light at 1.54 μm, a wavelength which is strategic in the telecommunication technology matching the window of maximum transmission for the optical fibers. The initial problems related to erbium incorporation and luminescence quenching have been now understood. In particular, it has been shown that Er excitation

161

L. Pavesi and E. Buzaneva (eds.), Frontiers of Nano-Optoelectronic Systems, 161–176.
© 2000 *Kluwer Academic Publishers. Printed in the Netherlands.*

occurs very efficiently in Si (with a cross section of 3×10^{-15} cm^2, to be compared with the cross section for direct photon absorption of 8×10^{-21} cm^2) demonstrating that, in principle, Er luminescence in Si can be very efficient [17]. The main limiting step has been recognized in the non-radiative decay channels, back-transfer (with the energy transferred back from excited Er to electron-hole couples) and Auger (with the energy released to free carriers) [16-17]. Nevertheless Er:Si devices operating at room temperature, with efficiencies of 0.1% and modulation speed of 10 MHz, have been fabricated [18-22].

A particularly interesting field of research concerns the coupling of Er and Si nanostructures. Indeed, erbium doping of Si nanocrystals (nc) has been recently recognized as a quite efficient method of obtaining 1.54 μm luminescence [23-29]. Indeed, the excited nc preferentially transfer their energy to the Er ions which subsequently de-excite radiatively. Several points of this process are of extreme interest. In particular, the transfer of energy from the nanostructure to Er is much more efficient than direct photon absorption and luminescence intensities 2 orders of magnitude higher than for Er in SiO$_2$ are observed [23-29]. Moreover, the non-radiative de-excitation processes typically limiting Er luminescence in Si, namely Auger with free carriers and energy back-transfer [17], are strongly reduced in this case further improving the luminescence yield [27]. In the present work we will review our recent work on Si nanostructures [32-34] and their interaction with Er ions [27-28,30-31].

2. Experimental

Si nc were produced by high temperature annealing of 0.2 μm thick substoichiometric SiO$_x$ thin films grown by plasma enhanced chemical vapor deposition (PECVD) on Si substrates. The deposition system consists of an ultra high vacuum chamber (base pressure 1×10^{-9} Torr) and a RF generator (13.56 MHz), connected through a matching network to the top electrode of the reactor. The source gases used were high purity SiH$_4$ and N$_2$O. The thermal annealing that follows the deposition induces the separation of the Si and SiO$_2$ phases and Si nc were formed as evidenced by plan view transmission electron microscopy (TEM).

Si/SiO$_2$ superlattices (SLs), consisting of 11 SiO$_2$ layers alternated with 10 Si layers, were also prepared by PECVD. The thickness of the SiO$_2$ layers was maintained constant at a value of 8.5 nm in all the deposited samples, while the thickness of the Si layers (D$_{Si}$) was varied between 0.9 and 2.6 nm. After deposition, the SLs were annealed for 1 hour at temperatures ranging between 1000 and 1250 °C in ultra-pure nitrogen atmosphere using a horizontal furnace.

Erbium ions were then implanted at doses in the range $2 \times 10^{12} - 3 \times 10^{15}$/cm^2. The energy was chosen in order to place the Er profile within the deposited layer. The same Er implants were also performed in SiO$_2$ layers not containing Si nc in order to have reference samples. All of these samples were finally annealed at 900 °C for 1 hour in order to remove the residual damage left over by the implantation process and to activate Er. Er-doped single crystalline Si samples were also prepared for comparison according to literature data [18].

Photoluminescence (PL) measurements were carried out by pumping with the 488 nm line of an Ar⁺ laser. The laser beam was mechanically chopped at a frequency of 55 Hz and the pump power was varied between 0.1 and 1500 mW over a circular area of about 1 mm in diameter. The luminescence signal was dispersed by a single grating monochromator, revealed by a photomultiplier tube or a liquid nitrogen cooled Ge detector and measured by a lock-in amplifier using the chopper frequency as a reference. Luminescence lifetime measurements were performed by monitoring with a digitizing oscilloscope the decay of the PL signal after pumping to steady state and mechanically switching off the laser beam by an acousto-optic system. The overall time resolution of our system is of 30 ns. All of the measurements, where not differently stated, were performed at room temperature. Low temperature measurements were performed by using a closed cycle liquid He cooler system with the samples kept in vacuum.

3. Structural characterization

Fig. 1 reports as an example the plan view TEM micrographs relative to a substoichiometric SiO$_x$ sample with 44 at.% Si after thermal annealing at 1100, 1200 and 1250 °C.

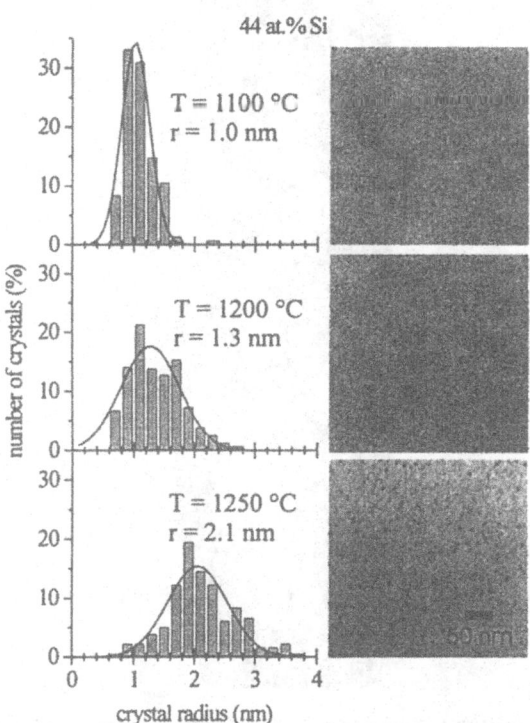

Figure 1. Plan view TEM micrographs and relative Si nanocrystals size distribution for a SiO$_x$ film having a Si concentration of 44 at.%, after annealing at (a) 1100 °C, (b) 1200 °C, and (c) 1250 °C.

164

All micrographs show that the amorphous SiO_x matrix contains a high density of small clusters, that, on the basis of the electron diffraction analysis, can be identified as silicon nanocrystals. Their formation is due to the thermal annealing, because they are completely absent in the as deposited sample. The statistical analysis of the crystal size distribution (reported in Fig. 1 as a histogram) obtained from the TEM micrographs indicates that the thermal process at 1100 °C (fig. 1a) induces the formation of crystals having a mean radius of 1.0 nm, as deduced by fitting with a Gaussian curve the size distribution. The distribution has a standard deviation (σ) of 0.2 nm, accounting for all the different crystal sizes detected in the micrograph. The Si nanocrystals are homogeneously distributed along all the film depth, as demonstrated by cross sectional TEM analysis (not shown).

From the analysis of Fig. 1 it is evident that, for the same silicon composition, the mean radius of the silicon crystals strongly depends on the annealing temperature; indeed, from the analysis of the micrographs relative to samples annealed at higher temperatures, it can be clearly noted an increase of the mean crystal size up to 1.3 nm at

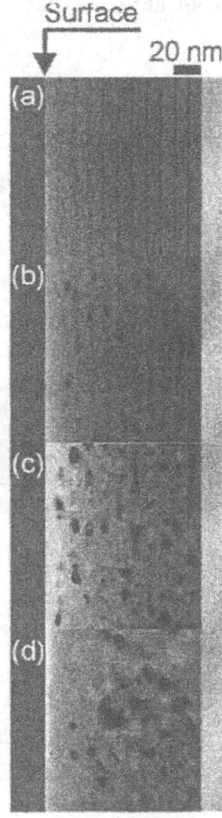

Figure 2. Cross sectional TEM images of ten-period Si/SiO$_2$ superlattices (SLs) with a silicon layer thickness of 2.6 nm and a SiO$_2$ layer thickness of 8.5 nm: (a) as deposited; (b) after annealing at 1100 °C for 1 h; (c) after annealing at 1200 °C for 1 h and (d) after annealing at 1250 °C for 1 h.

1200 °C (fig. 1b) and 2.1 nm at 1250 °C (fig. 1c). This phenomenon implies also an increase of the width of the distribution; indeed, the value of σ becomes 0.4 nm at 1200 °C and 0.5 nm at 1250 °C.

Also the superlattices were analized in details. Fig. 2a shows the cross sectional TEM image of the as deposited Si/SiO_2 superlattices (SLs), consisting of 11 SiO_2 layers ~ 8.5 nm thick, and 10 Si layers ~ 2.6 nm thick. Transmission electron diffraction (TED) measurements (not shown) confirmed the amorphous nature of the as deposited Si layers. The evolution of the SLs structure with the thermal treatment is shown in Figs. 2b-2d, where the cross sectional TEM micrographs of the SLs annealed at 1100 (b), 1200 (c) and 1250 °C (c) for 1 h are reported. In particular, the TEM images show that Si nanograins, whose crystalline nature was confirmed by TED measurements, have nucleated inside the Si layers already after a thermal process at 1100 °C (see fig. 2b). It is noteworthy that, in spite of the Si nanocrystals (Si-nc) formation, the layered structure still remains clearly visible, even if the thickness of the layers is changed. In particular, the Si layers thickness is increased up to ~ 5 nm, while the SiO_2 layers thickness is reduced to ~ 5.5 nm. This effect is probably due to the fact that the crystallization of the Si layers implies clustering with the formation of almost spherical grains, having a diameter larger than the thickness of the starting Si layers. Nanograins within a single layer are separated one another, the space between adjacent nc being filled by SiO_2, thus producing the observed decrease in the SiO_2 thickness. The Si-nc size increases by increasing the annealing temperature up to 1200 °C (see fig. 2c), however an ordered structure remains visible. As a consequence, the thickness of the Si layers further increases to ~6.5 nm (roughly following the increase of the Si-nc size) while the thickness of the SiO_2 layers is further reduced to 4.5 nm. Only after the 1250 °C annealing process (see fig. 2d), the initial ordered structure is almost completely lost due to the formation of very large Si nanograins. Therefore, by thermal annealing up to 1200 °C of amorphous Si/SiO_2 superlattices, we have been able to obtain Si nanocrystals arranged in ordered planar arrays, separated by about 5 nm of SiO_2 along the depth axis. The size distribution of the Si nanocrystals dispersed on the planes of the 10 layers can be evaluated by using plan view TEM analysis. Fig. 3a reports the dark field plan view TEM image of a SL having $D_{Si} = 2.6$ nm, after annealing at 1200 °C (see fig. 2c for the corresponding TEM cross sectional micrograph). A dense population of Si-nc embedded in an amorphous SiO_2 matrix can be seen; from a statistical analysis of the micrograph we estimate that the nc mean radius, obtained by fitting the histogram of the size distribution with a Gaussian curve, is about 2.3 nm (see left hand side in Fig. 3a). It is interesting to note that the size distribution results quite wide (FWHM of about 2.3 nm), accounting for a not negligible population of nc having mean radii remarkably larger (up to ~5 nm) and smaller than the mean value.

A similar behavior is observed also in SLs with different Si layer thicknesses. In particular the crystallization process leads to the formation of Si nanocrystals having a mean radius of ~1.7 nm after annealing at 1200 °C a Si/SiO_2 SL with $D_{Si} = 1.4$ nm (see fig. 3b), while Si nanocrystals having a mean radius of ~1.1 nm are formed by annealing at 1200 °C a Si/SiO_2 SLs with $D_{Si} = 0.9$ nm (see fig. 3c). Note also that the width of the size distribution becomes remarkably narrower as the Si layer thickness decreases.

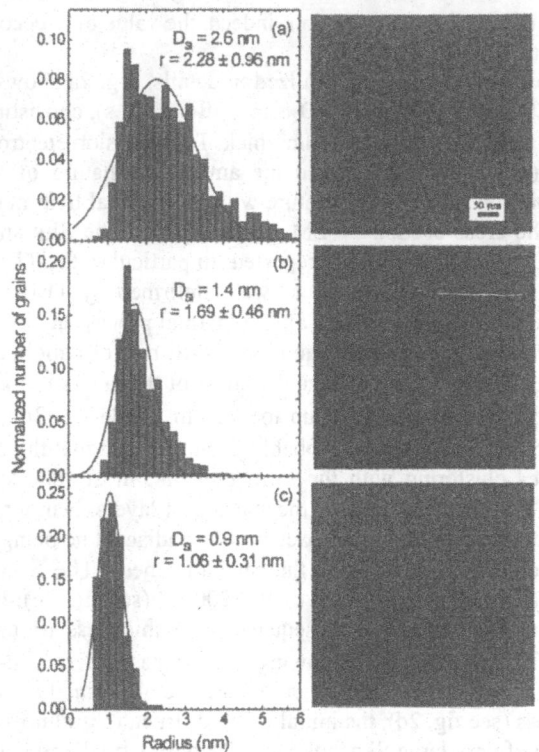

Figure 3. TEM plan view micrographs (right hand side) and relative nanocrystal size distributions (left hand side) for Si/SiO₂ SLs annealed at 1200 °C for 1 h, with (a) silicon layer thickness (D_{Si}) of 2.6 nm, (b) D_{Si} = 1.4 nm and (c) D_{Si} = 0.9 nm.

The TEM plan view analysis allowed us also to determine the effect of the temperature on the process of Si nc formation; indeed, for instance, for the sample with D_{Si} = 0.9 nm, the Si-nc mean size increases from the value reported in fig. 3c for a 1200 °C annealing (1.1 nm) to 1.7 nm for a 1250 °C annealing. On the other hand, TEM analysis is not able to detect any Si nc when the same sample is annealed at 1100 °C. This is probably due to the fact that only Si nanocrystals whose size is below the TEM resolution are present in this sample.

By summarizing, the data reported in this section show that nanocrystalline Si can be produced by plasma enhanced chemical vapor deposition. The size distribution can be properly tuned by modifying the excess Si concentration and/or the annealing temperature. In addition, nanocrystalline Si SLs with a tunable size distribution can be formed by sequential Si and SiO₂ deposition followed by annealing. The luminescence properties of these structures, prior and after Er doping, are reported in the following section.

4. Luminescence Properties

4.1 SILICON NANOCRYSTALS

All of the samples reported in the previous section present a strong room temperature luminescence with a wavelength which can be properly tuned by changing the dimensions. As an example in Fig. 4 the PL emission for a superlattice with a thickness of the Si layer of 0.9 nm is reported. From the structural changes reported above, i.e. the increase of the mean crystal radius with increasing annealing temperature, one expects that the wavelength of the PL signal should also vary by varying the annealing temperature for a fixed SL. Fig. 4 illustrates this effect for the SL with D_{Si}= 0.9 nm excited at a pump power of 50 mW. Indeed, the peak wavelength increases from 760 to 860 nm by increasing the annealing temperature from 1100 to 1250 °C. The maximum PL intensity can be obtained for an annealing temperature of 1200 °C, suggesting that the corresponding mean radius of 1.1 nm (with no appreciable nc population having sizes larger than 2 nm) should be the best one in order to maximize light emission. It is interesting to note that an intense light emission has been obtained also from the sample annealed at 1100 °C (only a factor of 2 less intense than the maximum observed value). This confirms our idea that the absence of any detectable Si nc in the TEM image of this sample can be due just to the small size of the nanograins.

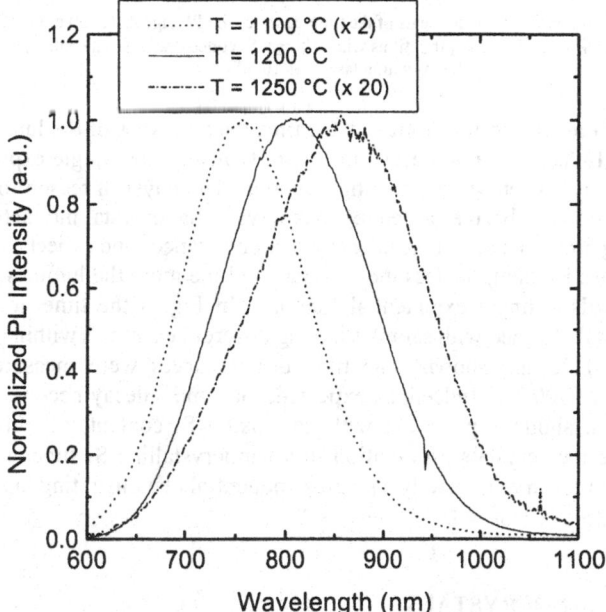

Figure 4. Normalized PL spectra of Si/SiO₂ SLs with D_{Si} = 0.9 nm after annealing at 1100, 1200 and 1250 °C for 1 h. Spectra were measured at room temperature, with a laser pump power of 50 mW.

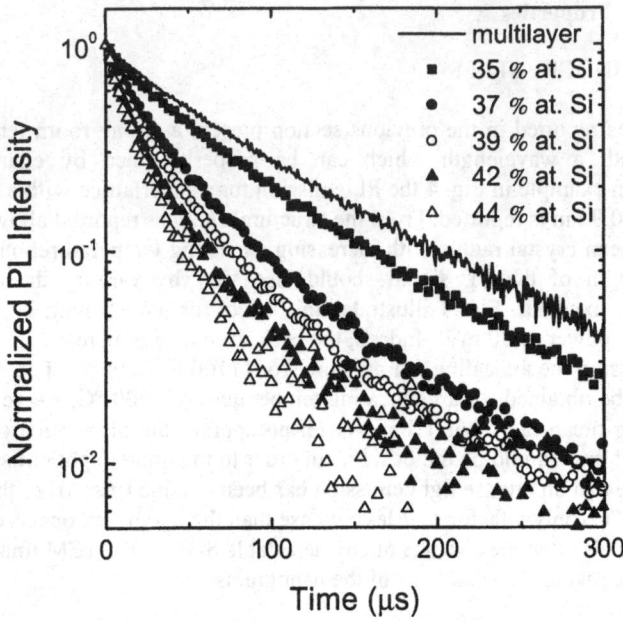

Figure 5. Room temperature measurements of the time-decay of the PL signals at 700 nm for Si/SiO₂ SLs with D_{Si} = 0.9 nm and for SiO$_x$ thin films with different Si concentrations. The pump power of the excitation laser was 10 mW.

A particularly interesting feature concerns the time decay curves of the luminescence in Si nanocrystals. Indeed the time decay curves are typically non-single exponential and can be described by a stretched exponential function. The physical reason for such non-exponential function has been ascribed to a nanocrystal-nanocrystal interaction with the energy travelling along the sample. Under these circumstances one expects that the more the nanocrystal are isolated, the less they interact and the more the luminescence decay time tends towards a single exponential function. In Fig. 5 the time decay data for several nanocrystals formed with samples having different excess Si within the SiO$_x$ and annealed at 1250 °C are shown. The time decay curves were measured at room temperature and at 700 nm. Indeed, as expected, the time - decay becomes longer and more similar to a single exponential with decreasing Si content, i.e. with a greater separation of the nanocrystals. In addition in a nanocrystalline Si superlattice sample (continuous line) the curve is exactly a single exponential demonstrating that in this case nanocrystals are totally isolated.

4.2 Er DOPED NANOCRYSTALS

Figure 6 shows the room temperature PL spectra obtained when pumping with a 200 mW laser beam the SiO₂ sample implanted with 3×10¹⁵ Er cm⁻² (continuous line) and the sample containing also Si nanocrystals (dash-dotted line). It should be noted that the

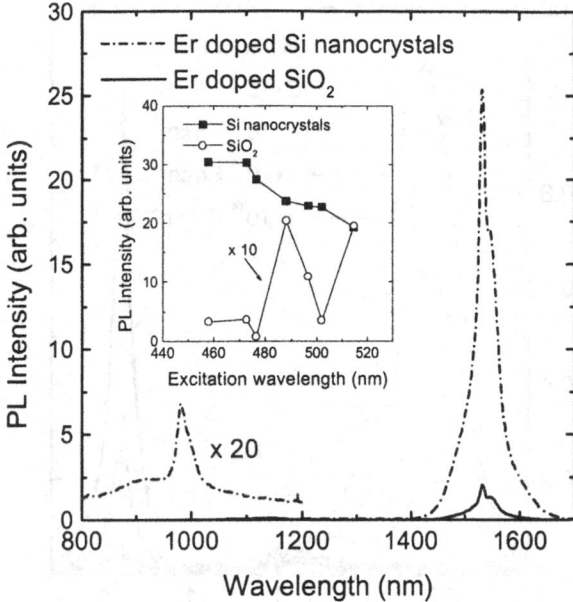

Figure 6. Room temperature photoluminescence spectra for SiO_2 (continuous line) and Si nanocrystals (dash-dotted line) samples implanted with 300 keV Er to a dose of 3×10^{15} cm^{-2}. Spectra were taken with a laser pump power of 200 mW. In the insert the photoluminescence signal at 1.54 μm for the two samples is reported as function of the excitation wavelength. Data are shown for both Er doped Si nanograins (■) and Er doped SiO_2 (o). The data for Er doped SiO_2 are multiplied by a factor of 10.

PL spectrum for the Er doped Si nanocrystals shows the typical features of the Er luminescence. Moreover, a strong enhancement of the 1.54 μm signal with respect to the Er doped SiO_2 is observed. In fact the PL signal is more than a factor of 10 higher with respect to the SiO_2 sample non containing Si nanocrystals. In addition, another PL peak at 0.98 μm, due to the radiative de-excitation of Er from the $^4I_{11/2}$ to the $^4I_{15/2}$ level, is observed in the sample containing Si nanocrystals while it is absent in the Er doped SiO_2 (at this Er concentration). All of these observations indicate that the presence of Si nanocrystals produces a strong enhancement of the Er luminescence.

In order to identify which is the mechanism responsible for the increase in the luminescence yield, we have measured the PL peak intensity at 1.54 μm at different pump wavelengths ranging between 457.9 and 514.5 nm, at a constant pump power of 200 mW. This wavelength scan was performed on Er implanted Si nanocrystals (•) as well as on the Er implanted SiO_2 film (o) and the results are reported in the insert of figure 6. The signal of the Er doped SiO_2 has been multiplied by a factor of 10 in order to do the comparison with the data of Er doped Si nanocrystals. For the SiO_2 sample two peaks at 488 nm and 514.5 nm are observed corresponding to the direct photon absorption from the ground state of Er ($^4I_{15/2}$) to the $^4F_{7/2}$ and $^2H_{11/2}$ levels, respectively. In contrast, for the Er doped Si nanocrystals no resonance is observed and the PL signal monotonically and slightly decreases with increasing the excitation wavelength (due to the different absorption coefficient at the different wavelengths). These data demonstrate

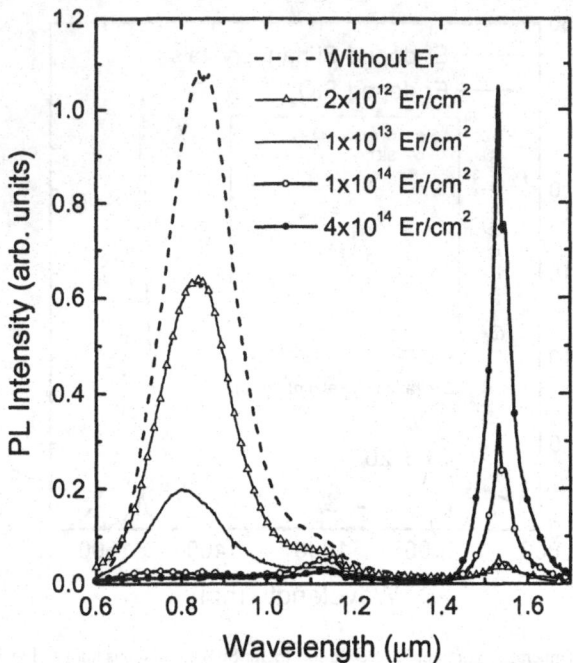

Figure 7. Room temperature PL spectra of Er - implanted Si nanocrystals at different Er doses. The pump power of the laser beam was 50 mW.

that the strong luminescence observed in the sample containing Si nanocrystals comes from Er ions that are pumped through an electron-hole mediated process in the Si nanocrystals.

If Er ions are excited through exciton recombination a competition between rare-earth ad nanocrystal intrinsic luminescence should be present. Figure 7 shows the room temperature PL spectra of Er-implanted Si nanocrystals at different Er doses and at a pump power of 50 mW. In absence of Er a well known signal at around 0.85 μm is observed coming from the Si nanocrystals (dashed line) as a result of confined exciton recombination, in agreement with literature data. As soon as Er is introduced, also at doses as small as $2 \times 10^{12}/cm^2$, the signal from Si nanocrystals at 0.85 μm is seen to decrease, while a new peak at around 1.54 μm, coming from the $^4I_{13/2} \rightarrow {}^4I_{15/2}$ intra 4f-shell Er transition, appears. With increasing Er dose this phenomenon becomes particularly evident with a quenching of the visible nanocrystal luminescence corresponding to a simultaneous enhancement in the Er-related luminescence. This effect is a demonstration of an energy transfer from the excitons confined in the nanocrystals to the erbium ions.

The observed behavior suggests that the two processes are in competition: in absence of Er the exciton recombines radiatively, in presence of Er the energy transfer to the Er 4f-shell takes over. If this picture is correct one might expect that, as the Er concentration increases, not only the nanocrystal PL signal decreases, but also its decay time. In fact, if the energy is preferentially transferred to Er, the energy transfer process

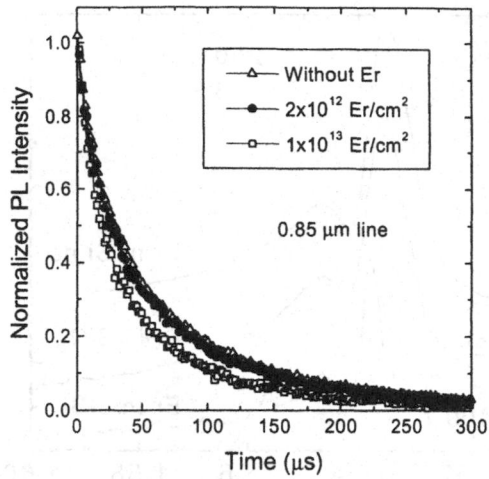

Figure 8. Time decay curves of the nanocrystal – related luminescence at 0.85 μm in absence of Er and for different Er contents. Data were taken by pumping to steady state at 50 mW and mechanically switching off the laser beam at t=0.

should be faster than the radiative recombination of the exciton. The exciton lifetime then decreases. The decay time of the signal at 0.85 μm should reflect the exciton lifetime and one can expect its decrease with increasing Er content. In this way, in principle, one should be also able to measure the average time of energy transfer to the rare earth. We have therefore performed lifetime measurements of the 0.85 μm nanocrystal signal in absence of Er and with increasing Er concentration. The results are reported in Fig. 8. In absence of Er the lifetime is around 130 μsec. Surprisingly, with increasing Er content, in spite of the decrease of the luminescence at 0.85 μm (see Fig. 7), no change in lifetime is observed (see Fig. 8). This result demonstrates that the energy transfer has characteristics very different from what expected.

A further important issue concerns the site of the Er ion. Fig. 9 shows the PL spectra measured at 16 K by pumping with a 250 mW laser beam three different samples: Er+O implanted crystalline Si (c-Si), Er implanted SiO_2 and Er implanted Si nc. The Er concentration is $1 \times 10^{20}/cm^3$ in all cases. In Er doped crystalline Si several sharp peaks, all Er related and due to the different transitions between the $^4I_{13/2}$ and $^4I_{15/2}$ manifolds, are present with the main emission occurring at ~1538 nm. In contrast the high resolution spectrum of Er doped SiO_2 consists of only two lines at 1535 and 1546 nm. These peaks are much wider than those observed in Er doped crystalline Si. In particular the width of the main peak is 8.7 nm to be compared with that of the main transition in c-Si that is 3.8 nm. The high resolution PL spectrum of Er doped Si nc is identical to that of Er implanted SiO_2 indicating that the environment of the emitting Er ions is the same in the two samples. Since photoluminescence excitation (PLE) measurements performed on Er doped Si nc demonstrated that Er is excited through an electron-hole mediated process (see insert in Fig. 6) (as it occurs in Er doped crystalline Si), these data, together with the PLE measurements, strongly suggest that the emitting Er ions in Si nc are pumped by the electron-hole pairs generated within the nc but are located in SiO_2 or at the Si-nc/SiO_2 interface.

172

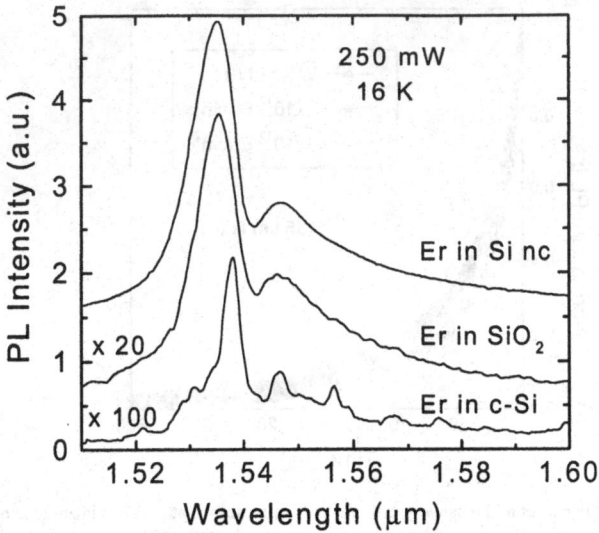

Figure 9: High resolution (~ 1 nm) photoluminescence spectra measured at 16 K by pumping with a laser pump power of 250 mW three different samples: Er + O implanted crystalline silicon; Er implanted SiO₂ and Er implanted Si nanocrystals.

More information on this system can be obtained by studying the decay time of the Er luminescence. Fig. 10 shows the luminescence time decay at 1.54 μm of Er at 17 K and at 300 K. It is interesting to note that this decay time is almost temperature independent. Indeed, the small temperature dependence of the lifetime demonstrates that thermally activated non-radiative processes are weak within the nanocrystals. Temperature-activated energy back-transfer is not observed as a result of the enlargement of the gap of Si nanograins, and also Auger processes are weak due to the absence of free carriers.

Figure 11 reports the PL signal at 1.54 μm measured with a laser pump power of 10 mW as a function of the reciprocal temperature for Er doped SiO₂ (Δ) and for the Er doped Si nanocrystals (•). The PL signal decreases by only a factor of ~3 in both nanocrystals and oxide samples by increasing the temperature from 17 K to RT. This is very different from the behaviour of Er-doped Si where the release of free carriers from donors levels and the excitation of phonons produce non-radiative Auger and back-transfer processes as the temperature is increased with a severe reduction in the luminescence intensity. Furthermore, by decreasing the pump power to 10 mW the luminescence signal in the Er doped Si nanocrystals is 2 orders of magnitude more intense with respect to Er doped SiO₂ and this difference is maintained at all temperatures.

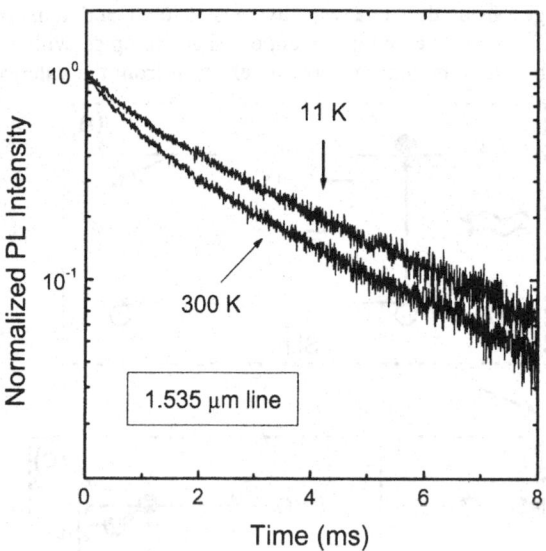

Figure 10. Time decay of the PL intensity in Er-doped Si nanocrystals. Data are shown for both room temperature and 17 K.

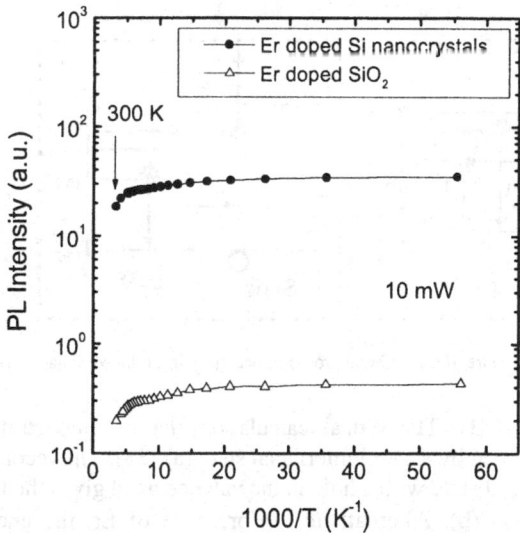

Figure 11. Temperature dependence of the photoluminescence signal at 1.54 μm for Er doped SiO₂ (Δ) and Er doped Si nanocrystals (•). Data were taken at a pump power of 10 mW.

174

A clear picture emerges from the data we have just shown and it is schematically summarized in Fig. 12. When we pump Er doped Si nc samples with a laser beam, photons are absorbed by the Si nc and promote an electron from the valence band (VB)

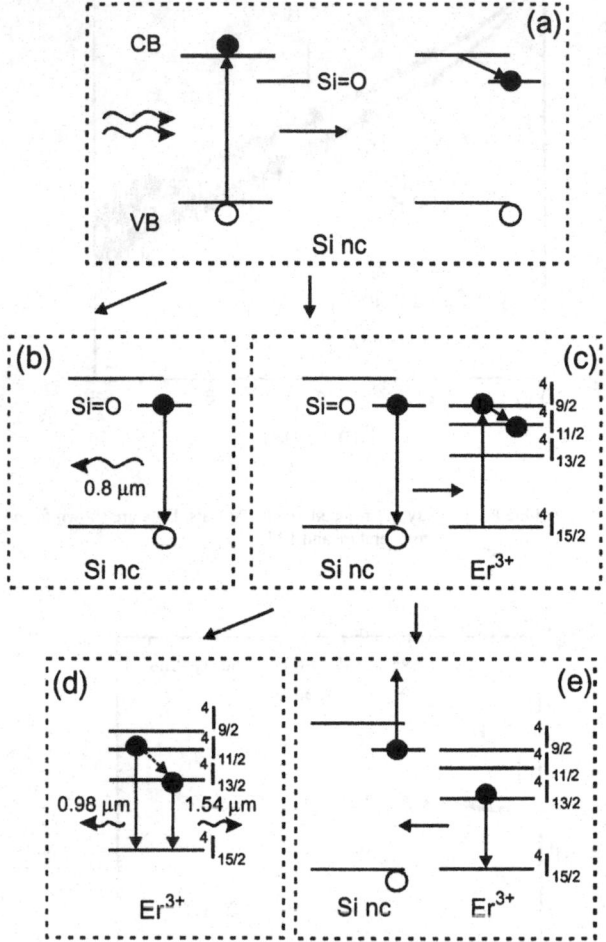

Figure 12. Schematic picture of the different processes occurring in Er doped Si nanocrystals.

to the conduction band (CB). Theoretical calculations have demonstrated that the electron in CB is then trapped by a Si=O interfacial state (a) [35]. The recombination of the electron in the interfacial state with a hole in the valence band gives the typical Si nc light emission at ~0.8 μm (b). Alternatively, in presence of Er, the energy can be transferred to the Er ion to excite it (c). Since the 0.8 μm nc wavelength (corresponding to ~1.5 eV) couples well with the $^4I_{9/2}$ level of the Er manifold we propose that this is indeed the Er level first excited by the nc. From this level a rapid relaxation occurs to the $^4I_{11/2}$ level with the subsequent emission of 0.98 μm photons or with a relaxation to the metastable $^4I_{13/2}$ level and emission of photons at 1.54 μm (d). Processes (b) and (c) are in competition one another. When an Er ion is close to a nc, that nc will become "dark" in the sense that the energy transfer to Er will be much more probable than the nc radiative emission. At N_{Er} ~1, all nc will give their energy to an Er ion. However, at 50 mW a nc is excited every ~ 0.1 ms. In presence of an Er ion ~ 5 ms will be needed to have a 1.54 μm photon out, i.e. most of the nc excitations (~ 40 excitations in 5 msec at 50 mW) won't produce any effect. This will decrease the emission efficiency of the sample. Indeed, these excitations might also be detrimental, in that an Auger effect between excited Er and the exciton in Si nc can occur (process (e) in Fig. 12). In fact, we have observed a decrease of the 1.54 μm decay time with increasing pump power demonstrating the presence of this effect. This decrease has also been observed at very low Er contents and, hence, cannot be attributed to cooperative upconversion processes [36] (which might also be present at higher concentrations). Finally, with increasing Er content several Er ions can be excited simultaneously by the same nc and the total number of emitted photons can hence become much higher than that in absence of Er. These data give a clear understanding of the Er – nc interactions.

5. Acknowledgments

The work presented in this paper is the result of the efforts of many people. We would like to acknowledge the contribution of C. Spinella, S. Pannitteri and A. Marino (CNR-IMETEM), V. Vinciguerra and D. Pacifici (University of Catania).

References

1. Canham, L.T. (1993) Progress towards crystalline-silicon-based light-emitting diodes, *MRS Bulletin* **18**, 22-28
2. Canham, L.T. (1990) Silicon quantum wire array fabrication by electrochemical and chemical dissolution of wafers, *Appl. Phys. Lett.* **57**, 1046-1048.
3. Takagi, H., Ogawa, H., Yamazaki, Y., Ishizaki, A. and Nakagiri, T. (1990) Quantum size effects on photoluminescence in ultrafine Si particles, *Appl. Phys. Lett.* **56**, 2379-2380.
4. Shimizu-Iwayama, T., Fujita, K., Nakao, S., Saitoh, K., Fujita, T. and Itoh, N. (1994) Visible photoluminescence in Si^+ - implanted silica glass, *J. Appl. Phys.* **75**, 7779-7783.
5. Zhu, J.G., White, C.W., Budai, J.D., Withrow, S.P., and Chen, Y. (1995) Growth of Ge, Si and SiGe nanocrystals in SiO_2 matrices, *J. Appl. Phys.* **78**, 4386-4389.
6. Min, K.S., Shcheglov, K.V., Yang, C.M., Atwater, H.A., Brongersma, M.L., and Polman, A. (1996) The role of quantum-confined excitons vs defects in the visible luminescence of SiO_2 films containing Ge nanocrystals, *Appl. Phys. Lett.* **68**, 2511-2513.
7. Min, K.S., Shcheglov, K.V., Yang, C.M., Atwater, H.A., Brongersma, M.L., and Polman, A. (1996) Defect-related versus excitonic visible light emission from ion beam synthesized Si nanocrystals in SiO_2, *Appl. Phys. Lett.* **69**, 2033-2035.
8. Brongersma, M.L., Polman, A., Min, K.S., Boer, E., Tambo, T., and Atwater, H.A. (1998) Tuning the emission wavelength of Si nanocrystals in SiO_2 by oxidation, *Appl. Phys. Lett.* **72**, 2577-2579.
9. Song, H.Z., and Bao, X.M. (1997) Visible photoluminescence from silicon-ion-implanted SiO_2 film and its multiple mechanisms, *Phys. Rev. B* **55**, 6988-6993.
10. Shimizu-Iwayama, T., Kunumado, N., Hole, D.E., and Townsend, P. (1998) Optical properties of silicon nanoclusters fabricated by ion implantation, *J. Appl. Phys.* **83**, 6018-6022.
11. Lockwood, D.J., Lu, Z.H., and Baribeau, J.M. (1996) Quantum confined luminescence in Si/SiO_2 superlattices, *Phys. Rev. Lett.* **96**, 539-541.
12. D'Avitaya, F.A., Vervoort, L., Bassani, F., Ossicini, S., Fasolino, A., and Bernardini, F. (1995) Light emission at room temperature from Si/CaF_2 multilayers, *Europhysics Lett.* **31**, 25-30.

176

13. Hirschman, K.D., Tsybeskov, L., Duttagupta, S.P., and Fauchet, P.M. (1996) Silicon-based visible light-emitting devices integrated into microelectronic circuits, *Nature* **384**, 338-341.

14. Ennen, H., Schneider, J., Pomrenke, G., and Axmann, A. (1983) 1.54-μm luminescence of erbium-implanted III-V semiconductors and silicon, *Appl. Phys. Lett.* **43**, 943-945.

15. Michel, J., Benton, J.L., Ferrante, R.F., Jacobson, D.C., Eaglesham, D.J., Fitzgerald, E.A., Xie, Y.H., Poate, J.M., and Kimerling, L.C. (1991) Impurity enhancement of the 1.54-μm Er^{3+} luminescence in silicon, *J. Appl. Phys.* **70**, 2672-2678.

16. Palm, J., Gan, F., Zheng, B., Mitchel, J., and Kimerling, L.C. (1996) Electroluminescence of erbium-doped silicon, *Phys. Rev. B* **54**, 17603-17615.

17. Priolo, F., Franzò, G., Coffa, S., and Carnera, A. (1998) Excitation and nonradiative deexcitation processes of Er^{3+} in crystalline silicon, *Phys. Rev. B* **57**, 4443-4455.

18. Franzò, G., Priolo, F., Coffa, S., Polman A., and Carnera, A. (1994) Room temperature luminescence from Er-doped crystalline silicon, *Appl. Phys. Lett.* **64**, 2235-2237.

19. Zheng, B., Michel, J., Ren, F.Y.G., Kimerling, L.C., Jacobson, D.C., and Poate, J.M. (1994) Room-temperature sharp line electroluminescence at λ=1.54 μm from an erbium-doped, silicon light-emitting diode, *Appl. Phys. Lett.* **64**, 2842-2844.

20. Stimmer, J., Reittinger, A., Nützel, J.F., Abstreiter, G., Holzbrecher, H., and Buchal, Ch. (1996) Electroluminescence of erbium-oxygen-doped silicon diodes grown by molecular beam epitaxy, *Appl. Phys. Lett.* **68**, 3290-3292.

21. Du, C-X., Ni, W-X., Joelsson, K.B., and Hansson, G.V. (1997) Room temperature 1.54 μm light emission of erbium doped Si Schottky diodes prepared by molecular beam epitaxy, *Appl. Phys. Lett.* **71**, 1023-1025.

22. Franzò, G., Coffa, S., Priolo, F., and Spinella, C. (1997) Mechanism and performance of forward and reverse bias electroluminescence at 1.54 μm from Er-doped Si diodes, *J. Appl. Phys.* **81**, 2784-2793.

23. Kenyon, A.J., Trwoga, P.F., Federighi, M., and Pitt, C.W. (1994) Optical properties of PECVD erbium-doped silicon-rich silica: evidence for energy transfer between silicon microclusters and erbium ions, *J. Phys.: Condens. Matter* **6**, L319-L324.

24. Fujii, M., Yoshida, M., Kanzawa, Y., Hayashi, S., and Yamamoto, K. (1997) 1.54 μm photoluminescence of Er^{3+} doped into SiO_2 films containing Si nanocrystals: evidence for energy transfer from Si nanocrystals to Er^{3+}, *Appl. Phys. Lett.* **71**, 1198-1200.

25. Fujii, M., Yoshida, M., Hayashi, S., and Yamamoto, K. (1998) Photoluminescence from SiO_2 films containing Si nanocrystal and Er: effects of nanocrystalline size on the photoluminescence efficiency of Er^{3+}, *J. Appl. Phys.* **84**, 4525-4531.

26. Komuro, S., Katsumata, T., Morikawa, T., Zhao, X., Isshiki, H., and Aoyagi, Y. (1999) Time response of 1.54 μm emission from highly Er-doped nanocrystalline Si thin films prepared by laser ablation, *Appl. Phys. Lett.* **74**, 377-379.

27. Franzò, G., Vinciguerra, V., and Priolo, F. (1999) The excitation mechanism of rare-earth ions in silicon nanocrystals, *Appl. Phys. A* **69**, 3-12.

28. Franzò, G., Iacona, F., Vinciguerra, V., and Priolo, F. (1999) Enhanced rare earth luminescence in silicon nanocrystals, *Mat. Sci. & Eng. B* **69/70**, 338-341.

29. Chryssou, C.E., Kenyon, A.J., Iwayama, T.S., Pitt, C.W., and Hole, D.E. (1999) Evidence of energy coupling between Si nanocrystals and Er^{3+} in ion-implanted silica thin films, *Appl. Phys. Lett.* **75**, 2011-2013.

30. Franzò, G., Pacifici, D., Vinciguerra, V., Priolo, F., and Iacona, F. (2000) Er^{3+} ions–Si nanocrystals interactions and their effects on the luminescence properties, *Appl. Phys. Lett.* **76**, 2167-2169.

31. Franzò, G., Vinciguerra, V., and Priolo, F. (2000) Room temperature luminescence from rare earth ions implanted into Si nanocrystals, *Phil. Mag.* **80**, 719-728.

32. Iacona, F., Franzò, G., and Spinella, C. (2000) Correlation between luminescence and structural properties of Si nanocrystals, *J. Appl. Phys.* **87**, 1295-1303.

33. Vinciguerra, V., Franzò, G., Priolo, F., Iacona, F., and Spinella, C. (1 June 2000 issue) Quantum confinement and recombination dynamics in silicon nanocrystals embedded in Si/SiO_2 superlattices, *J. Appl. Phys.* **87**, 8165-8173.

34. Franzò, G., Iacona, F., Spinella, C., Cammarata, S., and Grimaldi, M.G. (2000) Size dependence of the luminescence properties in Si nanocrystals, *Mat. Sci. & Eng. B* **69/70** 454-457.

35. Wolkin, M.V., Jorne, J., Fauchet, P.M., Allan. G., and Delerue, C. (1999) Electronic states and luminescence in porous silicon quantum dots: the role of oxygen, *Phys. Rev. Lett.* **82**, 197-200.

36. Polman, A. (1997) Erbium implanted thin film photonic materials, *J. Appl. Phys.* **82**, 1-39

NICKS, NODES, AND NEW MOTIFS FOR DNA NANOTECHNOLOGY

NADRIAN C. SEEMAN, CHENGDE MAO, FURONG LIU, RUOJIE SHA, XIAOPING YANG, LISA WENZLER, XIAOJUN LI, ZHIYONG SHEN, HAO YAN, PHISET SA-ARDYEN, XIAOPING ZHANG, WANQIU SHEN, JEFF BIRAC, PHILIP LUKEMAN, YARIV PINTO, XIAOJUN LI, JING QI, BING LIU, HANGXIA QIU, SHOUMING DU, HUI WANG, WEIQIONG SUN, YINLI WANG, TSU-JU FU, YUWEN ZHANG, JOHN E. MUELLER and JUNGHUEI CHEN
Department of Chemistry, New York University
New York, NY 10012, USA

Abstract

The properties that make DNA such an effective molecule for its biological role as genetic material also make it a superb molecule for nanoconstruction. One key to using DNA for this purpose is to produce stable complex motifs, such as branched molecules. Combining branched species by sticky ended interactions, leads to N-connected stick figures whose edges consist of double helical DNA. Zero node removal or reciprocal crossover, leads to complex fused motifs, such as rigid multi-crossover molecules and paranemic crossover molecules. Multi-crossover molecules have been used to produce 2D arrays and a nanomechanical device. Algorithmic assembly and the use of complex complementarities for joining units are goals in progress that are likely to produce new capabilities for DNA nanotechnology.

1. DNA as a Construction Material

Watson and Crick's DNA structure and replication model placed the phenomenology of genetics on a firm molecular basis.[1] It is now well known that genetic information resides in the complementarity of the bases: Adenine (A) forms two hydrogen bonds with thymine (T) and guanine (G) forms three hydrogen bonds with cytosine (C). In combination with backbone geometry, this specific pairing relationship leads to the double helical structure of the DNA polymer; the double helix is about 2 nm wide, and its repeat consists of about 10.5 nucleotide pairs, each separated by about 3.4 Å. The chemical polarities of the two strands are antiparallel, resulting in a series of twofold axes for the backbone structure, perpendicular to the helix axis, one axis roughly through the base pair planes, and another one halfway in between. It is important to recognize that the helix axis is linear in cellular DNA, not in the algebraic sense of

L. Pavesi and E. Buzaneva (eds.), Frontiers of Nano-Optoelectronic Systems, 177–197.

178

being a straight line, but in the topological sense of being unbranched. This feature characterizes not only naturally occurring molecules, but also recombinant DNA species produced by molecular biotechnology.[2]

The molecular properties of the DNA molecule that make it such an effective molecule for storing genetic information also confer on it the characteristics of an outstanding nanoscale synthon for a biomimetic nanotechnology. This fact has led to the development of both DNA nanotechnology[3,4]and the closely related field of DNA-based computation.[5] This article will describe the development of DNA nanotechnology by deriving the unusual motifs of DNA that have led to it. These motifs have been inspired by a variety of sources, and we have shown that they lead naturally to a generalization of Watson-Crick complementarity.[6] The basic feature shared by both unusual DNA motifs and generalized complementarity is discontinuous

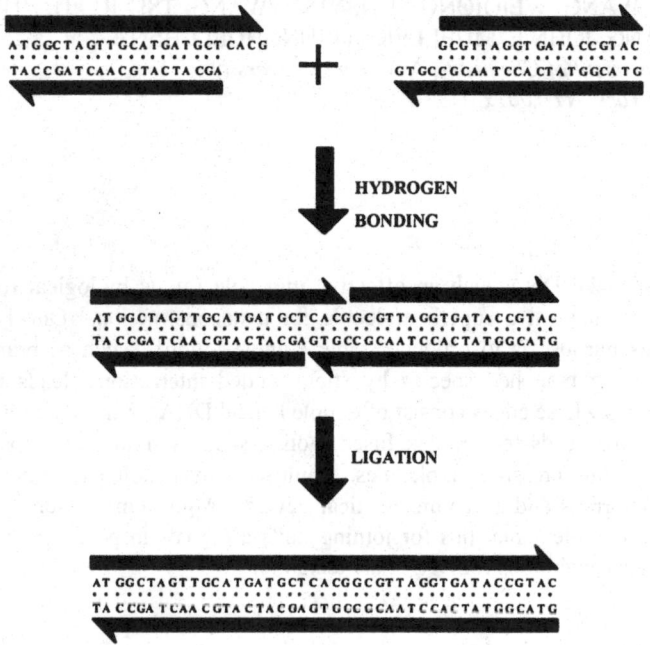

Figure 1. Sticky-Ended Cohesion and Ligation. Two linear double helical molecules of DNA are shown at the top of the drawing. The antiparallel backbones are indicated by the black lines terminating in half-arrows. The half-arrows indicate the 5'-->3' directions of the backbones. The right end of the left molecule and the left end of the right molecule have single-stranded extensions ('sticky ends') that are complementary to each other. The middle portion shows that, under the proper conditions, these bind to each other specifically by hydrogen bonding. The bottom of the drawing shows that they can be ligated to covalency by the proper enzymes and cofactors.

complementarity. Conventional nucleic acid complementarity entails two continuous polynucleotide backbones: The sequence on one strand is one-for-one complementary to the sequence on the other strand. It has long been known that the bases of nucleic acids can pair in many ways other than Watson-Crick complementarity (e.g., ref. 7), but our discussion here will be restricted to Watson-Crick interactions. Other investigators are using non-Watson-Crick pairing, such as G tetrads, in work on DNA nanotechnology.[8,9]

Figure 1 illustrates a prominent example of discontinuous complementarity, the cohesion of single-stranded overhangs, known as sticky ends. If, as shown, the two molecules contain complementary overhangs, they will cohere in solution. It is even possible to ligate them to form a longer covalent DNA double helix that is the sum of its two components. Sticky ended association makes DNA a remarkable synthon, not only because it leads to predictable and programmable intermolecular affinity, but also because it also leads to a predictable local product structure: The overlap region will be B-DNA.[10] However, the discontinuous complementarity illustrated in the middle panel of Figure 1 does not lead to new thinking about complementarity.

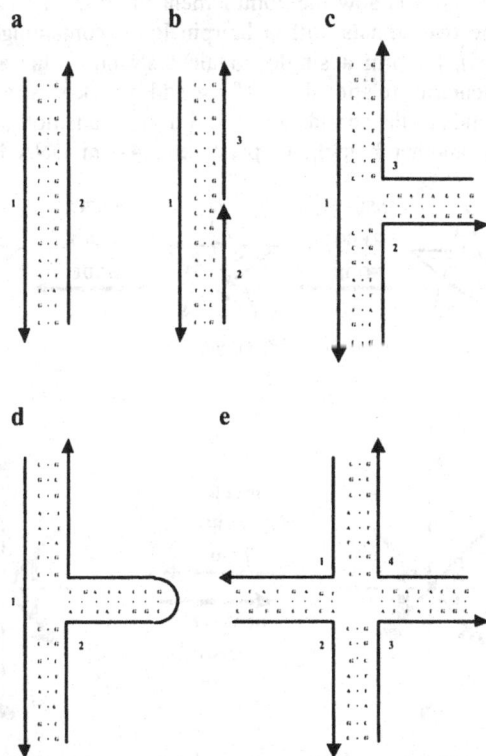

Figure 2. Generalizing Complementarity Through a Nick. Panel (a) illustrates a conventional Watson-Crick duplex. In (b), strand two has been broken into strands 2 and 3 through the insertion of a nick. In (c), the nick has been elaborated into a double helical arm by the 3' extension of strand 2 and the 5' extension of strand 3. Panel (d) shows that, in principle, stands 2 and 3 of (c) could be joined to make a new single-strand complement to strand 1. In (e), strand 1 of (c) has been broken into two lengthened strands to produce a four-arm junction; note that strands 2 and 3 of (c) have been renumbered here.

180

2. Generalized Complementarity Generates New DNA Motifs

The DNA backbone is an alternating copolymer of a sugar (deoxyribose) and phosphate, and the bases are attached to the sugar moieties. A duplex molecule DNA is shown unwound in Figure 2a; it is evident that strand 1 is the complement of strand 2, because it contains the nucleotides that complete all the base pairs of strand 2. Let us imagine a double helical DNA molecule lacking one of the phosphoester bonds, so that the molecule is nicked once, much like the nicks seen in the central panel of Figure 1. A molecule like this is shown in Figure 2b. Strand 2 of Figure 2a has been broken into two different strands, now numbered 2 and 3. The complement of strand 1 is now the sum of strands 2 and 3, not just a single strand.

We can extend the 3' end of strand 2 and the 5' end of strand 3. In Figure 2c, we show what the structure looks like when the two extensions are complementary. This structure is a 3-arm DNA branched junction. The structure consisting of the complex of strands 2 and 3 is now the complement of strand 1. As shown in Figure 2d, we could join the two strands with a hairpin loop (containing, say, four or five thymidine nucleotides), to form a single covalent strand. This alteration does not change the complementarity relationship. If we add a nick to strand 1 of Figure 1c, and then extend the ends of the strands as we did in the transition between Figures 1b and 1c, we will add another branch, to produce a 4-arm DNA branched junction,

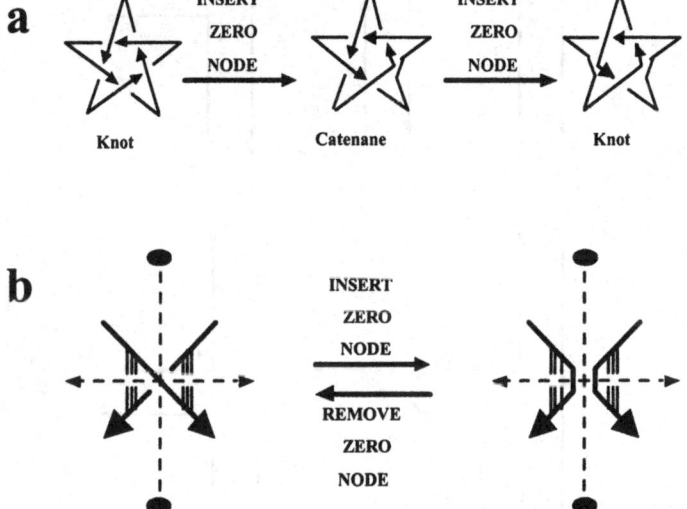

Figure 3. Zero Node Operations. (a) Inserting Zero Nodes in Knots and Catenanes. On the left is a 5-noded knot, with its polarity indicated by arrowheads. Passing to the middle, one zero node has been inserted, converting the knot to a catenane, drawn with lines of two different thicknesses. On the right, another zero node has been inserted, making a new 3-noded knot. *(b) Zero Node Insertion and Removal in a DNA Context.* Backbones are indicated by thick arrows, held together by three base pairs on each side. The helix axis is horizontal, and the dyad axis is vertical. The operation changes the strand segments connections, but maintains polarity. The reaction symbol replaces the right directional in *(a).*

shown in Figure 1e. The complement to any of the strands is now the set of the other three strands. Thus, in panel 1e, the complement to strand 1 in the four-arm branched junction context is the complex of strands 2, 3, and 4, even though there are no direct interactions between strand 1 and strand 3. Again, one could imagine covalent links between strands 2 and 3 and between strands 3 and 4, if these links help to clarify the concept.

One feature of generalized complementarity is that there is no longer a unique complement to a given strand. Strand 2 in Figure 2a is the unique complement of strand 1, but the complement to strand 3 of Figure 1c is different from the complement to the same strand (renumbered 4) in Figure 1e. Indeed, there exists a whole panoply of complements to any strand. This situation may prove useless or dangerous to biological systems, but it provides a wealth of new possibilities for DNA nanoconstruction.

The generalization of complementarity described above does not change the bases that are complementary to a given strand. However, it allows for the insertion of residues into the backbone between any two base pairs when defining a strand's complement. Below, we will describe several different DNA motifs that result from the generalization of DNA complementarity: DNA branched junctions, mentioned above, DNA multi-crossover (MX) molecules, and paranemic crossover (PX) molecules. To date, only DNA branched junctions and MX molecules have been used directly in DNA nanoscale constructions. PX molecules appear very promising, but are too new to have been applied successfully yet.

3. Zero Node Removal and Reciprocal Crossovers Produce New Motifs

If complements need not be continuous, it is useful to think about the generation of new motifs from the topological perspective of zero node removal.[11] Topologists have developed the concept of the zero node as an aid in the analysis of knotted and catenated structures (e.g., ref. 12). Figure 3a illustrates the creation of a zero node in a knot, leading to a catenane, and then creation of a zero node in the catenane leading to a simpler knot.

The initial five-noded knot has been assigned an arbitrary polarity. One can imagine each of the nodes (crossings) to consist of two pairs of two segments each; one segment pair before and after the crossing point on the strand above, and a second pair on the strand below. Inserting a zero node consists of destroying the node by disconnecting the existing pairs and reconnecting between layers, so as to maintain local polarity. Figure 3b puts the system in a DNA context, indicating helix axis, dyad axis and base pairs; a half-turn of duplex DNA is shown, the quantum of single-stranded DNA topology.[11] This drawing also shows the reverse operation, removing a zero node, and replacing it with an authentic node; is a synthetic operation of great utility in designing new DNA motifs.

Figure 4 shows that a branched junction can be generated from two adjacent helices by this operation. Thus, zero node removal is the same as a *reciprocal crossover* between two strands of adjacent double helices. Figure 4a illustrates this operation for a pair of antiparallel double helices, and Figure 4b shows it for two

182

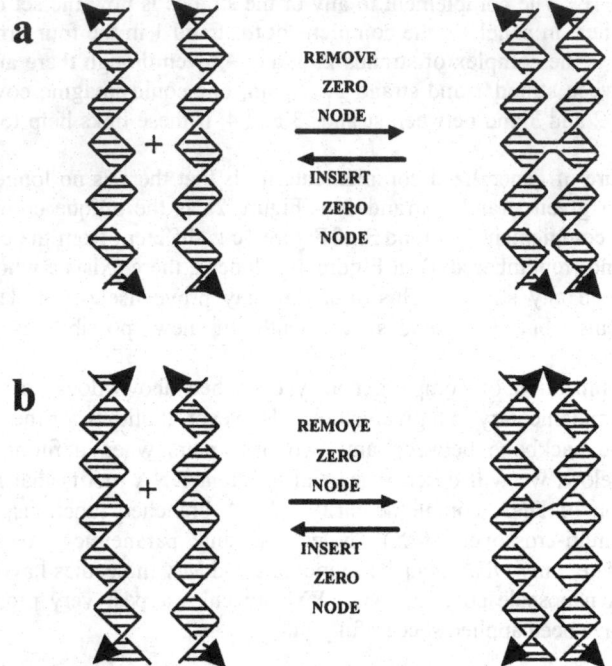

Figure 4. Zero Node Removal Forming Reciprocal Crossovers Between Two Double Helices with Parallel and Antiparallel Helix Axes. (a) A zero node between two antiparallel double helices is removed to produce an antiparallel branched junction with its arms stacked in pairs of helical domains. *(b)* A zero node between two parallel helix axes is removed to produce a parallel branched junction. These two molecules represent different conformations of the same molecule.

parallel helices. These two structures differ only in conformation, because rotation of one helix by 180° about the horizontal will interconvert them. It may appear from the drawing that a positive node is visible at the crossover point of the branched junction on the right side of Figure 4b, but that there is no node present in the comparable molecule in Figure 4a. However, this is misleading, because looking down the axis between the two helical domains would show that the strands are crossed over each other to form a node. The operations of nicking a duplex molecule and inserting a node that are shown in Figure 2 could equally well be shown in the context of zero node removal or reciprocal crossover. This process corresponds to what is known as a reciprocal exchange in genetic recombination. Thus, Figure 5 illustrates a reciprocal exchange the generates the 3-arm junction of Figure 2c.

The reciprocal crossovers generated by zero node removal in Figures 4 and 5 provide a general method of producing a large variety of unusual motifs. The operation in Figure 5 consists of removing a zero node between a double helix and another one whose helix axis is perpendicular to it. It is evident that repeating this operation would serve to convert the 3-arm junction on the right of Figure 5 to a 4-arm junction, such as the one shown in Figure 2e. Likewise, the operation shown in Figure 3b consists of fusing two hairpins with co-axial helix axes, to form a longer duplex molecule.

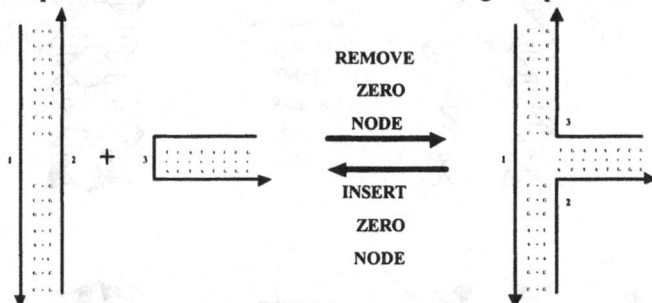

Figure 5. Zero Node Operations Converting a DNA Hairpin and a Duplex Molecule to a 3-Arm Junction.
The linear molecule on the left is the molecule shown in Figure 2a, except that the top and bottom portions have been separated by a space; thus, the backbone has been arbitrarily lengthened, for graphical purposes.. The top and bottom portions of strand 3, on the right, have been joined by a single backbone linkage. Both of these extensions are artificial. However, a reciprocal crossover (removal of a zero node) between the extended backbone linkage of strand 2 and the artificially long hairpin linkage of strand 3 results in the same 3-arm junction shown in Figure 2c.

The operations shown in Figure 4 demonstrate the fusion of two lateral helices to produce the same 4-arm junction generated by two reciprocal crosses (zero node removals) of the type shown in Figure 5. The lateral fusion reciprocal crossover is topologically similar to the result of the operations that produce Holliday junctions in living systems.[10]

A further series of motifs can be produced if one expands the operations shown in Figure 4 to more juxtapositions. Figure 6 illustrates the extreme cases of performing reciprocal crossovers at every point where strands are juxtaposed. The structure shown in Figure 6a illustrates an antiparallel fused helix motif that could be formed by a series of cyclic strands, although, owing to their linked character, they would have to be nicked to produce this topology. The system shown in Figure 6b is paranemic crossover (PX) DNA, which has been formed in the laboratory. It is likely to be used in constructions and devices, and, further, may play a role in the recognition of DNA homology by DNA in the process of homologous recombination.

184

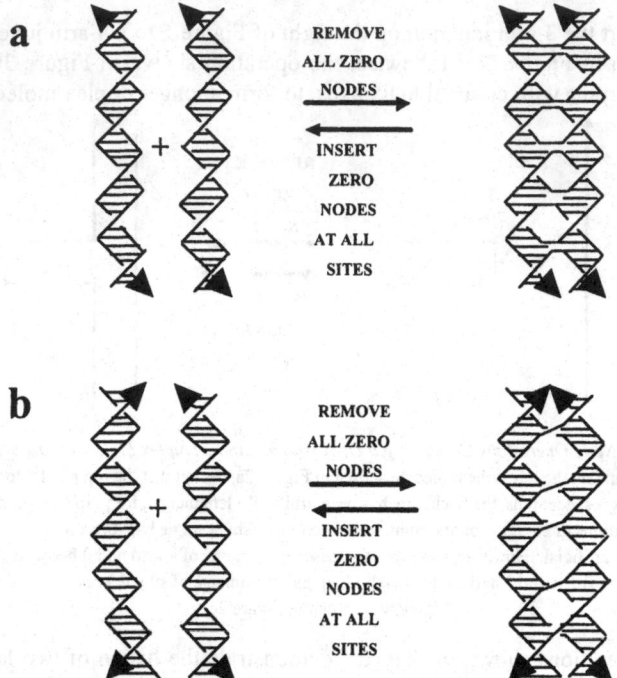

Figure 6. Generation of New Motifs by Reciprocal Strand Crossovers. (a) Antiparallel Helices. Each of the helices contain six half-turns of DNA, leading to five internal positions of strand juxtaposition. All five of these positions are converted to crossover sites, leading to the molecule on the right, which consists of a series of cyclic molecules. *(b) Parallel Helices.* Two parallel helices are shown forming all five possible reciprocal crossovers in the central zone of the molecule. This system is paranemic crossover (PX) DNA, wherein two double helices are wrapped around each other, as shown by the two different thicknesses of strands. This system offers the potential for sensitivity to homology between the two different pairs of strands.

A series of 'juxtaposed' (JX) molecules can be produced as well by refraining to remove every zero node in this system. These molecules have quite distinct topologies from those illustrated. Figure 7 illustrates the same processes illustrated in Figure 6, but only two zero nodes have been removed, rather than all of them. Figure 7 shows the formation of two classical double crossover (DX) molecules[13] from two double helices. Figure 7a illustrates the formation of a DAE molecule, a double crossover molecule with antiparallel helical domains, and with an even number of half-turns between the crossover points. Figure 7b shows, similarly the formation of a DPE molecule, with parallel double helical domains, and also with an even number of half-turns between the crossover points. These molecules are special cases of JX molecules, which incorporate both juxtaposition points and crossover points. JX molecules held together by a small ratio of crossovers to juxtapositions are called MX (multi-crossover) molecules. Applications of DX and triple crossover (TX) molecules already realized are described in greater detail below.

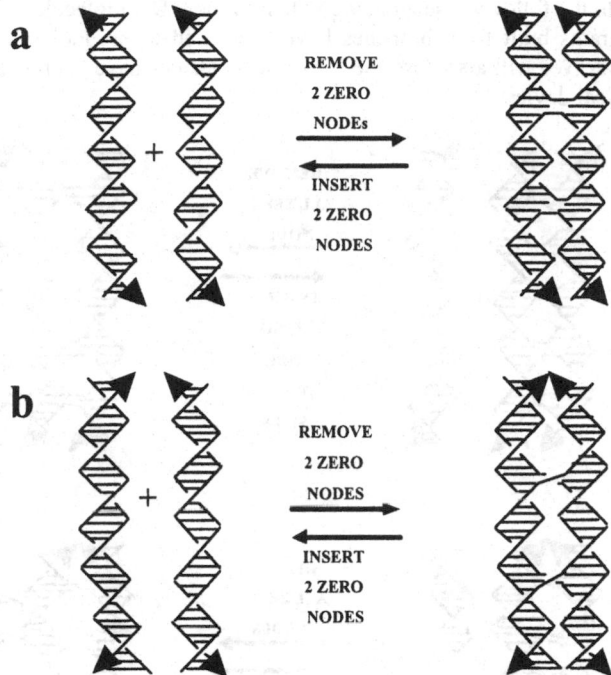

Figure 7. The Formation of Double Crossover Molecules from Two Helical Domains. (a) The Formation of a DAE Molecule. The two antiparallel double helices are combined by reciprocal exchanges at two sites separated by two half-turns (one full turn) of DNA. A different topology is seen when the separation is an odd number of half-turns. The symmetry axis of this molecule is perpendicular to the plane of the page. *(b) The Formation of a DPE Molecule.* The two double helices are parallel to each other, and are joined by the same reciprocal exchange operations that join the two molecule in *(a)*. The resulting molecule contains parallel helices separated by two half-turns of DNA. Two different topologies will result if an odd number of half turns separate the crossover points. The symmetry axis of this system is parallel to the helix axes, and in the plane of the page. It is evident that if the ends of the helices were closed by hairpin loops, the resulting single-stranded circles would be linked to each other.[11]

In principle, there is no reason that one need stop with two helical domains in any of these structures. Further motifs can be generated if one fuses more helices by reciprocal crossovers. This concept is illustrated in Figure 8, which shows the products of Figure 6 being fused with another helix to produce three-domain molecules with no juxtapositions, just crossovers.

This is similar to the combination shown in Figure 5, except that a DAE molecule replaces a linear duplex molecule. In general, we use a bulged junction for this operation, because it fits better into the duplex section of the molecule.[14] In this section and the previous section, we have shown how simple operations can lead to a variety of DNA motifs. In contrast to linear duplex DNA, the characteristic of all of these motifs is that they contain three or more ends associated with each unit.

These motifs, in turn, when stable, can be used in a variety of applications. In the next section, we will demonstrate how branched molecules can be used to build

186

stick figures. In the following section, we will show how DX molecules, TX molecules and parallelograms built from branches have been used to produce two-dimensional patterned arrays. We will also describe how DX molecules have been used to produce a nanomechanical device.

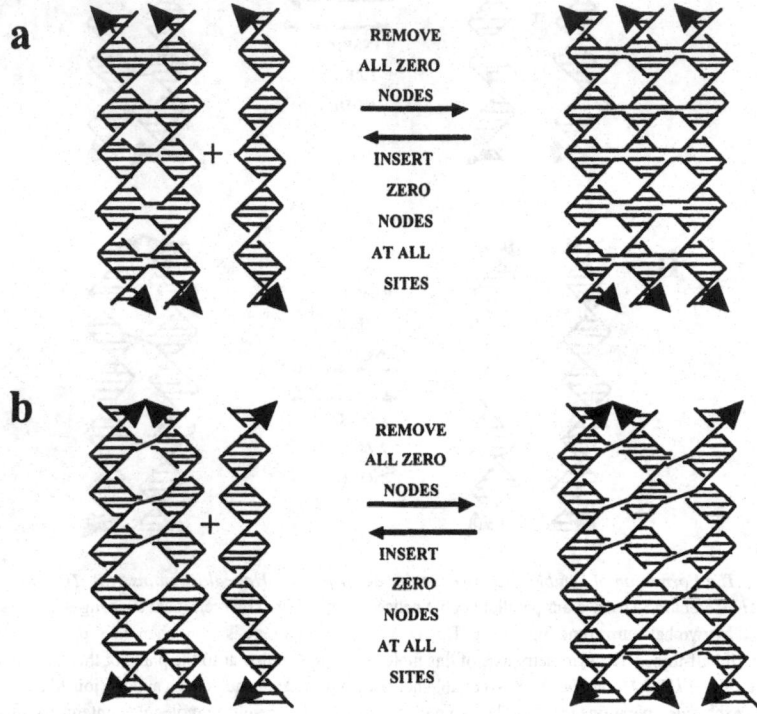

Figure 8. The Extension of Two-Domain Motifs to Three-Domain Motifs. (a) Antiparallel Molecules.
The three parallel molecules shown are formed from two (broken) strands, as indicated by the thickness of
the lines. *(b) Parallel Molecules.* The same extension is shown for the parallel system. The different line
thicknesses indicate adjacent pairs of DNA strands, with opposite polarities.

4.   DNA Branched Junctions With Sticky Ends Lead to DNA Nanotechnology

DNA nanotechnology entails the joining of complex motifs by complementary interactions. Branched junctions were the first motif to which this approach was applied. DNA branched junctions are stable branched molecules, such as the species shown in Figures 2c-e. The four-arm junction of Figure 2e is a stable analog of an intermediate in the biological process of genetic recombination known as the Holliday junction.[15] Junctions are produced readily from the self-assembly of synthetic strands whose sequences have been generated by the minimization of sequence symmetry.[3] Branched junctions containing three,[16] four,[17] five, and six[18] arms have been assembled experimentally, and there is no known limit to the number of arms that may flank a branch point.

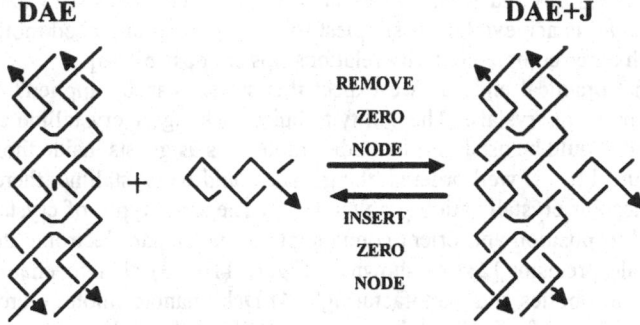

Figure 9. The Formation of a DAE+J Motif from a DAE Molecule and a Hairpin. The DAE molecule shown on the left is combined with a hairpin to produce the DAE+J motif. The 3-arm junction produced here is actually a bulged junction, containing an extra two thymidine residues. The junction is shown phased to have its helix axis coplanar with those of the other two domains, but in most cases it is used in another position, so that the helix axis comes out of the plane of the page, and serves as a topographic marker for the AFM.

The basic idea is shown in Figure 10. A four-arm branched junction is displayed in the left panel; the junction contains four sticky ends, **X** and its complement **X'**, and **Y** and its complement **Y'**. The arrangement on the right illustrates the assembly of four of these branched junctions into a quadrilateral. The quadrilateral has open sticky-ended 'valences' on the outside, so, in principle, it could be extended to form a two-dimensional periodic lattice. As described below, variations on this theme have been constructed successfully,[19] although, individual four-arm junctions could not be used, because they are too flexible.[20]

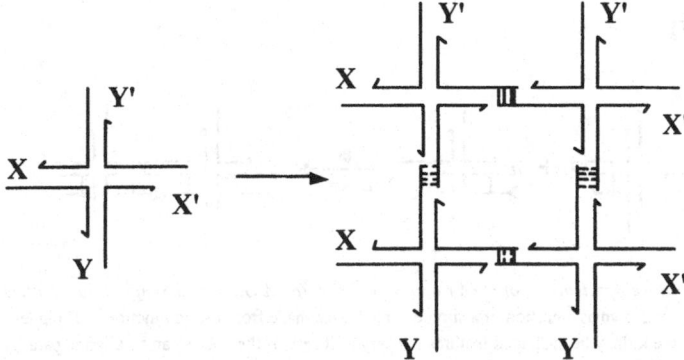

Figure 10. Formation of a Two-Dimensional Lattice from a Junction with Sticky Ends. X is a sticky end and X' is its complement. The same is true of Y and Y'. Four of the monomers on the left are complexed in parallel orientation to yield the structure on the right. DNA ligase can close the gaps left in the complex. The complex has open valences, so that it can be extended by the addition of more monomers.

188

Thus, the idea is to use the branch points of DNA junctions as the vertices of N-connected objects and lattices;[21,22] N-connected means that every vertex is connected to N other vertices through edges. In the DNA context, this means that we will make stick figures and stick-lattices in which the edges consist of double helical DNA. This target is achieved in its simplest form by joining branched motifs by sticky ends, although other complementarity relationships are possible.[6]

To what practical ends do we expect this system can be applied? 1) Spatially periodic networks are crystals. The ability to build stick-figure crystalline cages on the nanometer scale could be used to orient other molecules as guests inside those cages, as shown in Figure 11a. If well ordered, the guests would be crystalline, thereby solving the macromolecular crystallization problem.[3] 2) The same types of crystalline arrays could be used to position and orient components of molecular electronic devices with nanometer-scale precision,[23] as shown in Figure 11b. 3) Nanomechanical devices can lead to nanorobotics and 'nanofacturing'. 4) DNA nanotechnology creates motifs that are likely to be useful for DNA-based computation and the algorithmic assembly of

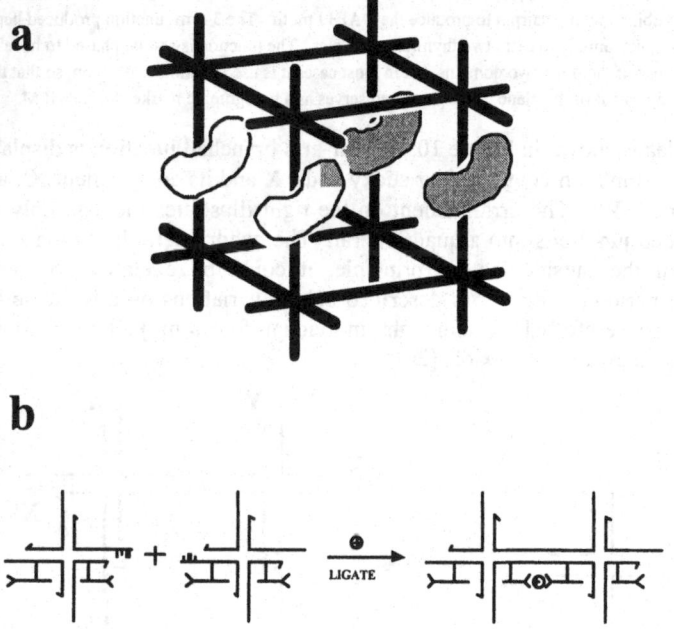

Figure 11. Future Applications of DNA Nanotechnology. (a) A Guest in a Simple Cubic Lattice. The DNA lattice is drawn as a portion of a simple cubic lattice made from 6-arm junctions. The guests are represented by the kidney bean-shaped features in every unit cell. If the guests can be aligned parallel to each other, their structures can be determined by X-ray crystallography. *(b) DNA as Scaffolding.* Two branched junctions are shown, and a molecular wire is attached to them. When the two junctions cohere with each other, so does the molecular wire, which forms a synapse. This drawing represents one way that DNA could be used to organize molecules whose properties are best suited to a particular molecular task.

materials.[24]

The flexibility of individual branched junctions led in the early days of DNA nanotechnology to the construction of topological targets, rather than geometrical targets. The first N-connected object to be produced this way was a DNA molecule whose helix axes are connected like the edges of a cube or a rhombohedron.[25] This molecule, shown in Figure 12a, has an estimated molecular weight of 150,000; it was synthesized in solution, and each of its twelve edges contains a unique recognition sequence for a restriction endonuclease that can cleave there. The vertices are separated by two complete turns of DNA, so this molecule is a hexacatenane of strands that correspond to each face. Each strand is linked twice to each of its neighbors, and proof of synthesis consists of digestion by specific restriction enzymes to produce target catenanes that can be synthesized independently.

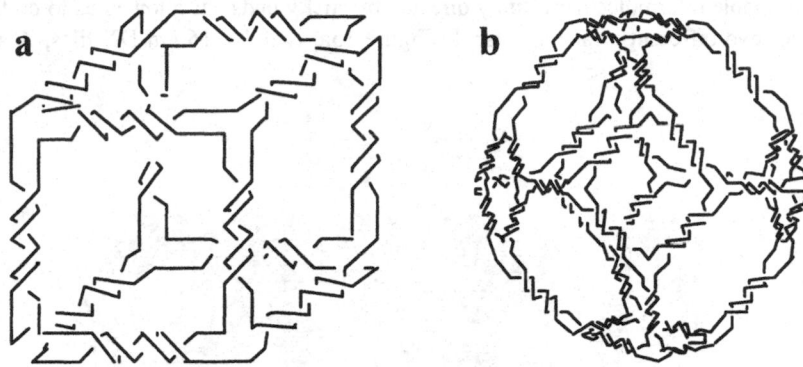

Figure 12. Ligated Products from Flexible DNA Components. (a) A Stick Cube and (b) a Stick Truncated Octahedron. The drawings show that each edge of the two figures contains two turns of double helical DNA. The twisting is confined to the central portion of each edge for clarity, but it actually extends from vertex to vertex. Both molecules are drawn as though they were constructed from 3-arm junctions, but the truncated octahedron has been constructed from 4-arm junctions, and the extra arms have been omitted for clarity.

Figure 12b illustrates a DNA truncated octahedron constructed by a solid-support based methodology[26] that uses the logic of sticky-ended ligation much more effectively than the solution synthesis that produced the cube. The truncated octahedron, a 3-connected figure, is a 14-catenane, of molecular weight ca. 790,000; it was prepared using four-arm junctions, but the fourth one has been omitted for clarity. Proof of synthesis was similar to that used for the cube.[27]

5. DNA Arrays and Devices Can Be Produced From Unusual DNA Motifs

We have shown above how to produce a variety of DNA motifs that contain two or more helical domains fused along their sides. The generic term for these molecules is multi-crossover molecules (MX). DX molecules (Figure 7) fall into this category. These molecules were derived from biology,[13] because they are intermediates in genetic recombination mediated by double strand breaks[28,29] and also they are known to be intermediates in meiosis.[30]

190

Parallel DX molecules are not well behaved when their separations are short,[13] however, not only are antiparallel DX molecules stable, but they also are quite stiff.[31] Their stiffness is not decreased markedly if they are expanded to be DAE+J motifs, as illustrated in Figure 9. The rigidity of DX molecules makes them useful for high-symmetry applications, such as array formation, as well as for components of nanomechanical devices. We have used DX motifs with sticky ends to form two-dimensional hydrogen-bonded arrays that tile the plane, as shown in Figure 13. As noted above, the extra helical domains of DX+J molecules can be directed out of the plane, to produce topographic markers visible in the atomic force microscope (AFM).[32]

In any microscopic study, it is important to have chemical correlates, to control for the fortuitous appearance of the desired image. Figure 13 illustrates that the power of programmable intermolecular affinity directed by sticky ends has allowed us to do this by using two different arrangements. In Figure 13a, two 4 x 16 nm DX tiles, A and

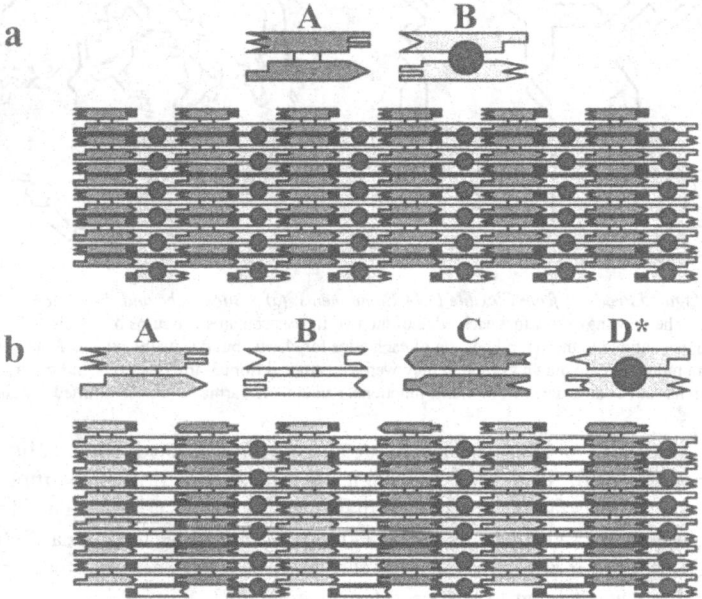

Figure 13. Arrays Assembled from Double Crossover Molecules. (a) A Two Component Array. Two DX molecules are illustrated schematically at the top of this panel, one labeled A and the other labeled **B***. The two helices are represented as rectangles, and the complementary sticky ends are represented by complementary geometrical shapes. A is a conventional DX molecule, but **B*** is a DX+J molecule, in which a DNA hairpin protrudes from the plane defined by the helix axes of the two antiparallel domains; this hairpin is shown as a filled black circle. Below these molecules is an array, drawn half-sized, that shows the two components fitting together to tile a plane. *(b) A Four Component Array.* The same conventions apply as in (a). The difference here is that the array has four components, A, **B**, C and **D***, where A, **B** and C are conventional DX molecules, and **D*** is a DX+J motif. It is clear from the bottom portion of the panel that the stripes should be separated by twice the distance seen in *(a)*.

B*, tile the plane; the topographic markers on the **B*** tiles, which are DAE+J molecules, should appear as stripes separated by 32 nm in the AFM, which does not resolve the 4 nm spacing. Figure 13b shows that switching to a set of four tiles, **A**, **B**, **C** and **D*** leads, as predicted, to a 64 nm spacing.[32] We have demonstrated that it is possible to edit the patterns by means of DNA modification: The 32 nm array can be converted to the 64 nm array by removing one set of stripes with a restriction enzyme; likewise, the 64 nm array can be converted to the 32 nm array be ligating or annealing DNA hairpins to sticky ends.[33]

It is evident from the section on reciprocal crossovers that the DX molecule is easily expanded to a triple crossover (TX) molecule, with three helical domains, each connected to the central domain by two crossover points. TX molecules have been made and characterized successfully.[34] These molecules have been used to tile the plane in the same way that DX molecules have been used, and TX+J motifs have been used to produce a striped array that has been visualized by AFM.[34] In addition, a TX array has been produced containing gaps that can be filled by rotated TX molecules.[34] It is likely that some variation of this approach ultimately will be used

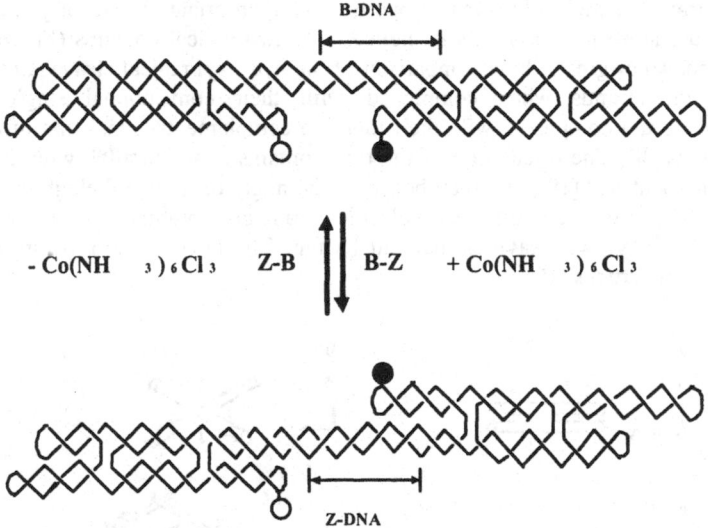

Figure 14. The Operation of a DNA Nanomechanical Device. The device consists of two DNA DX molecules (DAO motif) connected by 4.5 turns of DNA between the nearest crossover points. The upper portion of the drawing illustrates the molecule constructed entirely from right-handed B-DNA. Three strands are shown, one in the center drawn with a thick line, and two on the ends triply catenated to it. Fluorescent dyes are drawn schematically as filled (fluorescein) and unfilled (Cy3) circles attached to the free hairpins near the middle of the molecule. At the center of the connecting helix is a 20 nucleotide region of proto-Z DNA in the B-DNA conformation. When the B-Z transition is triggered by the addition of Co(NH3)6Cl3, this same portion becomes left-handed Z-DNA, as seen at the bottom of the drawing. When this transition occurs, the two double crossover molecules change their relative positions, increasing the separation of the dyes. It is possible to cycle this system in both directions.

to produce three-dimensional arrays.

The stiffness of the DX molecule makes it a valuable species to use in a nanomechanical device. Figure 14 illustrates that two DX molecules can be combined to produce a two-state device whose action is triggered by the B-Z transition of DNA.[35] Z-DNA is a left-handed conformation of conventional DNA molecules.[36] It has two requirements, a special sequence [(dCdG)$_n$ is a good one], and special solution conditions, such as the presence of high ionic strength or an effector such as $Co(NH_3)_6^{3+}$.[37] The sequence requirement allows us to control the transition in space, limiting it to a particular part of the molecule; the requirement for special conditions allows us to control it in time. We have used fluorescence resonance energy transfer to demonstrate that this device works, because the separation of the donor and acceptor dyes is different in the two states. The largely rotary motion moves atoms 20-60 Å, depending on their radius relative to the rotation axis. It took over ten years to produce this device, most of which was spent seeking a rigid motif.

We have managed to produce a rigid DNA component in another way. Although the individual branched junction is not rigid enough to be useful in producing DNA arrays, it is possible to combine four such junctions into a parallelogram-like molecule whose properties are appropriate. It has long been known that the four-arm junction stacks its helices into two helical domains (Figure 15a,b); this was shown originally by a comparison of the hydroxyl radical cleavage patterns of each of the strands when complexed with their conventional single-stranded complements and complexed with their junction complements, consisting of the other three strands.[38] The orientation of the two domains is antiparallel, with an angular separation about 60°,[19,39] which becomes the angle of the parallelogram in Figure 15c.[19] The parallelogram-like molecules can be combined to produce two-dimensional arrays with cavities that can be tuned by altering the separations of the parallelogram vertices.[19]

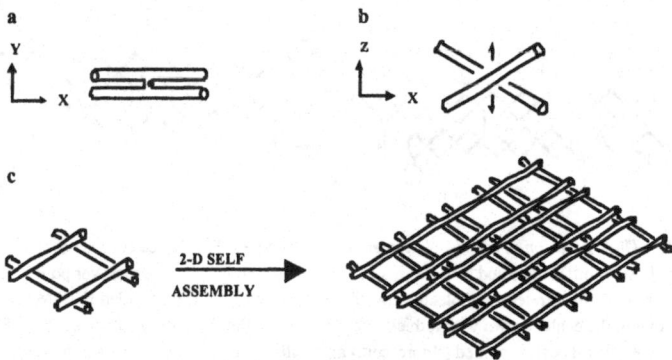

Figure 15. Parallelogram-Like Molecular Components and their Assembly. (a) A View Down the Dyad Axis of the Holliday Junction Solution Structure.

The dyad axis is indicated by the small lens-shaped figure. The upper helical domain is rotated 30° about the vertical so that its right end penetrates the page, and the lower helical domain is rotated 30° about the vertical so that its left end penetrates the page. The X and Y axes of a right-handed coordinate system are shown to help orient the reader. *(b) A View with the Dyad Axis Vertical.* The molecule has been rotated 90° about the X axis, as indicated. The dyad axis is indicated by the double arrows. *(c) The Combination of Four Junctions into a Parallelogram-Like Motif that Produces a Latticework of DNA by Self Assembly.* The array shows the 2-dimensional self-assembly product of the motif; the long separations between helices contain four helical turns, and the short separations contain two helical turns. Note that the latticework array contains two separate layers, an upper layer oriented from lower left to upper right, and a lower layer oriented from lower right to upper left.

6. DNA Paranemic Crossover Molecules Offer Link-Free Edge-Sharing

If one were to close the ends of DX, TX, or other MX molecules, one would obtain a set a of catenated strands.[40] It is also possible to devise DNA topologies in which the resulting capped molecules would not be linked at all, but would associate as DNA dumbbells. The product molecules shown in Figures 6b and 8b are molecules of this sort. We call this system paranemic crossover (PX) DNA.[41] The triple domain molecules of Figure 8b have not yet been constructed successfully, but the double domain molecules of Figure 6b have been. This is a remarkable and stable system,[41] in which a four-stranded structure can be assembled from two dumbbell molecules. We have just begun to explore the nanotechnological properties of PX molecules. In

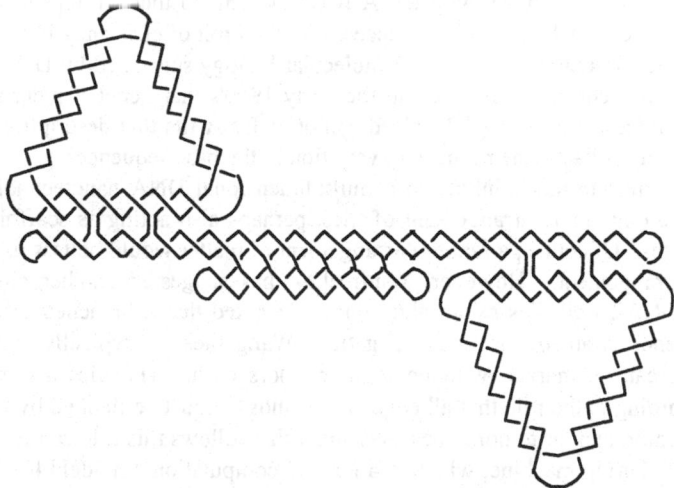

Figure 16. Cohesion by the PX Motif. Two DNA triangles are shown cohering by a PX-directed cohesion. These molecules are closed, because the PX motif does not generate links between individual strands, although the participating double helices are wrapped plectonemically around each other.

our first experiments, they appear to be as rigid as DX molecules (X. Zhang and NCS, in preparation). In addition, their properties suggest a general method for the recognition of DNA homology, although that remains to be demonstrated.[41]

On a practical level, PX DNA offers the extension of assembly motifs beyond simple sticky ends. The production of long sticky ends is difficult or awkward,[42] although a new development by Zhang and Taylor[43] may simplify it markedly. Figure 16 illustrates DNA triangles[44] held together by PX complementarity, rather than sticky ends. This approach takes advantage of the unlinked nature of PX DNA, and could link topologically closed species. The drawing shows triangles that cohere through the cohesion of two helices that are complementary to each other in the PX sense, even though they would not seem complementary in the conventional sense. It is very useful to be able to work with topologically closed species in DNA systems, so this linking function ultimately may prove to be extremely valuable in DNA nanotechnology.

7. Algorithmic Assembly May Lead to Complex Patterns

We have indicated above that the deliberate assembly of target periodic DNA arrays is a key goal of our work, particularly if the DNA acts as scaffolding for other molecules. Of course crystals are necessary for crystallographic diffraction analysis, but they also represent the simplest and likely cheapest way to organize matter from the bottom up in a specific and known fashion, with well-defined spatial relationships between the components. Furthermore, it is possible to impose a certain level of complexity on the system, such as we did above, with the ABCD* system, so that the repeat may be a complex unit. Nevertheless, this is not necessarily the limit of efficiency in DNA array assembly. One can return to the root of molecular biology suggested by Delbrück and Schrödinger, who correctly predicted in the early 1940's that genetic material would likely be 'an aperiodic crystal'.[45] Indeed, genomic DNA fits that description: There is a repeating linear backbone motif with variation in the base sequences.

If we translate this thinking to a multidimensional DNA nanotechnology, we can imagine an aperiodic arrangement of tiles, perhaps also acting as scaffolding for other molecules. Of course, a random arrangement of matter would be less useful than a periodic arrangement. However, Winfree[24] has suggested another alternative, assembling DNA-based tiles by an algorithm. He noted that a branched DNA motif with sticky ends is analogous to a Wang tile. Wang tiles are typically square tiles whose edges can be marked with up to four colors each. The tiles assemble into mosaics according to the rule that all edges in the mosaic must be flanked by tile edges of the same color, similar to dominoes. Assembly that follows this rule can emulate the operation of a Turing machine, which is a general computational model.[46] A simple computation adapted from ref. 47 is illustrated in Figure 17. Thus, one can imagine the assembly of DNA tiles, such as DX or TX molecules, to do computation or to produce programmed patterns in which each component need not be separately specified. If reduced to practice, this possibility holds out truly exciting possibilities for self-assembled systems. Recently, we have used this approach to perform a four-step cumulative XOR calculation in one dimension (C. Mao, T.H. LaBean, J.H. Reif and N.C. Seeman, submitted for publication.)

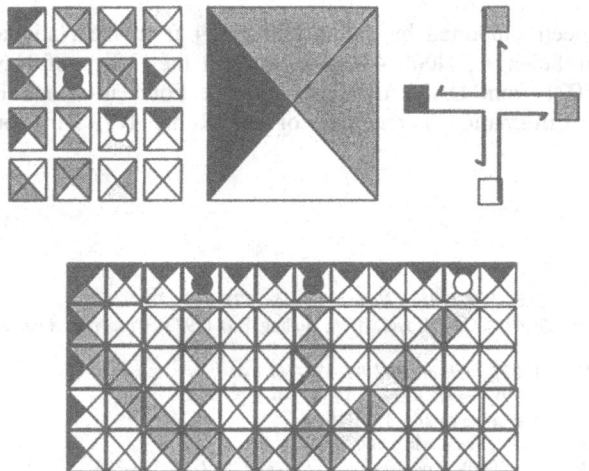

Figure 17. Wang Tiles. The upper left corner contains a set of 16 Wang tiles. The triangles that flank each edge are shaded differently to represent colors on the edge. A magnified tile is shown at the top center, and a four-arm branched junction is at its right. The shadings are the same in both the large tile and the sticky ends of the junction, indicating that the sticky ends on a branched junction can emulate a Wang tile. At the bottom is an assembly that obeys the rule that all edges are flanked by edges of the same color. The assembly represents the addition of four and seven to produce eleven. The fourth tile from the left in the top row has a dot indicating that it is one of the addends and the seventh tile is similarly marked; the eleventh tile has a white dot, indicating it is the answer tile. The 16 tiles shown would do a large set of additions in parallel assemblies; to do this particular calculation specifically, the second through sixth tiles in the top row would also have to be unique. The tiles within the calculation diagonals are specified by the assembly rule. Thus, not counting border tiles that could extend infinitely, 21 tiles can specify the positions of 46 elements in a directed fashion.

8. Concluding Remarks

Biology is nanotechnology that works. We have tried to show how the biomimetic system of DNA nanotechnology is predicated on a few simple concepts, particularly generalized complementarity and the reciprocal crossover structures that result from the removal of zero nodes. This approach leads in a general fashion to new motifs that have proved useful or are very promising. The immediate challenges in this area consist of extending the system to three-dimensions, to combine the chemistry described here to scaffold other nanoscale systems, such as carbon nanotubes[48] or proteins, to generalize the action of nanomechanical devices to programmability, to extend the system from periodicity to algorithmic assembly and to investigate the value of generalized complementarity in linking motifs. Ultimately, we strive to produce a non-living but life-emulating machinery that responds to instructions and can do useful

196

chemistry. As exciting as the past has been, we expect the future to be even more thrilling.

Acknowledgments

This work has been supported by grants GM-29554 from the National Institute of General Medical Sciences, N00014-89-J-3078 from the Office of Naval Research, NSF-CCR-97-25021 from DARPA/National Science Foundation and F30602-98-C-0148 from the Information Directorate of the Rome NY Air Force Research Laboratory.

References

1. Watson, J.D.; Crick, F.H. (1953), *Nature (London)* **171**, 737-738.
2. Cohen, S.N.; Chang, A.C.Y.; Boyer, H.W.; Helling, R.B. (1973), *Proc. Nat. Acad. Sci. (USA)* **70**, 3240-3244.
3. Seeman, N.C. (1982), *J. Theor. Biol.* **99**, 237-247.
4. Seeman, N.C. (1999), *Trends Biotech.* **17**, 437-443.
5. Adleman L. (1994), *Science* **266**, 1021-1024.
6. Seeman, N.C. (2000), *Synlett*, in press.
7. Voet, D.; Rich, A. (1970), *Prog. Nucl. Acid Res. Mol. Biol.* **10**, 183-265.
8. Fahlman, R.P.; Sen, D. (1999), *J. Am. Chem. Soc.* **121**, 11079-11085.
9. Protozanova E.; Macgregor R.B. Jr. (1996), *Biochem.* **35**, 16638-16645.
10. Qiu, H; Dewan, J.C.; Seeman, N.C. (1997), *J. Mol. Biol.* **267**, 881-898.
11. Seeman, N.C.; Chen, J.; Du, S.M.; Mueller, J.E.; Zhang, Y.; Fu, T.-J.; Wang, H.; Wang, Y.; Zhang, S. (1993), *New J. Chem.* **17**, 739-755.
12. White, J.H.; Millett, K.C.; Cozzarelli, N.R. (1987), *J Mol. Biol* **197**, 585-603.
13. Fu, T.-J.; Seeman, N.C. (1993), *Biochem.* **32**, 3211-3220.
14. Liu, B.; Leontis, N.B.; N.C. Seeman (1994), *Nanobiol.* **3**, 177-188.
15. Holliday, R. (1964), *Genet Res.* **5**, 282-304.
16. Ma, R.-I.; Kallenbach, N.R.; Sheardy, R.D.; Petrillo, M.L.; Seeman, N.C. (1986), *Nucl. Acids Res.* **14**, 9745-9753.
17. Kallenbach, N.R.; Ma, R.-I.; Seeman, N.C. (1983), *Nature (London)* **305**, 829-831.
18. Wang, Y.; Mueller, J.E.; Kemper, B.; Seeman, N.C. (1991), *Biochem.* **30**, 5667-5674.
19. Mao, C.; Sun, W.; Seeman, N.C. (1999), *J. Am. Chem. Soc.* **121**, 5437-5443.
20. Petrillo, M.L.; Newton, C.J.; Cunningham, R.P.; R.-I. Ma; Kallenbach, N.R.; Seeman, N.C. (1988), *Biopolymers* **27**, 1337-1352.
21. Wells, A. F. (1977), *Three-dimensional Nets and Polyhedra*, John Wiley & Sons, New York.
22. Williams, R. (1979), *The Geometrical Foundation of Natural Structure*, Dover, New York.
23. Robinson, B.H.; Seeman, N.C. (1987), *Prot. Eng.* **1**, 295-300.
24. Winfree, E. (1995), In: *DNA Based Computers*, ed. by R. Lipton and E. Baum, Am. Math. Soc., Providence, pp. 199-215.
25. Chen, J.; Seeman, N. C. (1991), *Nature (London)* **350**, 631-633.
26. Zhang, Y.; Seeman, N.C. (1992), *J. Am. Chem. Soc.* **114**, 2656-2663.
27. Zhang, Y.; Seeman, N.C. (1994), *J. Am. Chem. Soc.* **116**, 1661-1669.
28. Thaler, D.S.; Stahl, F.W. (1988), *Ann. Rev. Genet.* **22**, 169-197.
29. Sun, H.; Treco, D.; Szostak, J.W. (1991), *Cell* **64**, 1155-1161.
30. Schwacha, A.; Kleckner, N. (1995), *Cell* **83**, 783-791.
31. Li, X.; Yang, X.; Qi, J.; Seeman, N.C. (1996), *J. Am. Chem. Soc.* **118**, 6131-6140.
32. Winfree, E.; Liu, F.; Wenzler, L.A.; Seeman, N.C. (1998), *Nature* **394**, 539-544.
33. Liu, F.; Sha, R.; Seeman, N.C. (1999), *J. Am. Chem. Soc.* **121**, 917-922.
34. LaBean, T.; Yan, H.;. Kopatsch, J.; Liu, F.; Winfree, E.; Reif, J.H.; Seeman, N.C. (2000), *J. Am. Chem. Soc.* **122**, 1848-1860.
35. Mao, C.; Sun, W.; Shen, Z.; Seeman, N.C. (1999), *Nature (London)* **397**, 144-146.

36. Rich, A.; Nordheim, A.; Wang, A.H.-J. (1984), *Ann. Rev. Biochem.* **53**, 791-846.
37. Behe, M.; Felsenfeld, G. (1981), *Proc. Nat. Acad. Sci. (USA)* **78**, 1619-1623.
38. Churchill, M.E.A.; Tullius, T.D.; Kallenbach, N.R.; Seeman, N.C. (1988), *Proc. Nat.Acad.Sci. USA* **85**,4653-56.
39. Murchie, A.I.H.; Clegg, R.M.; von Kitzing, E.; Duckett, D.R.; Diekmann, S.; Lilley, D.M.J. (1989), *Nature (London)* **341**, 763-766.
40. Fu, T.J.; Tse-Dinh, Y.C.; Seeman N.C. (1994), *J. Mol. Biol.* **236**, 91-105.
41. Shen, Z. (1999), Ph.D. Thesis, New York University.
42. Liu, F.; Wang, H.; Seeman, N.C. (1999), *Nanobiol.* **4**, 257-262.
43. Zhang, K.; Taylor, J.-S., (1999) *J. Am. Chem. Soc.* **121**, 11579-11580.
44. Yang, X.; Wenzler, L.A.; Qi, J.; Li, X.; Seeman, N.C. (1998), *J. Am. Chem. Soc.* **120**, 9779-9786.
45. Schrödinger, E. (1967),*What is Life? and Mind and Matter*, Cambridge Univ. Press, Cambridge, pp. 60-71.
46. Wang, H. (1963), In *Proc. Symp. Math. Theory of Automata*, Polytechnic Press, New York, pp. 23-26.
47. Grünbaum, B.; Shephard, G.C. (1987), *Tilings & Patterns*, W.H. Freeman & Co., New York, pp. 583-608.
48. Colbert, D.T.; Zhang, J.; McClure, S.M.; Nikolaev, P.; Chen, Z.; Hafner, J.H.; Owens, D.W.; Kotula, P.G.; Carter, C.B.; Weaver, J.H.; Rinzler A.G.; Smalley, R.E. (1994), *Science* **266**, 1218-1222.

INTERACTION OF BIOMATERIALS WITH POROUS SILICON

S.C. BAYLISS, L.D. BUCKBERRY and A. MAYNE
Solid State Research Centre and Biomaterials Research Group
De Montfort University
Leicester LE1 9BH UK

1. The growing use of porous silicon in bioapplications

Porous silicon has now been used in a range of *in vitro* and *in vivo* bioapplications, from implant coatings to cell culture substrates and biosensors [1-5]. These widening applications are in part due to the large specific surface of porous silicon, the nano and micro-machinability of silicon, and the potential of nano and microstructured silicon to be used in novel devices due to the greatly modifiable optoelectronic properties. However there is also a growing body of literature which testifies the biocompatibility of porous silicon. Thus there appears to be many exciting biological areas in which porous silicon could play a major role. These include medical, environmental and computing technologies, and generally will require the immobilization of cells and enzymes in specific regions on porous silicon substrates. Once this has been achieved, the porous silicon can be built into, or supplied with, the relevant device architectures. However much background work needs to be done for these first steps to be realised. This itself requiring an understanding at the molecular level of interactions at the porous silicon surface. This interface between a cell and an inorganic material may seem straightforward at the micron level. It is in fact a dynamic system at the molecular level. Firstly, the substrate usually exhibits an oxide on the surface, due to exposure to air and the sterilization pre-treatment. Secondly, it is in a physiological environment, and generally proteins in the serum will coat the surface. In addition to this, it is necessary to ensure that the correct biosystem is adhered to the surface at the correct location. To go further in the deployment of porous silicon, therefore, it is necessary first to gain an understanding of the current techniques used in the research and development of biomaterials. And it is necessary to assess their applicability for the case of porous silicon. In particular the requirements for biomaterials need to be described, and the methods for controlling the interactions of biomolecules with biomaterials need to be clarified.

199

L. Pavesi and E. Buzaneva (eds.), Frontiers of Nano-Optoelectronic Systems, 199–207.
© 2000 *Kluwer Academic Publishers. Printed in the Netherlands.*

200

2. Biomaterials and toxicity

Biomaterials are materials which show biocompatibility. As well as for inert implants, such materials can be used in devices for restoring sensory and/or motor function, living skin equivalents, drug delivery systems, sensors, and for supporting matrices [6]. The term biocompatible includes the effect of the substrate on the function of the biosystem, including lack of toxicity, and the effect of the biosystem on the function of the biomaterial, including its sensitivity and stability. Along with the growing body of literature describing use of porous silicon for bioapplications, a recent study of toxicity of the silica and silicic acid forms of silicon following the MTT assay method of Buckberry has been reported [7]. Respiring cells metabolise MTT to its formazan derivative, a reaction catalysed by succinate hydrogenase. The derivative is purple, and the dye released from adhered cells cultured in the MTT spiked media can be assessed spectrophotonically. A sigmoid response is expected at a specific concentration, and the concentration which kills 50% of the cells (corresponding to 50% absorbance of untreated cells) can be used to calculate toxic efficacy. The B50s have, however, shown a definite lack of toxicity (Figure 1). The cell line chosen for our studies is the immortalised B50 rat hippocampal neuron cells, as these cells produce neurotransmitters necessary for the neuron-like functions, as would normal glial cells. The B50s were exposed to a range of concentrations from 0.0001 to 100 mM of either chemical for a period of 24hrs. The substrates used were coverslips since it was known that these substrates would not produce metabolites which could compromise the data. No toxicity was observed from either the silica or the silicic acid: cell cultures remained 100% viable (as compared with a control culture) over the exposure period of 24 h (Figure 1) even at the highest concentration levels used. Of course the *in vivo* or *in vitro* environments exposed to PS wafer would not include concentrations as high as 100 mM of silicic acid and silica: it is at these concentrations that salts precipitate out.

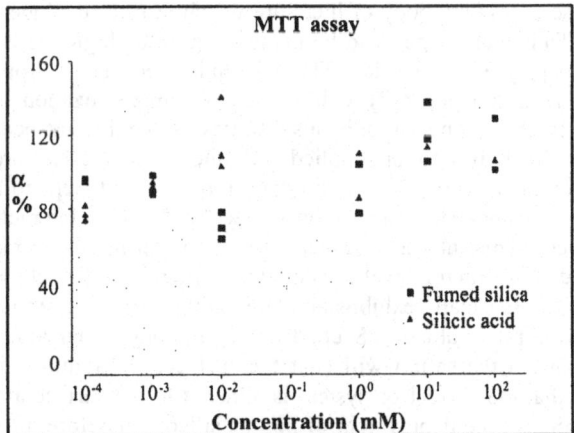

Figure 1. Toxicity of silicic acid and silica in culture of B50 cells on silica, using MTT assay.

The observation of lack of toxicity is supported by the anecdotal evidence of cell growth directly onto PS [8] and by other works [9].

3. Culturing cells *in vitro*

Recently there have been major advances in the techniques available for the *in vitro* culture of cells derived from mammalian tissues. Along with improvements in cell handling methods, the developments which have given rise to this progress include cell line characterisation and isolation, and identification of specific cellular growth factors. This has led to the situation where it is now possible to culture almost any type of mammalian cells.

Since there are 2 different types of cell systems, cells which grow in suspension and those which are anchorage-dependent, different methods are required for cell culture. For applications where interfaces between cells and inorganic materials are at issue, as in the deployment of porous silicon for implants, sensors and cell culture substrates, we are interested in the latter type. Individual cells from such cultures adhere to the substrate and also make contact with neighboring cells. Apart from ensuring that the correct physiological environment, including adhesion proteins (usually in the serum, or applied to the substrate surface), is in place, that the substrate is compatible with the biological system, and that the cell culture is healthy, no further chemical modification is required. The serum contains the correct concentration of nutrients and oxygen for the cells to grow and reproduce. Note that certain substrates (and certain cell lines) do not require adhesion proteins ~ in particular porous silicon has been successfully used to culture 4 cell line types without the need for such adhesive. In the case of enzymes, microbial cells and antibodies, however, the process of immobilization requires chemical modification.

4. Controlling cell growth and division

The growth of cells on surfaces has led to an increased understanding of the cell function and life cycle, and of the differentiation between healthy and cancerous cells. Furthermore the effect of drugs, toxins, and growth hormones on cell function can be investigated. In the case of porous silicon substrates, neuronal growth factor (NGF) has been tested on B50 cell cultures [10]. The reason for choice of NGF is that we hoped to slow down the cell division rate, thereby allowing individual cells to have increased longevity. This is of importance when one is trying to produce devices based on addressing specific cells ~ the problem with immortalized lines such as B50s is that they thrive very well on certain surfaces.

This means that they become confluent after a few days, eventually leading to the inability of the cells on the support surface to acquire the necessary nutrients to support their metabolism. This in turn leads to apoptosis, a form of internally-programmed cell death, effectively destroying the device characteristics. As an alternative to addition of factors such as NGF, a primary cell line could be used, but use of NGF is much simpler.

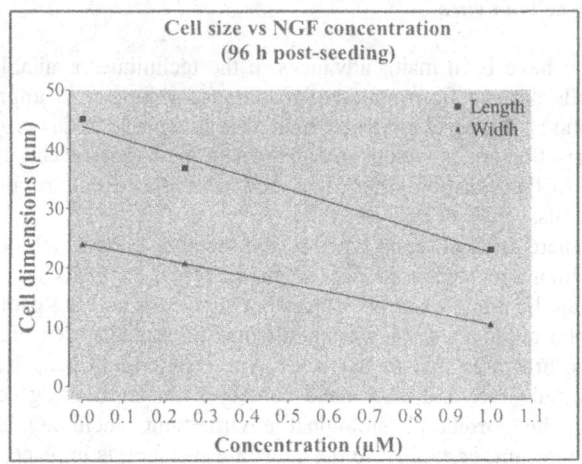

Figure 2: Length and width of B50 neurons cultured with 3 concentrations of NGF.

NGF-supplemented DMEM with a range of NGF concentrations from 0.1 to 0.75mM was supplied to cells in culture on porous silicon and on coverslips over a period of 96 hours. The cells were then fixed in glutaraldehyde at 24, 48, 72 and 96 hours, goldcoated, and imaged using SEM. The effect of NGF on the cell morphology has been assessed through the average length and width of the cell body (Fig. 2), and by cell counting.

There is a monotonic decrease in average cellular length and width with increase in NGF concentration but at present it is not known whether these effects are accompanied by changes in cell vitality and function, and in particular, the ability to process signals. In other words the biochemical effect of NGF on B50 cells is unclear.

The cell count data show that compared to a standard the NGF produced no increase in cell number, but also no decrease even with the highest concentration supplied. Once again an understanding at the molecular level is required. Studies are underway which seek to (i) confirm that cell division is suspended, and (ii) confirm if the changes in morphology are due to a decrease in cell viability. This would ultimately result in apoptopic cellular 'rounding', release from the substrate and death. Alternatively, this may be an intermediate morphological state prior to redifferentiation.

5. Effect of porous silicon surface features on cell anchorage

In addition to the nutrient requirements for cells to be viable on a surface, surface topology is of great importance [11-14]. These effects are found to be more dramatic for feature sizes around and below the sizes of the external cell structure (some, axons etc.). Thus, although it is clear that B50 cells are viable on porous silicon without the usual requirement of an adhesive layer, we need to determine whether such cells have any preference for surface features on a nm-micron scale. As a preliminary investigation, by lithography we have produced a regularly array of plateaux and troughs in porous silicon separated by 100 micron (Figure 3), and have estimated the preference for adhesion points through cell counts: for an average of 15 images, cell counts numbers were- plateaux 65, troughs 32, ridges 98. Thus B50 cells appear to prefer to adhere to the ridges, possibly due to the presence of increased surface charges at these points.

Using fluorescence interference contrast microscopy, it has been estimated by the Fromherz group [15] that cells do not attach directly to a surface support, but rather sit up to 100nm above the surface. This has great implications for direct electrical addressing of neuron cells, and will be discussed below.

 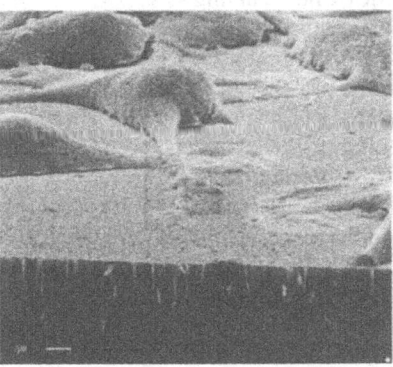

Figure 3: left: B50s cultured on an array of porous silicon plateaux and troughs; right: side view, showing good interconnections between B50s cultured on porous silicon, and the presence of anchorage supports from the cell to the surface.

6. Immobilization techniques

There are at present a range of general techniques used for immobilization of cells and enzymes. These are

1. Adsorption and covalent coupling (used for enzymes) [16]
2. Metal link/chelation processes (used for biocatalysts) [17,18]
3. Gel entrapment (used for cells and enzymes)
4. Microencapsulation (used for enzymes)

These methods make possible a wide range of applications, such as production of immobilized enzyme electrodes [19] and electrochemical techniques [20], which could be used to drive specific biochemical reactions. The ability to immobilize *cells* could give rise to completely different and important developments, such as the transforming of steroids, manufacture of cell factories for synthesising specific products, and novel bioassays.

The technique chosen for immobilisation is based on whether any one mode of attachment will disrupt the structure of the protein and hence its activity. It is also important to avoid overloading to prevent steric hindrance which will reduce the activity by reducing the binding.

(a) Adsorption and covalent coupling
Adsorption is the simplest method. The procedure for this is
i.   Mix enzyme or other protein and support
ii.   Incubate
iii.   Centrifuge to separate soluble from insoluble

The disadvantage is that the enzyme is not bonded to the support but adsorbed (via salt-linkages), so changes in pH, ionic strength and temperature could debind the enzyme.

Examples of current adsorption solid supports are alumina, calcium carbonate collagen, cellulose, glass (porous), hydroxyapatite, silica gel. It is possible that porous silicon could be used as the solid support in this method.

(b)Covalent binding
It is important that the amino acids involved in the activity of the protein are not bound to the solid support (binding of this type is minimised if the enzyme involved is bound in the presence of its substrate). To ensure this, covalent modification is used. Typical supports for covalent modification are agarose (sepharose), dextran (sephadex, a sugar polymer), glass and polyacrylamide co-polymers.

For this the support must first be activated, and the most widely used activated agent is cyanogen bromide (which is bound to the support) as no activation step is required. The cyanogen bromide then reacts with free amino groups of proteins.

(c) Metal link/chelation processes
Using chelation processes [17,18] should be simpler and quicker than using covalent modification as this latter method requires matrix activation as an extra step. Good candidate metals for this have been titanium and zirconium (particularly good because the oxides are non-toxic). Increased activation of a range of enzymes on various supports have been found including the transition metal chloride compared to that in its absence.

The method generally used is:
i. titanium (III) chloride is used as a reducing agent for the nitroaryl derivative of cellulose.
ii. the resulting aminoaryl derivative promotes the diazo coupling of the protein.

The result should be direct binding between the hydroxide and hydrogen terminations on the porous silicon and those in biomolecules in the cell membrane. Thus this will reduce the physical gap between the cell/enzyme and the porous silicon to a few well-defined monolayers.

7. Biocomputing

The possibility of the production of a well-defined interface between silicon and neurons could be used as the basis for novel computing systems based on conventional binary data processing interfaced to living parallel computational units. For this the dynamic nature of the interface must be taken into account.

By chance the sizes of the comparable units involved in these 2 types of data processing, the soma of the neuron and the transistor chip, are rather similar, being on the order of microns. Even the relative signals are similar and of the same order of 1V. The main difference between the systems is that neural systems are capable of training and adaptation. In addition, neural systems exhibit low power consumption, (40W for a sleeping human including cell regeneration). Although each action potential travels relatively slowly, there are upto thousands of connections to each neuron in the brain, and over hundreds of thousands of neurons. Thus there exists in the brain a highly complex 3D architecture, which would be impossible to reproduce in an artificial network in the foreseeable future. This enables a living system to be extremely efficient at pattern recognition, even in the presence of highly damaged input data.

The analogue computer, on the other hand, is capable of highly precise calculations, but if parts of the processor are damaged, the whole system breaks down. There are great gains to be had, therefore, by a combined approach to computational tasks.

Towards this end there have been many attempts to address individual neurons using conventional semiconductor structures. Very exciting results have been obtained by the Fromherz group in Munich [21] on silicon, and the Jerry Pines group in Caltech [22] on gold/silica. Both of these groups have demonstrated electrical stimulation and recording. The Pine group have further packaged individual neurons in units arranged in 4x4 arrays with nutrient supply available. However, even with a comprehensive enclosure and food supply, the neurons behave plastically, attempting to grow out of their containers. Furthermore the signals recorded are much weaker than would be expected (of the order of μV rather than mV). Once these problems have been solved, then much study on neural signal processing can take place, and a wide range of computational devices can be realised, such as the pRAM developed by the Clarkson group in London[23].

As an alternative to the approaches above, we have carried out mechanical stimulation of B50 cells using a microprobe [24]. The cells were cultured on a 22mm coverslip, and were loaded with Ca+ dye. It was possible to record the passage of a signal through the cell network 15 mm away from the stimulation point. The signal was recorded as a function of time, and the signal pulse peak was obtained 30 microseconds after stimulation. No signal was recorded when an artificial break was made in the network, showing that the signal was propagating through the network.

8. Final remarks

Earlier work on cell stimulation used the patch-clamp method to stimulate and probe a cell response [25]. This is not suitable for long-term studies, as the neuron quickly dies, and furthermore it is not investigating the neuron under natural conditions. The procedures described above [21,22] are an improvement as they use external electrodes. Despite the additional progress in the cell culture procedures and the design of neuron addressing devices such as the Neurochip, it is surprising that good signal to noise data on stimulation and recording has not been achieved. It may be that the neuron's apparent resistance to toxicity is linked to its inability to transmit artificial signals. This is perhaps a self-protection mechanism which allows only the correctly-produced information to be passed. Thus either mechanisms other than electrical may need to be developed, or the functionality itself may need to be adapted at a biomolecular level.

Finally it should be noted that cells can adapt to their environment over time, sometime involving a change in functionality. Although it is possible to revive some of these functions, using the appropriate physiological environments, it is possible that in a living system, such as is envisaged for biocomputing, the neurons will be modified over time. In this case it may not be possible to predict how the system will behave, and this leads to many questions concerning the use of living systems for technological reasons.

Thus, despite the problems highlighted on signal processing and cell localization, it appears porous silicon has a lot to offer to the area of neurobiology as it so far given no indication of toxicity to neurons, and good adherence when cultured on porous silicon. In addition it can be patterned using conventional methods and has optoelectonic emission and detection properties. The development of a silicon to neuron interface is of particular importance for the development of energy efficient, powerful 3D architectures, complementing or even replacing the need for artificial intelligence. We apologize for not including in this paper more of the vast body of work in the area of neural stimulation.

References

1. Laurell, T., Drott, J., Rosengren, L. and Lindstrom, K., *Sensors and Actuators* **B31** (1996) 161
2. Wei, J., Buriak, J. and Siuzdak, G. *Nature* **399**(6733) (1999) 243
3. Van Noort, D., Welin-Klintstrom, S., Arwin, H., Zangooie, S. Lundstrom, I., and Mandenius, C.-F., *Biosensors and Bioelectronics* **13** (1998) 439
4. Bateman, J.E., Worrall, D.R., Horrocks, B.R. and Houlton, A., Angew.Chem.Intl.Ed. 37 (1998) 2683
5. Zangooie, S., Bjorkland, R. and Arwin, H., *J. Electrochem. Soc.* **144** (1997) 4027
6. Collings, A.F. and Caruso, F., *Rep. Prog. Physics* **60** (1997) 1397
7. Buckberry, L.D., Adcock, H.J., Gaskin, P.J., Adler, J. and Shaw, P.N., *A.T.L.A.- Altern. Lab. Anim.* **22**, 72 (1994).
8. Bayliss, S.C., Harris, P.J. and Buckberry, L.D., *J. Porous Mater.* **7**, 191, (2000)
9. Tanaka, T., Tanigawa, T., Nose, T., Imai, S. and Hayashi, Y., *J. Trace Elem. Exp. Med.* **7**, 101, (1994); Canham, to be presented at the ARW Kiev 2000
10. Mayne, A.H., Bayliss, S.C., Barr, P., Tobin, M. and Buckberry, submitted to *Phys. Stat. Sol.* April 2000
11. Eisenbarth, E., Meyle, J., Nachtigall, W. and Breme, J., *Biomaterials* **17** (1996) 1399
12. Curtis, A. and Wilkinson, C., Biomaterials, **18**, 1573, (1998)
13. Brunette, D.M. and Chehroudi, B., *J. Biomech. Eng.* **121** (1999) 49
14. Ito, Y. *Biomaterials* **20** (1999) 2333

15. Braun, D. and Fromherz, P., *Phys. Rev. Lett.* **81** (1998) 5241
16. Kennedy, J.F. and Cabral, J.M.S. in *Applied Biochemistry and Bioengineering* Vol.4, eds. Chibata, I. and Wingard, L.B. Academic Press New York p190 (1983)
17. Novais, J.M., PhD Thesis, University of Birmingham UK (1971);
18. Clark, L. and Lyons C., *Ann. N. Y. Acad. Sci.* **102** (1962) 29
19. Guilbault, G.G. *Handbook of Immobilised Enzymes*, pub. Marcel Dekker, New York (1984)
20. Castner, J.F. and Wingard, L.B. Jr, *Biochemistry (Wash.)* **23** (1984) 2203
21. Fromherz, P., *Eur. Biophys. J.* **28** (1999) 254
22. Maher, M.P., Pine, J., Wright, J. and Tai, Y.C., *J Neurosci. Methods*, **87** (1999) 45
23. Clarkson, T., *IEEE Trans.* (1999)
24. Mayne, A. and Tobin, M., unpublished data
25. Fromherz, P., Muller, C. and Weis, R. *Phys. Rev. Lett.* **75** (1993) 1670

POLYDIACETYLENE PTS: A MOLECULAR QUANTUM WIRE WITH EXCEPTIONAL OPTICAL PROPERTIES

CARLOS G. TREVIÑO-PALACIOS*, GEORGE STEGEMAN,
MINGGUO LIU, FUMIYO YOSHINO, SERGEY POLIAKOV, AND
LARS FRIEDRICH
*School of Optics and CREOL,
University of Central Florida, Orlando, Florida, USA*

STEVEN R. FLOM, J.R. LINDLE AND F.J. BARTOLI
*Optical Sciences Division, U.S. Naval Research Laboratory, Washington,
DC 20375*

♦*Permanent address: INAOE Apdo. Postal 51 y 216 Puebla 07200 Pue.
Mexico*

Abstract

The conjugated polymer 2,4-hexadiyne-1,6-diol-bis-(para-toluenesulfonate) or PTS is a classical example of a one-dimensional quantum wire. The π-electrons can move more or less freely in one dimension. There are large repercussions to this effect in the photonics field, including photoconductivity, and the linear and nonlinear optical response. Growing techniques of crystals with high optical quality as well as measurements on a variety of nonlinear optical effects of PTS are reported in this chapter. These include measurements on the large magnitude of the exciton absorption line, the well-defined vibrational side bands at room temperature, massive two and four photon absorption coefficients, very large Raman gain coefficients, minimal excited state absorption and a large nonlinear refractive index.

1. Introduction

Quantum wires are quasi one-dimensional systems of electrons, which by virtue of their one dimensionality have unique properties. They are now under intense investigation because it has become technically feasible to fabricate quantum wires in semiconductors. [1] Many of these unique properties hold considerable promise for new electronic and photonic devices in the semiconductor case. [2]

One form of a quantum wire is a conjugated polymer, the backbone of which consists of multiple sequential carbon single, double and triple bonds. The key point is

209

L. Pavesi and E. Buzaneva (eds.), Frontiers of Nano-Optoelectronic Systems, 209–226.
© 2000 *Kluwer Academic Publishers. Printed in the Netherlands.*

210

that a carbon atom has four valence electrons that occupy the 2s, $2p_x$, $2p_y$ and $2p_z$ orbitals. Upon carbon-carbon bond formation the s and p electrons hybridize to form σ and π orbitals. In the case of alternating double (or triple) bonds, the $2p_z$ (and $2p_y$) orbitals that lie orthogonal to the σ bond, form a continuous orbital in which electrons can move more or less freely parallel to the carbon-carbon axis, the polymer "backbone", see Figure 1. The facile movement of the π–electrons results in the large polarizabilities (and hyperpolarizabilities) observed in conjugated polymers.

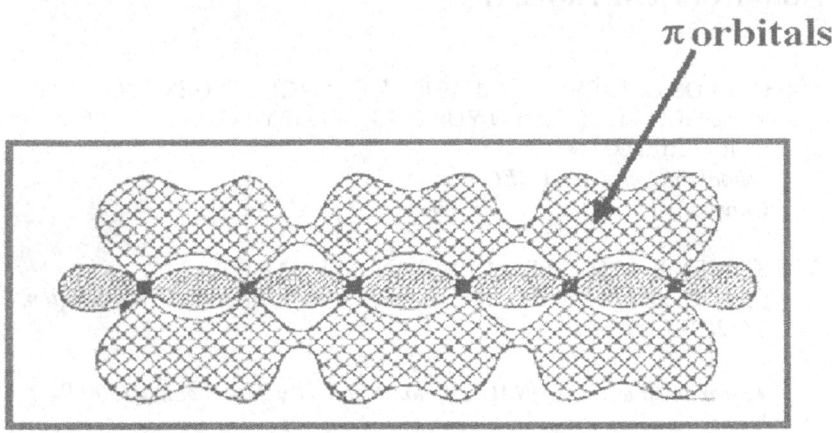

Figure 1. π-electron cloud along a conjugated polymer backbone.

These delocalized π–electrons in conjugated polymers give rise to the fundamental nature of the polymeric quantum wire. When the polymer is in its non-crystalline or polycrystalline, amorphous state, i.e. not kept straight in a rigid lattice, the effective delocalization length is short. More importantly, the individual quantum wires are not well aligned yielding a net reduced bulk polarizability. In this case the properties of such quantum wires are not fully optimized. However, conjugated polymers can be grown as single crystals and presumably their "quantum wire" properties are then significantly improved.

The special properties of conjugated polymers were recognized in the late 1960's and early 1970's by Wegner and colleagues.[3] Many interesting properties were investigated over the succeeding years, including the excitonic nature of the dominant and very intense absorption peak in the optical absorption spectrum, the high electrical conductivity and the photo-activated conductivity. The conjugated polymer with optimized quantum wire properties is polyacetylene which has a sequence of alternating carbon single and double bonds. [4] However, the material is not processable and single polyacetylene crystals have proven challenging to grow. On the other hand, it has been possible to grow polydiacetylene crystals that consist of single, double and triple carbon bonds. [5-6] The parallel polymer backbones which constitute the crystal are held together through Van der Waals interactions. The side groups that give rise to the monomer solubility are only weakly coupled to the polymer backbone. Indeed, investigations have shown that the many polymer properties are to a large degree independent of the details of these side groups. [7]

Although thin single crystals of polydiacetylenes have been grown successfully by many groups, it has proven to be challenging problem to grow thick high optical quality crystals suitable for photonics applications. [8-12] They are typically grown as monomer single crystals, and then polymerized by heating, electron bombardment etc. [13-16] Typically the crystals exhibit cracks, twining etc. [17-20] The principal cause is the shrinkage that occurs along the b-axis on polymerization. [21] As a result the crystals have scattered light due to these imperfections, making them unsuitable for photonics applications. The situation, however, has been different for thin films. Thakur and co-workers have produced optical quality films by a shear growth method. [22]

In this chapter we describe our research into the growth and measurements of the optical properties of a specific polydiacetylene, namely polydiacetylene PTS, 2,4-hexadiyne-1,6-diol-bis-(para-toluenesulfonate). Although good quality polydiacetylene thin crystals have been made from this polymer, the optical quality of thicker crystals has not been adequate. Nevertheless, many of the interesting features of this polydiacetylene have been studied on such crystals, for example the absorption spectrum, the photoconductivity, two-photon absorption etc. We have invested a number of years investigating the growth techniques for the monomer diacetylenes, optimizing the handling and polymerization conditions, and significantly improving their optical properties. In these crystals we have found remarkable nonlinear optical properties which are reported here.

2. Polydiacetylene Single Crystals

Standard techniques were used to synthesize the PTS starting material and purify it by multiple re-crystallizations. [23] One of the keys to high optical quality crystals is definitely the purity of the starting material. The final monomer crystals were grown in the closed apparatus shown schematically in Figure 2. The rate of nitrogen flow controlled the rate of evaporation of the solvent, and hence the rate of crystal growth. Since no seeds were used, multiple crystals were obtained with volumes varying from a few mm^3 to a cm^3. The as-grown monomer crystals were light pink due to the partial

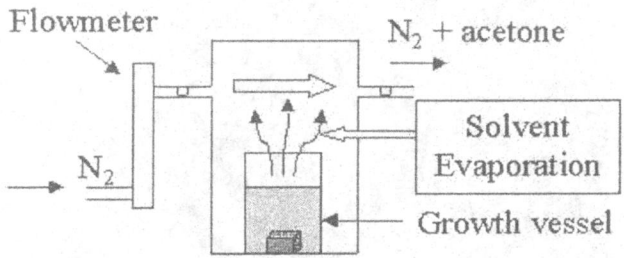

Figure 2. Experimental apparatus for monomer PTS crystal growth.

polymerization of the monomer during the crystal growth, and had the chemical structure shown in Figure 3a.

These crystals were cleaved into platelets with thicknesses varying from about 50 microns to about a millimeter. The final surface quality depends crucially on the quality of the cleave and this step requires great care. A SEM picture of the surface,

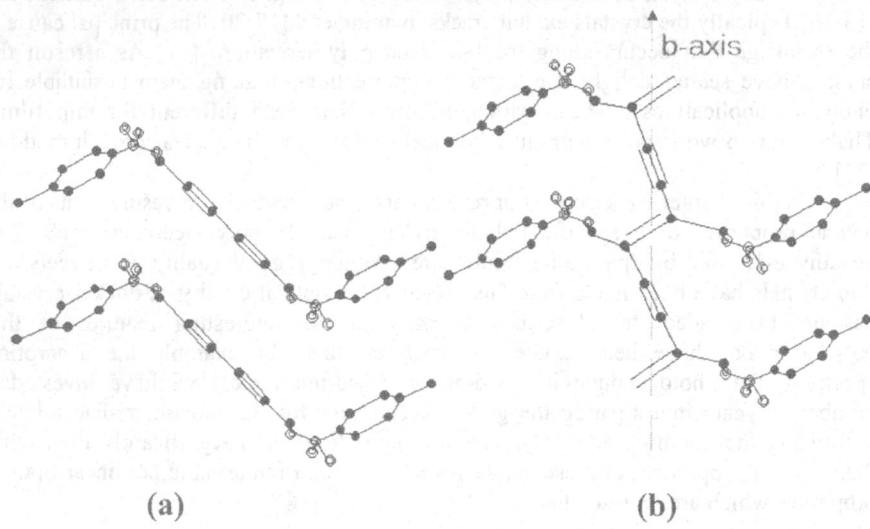

(a)   (b)

Figure 3. PTS structure after crystal growth (a) before polymerization and (b) after polymerization. The main optical axis is shown as the b-axis.

after polymerization, is shown in Figure 4.

The crystals can be polymerized either by heating, or by UV light irradiation. The chemical structure changes resulting in the alternating double and triple bonds

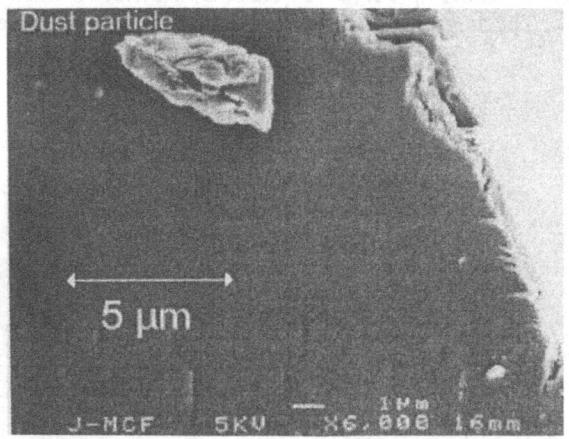

Figure 4. SEM of the (100) facet of a PTS polymer crystal. The dust particle in the upper third of the surface was used for focusing the electron beam.

shown in Figure 3b. It is during this step that the large shrinkage (□4.5%) occurs along the b-axis of the final crystal [24]. If the polymerization is allowed to occur in an uncontrolled fashion multiple defects are created. By controlling the polymerization rate and effectively annealing the crystal during polymerization, crystals with very low defect concentration were obtained. The crystal morphology after polymerization and the direction of the optical b-axis along which the conjugation lies is shown in Figure 5a. The final crystal platelet has a dark gray metallic sheen. A picture of one is shown in Figure 5b.

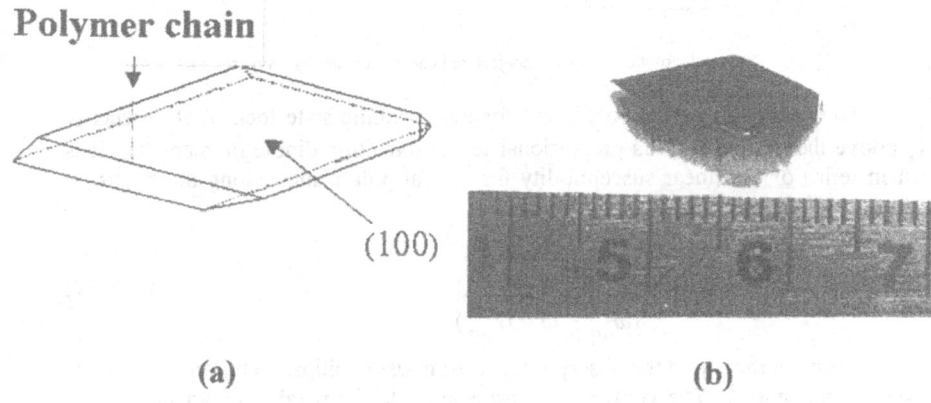

Polymer chain

(100)

(a)   (b)

Figure 5. (a) The morphology of a PTS single crystal. (b) PTS crystal.

3.   Linear Optical Properties

A good way of estimating the quality of polydiacetylene crystals as conjugated polymers is to measure the location and strength of the exciton absorption feature. The known and conjectured electronic energy levels of PTS are shown in Figure 6. Our determination of the two photon excited states (even symmetry A_g) will be discussed later in this chapter. The ground state has even symmetry ($1A_g$) and the exciton state is $1B_u$. Each electronic state has associated with it a manifold of vibrational sub-levels. The quasi-continuum for wavelengths shorter than about 0.55 µm consists of alternating even and odd symmetry states. As will be shown later, the dominant even symmetry excited state lies within this continuum. The onset of the continuum has been measured by electro-reflectance. [25] There is also a triplet set of energy levels about which very little is known other than that they can be accessed at very high intensities and at high photon energies in the visible and UV regions of the spectrum.

214

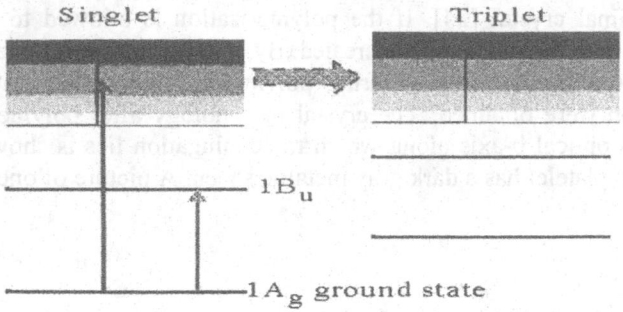

Figure 6. Schematic of the known and conjectured one photon energy levels for PTS.

The linear absorption coefficient for the excitonic state located at an energy $\Sigma\omega_{ug}$ above the ground state is proportional to the transition dipole moment μ_{gu}. It is given in terms of the linear susceptibility for optical polarization along the b-axis by $\alpha_1 \propto \Im mag\{\chi^{(1)}(-\omega; \omega)\}$ and

$$\chi^{(1)}(-\omega; \omega) \propto \frac{|\mu_{ug}|^2}{(\omega_{ug} - \omega - i\Gamma_{ug})}. \tag{1}$$

We have measured the absorption spectrum using ellipsometry and the results are shown in Figure 7. The vertical line corresponds to the parallel polarized exciton line, centered at 630 nm. The horizontal line indicates the spectrum which can be attributed to vibronic sub-bands, and in fact the two peaks have been successfully associated with vibrational transitions involving the strongest lines in the Raman spectrum, i.e. the 1493 cm^{-1} and 2085 cm^{-1} shifts.

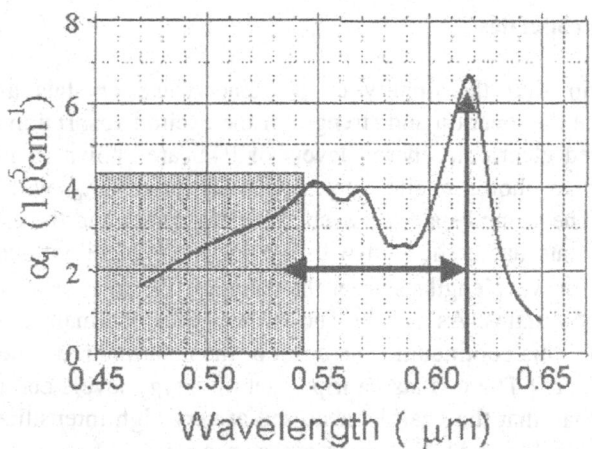

Figure 7. PTS absorption spectrum for light polarized along the b-axis. The vertical line corresponds to the exciton for the transition $1A_g \,|\, 1B_u$. The horizontal line represents vibronic sub-bands of this transition. The shaded area represents transition to the quasi-continuum even and odd symmetry states.

The shaded region is believed to be due to transitions involving the quasi-continuum, as well as probable overlap with the vibronic sub-structure associated with the exciton line. Given the magnitude of the absorption maximum ($\alpha_1 = 7 \times 10^5$ cm^{-1}), and the amount of oscillator strength associated with the vibronic sub-bands, the total oscillator strength for the $Ag_1 \parallel 1B_u$ is very large indeed!

Ellipsometry has also been used to measure the dispersion in the refractive index for polarization along the b-axis, and the results are shown in Figure 8. What is important here is the large refractive index of about 2.0 at 1 micron, much larger than the 1.58 measured for the orthogonal polarization (not shown here). In an independent experiment, the parallel and perpendicular refractive indices have been measured interferometrically at 2 microns to be 1.87±0.02 and 1.62±0.04. Both are among largest birefringences known. All of these linear optical characterizations indicate very high quality crystals.

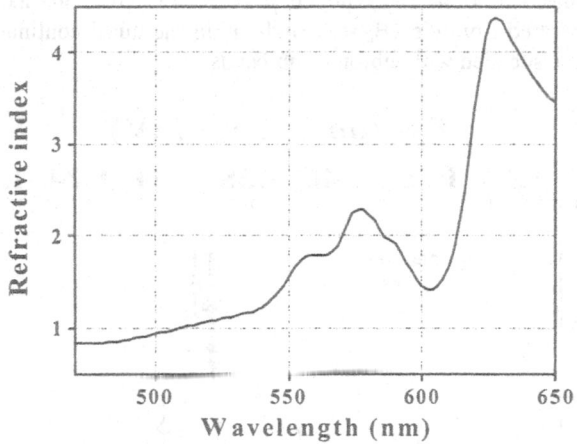

Figure 8. Refractive index dispersion for PTS measured using ellipsometry.

4. Two-Photon Absorption Spectrum

The PTS molecular unit is centro-symmetric which results in electronic states with either even or odd symmetry, but not mixed symmetry. Because a one-photon electric dipole transition involves a change in the symmetry between the initial and final states, the even symmetry excited states cannot be studied by linear absorption involving the ground state. However, the even symmetry excited states can be identified by two-photon spectroscopy which involves the simultaneous absorption of two photons.

The Two-photon absorption (TPA) coefficient, α_2, is given by $\Delta\alpha = \alpha_2 I$ with $\alpha_2 \propto \Im\text{mag}\{\chi^{(3)}(-\omega; \omega, -\omega, \omega)\}$ and

$$\chi^{(3)}(-\omega; \omega, -\omega, \omega) \propto \frac{|\mu_{gu}|^2 |\mu_{ug'}|^2}{(\omega_{gu} - \omega - i\Gamma_{gu})^2 (\omega_{gg''} - 2\omega - i\Gamma_{gg'})}. \tag{2}$$

here $\mu_{ug'}$ is the transition dipole moment between the $1B_u$ and the even symmetry excited state g', and $\Sigma\omega_{gg'}$ is the energy difference between the ground state and the even symmetry excited state. Also, the intensity-dependent refractive index coefficient $n_2 \propto \Re\text{eal}\{\chi^{(3)}(-\omega; \omega, -\omega, \omega)\}$. Therefore by scanning the frequency ω, maxima in the intensity-dependent absorption are indicative of the spectral location of excited even symmetry states.

The technique well suited for measuring two-photon absorption is Z-scan spectroscopy. It involves moving a sample through the focal point and detecting the full-transmitted signal (for measuring α_2) and through a limited aperture for evaluating n_2. The results for the wavelength range 700-1000 nm obtained with multiple laser pulse widths are shown in Figure 9. The most remarkable result is the magnitude of the peak at 925 nm, 700 cm/GW. This should be compared to the value in semiconductors, which are 10's of cm/GW. [26] Therefore this state, labeled here as $3A_g$, also has a large transition dipole moment to the $1B_u$ state, i.e. $\mu_{ug'}$. Furthermore, this state is located at 460 nm, well-separated from the $1B_u$ state and within the quasi-continuum. The short wavelength tail is associated with vibronic sub-bands.

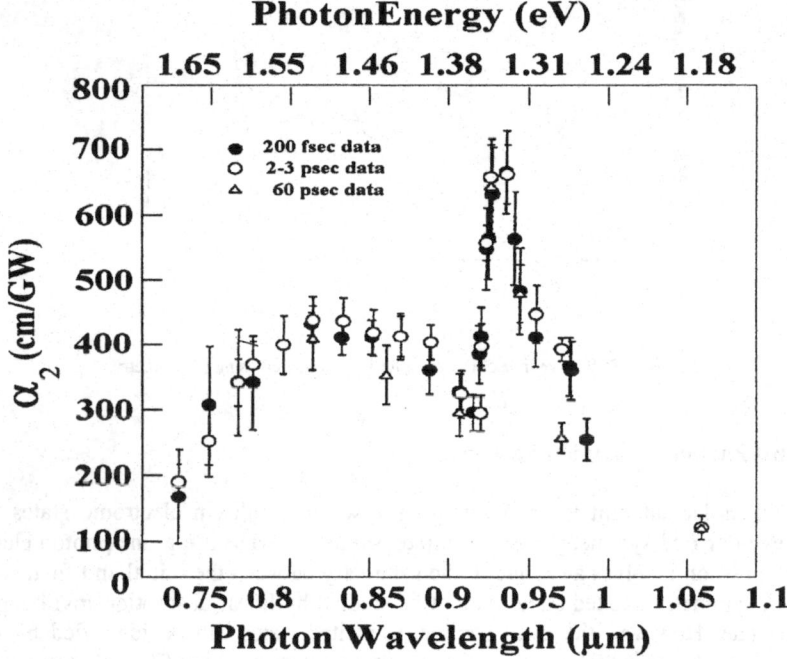

Figure 9. Two-photon absorption (α_2) spectrum of PTS from 700 nm to 1100 nm measured with 60 psec, 2 psec, and 200 fsec pulses.

Additional experiments (not shown here), with a single laser pulse width of 60 psec, extended the wavelength range studied right out to 1600 nm and allowed a lower lying even symmetry excited state, labeled $2A_g$, to be identified just below the $1B_u$

exciton state at about 630 nm. The maximum α_2 was much smaller in this case, 30 cm/GW.

The experiments in the 700-1000 nm range were actually performed as a function of intensity. When $\Delta\alpha$ is not strictly proportional to I, this means that additional processes that will be discussed in greater detail below, are active. The difference between $\alpha_2 I$ where α_2 was obtained by extrapolating $\Delta\alpha$ to low intensities, and the measured $\Delta\alpha$, i.e. $\Delta\alpha - \alpha_2 I$ was interpreted as $\alpha_3 I^2$ and the deduced value for α_3 is shown in Figure 10 as a function of intensity.

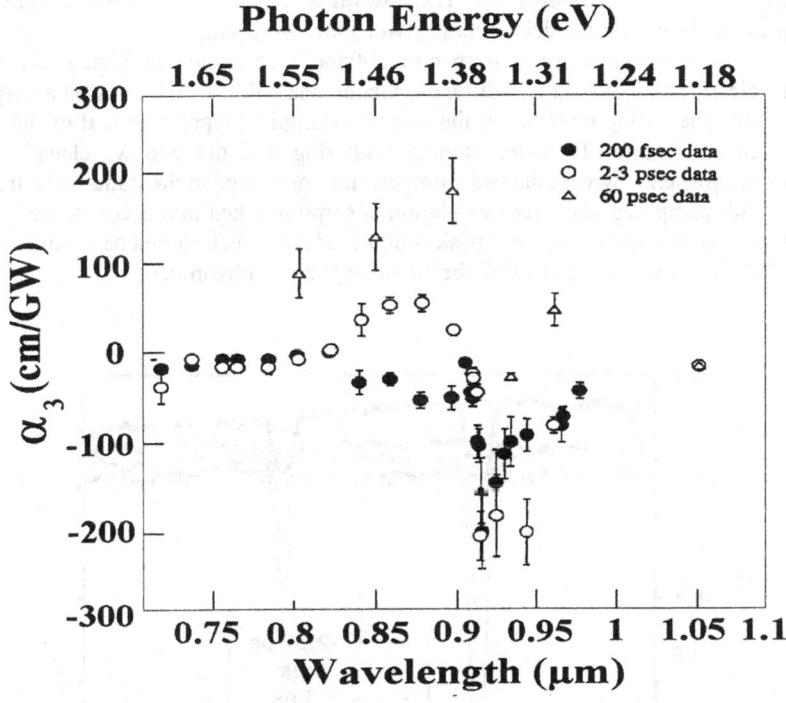

Figure 10. α_3 spectrum of PTS from 700 nm to 110 nm measured with 65 psec, 2 psec, and 200 fesc pulses. The values for α_3 were deduced as described in the text.

The results show two distinct kinds of behavior. The minimum at 925 nm is interpreted as due to saturation of the two-photon state. The pulse-width dependent increase in α_3 at shorter wavelengths indicates that there is additional absorption from a state accessed through the two-photon transition. However, it is not clear whether this additional absorption is from the $3A_g$ state, or from the $1B_u$ state to which the $3A_g$ is believed to relax very rapidly. However, it is clear that there is some additional excited state absorption.

5. Transient Absorption Spectroscopy

Roughly the same wavelength range as discussed in detail above was probed by transient absorption spectroscopy. In these experiments, an intense 1.2 ps pump laser pulse of tunable wavelength was used to both illuminate the sample and to generate a white light continuum. First the two-photon absorption data discussed above was verified by measuring the nonlinear transmission of the pump pulse as a function of wavelength. Good agreement was obtained between the two sets of data. The pump pulse was then fixed at 725 nm and used to create a population in an excited state accessed by two-photon absorption. The absorption of a continuum probe beam was then measured for different delay times between the two beams.

The transient absorption spectra obtained are shown in Figure 11. These spectra clearly show at least two different phenomena, all of which relax on a very fast time scale. The strong emission at the two wavelengths longer than that of the pulse beam corresponds to stimulated Raman scattering and the two wavelength shifts measured agree exactly with the two strongest lines measured in the Raman spectrum of PTS. With pump depletion via two-photon absorption taken into account, the Raman gain factor for the more prominent peak is 600 cm/GW, which should be compared with 24 cm/GW for CS_2 and 1.5 cm/GW for 10 atmospheres of hydrogen.

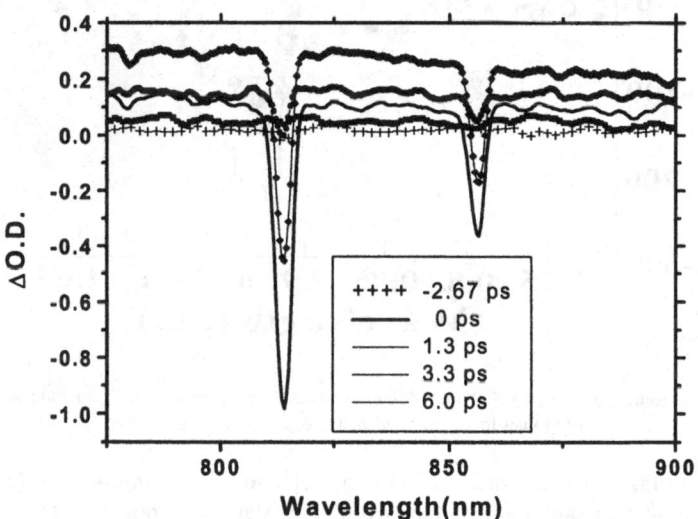

Figure 11. Transient Absorption Spectra of a 57 μm thick single crystal of pTS. The excitation wavelength is 725 nm and the intensity is 5.5 GW/cm².

The broad induced absorption observed throughout the spectra can be attributed largely to population of excited states through two-photon excitation. The lifetime of the state is consistent with that observed under one-photon excitation. [27] Detailed analysis of the time and intensity dependences yields evidence that in addition to excited state absorption, nondegenerate two-photon absorption is contributing to the observed signal.

Further experiments were performed using degenerate four-wave mixing (D4WM). In this technique, three laser pulses are overlapped in the sample. When all three arrive simultaneously the response is proportional to $|\chi^{(3)}(-\omega; \omega, -\omega, \omega)|^2$. When one pulse is delayed with respect to the other two, the two pump beams interfere inside the sample to create a grating. If components of this grating create a population in an excited state, then a later beam may be Bragg deflected by this grating provided that the population has not decayed. The intensity dependence of the zero time peak shown in Figure 12 yields $\chi^{(3)}(-\omega; \omega, -\omega, \omega)$ values consistent with those previously reported. [28] In addition, there is clear evidence for a long lived excited state, probably the aforementioned triplet, as a very weak background level in the grating response shown in Figure 12. Also in evidence is scattering from an oscillating grating which corresponds to an acoustic wave launched by the impulsive heating at the peaks of the interference grating.

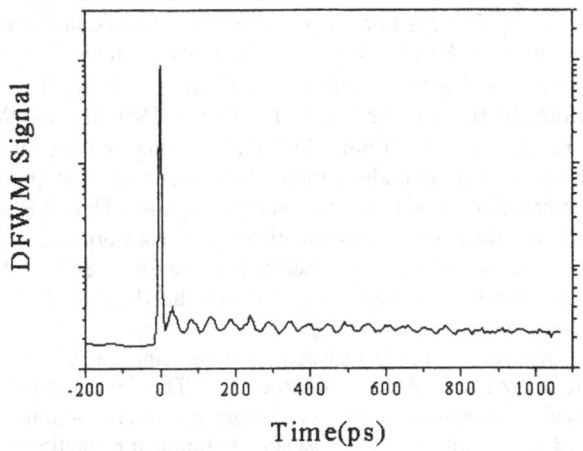

Figure 12. Degenerate four wave mixing (D4WM) deflection of the probe pulse on PTS.

6. Higher Order Absorption

The very large one- and two-photon absorption coefficients measured suggest that higher order absorption processes may also be strong. It is necessary to distinguish between processes which involve the simultaneous absorption of three or four photons, versus processes in which multiple, sequential, one- or two-photon process are involved.

Multiple one-photon absorption processes have been reported previously, usually in semiconductors where carriers are excited by or two photon processes and then subsequently free carrier absorption occurs.

A detailed calculation for the three-photon absorption coefficient α_3 to be published elsewhere gives

$$\alpha_3 \propto \frac{|\mu_{gu}|^6}{(\omega_{gu} - 3\omega - i\Gamma_{gu})} f(\omega, \omega_{gu}) \qquad (3)$$

where $\Delta\alpha = \alpha_3 I^2$ and $f(\omega, \omega_{gu})$ is a slowly varying function of its variables. Note that the three-photon absorption spectrum peaks at $\omega_{gu}/3$. In terms of wavelength, the absorption spectrum in Figure 7 suggests that three-photon should occur over the wavelength range $3 \times 500 = 1500$ nm to $3 \times 650 = 1950$ nm, peaking around 1860 nm. The same calculation extended to the next highest order for α_4 where the increase in the absorption coefficient is $\Delta\alpha = \alpha_4 I^3$ gives

$$\alpha_4 \propto \frac{|\mu_{gu}|^6 |\mu_{ug'}|^2}{(\omega_{gu} - 3\omega - i\Gamma_{gu})^2 (\omega_{gg'} - 4\omega - i\Gamma_{gg'})} F(D, \omega, \omega_{ug}, \omega_{gg'}) \qquad (4)$$

where $\Sigma\omega_{gg'}$ is the energy difference between the ground and excited two-photon state, $D = |\mu_{ug'}/\mu_{gu}|$, $\mu_{ug'}$ is the dipole transition moment between the $1B_u$ and even symmetry excited states and the function F varies only slowly with its variables. The denominators imply that α_4 is resonantly enhanced at both $\omega \cong \omega_{gu}/3$ and at $\omega \cong \omega_{gg'}/4$.

As noted before, the first corresponds to 1500 nm to 1950 nm, and the second to $2 \times 700 = 1400$ nm to $2 \times 1000 = 2000$ nm. This double enhancement for α_4 is called three-photon enhanced, four-photon absorption. Note that three-photon absorption is not enhanced by the coincidence with four-photon absorption. Thus it is possible that four-photon absorption could actually dominate three-photon absorption, despite being a higher order effect. However, there is another pathway with an end product of four-photon absorption, namely three-photon absorption into the $1B_u$ state, followed by one-photon absorption from one $1B_u$ to $3A_g$.

There are multiple complications associated with using the Z-scan technique to separate out different nonlinear absorption processes. The first has to do with the complicated interplay between strong nonlinear absorption and refraction. An example is shown in Figure 13. For example, in the absence of nonlinear refraction, a transmitted beam's spatial profile is flattened out by three- (or four-) photon absorption, as shown in Figure 13a. However, if $n_2 > 0$ (self-focusing case) with a value comparable to that actually measured in our experiments (discussed later), the beam actually collapses and produces a ring like structure as observed in Figure 13b. Because this leads to a higher local intensity than in the absence of self-focusing, a larger fraction of the beam is absorbed, resulting in a measurement of a larger effective multi-photon absorption coefficient if only a first order analysis of the data is made. Since self-focusing proceeds with increasing distance, this implies that thin samples and low input intensities are preferable for these measurements. (And in fact it was found that thick samples did indeed give inconsistent results.)

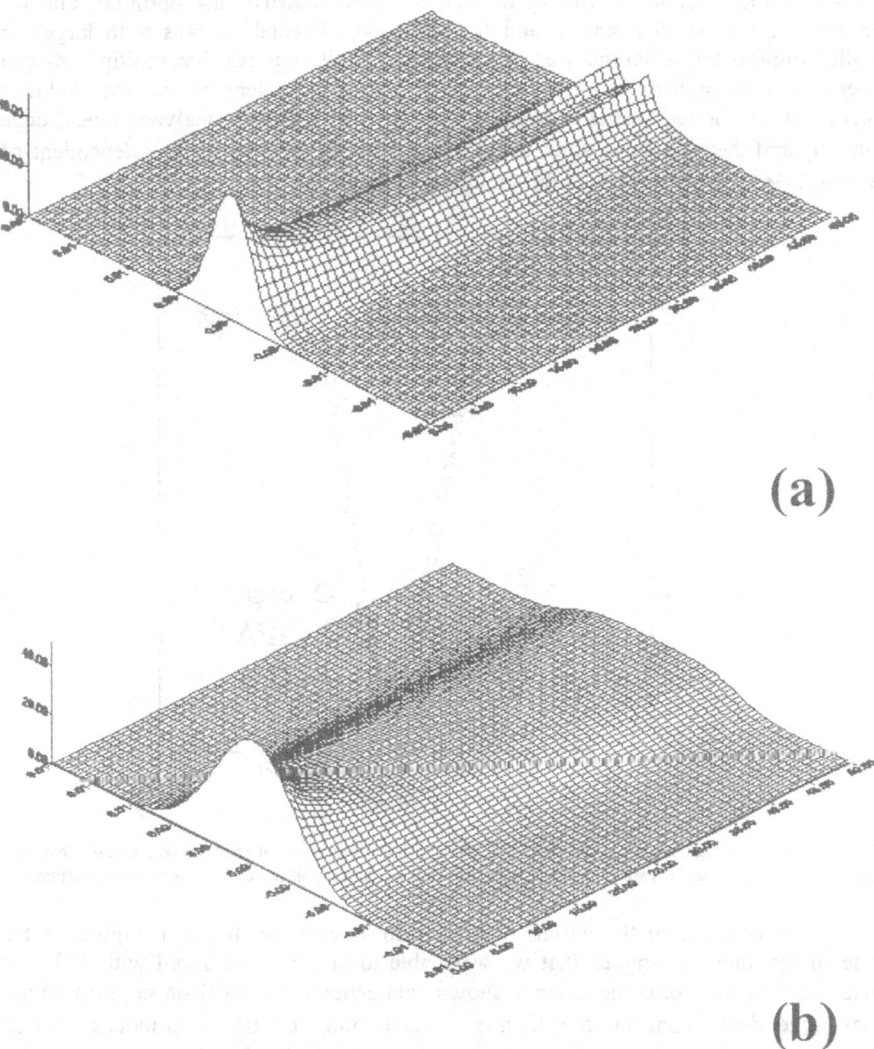

(a)

(b)

Figure 13. Computer simulation of beam propagation in a nonlinear medium in the presence of (a) pure nonlinear absorption in the absence of nonlinear refraction and (b) nonlinear absorption and nonlinear refraction ($n_2 > 0$).

The second complication is the insensitivity of the shape of the Z-scan to the order of the nonlinear absorption coefficient. This is illustrated in Figure 14 with actual data. Note that the small differences in the predicted shapes of the "valley" between three- and four-photon absorption. In fact, the case shown is the optimum one for demonstrating these differences, and for other experimental Z-scans with larger or smaller minima the differences are even smaller. This requires that multiple Z-scan experiments be performed as a function of increasing intensity and the deduced coefficients be plotted versus intensity. That is, each Z-scan is analyzed twice, once using α_3 and then a second time using α_4. For an α_3 process, α_3 is independent of intensity. An α_4 process will exhibit an intensity-independent α_4.

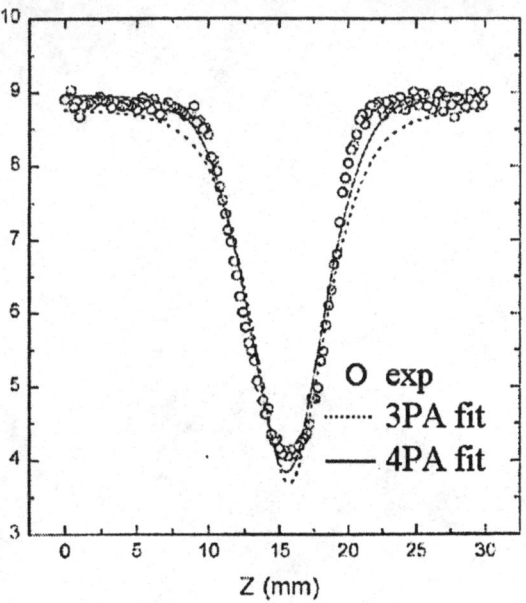

Figure 14. Open aperture Z-scan spectrum of PTS taken with 100 fsec pulses at 1600 nm with a peak intensity of 83 GW/cm². Two best fits are shown, one to the three photon absorption and to the four photon absorption.

An example of the variation in α_4 with intensity is shown in Figure 15 for some of the thinnest samples that we were able to produce, measured with 100 fsec pulses at 1600 nm. For comparison is shown data generated in 1997 on samples with a much larger defect concentration than is available now. Clearly α_4 is independent of intensity. When the calculated α_3 was plotted in a similar fashion, it grew rapidly with increasing intensity. Therefore the higher order process is clearly four-photon absorption. However, it is not known whether this is three-photon absorption followed by one-photon absorption, or the simultaneous absorption of four photons. Because 100 fsec pulses are expected to produce minimal population of the $1B_u$ state because of their small energy, we surmise that it is simultaneous four-photon absorption that is dominant. In either case, the process is three-photon enhanced and the deduced value is $\alpha_4 = 0.24 \pm 0.06$ cm⁵/GW³.

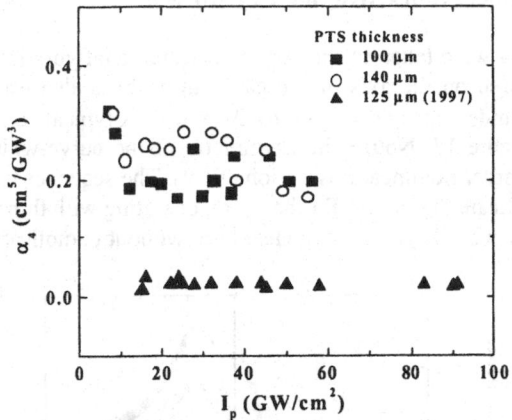

Figure 15. Variation of the four photon absorption (α₄) coefficient with peak input intensity
for PTS taken with 100 fsec pulses at 1600 nm.

Preliminary measurements on the wavelength dispersion of α_4 are shown in Figure 16. The data was taken primarily on the old samples and is probably an underestimate for the values available from the new, almost defect-free samples. Nevertheless a large enhancement is clear around 1800 nm relative to the 1600 nm data. Such experiments need to be repeated with the new samples.

There are interesting repercussions to such large four-photon coefficients. The absorption increases rapidly with increasing intensity, i.e. as I^3. An increase by only a factor of two in intensity results in an increase in $\Delta\alpha$ by a factor of 8, and of course the transmission decreases slower than exponentially with this factor. Basically PTS acts like an optical shutter in this wavelength range.

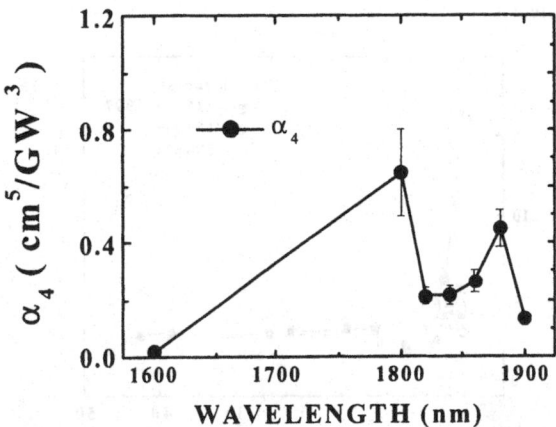

Figure 16. Wavelength dependence of the four-photon absorption (α₄) coefficient for PTS. The values of α₄
were deduced by fitting the open aperture Z-scan data to just the four photon absorption.

224

7. Intensity-Dependent Refractive Index Coefficient

All of the above data were taken in the "open aperture" configuration. Z-scan with a small aperture centered on the axis of the gaussian beam is also used to measure n_2 where the nonlinear index change is given by $\Delta n = n_2 I$. A typical "closed aperture" Z-scan is shown in Figure 17. Notice the asymmetry of the curve-width respect to the focus due to higher order nonlinear absorption effects. The sequence of peak and valley with increasing z indicate that $n_2 > 0$. Furthermore, operating with thin samples and low intensities, it proved possible to measure cleanly n_2 without complications due to four-photon absorption.

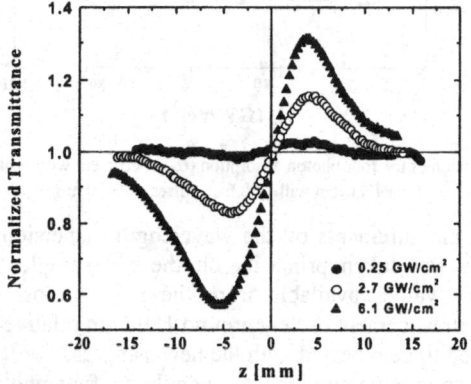

Figure 17. Typical closed aperture Z-scans. Curves were taken on a 380 μm thick PTS crystal with 100 fsec pulses at 1600 nm for different peak intensities.

The results in Figure 18 show an intensity-independent $n_2 = 3 \times 10^{-13}$ cm²/W, with perhaps a small increase in n_2 at very low intensities. However, the increase is much smaller than that initially found in the old samples. This is one of the largest known non-resonant nonlinearities.

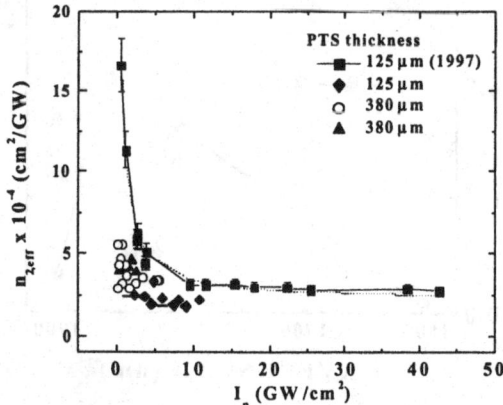

Figure 18. Intensity dependence of the nonlinear refractive index (n_2) coefficient for PTS at 1600 nm. The values of n_2 were deduced by fitting the close aperture Z-scan data to theory.

8. Summary

The conjugated polymer PTS is a classical example of a one-dimensional quantum wire. The π-electrons can move more or less freely in one dimension. There are large repercussions to this effect in the photonics field, including photoconductivity, and the linear and nonlinear optical response. Some of these are reported in this paper.

In summary, techniques were developed to grow single crystals of the polydiacetylene PTS. These crystals are of very high optical quality and have a minimum of defects and scattering centers. As a result it proved possible to measure a variety of nonlinear optical effects, all of which turned out to be large in PTS. Examples are the large magnitude of the exciton absorption line, the well-defined vibrational side bands at room temperature, massive two- and four-photon absorption coefficients, very large Raman gain coefficients, minimal excited state absorption and a large nonlinear refractive index.

This research was supported at CREOL by the National Science Foundation and at the Naval Research Labs by the Office of Naval Research and the Ballistic Missile Defense Organization.

References:

1. Also known as nanowires, atomic or molecular wires.
2. Joachim, C. and R oth, S. (eds.) (1996) *Proceedings of the NATO Advanced Research Workshop in Atomic and Molecular Wires*, Vol 341, Kluwer Academic Publisher.
3. Wegner, G. (1979) in W.E. Hatfield (ed.) *Molecular Metals*, Plennum Press, New York, p209.
4. Ito, T., Shirakawa, H., and Ikeda, S., (1974) *J. Polym. Sci.* **12**, 11.
5. Wegner, G., (1969) Z, *Naturforsch* B 74, 824.
6. Baughman, R.H. (1974), *J. Polym. Sci.*, **12**, 1511.
7. Mukhopadhyay, D. and Soos, Z.G. (1996) Nonlinear optical and electroabsorption spectra of polydiacetylene crystals and films, *J. Chem. Phys.* **104**, 1600-1610.
8. Townsend, P.D., Jackel, J.L., Baker, G.L., Shelburne, J.A., and Etemad, S, (1989) Observation Of Nonlinear Optical-Transmission And Switching Phenomena In Polydiacetylene-Based Directional-Couplers, *Appl. Phys. Lett*, **55**, 1829-1831
9. Sasaki, K., Fujii, K., Tomioka, T., and Kinoshita, T., (1998) All-Optical Bistabilities Of Polydiacetylene Langmuir-Blodgett Film Wave-Guides, *J. Opt. Soc. Am. B* **5**, 457-461
10. Winful, H.G., Marburger, J.H., and Garmire E., (1979) Theory of bistability in non-linear distributed feedback structures, *Appl. Phys. Lett.* **35**, 379-381
11. Sasaki, K, Sasaki, S, and Furukawa F. (1992) All-Optical switches and all-optical bistability by nonlinear optical materials, in Chiang, L.Y., Garito, A.F., and Sandman, D.J. (eds.) Materials Research Society Symposium Proceedings on *Electrical Optical and Magnetic Properties of Organic Solid State Materials*, **247**, pp 141-149.
12. Jensen, S.M. (1982) The Non-Linear Coherent Coupler, *IEEE J Quantum Elect.* **QE-18**, 1580-1583.
13. Albouy, P.A, Keller, P., and Pouget J.P. (1982) Structural and optical studies of the topochemical polymerization mechanism of the bis(para-toluenesulfonate) of 2,4-hexadiyne-1,6-diol, *J. Am. Chem. Soc.* **104**, 6556-6561
14. Chance, R.R., and Patel, G.N., (1978) Solid-state polymerization of a diacetylene crystal - thermal, ultraviolet, and gamma-ray polymerization of 2,4-hexadiyne-1,6-diol bis-(para-toluene sulfonate), *J.Pol.Sci. B* **16**, 859-881.
15. Dudley, M., Sherwood, J.N., Ando D.J., and Bloor, D. (1983) SRS radiation-induced polymerization of 2,4-hexadiynediol-bis-(para-toluenesulphonate) (PTS), *Mol.Cryst.Liq.Cryst* **93**, 223-237.

16. Bloor, D, and Stevens G.C. (1977) Solid-state thermal polymerization of 2,4-hexadiyne-1,6-diol, *J.Pol.Sci. B* **15**, 703-714
17. Lequime, M., and Hermann J.P. (1977) Reversible creation of defects by light in one dimensional conjugated polymers, *Chem. Phys.* **26**, 431-437.
18. Young, R.J., Read, R.T., and Petermann, J. (1981) Defects in polydiacetylene single-crystals. I. the perfect crystal and stacking-faults, *J. Mat Sci.* **16**, 1835-1842
19. Young, R.J., and Petermann, J., (1982) Defects in Polydiacetylene Single Crystals. II. Dislocations in pTS, *J. Mat Sci.* **20**, 961-974
20. Krug, W., Miao, E., Derstine M, and Valera, (1989) Optical-absorption and scattering losses of pts and poly(4-bcmu) thin-film wave-guides in the near-infrared, *J.Opt.Soc.Am. B* **6**, 726-732.
21. Bloor, D., Koski, L., Stevens, G.C., Preston, F.H., and Ando, D.J., (1975) Solid state polymerizartion of bis-(p-toluene sulphonate) of 2,4-hecadiyne-1,6-diol, *J. Mat. Sci.* **10**, 1678-1688.
22. Thakur, M., Xu, J.J., Bhowmik, A., and Zhou, L.G., (1999) Single-pass thin-film electro-optic modulator based on an organic molecular salt, *Appl.Phys.Lett,.***74**, 635-637.
23. G. Wegner (1971) *Makromol. Chem.* **145**, 85-94.
24. Monomer pTS crystallizes in the monoclinic space group P2₁/c with cell parameters: a=1.460 nm, b= 1.515 nm c = 1.502 nm and β = 118.4°, after polymerization the space group remains the same with the cell parameters change to: a = 1.448 nm, b = 0.493 nm, c = 1.491 nm and 118.0°.
25. Soos, Z.G., Mukhopadhyay, D., and Henessy, M.H. (1996) Stark profiles of singlet excitons in conjugated polymers, , *Chem. Phys.* **210**, 249-257.
26. Reviewed in (1995) *Section 8: Nonlinear Optics* in M.J. Weber (ed.) *CRC Handbook of laser science and Technology, Supplement 2: Optical Materials*, CRC Press, Boca Raton, Fl, pp 308-320.
27. Greene, B.I., Orenstein, J., Millard, R.R., Williams, L.R. (1987) Picosecond relaxation dynamics in polydiactylene-pTS, *Chem. Phys. Lett.*, **139**, 381.
28. Carter, G.M., Thakur, M.K., Chen, Y.J., Hryniewicz, (1985) Time and wavelength resolved nonlinear optical spectroscopy of a polydicetylene in the solid state using picosecond dye laser pulses, *App. Phys Lett.*, **47**, 457.

THE ELECTRONIC STRUCTURE OF CARBON-BASED NANOSTRUCTRURES: FULLERENES, ONIONS AND TUBES

J. FINK, M. KNUPFER, T. PICHLER AND M. S. GOLDEN
Institut für Festkörper- und Werktofforschung
Postfach 270016 D-01171 Dresden Germany

1. Introduction

Materials which are based upon sp^2-hybridized carbon atoms are becoming increasingly important in solid state physics, chemistry and in materials science. They have in common that their electronic properties are predominantly determined by π-orbitals which are formed from C2p electrons, while their mechanical properties originate from the sp^2-hybrid σ-orbitals having three bonds within a plane. These systems comprise graphite, conjugated polymers and oligomers, fullerenes, carbon onions and carbon nanotubes. Many of these systems can be doped or intercalated. In this way new materials can be tailor-made to have interesting and potentially useful properties. These new materials have often become model compounds in solid state physics since they show semiconducting and metallic behaviour, superconductivity and magnetism. Correlation effects, which result from electron-electron interactions and the electron-phonon interaction are also important in many of these systems. In addition, dimensionality plays an important role, too. Many of the fascinating open questions in present day solid state research are encountered again in sp^2-hybridized carbon systems.

On the other hand, a number of these sp^2-hybridized carbon systems has high technical potential. They show remarkable mechanical properties, e.g., a record-high elastic modulus. In addition, electronic devices from conjugated carbon systems are coming close to their realization and large-scale commercialization. Organic transistors and light emitting diodes based on conjugated polymers or molecules are already on the market. Industry is strongly interested in the field emission properties of carbon nanotubes to fabricate bright light sources and flat-panel screens. Finally, future nanoscale electronic devices could be realized using carbon nanotubes.

The fullerenes, a third allotrope of carbon, were discovered in 1985 by Kroto et al. [1]. The prototype fullerene is C_{60}, a molecule formed out of 60 carbon atoms distributed on a sphere with a diameter of 7 Å. Multi-wall carbon nanotubes, a new member of the growing family of novel fullerene materials were discovered in 1991 by Iijima [2]. In 1992, D. Ugarte first observed carbon onions, which consist of concentric spherical layers of carbon with an interlayer distance of 0.34 nm [3]. In 1996, the group at Rice University successfully synthesized macroscopic quantities of bundles of single-wall nanotubes with a small diameter distribution [4]. This historical listing illustrates that many new carbon nanostructures have been discovered in the 15 years since 1985.

227

L. Pavesi and E. Buzaneva (eds.), Frontiers of Nano-Optoelectronic Systems, 227–242.
© 2000 *Kluwer Academic Publishers. Printed in the Netherlands.*

228

The exceptional properties of these carbon nanostructures are strongly related to their electronic structure. In this contribution we review recent high-energy spectroscopic studies of the electronic structure of the quasi zero-dimensional fullerene solids and the quasi one-dimensional carbon nanotubes. In particular, we show investigations using photoemission spectroscopy (PES) and electron energy-loss spectroscopy (EELS). We compare these results with the better known results from the quasi two-dimensional graphite.

2. Electronic structure of graphite

For the understanding of the electronic structure of carbon nanostructures we first present some spectroscopic results on graphite as a representative of a quasi two-dimensional conjugated carbon system. In Fig 1 we show angular resolved photoemission and inverse photoemission data from graphite together with the reults of a bandstructure calculation [5].

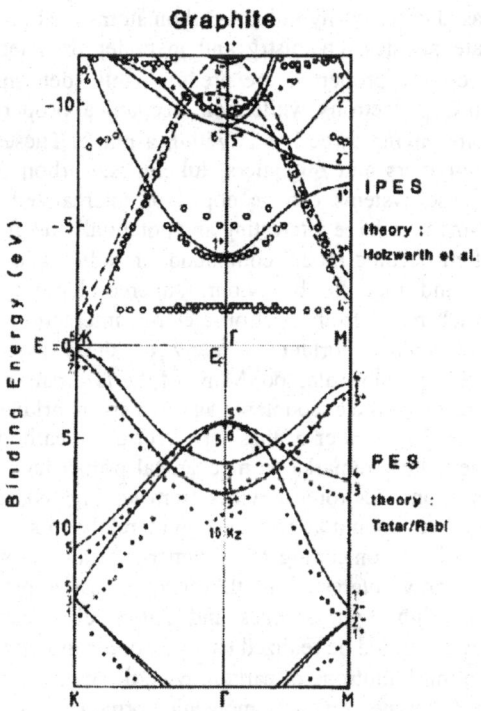

Figure 1. The bandstructure of graphite as determined from angular resolved photoemission and inverse photoemission spectroscopy. The inverse photoemission data have been obtained from highly oriented pyrolytic graphite. Therefore, the ΓM and ΓK directions are indistinguistable. Solid line: data from bandstructure calculations (from Ref. 5).

In graphite one s-electron and two p-electrons form the sp^2-hybrid which has trigonally directed σ-bonds in a plane. In the solid, these σ orbitals form strong covalent bonds with the σ-orbitals from neighboring carbon atoms, resulting in the formation of occupied σ and unoccupied σ* bands. These can be seen in the data of Fig. 1 for binding energies, E_B, either greater than 4 eV or less than -4 eV. The third C2p valence electron occupies a $2p_z$-orbital perpendicular to the plane, and forms a weaker π-bond with the $2p_z$-orbitals of neighboring C atoms. The electrons from the $C2p_z$ orbitals in this configuration are usually called π-electrons. Due to the weaker bonding, the splitting between the occupied π-bands and the unoccupied π*-bands is smaller. In Fig. 1 the bonding π-band can be seen between the Fermi level, E_F, and $E_B = 8$ eV and the antibonding π* band between E_F, and $E_B \sim -12$ eV. The band structure indicates that graphite is a zero gap semiconductor with the zero gap located at the K-point of the Brillouin zone. A simple tight-binding calculation of a single graphene sheet yields a total width of the π-bands of $6\gamma_0$ where γ_0 is the ppπ hopping integral between two carbon sites. This calculation also yields a flat band region near the M-point at $E_B = \pm \gamma_0$ for both the occupied π and the unoccupied π* states. In a real graphite crystal, there is a small splitting of the π-bands due to the weak van der Waals interaction between the graphene layers, which is also detectable in the data of Fig. 1.

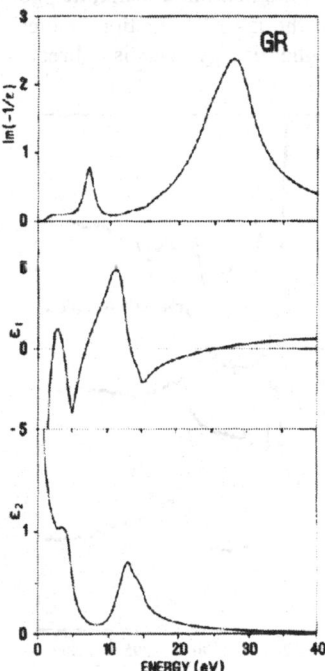

Figure 2. Electron energy-loss function Im[-1/ε] and the real (ε_1) and the imaginary (ε_2) parts of the dielectric function of graphite [6]. The momentum transfer was 0.1 Å^{-1}

The flat band region causes maxima (van Hove singularities) in the electronic densitiy of states near $E_B \sim \pm 2.5$ eV. The binding energy E_B of the flat band regions and the total width of the π-bands $W = 15$ eV $= 6\gamma_0$ yields a ppπ hopping integral between two carbon sites of $\gamma_0 \sim 2.5$ eV.

In Fig. 2 we come to the valence band excitations of graphite. The loss function, Im [-1/ε], where ε is the frequency (ω) and momentum (q) dependent dielectric function was measured using EELS in transmission. The real (ε_1) and the imaginary (ε_2) part of the dielectric function were then derived by a Kramers Kronig analysis. The data were taken at low momentum transfer q = 0.1 Å^{-1} (small scattering angle) parallel to the graphene sheets. The chosen momentum transfer is much smaller than the extension of the Brillouin zone, meaning that the data are comparable to optical data derived from reflectivity measurements [7]. The imaginary part of the dielectric function, ε_2, which is related to the absorption, shows a Drude-like tail at low energy due to the small concentration of free carriers (electrons and holes). The tail is followed by an oscillator resonance at 4 eV. The latter corresponds predominantly to a π - π* transition between the flat band regions of the π and π* bands at the M-point. The second peak in ε_2 near 12 eV is predominantly caused by σ - σ* transitions. The π-resonance at 4 eV causes a zero-crossing of ε_1 near 6 eV where ε_2 is small and, therefore, the loss function Im[-1/ε] = $\varepsilon_2/(\varepsilon_1^2+\varepsilon_2^2)$ shows a maximum there, i.e. a plasmon is observed. Since this plasmon is related to a π - π* interband transition, we call this plasmon an interband plasmon or a π-plasmon. When turning to higher momentum transfer, i.e., going from vertical to non-vertical transitions, the energy of the π - π* transitions increases and therefore the π-plasmon shows a dispersion to higher energy. This is a direct consequence of the large width of the π-bands.

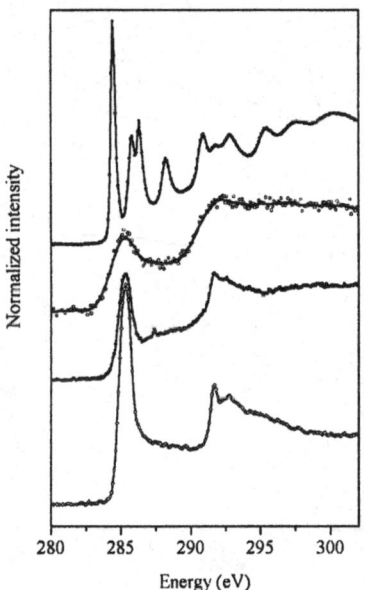

Figure 3. C1s excitation spectra of various carbon modifications recorded using electron energy-loss spectroscopy. From top to bottom: C$_{60}$ fullerene, carbon onions, single-wall carbon nanotubes, graphite.

The second peak in the loss function of graphite at 27 eV is caused by the zero-crossing of ε_1 near 25 eV. This zero-crossing is related both to the number of valence electrons and to the energy of the π - π* and, predominantly, of the σ - σ* transitions. Since this plasmon involves all the valence electrons, it is called the π + σ plasmon.

In Fig. 3 we show C1s excitation spectra of the various carbon modifications discussed in this paper. These spectra were recorded using EELS in transmission. In the spectrum of graphite, a maximum is observed at 285 eV, corresponding to transitions from the C1s level to unoccupied π* states. The width of this resonance is considerably reduced compared to the width of the unoccupied π* band. This comes from an excitonic enhancement of the spectral weight at the bottom of the π* band [8]. Above 291 eV, core level excitations to the unoccupied σ* bands take place, also resulting in the formation of a core-exciton.

3. Fullerenes

3.1. UNDOPED FULLERENES

As already mentioned above, the prototype fullerene - C_{60} - is comprised of 60 C atoms distributed on a sphere with a diameter of 7 Å. The C atoms are assembled in the form of a truncated icosahedron. Thus the C atoms outline 20 hexagons and 12 pentagons, the latter of which give rise to the curvature and thus enable the closed, quasi-spherical structure of the molecule. The valence electrons of the carbon atoms are predominantly sp^2-hybridized. Due to the finite curvature of the molecule there is some admixture of sp^3 hybridization which occurs itself in a pure form in diamond. A simple tight-binding calculation of the π-derived molecular electronic levels of C_{60} is shown in Figure 4.

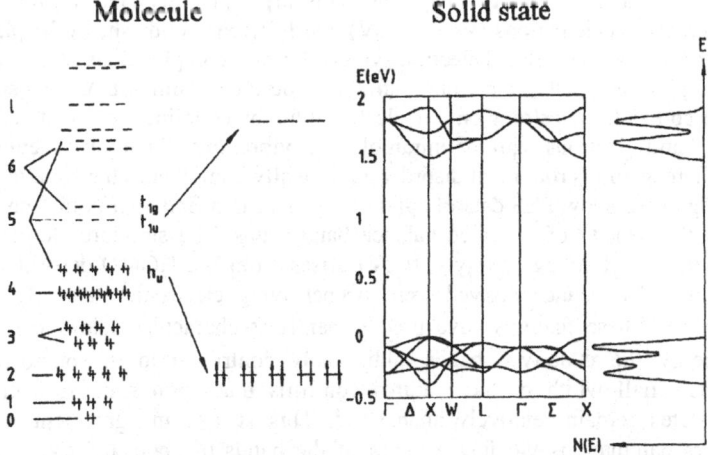

Figure 4. Schematic representation of the π-derived molecular electronic levels of C_{60} and the formation of the bandstructure in the solid state as predicted from LDA bandstructure calculations (from Ref. 9). Additionally shown is the electronic density of states N(E).

As the hopping integral of the π-electron between C sites in C_{60} is again $\gamma_0 \sim 2.5$ eV, the total width of the π-electron system is again of the order of 15 eV. The highest occupied molecular orbital (HOMO) has h_u symmetry while the triply degenerale lowest unoccopied molecular orbital (LUMO) has t_{1u} symmetry. Upon condensation into a solid the C_{60} molecules form a close-packed structure with face-centered cubic (fcc) symmetry and a lattic constant of $a_0 = 14.198$ Å. Similar as between graphene sheets in graphite, there is only a relatively weak van-der-Waals-type interaction between the C_{60} molecules. The hopping integral of the π-electron *between* C_{60} molecules is 50 times weaker than the $pp\pi$-intra-ball hopping integral. Therefore, there is only a small broadening of the molecular levels when going from the molecule to the solid. The width of the h_u-derived valence bands and the t_{1u}-derived conduction bands is of the order of 0.5 eV, as predicted from LDA bandstructure calculations [9] (see Fig. 4). From these calculations, also a direct band gap of about 1.5 eV is predicted.

This picture of the electronic structure of solid C_{60} indicates that the molecular character of the electronic structure is scarcely altered. This also indicates that the charge distribution of the electrons is rather inhomogeneous with the charge mainly confined to a shell with a radius of about 3.5 Å. This charge confinement is reminescent of the situation found in many solids with partially filled 3d or 4f electronic levels, i.e. one could expect that the on-site Coulomb energy, U, between two electrons to play an important role in a complete description of the electronic structure of fullerenes. This situation causes a deviation of the electronic properties from the predictions within LDA bandstructure calculations, as the latter treat exchange and correlation effects only within a free-electron approximation. Indeed, for solid C_{60} the Coulomb repulsion, U, between two holes in the valence bands has been determined from a comparison of the C KVV Auger spectrum with a self-convoluted photoemission valence band spectrum and has been found to be ~ 1.5 eV [10]. Since U is larger than the band width predicted from bandstructure calculations (W ~ 0.5 eV), the fullerene solids should be placed in the regime of strongly correlated electron sytems. Further complications with respect to the simple picture of the electronic structure derived from LDA bandstructure calculations comes from a strong on-ball electron-phonon coupling, i.e. an interaction of the valence band electrons with the intramolecular vibrations. Finally we mention that in many fullerene solids rotational disorder additionally complicates the situation.

In Fig. 5 we show PES data of solid C_{60}, which, in a first approximation, give a measure of the density of occupied valence band states. The structure closest to the chemical potential (binding energy = 0 eV) arises from the HOMO-derived valence bands followed by bands derived from deeper lying electronic states (HOMO-1, HOMO-2 etc.). These features have predominantly π character. Below about 6 eV binding energy the σ-derived bands additionally contribute to the photoemission spectra. The small width of the features confirms that upon solid formation the electronic states remain relatively unchanged. This is also in agreement with the bandstructure calculations yielding a width of the bands of about 0.5 eV. A further broadening observed in the PES spectrum is due to excitations of intramolecular phonons and possibly also due to disorder effects.

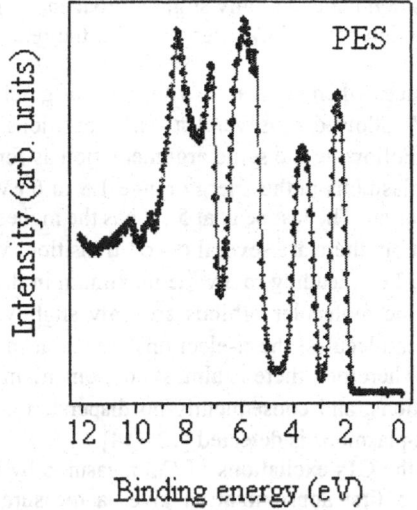

Figure 5. Photoelectron spectrum of solid C_{60} (from Ref. 11)

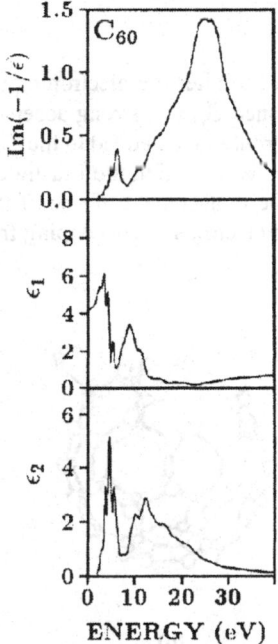

Figure 6. Electron energy-loss function Im[-1/ε] and the real (ε_1) and the imaginary (ε_2) parts of the dielectric function of solid C_{60}. The momentum transfer was chosen 0.1 Å^{-1} [12]

A comparison to the angular resolved PES data of graphite (see Fig. 1) and the C_{60} PES data clearly demonstrates the differences between a quasi two-dimensional and a quasi zero-dimensional system. In the former case, wide π-bands are observed, while in the latter, the density of states is dominated by only slightly broadened molecular levels. In Fig. 6 we show the loss function of solid C_{60} together with the real and imaginary part of the dielectric function.

The data show clear differences when compared with those of graphite (see Fig. 2). In ε_2, there is a gap of 1.8 eV followed by several π - π^* transitions between the well-separated molecular levels. Following the same argumentation as for graphite, these π-oscillators cause several π-plasmons in the energy range 1.8 to 6 eV, although ε_1 does not vanish in this energy interval. The last peak at 5 eV has the highest intensity because ε_2 is smallest there. In addition, there are several σ - σ^* transitions which almost cause an zero-crossing of ε_1 near 22 eV, leading to a wide maximum in the loss function, i.e. the π + σ-plasmon. Since the molecular orbitals are only slightly broadened by the interaction between the C_{60} molecules, the π-electrons are, even in the solid, strongly localized on the molecules. Therefore, there is almost no momentum dependence of the energy of the π - π^* transitions, and consequently no dispersion as a function of the momentum transfer of the π-plasmons is detected [12 - 14].

Finally, we mention the C1s excitations of C_{60} measured by EELS (see Fig. 3). In this spectrum, which in a first approximation gives a measure of the density of unoccupied C2p-related states, four features are present below the σ^* onset at 291 eV. They correspond to transitions into unoccupied π^*-bands (LUMO, LUMO+1 etc).

3.2 DOPED FULLERENES

The main strategy that can be used to alter the electronic properties of fullerenes in a controlled way is to dope them. Since C_{60} is a strong acceptor, only n-type doping has been achieved until now - i.e. electrons can be added to the C_{60}'s conduction bands. This can be practically realized in three ways which are illustrated in Figure 7: (i) doping from outside the molecules; (ii) the replacement of one of the carbon atoms by atoms with a different number of valence electrons or (iii) doping from inside.

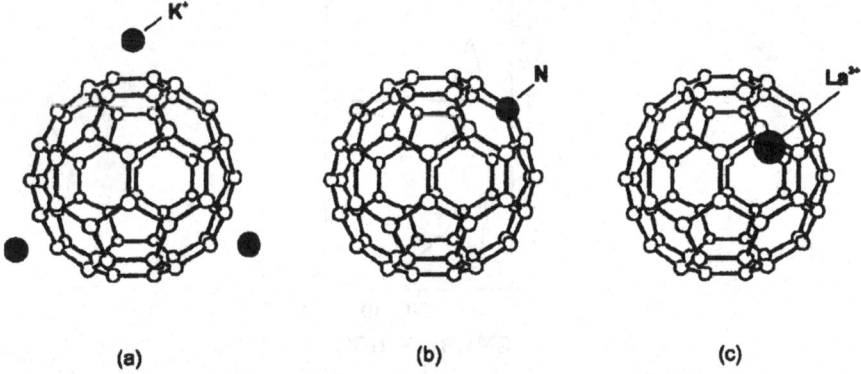

(a)   (b)   (c)

Figure 7. The three possibilities to dope fullerenes: (a) from the outside (exohedrally via intercalation), (b) on-ball (heterofullerene formation) or (c) from the inside (endohedrally)

All these routes have been successfully followed, resulting in a large number of fullerene intercalation compounds, heterofullerenes in which, e.g., a C-atom is substituted by a N atom or endohedrally-doped fullerenes, where the counter ion is inside the cage.

Here we focus on experimental studies of potassium intercalated C_{60}, which form a series of salts. In Fig. 8 we show PES results [15,16] and the C1s excitation edges [17,18] for C_{60}, K_3C_{60}, K_4C_{60} and K_6C_{60}.

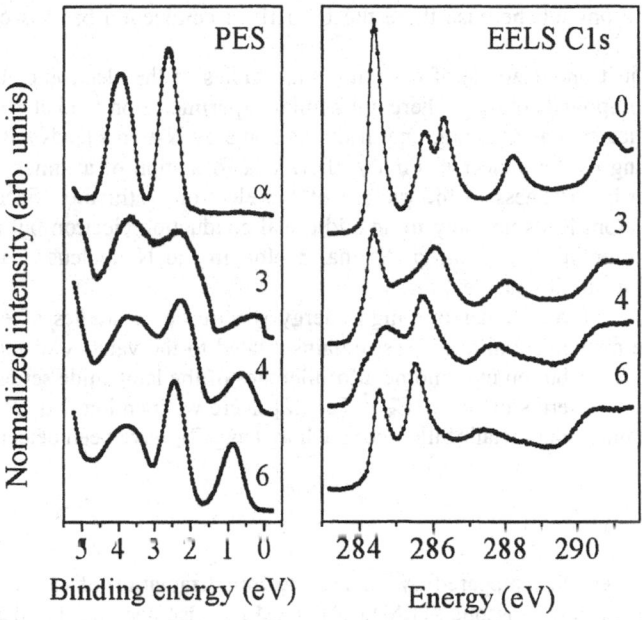

Figure 8. Left panel: photoemission spectra of K_xC_{60} compounds (x = ~0 (α), 3,4,6). Right panel: C1s excitation edges of the same compounds (from Refs. 15-18)

The spectrum labelled α was recorded from a sample which had been exposed to a very low flux of K atoms. In this case the Fermi level is pinned at donor levels near the bottom of the LUMO-derived conduction bands. Upon intercalation the charge transfer of electrons from K to C_{60} is clearly revealed by the presence of a new feature located in the former gap. This is due to the partly or completely filled conduction band formed from the triply degenerate molecular LUMO. For the superconducting compound K_3C_{60} (T_c = 18 K), a considerable density of states at the Fermi level and a clear Fermi cut-off are visible indicating the metallic ground state of this compound. On further intercalation to x ~ 4 the LUMO-derived peak has grown further but has shifted away from the Fermi level, resulting in a small or zero intensity at E_F. This indicates an insulating behaviour for K_4C_{60}, despite the fact that LDA bandstructure calculations predict a metallic ground state for this system. This discrepancy can be explained by correlation effects which are probably larger for K_4C_{60} than for K_3C_{60}. This puts the

K_4C_{60} compound firmly into the class of Mott-Hubbard insulators [19]. By x = 6 the LUMO-derived bands are completely full and there is a clear gap between the occupied and unoccupied states straddling E_F.

The filling of the t_{1u}-derived conduction band states upon K-intercalation is also clearly visible in the C1s excitation spectra also shown in Fig. 8. For K_3C_{60} the first peak at 284.5 eV is reduced by a factor of 2 because the LUMO-derived bands are half-filled (with 3 electrons). For K_4C_{60}, the intensity of the first peak is additionally reduced, consistent with further charge transfer to C_{60}. In K_6C_{60}, there are only three C1s - π^* transitions left because there the t_{1u}-derived conduction band is completely filled.

This is just one example of the numerous studies of the electronic structure of intercalation compounds of C_{60}. There are similar experiments on the heterofullerenes [20] which indicates that replacing one of the C atoms by N-atoms leads not simply to an n-type doping of the molecule. Firstly, there is a formation of a dimer, $(C_{59}, N)_2$, which adds further richness to the picture of the electronic structure. Secondly, the presence of N atom leads not only to an additional conduction electron but also to an altered local potential due to the additional proton in the N nucleus, leading to a localization of the additional electron.

Finally, there are numereous high-energy spectroscopic studies of endohedral fullerenes. One of the central questions remains related to the valency of the encaged ion. The charge distribution in monometallofullerenes of the lanthanide series has been discussed mainly in terms of $M^{3+}@(C_{82})^{3-}$ [21,22]. Here we mention that recently also divalent lanthanide monometallofullerenes such as $Tm@C_{82}$ have been detected [23].

4. Carbon onions

Carbon onions were first prepared by intense electron irradiation of carbon soot in a transmission electron microscope (TEM) [24]. TEM has also been used to demonstrate an interesting application of these systems as nanoscopic pressure cells to produce nanodiamond in their core [24]. A further method to produce carbon onions with a narrow diameter distribution is high dose carbon ion implantation into silver substrates held at elevated temperature [25]. The silver can be subsequently be removed by annealing the sample in high vacuum for 10 h at 850 ^{0}C.

In this section we review recent studies of the electronic structure of such carbon onions produced by ion implantation [26]. In Fig. 9 we show momentum dependent valence band excitations of carbon onions with a mean diameter of 50 Å recorded using EELS in transmission. For small momentum transfer a π-plasmon is observed at 6 eV with the $\pi + \sigma$ plasmon appearing at 24 eV. Contrary to the fullerenes and the carbon nanotubes there is only a single π-plasmon. Both the π and $\pi + \sigma$ plasmons show a strong dispersion as a function of momentum transfer similar to that observed in graphite. The reason for this similarity to graphite is the following. In the measured momentum range q = 0.15 - 0.8 Å^{-1}, the wavelength of the excitations $\lambda = 2\pi/q$ are between ~ 40 and 8 Å which is in all cases smaller than the diameter of the onions.

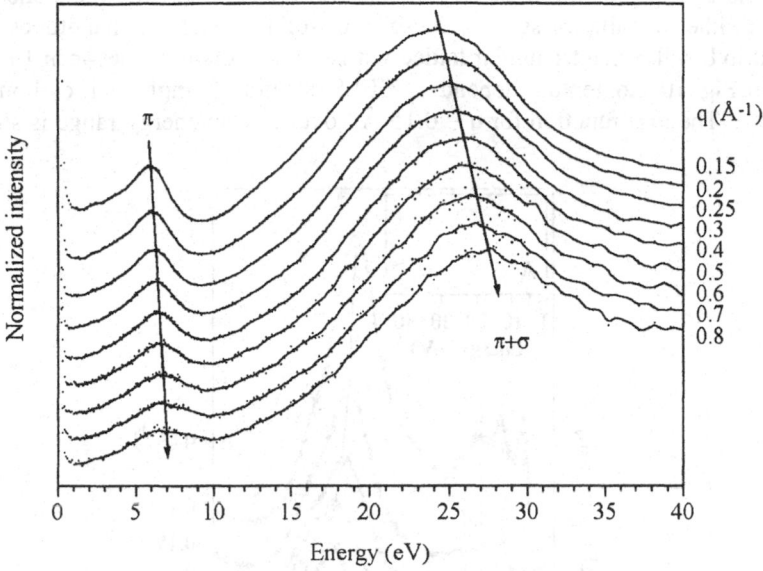

Figure 9. Momentum dependent loss functions of carbon onions (from Ref. 26)

Therefore, on this small wavelength scale, the electronic structure of the onions should be similar to that of graphite. Consequently, in this context it would be interesting to study the loss function at much smaller momentum transfer where the wavelength of the excitations should be larger than the diameter of the onions, although such experiments are beyond the cpabilities fo the present day instrumentation.

In Fig. 3, the C1s core level excitations spectrum of these carbon onions is also shown. A broad resonance related to transitions into π^* derived states at 285.2 eV and a threshold at 291 eV due to excitions into σ^* states is observed. The overall shape of the spectrum is very similar to that of a strongly broadened spectrum of graphite. Compared to C_{60}, no fine-structure related to transistions into narrow molecular-like states is observed. This is a further indication that, for localized probes - to which also the core excitations belong - the onions show an electronic structure which is closely related to that of graphite.

5. Carbon nanotubes

5.1 UNDOPED CARBON NANOTUBES

In this Section we review electronic structure studies of single-wall carbon nanotubes produced by laser evaporation of graphite. These nanotubes have a diameter of about 1.2 nm and can be envisaged as rolled-up graphene sheets which are capped with fullerene-like structures. Macroscopic nanotube samples generally contain a distribution of tubes with different diameters and chiralities. Depending on the diameter

238

and on the chirality or more exactly on the roll-up vector of the graphene sheets, the tubes are either metallic or semiconducting. In solid samples the nanotubes are usually arranged in bundles in a triangular lattice with a bundle diameter between 10 and 20 nm.

In Fig. 10 momentum dependent EELS spectra of single-wall carbon nanotubes are shown. The loss function for q = 0.15 Å^{-1} over a wide energy-range is shown in the inset.

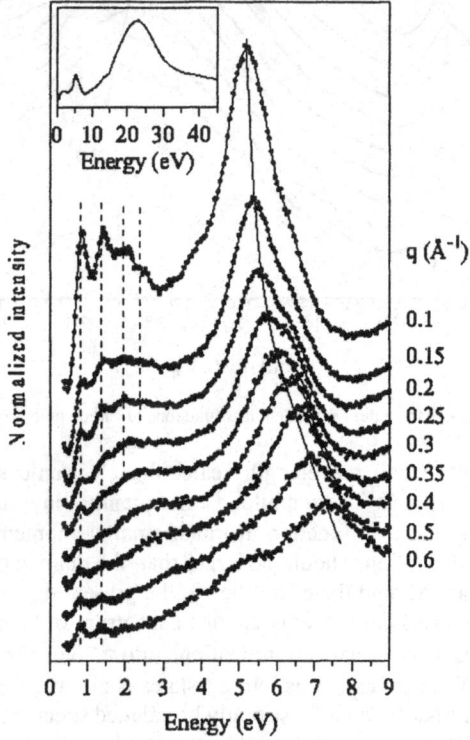

Figure 10. The loss function of purified single-wall carbon nanotubes for various momentum transfers, q. The inset shows the loss function over a larger energy range for q = 0.15 Å^{-1} (from Ref. 27)

Similarly to graphite, a π-plasmon is observed at 5.2 eV and a π + σ-plasmon appears at 21.5 eV. The momentum-dependent measurements (see Fig. 10) show a strong dispersion of the π-plasmon, similar to that of graphite or carbon onions. As discussed in Section 2 this plasmon is related to π - π* interband transitions near 5 eV, and therefore the dispersion indicates dispersive bands of delocalized π-electrons as in graphite. This dispersion can only occur along the nanotube axis and therefore the π-plasmon at 5.2 eV is related to a collective oscillation of π-electrons along the tubes. On the other hand, the low-energy peaks at 0.85, 1.45, 2.0, and 2.55 eV show no dispersion as a function of the momentum transfer, similar to the π-plasmons in C$_{60}$. This indicates that the low-energy maxima are related to excitations of localized electrons. It is tempting to attribute these excitations to collective excitations around the circumference of the tubes. The interband transitions which cause these peaks in the loss function are related to transitions between van Hove singularities in the density of states. These occur as a result of the quantum confinment perpendicular to the axis of the tubes due to the satisfaction of periodic boundary conditions for the wave functions of the rolled-up graphene sheet around the tube circumference. Hence the component of the Bloch wave vector perpendicular to the nanotube axis, $k_\perp$, can only assume discrete values, thus bringing with it the existence of discrete energy values. Therefore, the electronic structure of the carbon nanotubes might be regarded as molecular-like in the circumferential direction leading to van Hove singularities in the density of states.

A comparison with calculations of the electronic structure of nanotubes with the EELS spectra indicates that the lowest two peaks are due to semiconducting tubes while the next highest peak at 2 eV can be assigned to metallic tubes. From the intensitiy of the transitions, one can conclude that roughly two third of the tubes are semiconducting while one third is metallic. After these first EELS measurements, the same transitions between the van Hove singularities have also been observed using optical absorption measurements [28,29].

The momentum dependent loss data on one-dimensional single wall nanotubes lead to a picture of the electronic structure which is intermediate between the quasi zero-dimensional fullerenes and the quasi two-dimensional graphite. Perpendicular to the nanotubes these are localized π-electrons like in molecular systems while along the tubes these are delocalized π-electrons leading to wide π-bands like in graphite.

The C1s excitation spectrum of single-wall carbon nanotubes is shown in Fig. 3. As in graphite the first peak at 285 eV corresponds to transitions into unoccupied π* states while above 292 eV mainly unoccupied σ*-states are detected. Probably due to excitonic effects no details of the unoccupied density of states near the Fermi level and in particular no peaks due to the van Hove singularities could be detected in the EELS spectra, although the energy resolution in these experiment was 0.1 eV.

5.2 DOPED CARBON NANOTUBES

Right after the discovery of single-wall nanotubes it was clear that one should try to intercalate these materials, in analogy with the well-known examples of the graphite intercalation compounds (GICs) and intercalated fullerenes. Indeed, a decrease of the electrical resistivity by one order of magnitude was detected when single-wall nanotubes were exposed to potassium or bromine vapors [30,31]. The intercalation of the bundles by K can be followed by electron diffraction [32]. Upon successive intercalation, the first Bragg peak, the so-called rope-lattice peak, characteristic of the nanotube triangular lattice, shifts to lower momentum. This is consistent with an expansion of the inter-nanotube spacings concomitant with intercalation in between the tubes in the bundle. The maximal intercalation can be derived from the intensitiy of the C1s and K2p excitations recorded by EELS [32]. A comparison with data of KC_8 GIC yields the highest concentration of C/K ~ 7 for single-wall nanotubes intercalated to saturation, which is essentially a similar value as for stage I GIC KC_8.

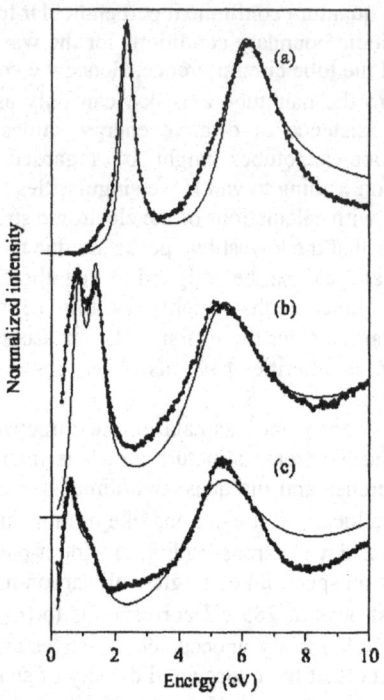

Figure 11. Loss function at a momentum transfer if 0.15 Å^{-1} for (a) GIC KC$_8$; intercalated single wall nanotubes with C/K ratios (b) 7 ± 1 and (c) 10 ± 2. The solid lines represent a fit of a Drude-Lorentz model (from Ref. 32)

To discuss the changes of the conduction-band structure upon intercalation, we show in Fig. 11 the loss function of K-intercalated single carbon nanotubes for two different concentrations together with that of GIC KC$_8$ [3]. For the latter compound, besides the π-plasmon, a charge carrier plasmon due to the filled π* bands is observed at about 2.5 eV. In the case of the intercalated nanotubes, the low-energy features cannot be satisfactorily described by a charge carrier plasmon alone. The introduction of an additional interband excitation located at 1.2 eV is necessary. Taking into account the previous discussion on undoped nanotubes, this excitation corresponds to the second transition between the van Hove singularities of the semiconducting nanotubes. In this view, the first transition at 0.6 eV has disappeared as a result of the filling of the lowest unoccupied van Hove singularity due to the K4s electrons transferred to the nanotubes, which leads to a shift of the Fermi level. Within a Drude-Lorentz model, the low-energy features can be explained by a charge carrier plasmon in addition to the remaining interband transitions between van Hove singularities, and the π-plasmon. Additionally, from the fit of the model to the data, the effectiv mass and the intrinsic DC conductivity for the intercalated nanotubes could be estimated. These studies gave a first information on the electronic structure of an new class of intercalation compounds.

6. Conclusions

Since the discovery of the fullerenes in 1985 a wide variety of new carbon nanostructures have been discovered. These new modifications of carbon form now a new class of undoped and doped materials. However, not only the macroscopic compounds from these carbon systems have attracted the attention of solid state physicists, chemists and material scientists. Additionally, the individual nanostructures themselves are of great interest for those who want to develop new nanoscaled mechanical and electronic devices. Since the properties of these nanostructure strongly depend on their electronic structure, the study of their electronic structure and properties represents a first, important step into the future for the application of these remarkable nanostructures.

References

1. Kroto, H. W., Heath, J. R., O'Brien, S. C., Curl, R. F., and Smalley, R. E. (1985) *Nature* **318**, 162
2. Iijima, S. (1991) *Nature* **354**, 56
3. Ugarte, D. (1992) *Nature* **359**, 707
4. Thess, A., Lee, P., Nikolaev, H. Dai, Petit, P., Robert, J. Xu, C., Lee, Y. H., Kim, S. G., Rinzler, A. G., Colbert, D. T., Scuseria, G. E., Tomanek, D., Fischer, J. E., and Smalley, R. E. (1996) *Science* **273**, 483
5. Hüfner, S. (1995) *Photoelectron Spectroscopy*, Springer, Berlin
6. Fink, J. (1989), *Adv. Electron. Electron Phys.* **75**, 121
7. Taft, E. A., and Phillip, H. R. (1965) *Phys. Rev.* **138**, A 197
8. Mele, E.J., and Ritsko, J. J. (1979) *Phys. Rev. Lett.* **42**, 68
9. Saito, S. and Oshima, A (1991) *Phys. Rev. Lett.* **66**, 2637
10. Lof, R. W., van Veenendaal, M. A., Koopmans, B., Jonkman, H. T., and Sawatzky, G. A. (1992) *Phys. Rev. Lett.* **68**, 3924
11. Golden, M. S., Knupfer, M., Fink, J., Armbruster, J. F., Cummins, T. R., Romberg, H. A., Roth, M., Sing, M., Schmitt, M., and Sohnen, E. (1995) *J. Phys. Condens. Matter* **7**, 8219
12. Sohmen, E., Fink, J., and Krätschmer, W. (1992) *Z. Phys. B-Condensed Matter* **86**, 87
13. Romberg, H., Sohmen, E., Merkel, M., Knupfer, M., Alexander, M., Golden, M., Adelmann, P., Pietrus, T., Fink, J. Seemann, R., and Johnson, R. L. (1993) *Synth. Met.* **55-57**, 3038; Sohmen, E., unpublished results
14. Knupfer, M., and Fink, J. (1999) *Phys. Rev. B* **60**, 10731
15. Merkel, M., Knupfer, M., Golden, M. S., Fink, J., Seemann, R., and Johnson, R. L. (1993) *Phys. Rev. B* **47**, 11470
16. Knupfer, M., Merkel, M., Golden, M. S., Fink, J., Gunnarsson, O. and Antropov, V. P. (1993) *Phys. Rev. B* **47**, 13944
17. Knupfer, M., Armbruster, J. F., Romberg, H. A., and Fink, J. (1995) *Synth. Met.* **50**, 1321
18. Sohmen, E., Fink, J., and Krätschmer, W. (1992) *Europhys. Lett.* **17**, 51
19. Knupfer, M., and Fink, J. (1997) *Phys. Rev. Lett.* **79**, 2714
20. Pichler, T., Knupfer, M., Golden, M. S. Haffner, S., Friedlein, R., Fink, J., Andreoni, W., Curioni, A., Keshavarz-K, M., Bellavia-Lund, C., Sastre, A.,

242

Hummelen, J.-C., and Wudl, F. (1997) *Phys. Rev. Lett.* **78**, 4249

21. Poirrier, D. M., Knupfer, M., Weaver, J. H., Andreoni, W., Laasonen, K., Parrinello, M., Bethune, D. S., Kikuchi, K., and Achiba, Y. (1994) *Phys. Rev. B* **49**, 2289

22. Kessler, B., Bringer, A., Cramer, S., Schlebusch, C., Eberhardt, W., Suzuki, S., Achiba, Y., Esch, F., Barmaba, M., and Cocea, D. (1997) *Phys. Rev. Lett.* **79**, 2289

23. Pichler, T., Golden, M. S., Knupfer, M., Fink, J., Kirbach, U., Kuran, P., and Dunsch, L. (1997) *Phys. Rev. Lett.* **79**, 3026

24. Banhardt, F., and Ajayan, P. M. (1996) *Nature* **382**, 443

25. Cabioc'h, T., Girad, J. C., Jaonen, M., and Denanot, M. F. (1997) *Europhys. Lett.* **38**, 471

26. Pichler, T., Knupfer, M., Golden, M. S., Fink, J., and Cabioc'h, T. *Phys. Rev. B submitted*

27. Pichler, T., Knupfer, M., Golden, M. S., Fink, J., Rinzler, A., and Smalley, R. E. (1998) *Phys. Rev. Lett.* **80**, 4729 (1998)

28. Kataura, H., Kumazawa, Y., Haniwa, Y., Umezu, I., Suzuki, S., Ohtsuka, Y., and Achiba, Y. (1999) *Synth. Met.* **103**, 2555

29. Jost, O., Gorbunov, A. A., Pompe, W., Pichler, T. Friedlein, R., Knupfer, M., Reibold, M., Bauer, H.-D., Dunsch, L., Golden, M. S. and Fink, J. (1999) *Appl. Phys. Lett.* **75**, 2217

30. Lee, R. S., Kim, H. J., Fischer, J. E., Thess, A., and Smalley, R. E. (1997) *Nature* **388**, 255

31. Rao, A. M., Eklund, P. C., Bandow, S., Thess, A., and Smalley, R. E. (1997) *Nature* **388**, 257

32. Pichler, T., Sing, M., Knupfer, M., Golden, M. S., and Fink, J. (1999) *Solid State Commun.* **109**, 721

CARBON NANOSTRUCTURE CHARACTERIZATION BY OPTICS AND RESONANCE RAMAN SCATTERING

H. KUZMANY, M. HULMAN, W. PLANK
Institut fuer Materialphysik, Universitaet
Wien, Strudlhofgasse 4, A-1090 Wien, Austria

Abstract: A review is presented on optical and Raman characterization of carbon nano-structures. Particular attention is paid to the new phases of carbon such as fullerenes, poly-fullerenes and carbon nanotubes. For the case of the polyfullerenes the two isostructural dimeres $(C_{59}N)_2$ and $(C_{60}^-)_2$ are discussed in detail. Both structures exhibit a very strong resonance for red laser excitation of the radial modes. The frequency of the radial breathing mode of single wall carbon nanotubes is shown to exhibit an oscillating behavior as a function of the excitation energy. A quantitative analysis performed for a quasi-continuous distribution of diameters revealed 2.9 eV as the best value for the π-overlap γ_0 and an 8% up shift of the mode due to the tube-tube interaction in the bundle.

1. Introduction

Carbon nanostructures are attracting considerable interest in science due to their novel cha-racter and their application potential. This contribution summarizes in the first part our knowledge in optics and Raman from carbon nanophases in general, then discusses recent results from the various phases, and finally reports on fullerene dimers and single wall carbon nanotubes which have been the target of recent research work in the group of the authors.

1.1 CARBON NANOPHASES

The small size of the carbon atom and the four valence electrons provide this element with a unique potential for forming a large variety of solids. Such solids can be highly crystalline or highly disordered, extreme soft or extreme hard, highly insulating or quasimetallic, all depending on the type of bonding the carbon atoms undergo or the type of structure the material accepts. Table 1 summarizes data about the most important carbon phases, including the new structures like fullerenes, polyfullerenes, onions, and carbon nanotubes.

243

L. Pavesi and E. Buzaneva (eds.), Frontiers of Nano-Optoelectronic Systems, 243–258.
© *2000 Kluwer Academic Publishers. Printed in the Netherlands.*

After a short introduction into the electronic structures of fullerenes, to be elucidated for the case of C_{60}, and to single wall carbon nanotubes (SWCNTs), particular attention will be paid to the latter and to special forms of polyfullerenes.

1.2 ELECTRONIC STRUCTURE

TABLE 1. Optical and Raman signatures for several carbon phases. $N(0)$ is the density of states at the Fermi level per unit length a_0. Column 5 lists most characteristic Raman lines. References address review articles or recent books for orientation, rather than original work. loc stands for localized states.

Carbon phase	Gap (eV)	$N(0)$ (eV^{-1})	Hybridi- zation	Raman lines (cm^{-1})	Ref.
Crystallin					
Diamond	5.5	-	Sp^3	1335	[1]
Graphite	-0.04	finite	Sp^2	1570, 1350	[2]
Singlewall nanotubes	0, 0.7	0, 0.6	Sp^{2+x}	1570, 180	[3]
Disordered					
NC-diamond	5	loc.	Sp^2/sp^3	1140	[4]
DL-carbon	1-2	loc.	Sp^3+sp^3	1520	[5]
A-carbon		loc.	Sp^2 dist.		[6]
Ultrahard carbon			Sp^3 dist.		
Molecular					
Fullerenes (C_{60})	1.9	-	Sp^{2+x}	1469, 273	[7]
Higher fullerenes	< 1.9	-	Sp^{2+x-y}	< 273	[8]
Polyfullerenes	0-1.9	-	$Sp^{2+x}+ sp^3$		[9]
Onions			Sp^{2+x-y}		[10]
Carbynes			Sp or sp^2	1800	

C_{60} is a molecular crystal where the electronic structure is widely determined by the electronic structure of the molecule. The carbon atoms in the football like molecule with I_h symmetry are nearly sp^2 bonded. Thus, the 60 p_z electrons form a rather independent and separated set of bands according to their accommodation in an fcc lattice. Figure 1 depicts the energy levels of the 60 molecular orbitals and the corresponding band structure. Occupation of orbitals is up to the 2nd h_u levels so that the h_u derived and the t_{1u} derived relations for E(k) represent the valence band and the conduction band, respectively. The width of the bands is only about 0.5 eV and the gap energy is 1.9 eV. Optical transitions between valence band and conduction band are symmetry forbidden. The first allowed transitions are between the h_u and t_{1g} derived bands. However, excitonic transitions are possible so that optical absorption starts at 1.6 eV. Absorption spectra exhibit several characteristic peaks corresponding to the various allowed band to band transitions. With increasing size of the fullerene molecules the gap energy decreases and approaches zero for infinitely large cages.

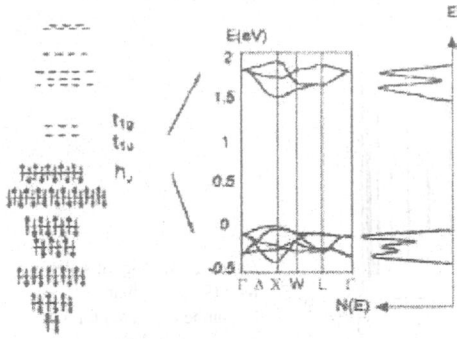

Figure 1. Energy schedule for molecular orbitals and π–electron band structure of C_{60}.

The situation for SWCNTs is different. Tubes can either be considered as very large elongated fullerenes or as rolled up graphene sheets. The process of rolling up is best described by an arbitrary lattice vector (n,m) of the graphene plane, where n and m are two integers. This vector is called the *folding vector* or the *Hamada vector* and determines the chirality, the diameter, and the electronic structure of the tube. For n = m the chiral angle is zero and the tubes are called armchair. Such tubes have always metallic character. For either m or n equal to zero the chiral angle is 30^0 and the tubes are called zigzag. All tubes where n-m is a multiple of three are quasimetallic with an almost zero gap energy. The SWCNTs exhibit an exactly one-dimensional periodicity along the tube axis. The unit cells are again determined by the components of the Hamda vector and can contain as many as several hundred carbon atoms. Only armchair and zigzag tubes have reasonably sized unit cells which allow *ab initio* and LDA based evaluation of physical properties.

Band structure and density of states for the tubes are to a first approximation obtained from a zone folding procedure applied to the graphen sheet.The one-dimensional character of the tubes leads to a condensation of the states into van Hove singularities. An example is given in Fig.2 for a (9,9) tube. The left side represents the valence band, the right side the conduction band. The density of states at the Fermi level is obviously ≠0. Similar structures in the density of states are obtained for the other tubes. The number of van Hove singularities is almost the same for tubes with varying chirality. Optical transitions are only allowed between states symmetric to the Fermi level. This selection rule determines the

246

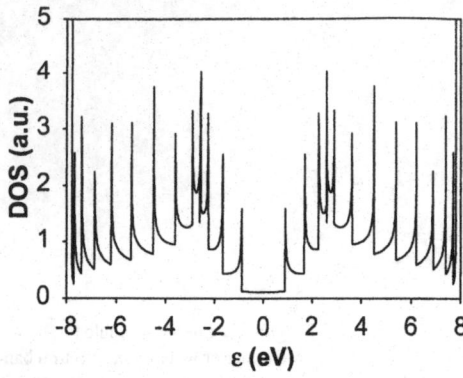

Figure 2. Density of states
for a (9,9) armchair
Nanotube evaluated for a π
-overlap of 2.7 eV

shape of the optical absorption spectra. The transition energies between the symmetric
van Hove singularities scale down with increasing tube diameter.

2. Carbon nanophase photonics

Optical and electronic properties of the new carbon nanophases have attracted the
interest of scientists working in research and application due to several findings. Some
of these are still in the process of exploration but others have already reached a high
degree of technological perfection.

In the early days of fullerene research photo-polymerization was one of the first
photo-activities reported for these phases. C_{60} and to a lesser amount also C_{70} is
photosensitive and undergoes a photo-polymerization by a cyclo-addition reaction [11].
To activate this process the quantum energy of the light must be higher than the
transition energy to the first excitonic state with singlet character. This state decays
spontaneously into a triplet exciton with a long life time and high reactivity. Since
oxygen reacts as a triplet quencher only oxygen free samples exhibit high photo-
reactivity. Also, the sample temperature must be high to guarantee free rotation of the
molecules which is a presumption for the cyclo-addition reaction. On the other hand
the temperature must not be too high in order to stabilize the resulting cyclo-butene
bonding. Typical dissociation temperatures for the polymers are 400 to 450 K. This
finally allows polymerization for a temperature window between 260 K and 400 K [12].
Photodimeres can be grown by irradiation just below 400 K [12]. The very fast and
sensitive process of photo-polymerization was suggested to be applied for
photolithography [13].

Nonlinear optics is an other phenomenon of interest in the field. The high cross
section for a further excitation of electrons from the singlet excited state leads to a light
limiting behavior. Since this process is very fast it is applicable even for light pulses in
the ns range. Prototypes of light limiters have been prepared from water soluble
fullerenes. This was done by growing glasses from a sol-gel technique which contained
the optically nonlinear molecules to a certain concentration [14].

Photoluminescence and ultrafast photo-electronic charge transfer between
fullerenes and conducting polymers are other fields where photonic properties are

promising for application. On the basis of the latter solar cells have been grown. Even though the overall efficiency is not yet very high the ratio of production costs to output power is promising [15].

Finally, in the field of carbon nanotubes, field emission for flat panel displays have reached an almost commercialized state. SWCNTs as well as multiwall tubes are considered in various laboratories as the sources for the electron emission. The low voltage requested for the field emission is a consequence of the small radius of the tip at the end of the nanotubes [16,17].

3. Doping of fullerenes from the cage and fullerene dimers.

Doping of pristine fullerenes was one of the central issues of fullerene research since the discovery of these new phases. In the solid only doping by reduction of the cage by introducing donor atoms or molecules was possible so far. Three basic techniques have been used: doping by intercalation of donor atoms [18], by substitution of carbons on the cage [19] and by inclusion of donor atoms or donor molecules into the cage [20,21]. Doping by the first two techniques were almost exclusively applied to C_{60} whereas doping by inclusion of atoms or molecules needs in general higher fullerenes. Intercalation doping was demonstrated to yield metallic states for one and three extra electrons on the cage. For the latter case even superconductivity with rather high transition temperatures were observed. The systems doped with one extra electron exhibit various forms of spin ordering leading to unusual magnetic phases and magnetic phase transitions.

The doping process was demonstrated to result in a polymerization process where the bonding between the cages is either established by a cylobutene ring [9] or by single bonds. Typical representatives of the former are RbC_{60} or KC_{60} and for the latter Rb_4C_{60} [22].

We will discuss in the following optical properties and Raman scattering for cage doped material of the form $(C_{59}N)_2$ and compare this compound with the isostructural and isoelectronic dimer $(C_{60}^-)_2$. For both system the intercage bond is established by a single covalent bond in trans geometry, rendering the system in a C_{2h} geometry. For the heterofulleren $(C_{59}N)_2$ the bonding connects the carbons next to the nitrogens.

3.1 OPTICAL AND RAMAN CHARACTERIZATION OF $(C_{59}N)_2$.

The biazafullerene samples were prepared and purified as described previously [23]. For the determination of the optical absorption the dimer was UHV-deposited on a single crystal KBr substrate and subsequently transmission was measured. Raman spectra were excited with various lasers extending from the deep blue to the near IR. The thermal stability of the dimer was checked from Raman and from IR experiments recorded at elevated temperatures.

Optical spectra from a thin film and from $(C_{59}N)_2$ dissolved in CS_2 are depicted in Fig.3. The absorption in the film starts just below 1.5 eV but tails down to about 1.3 eV.

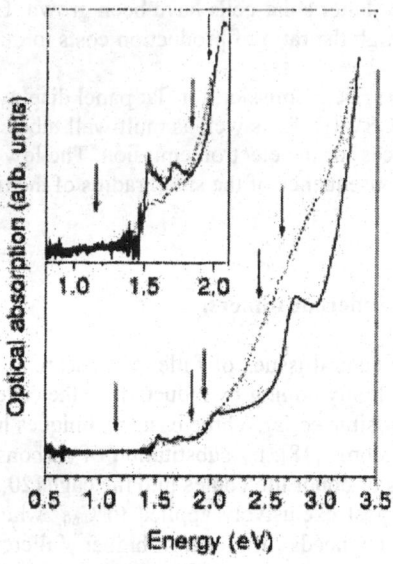

In solution the cut on of the absorption is even sharper. It starts at 1.4 eV and exhibits several characteristic peaks at 1.5, 1.7, 2, and 2.7 eV. The laser line positions used for the excitation of the Raman spectra extended from below the absorption edge to well above. The results for the optical absorption are in good agreement with reports from EELS where a red shift of electronic transition energies was observed as compared to C_{60} [25].

Raman spectra of $(C_{59}N)_2$ as excited with three different laser lines are depicted in Fig.4, together with the spectrum of C_{60}. Two things are striking at a first glance. The almost identical response of the C_{60} monomer and of the diazafullerene for the spectral range of the tangential modes, i.e. above about 1200 cm^{-1} and the strong resonance for the radial modes of $(C_{59}N)_2$ for excitation with the red laser. However, for excitation below the edge this resonance is completely quenched.

Figure 3. Optical absorption for a $(C_{59}N)_2$ thin film (dotted line) and for $(C_{59}N)_2$ dissolved in CS$_2$ (full line). The arrows indicate positions of laser lines used for excitation of Raman spectra; after [24].

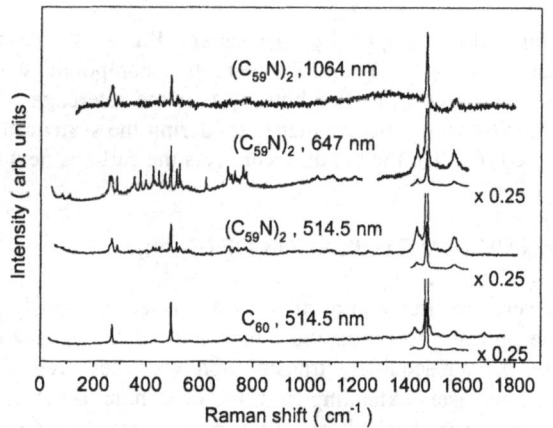

Figure 4. Overall Raman spectrum for C_{60} and for $(C_{59}N)_2$ as excited with the laser lines indicated.

At the low frequency end of the spectrum excited for resonance conditions three well expressed lines are located at 82, 103, and 111 cm^{-1}. These lines can be assigned immediately to the three Raman active intercage modes of the dimer.

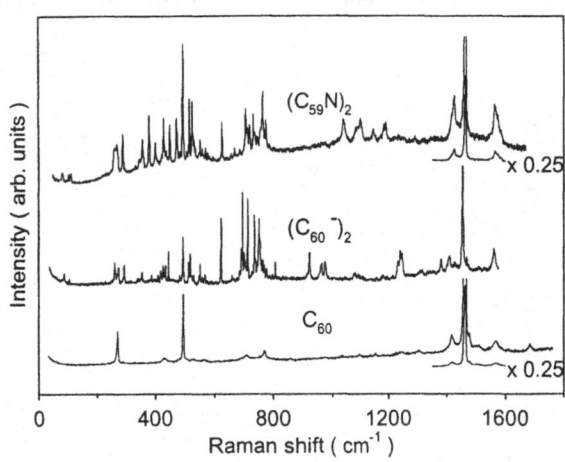

Figure 5. Overall Raman spectra of $(C_{60}^-)_2$ in comparison to spectra for C_{60} and $(C_{59}N)$

A rough assignment of the lines observed for the dimer was possible by comparison with C_{60} and from a calculation on a semi-empirical level [23]. Details of the assignment remained, however, difficult from this analysis since resonance enhancement and splitting appeared too strong to discriminate between split components of I_h degenerated modes and activation of I_h forbidden modes. Progress at this problem came from the analysis of the Raman spectra from the doped C_{60} dimers as demonstrated below.

3.2 RESONANCE EXCITATION AND THERMAL STABILITY OF $(C_{60}^-)_2$.

The dimeric phase of C_{60} was prepared as a thin film on a single crystal in the following way. A single crystal of C_{60} was first doped with Rb at 500 K to a phase RbC_{60}. Then the crystal was then rapidly quenched from 500 K to liquid nitrogen temperature. Quenching rates of about 10 K/s were achieved. In this way monomeric RbC_{60} was obtained which otherwise would have been transformed immediately to a polymeric phase during the cooling. Finally, the crystal was slowly warmed up until the dimeric phase was observed. This phase was found to be stable between 160 and 260 K. Details of the preparation process will be reported elsewhere [24].

Overall Raman spectra for the doped dimer are depicted in Fig.5, together with spectra for C_{60} and for $(C_{59}N)_2$ for comparison. Inspection indicates immediately that $(C_{60}^-)_2$ is again in resonance for excitation with the red laser. The center of the resonance is now shifted to the $H_g(3)$ and $H_g(4)$ derived modes around 690 and 770 cm^{-1}, respectively. Assignment is now easier since the two modes appear as separated groups of lines.

Also the assignment of the $H_g(2)$ derived mode around 430 cm^{-1} is more evident. Instead of the broad hump observed for C_{60} five almost equidistantly split features are observed. The rather strong line at 440 cm^{-1} has been assigned to a G_g derived mode which remains unsplit in spite of the distortion due to the intercage bond.

In the low frequency part of the spectrum again the intercage modes can be seen at 88, 98, and 106 cm^{-1}. These modes exhibit a dramatic change with temperature and are very good indicators for the transitions from the monomeric state after quenching to the

250

dimeric phase at 150 K and finally to the polymeric phase beyond 260 K. This behavior is demonstrated in Fig.6 which depicts the low frequency part of the spectrum recorded

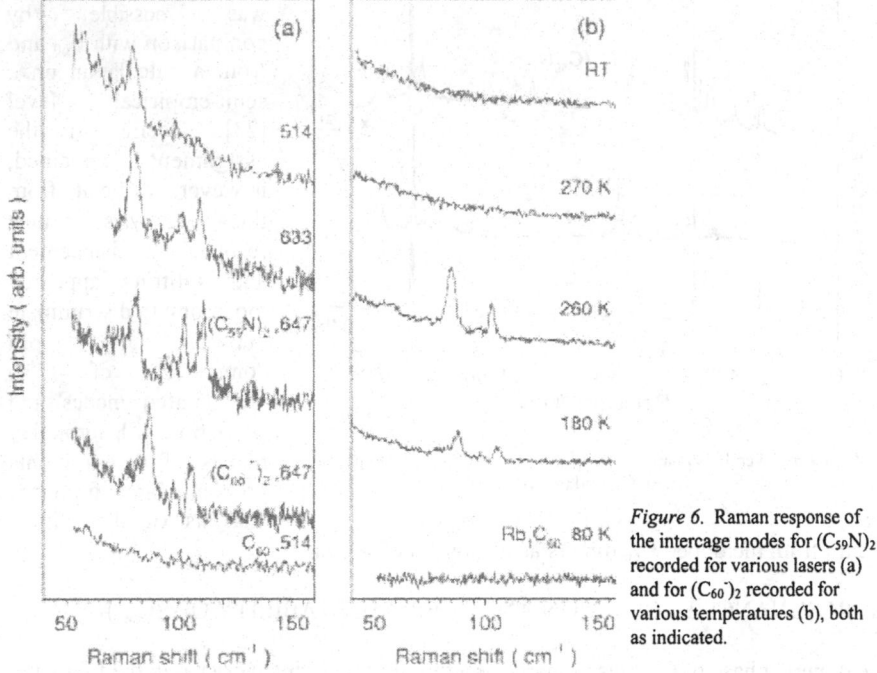

Figure 6. Raman response of the intercage modes for $(C_{59}N)_2$ recorded for various lasers (a) and for $(C_{60})_2$ recorded for various temperatures (b), both as indicated.

at various temperatures from 80 K to 270 K, together with the response of the intercage modes from the heterofullerene recorded for various laser lines. The response from the intercage modes of the C_{60} dimer increases with increasing temperature but disappears suddenly at the transition to the intermediate monomeric and to the final orthorhombic polymeric phase. This is in contrast to the behavior of the response of the intercage modes in $(C_{59}N)_2$ which remain visible until the dimer degrades at elevated temperatures. For the polymeric orthorhombic phase the intercage mode should also appear in the Raman spectrum but at higher frequencies, probably around 150 cm^{-1}. The lack of a signal in this frequency range may be due to the lack of a strong resonance in the polymeric phase.

4. Resonance Raman of single wall carbon nanotubes and optical response

Single wall carbon nanotubes have been grown either by laser evaporation of graphite [26], by gas discharge in a Kraetschmer-Huffman reactor [27], or most recently by chemical vapor deposition [28]. Typical tube diameters are 1.35 nm which would correspond to a (10,10) tube. During deposition of the tubes bundles or ropes are established in which individual tubes are compacted into a hexagonal like lattice. The determination of the lattice constant in such bundles has been one of the methods to find the average tube diameter.

In several recent work it was demonstrated that SWCNTs with smaller and larger diameters can also be grown [29]. Average tube diameters between 0.7 nm and 1.5 nm were reported.

Raman scattering was considered to be another possibility to determine the tube diameters since the frequencies for some of the Raman active vibrational modes depend strongly on the diameter of the tubes. Similarly, the optical transition energies depend on the tube diameters and thus allow as well to draw information on the latter.

4.1 THE RADIAL BREATHING MODE AND PHOTOSELECTIVE RESONANCE SCATTERING

According to the large size of the unit cell single wall carbon nanotubes can exhibit a very large number of optical modes. However only 15 or 16 modes are Raman active, independent from the chirality and from the diameter of the tube. Experimentally the Raman spectrum of SWCNTs is dominated by two lines, the radial breathing mode (RBM) in the low frequency region around 190 cm^{-1} and the graphitic lines in the C=C stretch region around 1590 cm^{-1}. The other 13 or 14 modes have not been identified so far. Figure 7 depicts a spectrum of unpurified SWCNTs as excited with a green line from an argon ion laser. The two dominating lines exhibit a split pattern which at least for the RBM mode originates from the resonance excitation of different tubes.

Tube frequencies have been calculated for various degrees of sophistication. In all cases several of the modes were found to change dramatically with the diameter of the tube. Such modes are called *dispersive.*

The RBM, where the normal coordinate has purely radial character, exhibits a particularly strong dispersion and is therefore most appropriate for an analysis of the tube diameter. The frequency for this mode was evaluated recently for armchair and zigzag tubes on an *ab initio* level with high accuracy [30]. The result is well described by a scaling law of the form B/d where B is 239 and 234 for armchair and zigzag tubes, respectively, and d is the diameter in nm. The estimated error for the calculated frequency was claimed to be lower than 2%.

Experimentally the RBM exhibits a characteristic fine structure and a strong photo-selective resonance scattering. This means line position and fine structure depend significantly on the quantum energy of the laser used for excitation of the spectrum. An example is depicted in Fig.8a,b. The spectra were recorded for two different samples with the laser lines indicated in nm. In the left part of the figure a dramatic change of the

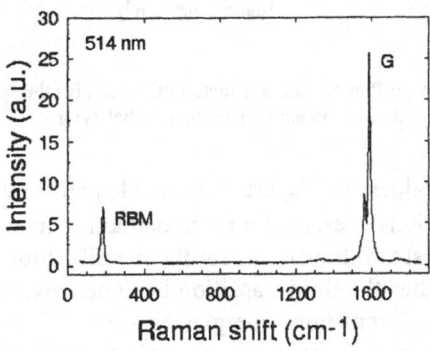

Figure 7. Raman spectrum of unpurified SWCNTs for excitation with 514 nm

line pattern is observed whereas the right part of the spectra indicates an oscillatory behavior for the peak positions. The photo-selective resonance scattering is in this case a consequence of the mode dispersion.

Photo-selective resonance scattering is often observed if the material under consideration is inhomogeneous with respect to the scaling parameter which determines the vibrational frequency and the optical transition energies. The laser then depicts resonantly this part of the material where its quantum energy matches best to the optical transition. In the above example the diameter of the tubes must be considered as the scaling factor which drives the dispersion and the energy shift of the transitions. Figure 9 has a schematic presentation. The dotted arrows connect the resonance energies between the van Hove singularities on the right side of the x-axis with the frequencies of the RBM on the left side of the x-axis. The distance between the van Hove singularities decreases with increasing tube diameter, approximately as $\gamma_0 A_n/d$, where γ_0 is the π-electron overlap integral of the graphene sheet and A_n are constants.

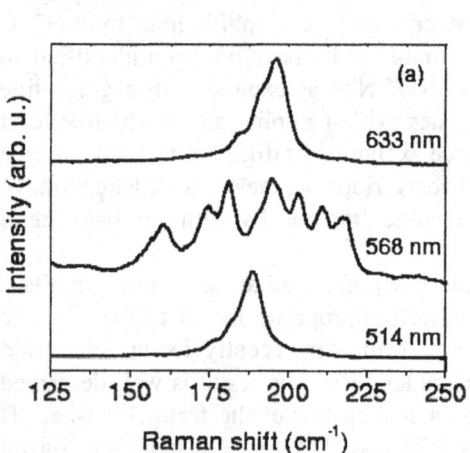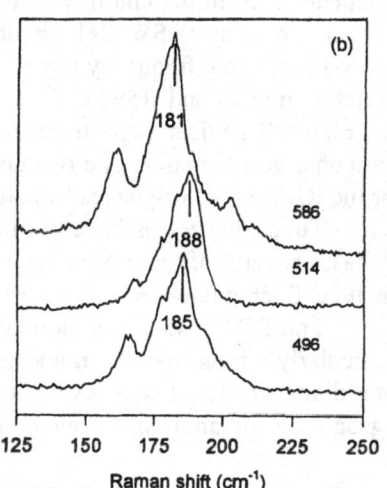

Figure 8. Change of Raman response for the RBM for two different nanotube materials. In part (a) the change of the fine structure dominates, in part (b) the peak position exhibits oscillatory behavior.

For tube diameters of 1.35 nm the first three values of A_n are 0.27, 0.54, and 0.99, respectively which yields, for $\gamma_0 = 2.9$ eV, transitions energies for semiconducting tubes of the order of 0.6 eV, 1.2 eV and 2.2 eV. The first two transitions for the metallic tubes are at 1.8 and 3.1 eV. For the semiconducting tubes the third transition becomes highly dependent on the chirality so that a wide range of transitions energies between 2.2 eV and 3 eV becomes available. Thus, the resonant transitions cover well the visible spectral range.

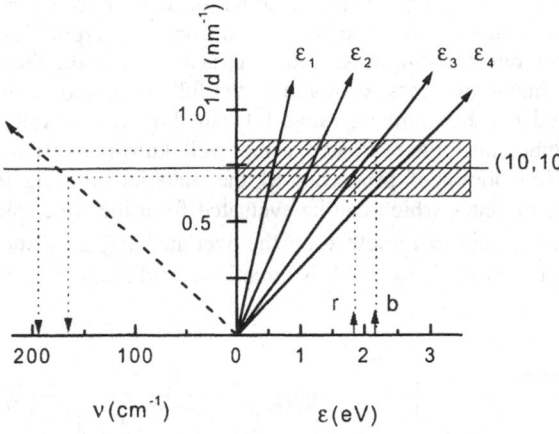

Figure 9. Schematic diagram for photoselective resonance scattering for SWCNTs. The radial arrows indicate the increase of transition energy (full lines) and the increase of RBM frequency (dashed line) with 1/d. The hatched area covers typical tube diameters.

4.2 QUANTITATIVE EVALUATION FOR THE DISPERSION OF THE RBM MODE

Whereas the general shift of the line for the RBM is straight forwardly understood from the 1/d scaling of the frequency, the interpretation of the fine structure of the line pattern and its dramatic change with excitation energy is not so simple. Since the peak positions in the pattern for the various laser excitations occur at almost the same frequencies it would be standard praxis in spectroscopy to fit the observed results with a set of Lorentzian or Gaussian lines of constant width. The intensities of each line would have to be varied with the laser excitation. This procedure was indeed demonstrated to be appropriate for standard SWCNT material [31]. All observed spectra could be fitted with a set of 14 different oscillators. Considering the diameter distribution known for the investigated material from TEM analysis 14 oscillators mean that not only armchair or zigzag tubes are relevant but a certain number of chiral tubes must be included as well. This raises of course the question which of the chiral tubes are more relevant and which are less relevant.

To solve the problem recently a much larger number of different laser energies were used for excitation [32]. Looking into the details of such spectra revealed a systematic oscillation in the pattern. Starting from the long wavelengths (837 nm) the response decreases with decreasing laser wavelength and simultaneously the peak shifts to lower wave numbers. For a further reduction of the laser wavelength the peak response shifts up again until it exhibits an other maximum value at 647 nm. Beyond this again a minimum in intensity is reached for 633 nm and the peak position switches back to lower wave numbers, and so on. This oscillatory behavior could be demonstrated explicitly by plotting the peak position of the response versus the excitation energy as demonstrated in Fig. 10a.

The interpretation of the experimental observations needs a more detailed quantitative analysis. For this, use has been made of the *ab initio* calculation for the

RBM. Even though calculations of this type can only be performed for armchair and zigzag tubes results for arbitrary helicity are reliably obtained from an interpolation. To avoid the question of any preference in chiralities and in agreement with the fact that theory does not provide any magic numbers for n and m the full set of geometrically allowed tubes was considered for the diameter range relevant for the material. The Raman intensity for each tube must be evaluated using well known relations for resonance scattering [33]. The dominant properties of the material entering these relations are the joint density of states which can be evaluated from the zone folding procedure for all tubes. Other crucial parameters are the overlap integral γ_0 and the width of the excited electronic state α. This width is directly related to the lifetime of the state.

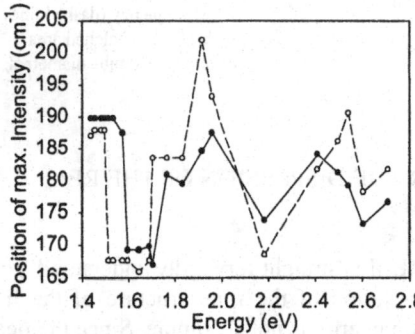 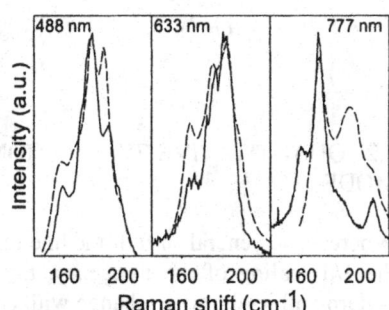

Figure 10. Peak position for the RBM in SWCNTs versus energy for the exciting laser. Full points are experimental results, open circles are as calculated and downshifted in frequency by 8%. Lines are guides for the eye (a), and comparison between observed (full line) and calculated (dashed line) line pattern for the RBM for excitation with three different lasers. The matching of the peak position was obtained for a 3%, 8% and 13% up shift of the calculated spectra, respectively (b); after [32].

For the sample under consideration a Gaussian distribution of tube diameters peaking at 1.35 nm with a variance $\sigma = 0.01$ nm^2 was assumed. This means 80 different types of tubes must be considered. Best agreement with the experiment was obtained for a value of the overlap integral $\gamma_0 = 2.9$ eV and for $\alpha = 0.01$ eV. Superposition of the response from all contributing tubes yields a line pattern which is rather similar to the observed results. In contrast to what one might have expected from the superposition of the response from the 80 tubes the fine structures in the line pattern are retained. and change very sensitively with changing laser energy. Examples are depicted in Fig. 10b and compared with experimental pattern. If the frequencies for the calculated spectra are up scaled as described in the caption even semi-quantitative agreement is evident.

A more rigorous comparison between experiment and calculation may be obtained from an evaluation of the peak position of the response as it was shown for the experimental data in Fig. 10a. The calculated peak positions are depicted in the figure as open circles connected by the broken line. Again, to obtain the good agreement between experiment and calculation the overall scale of the calculated frequencies was up shifted by 8%.

The good agreement between calculation and experiment supports suggestions that tubes grow with all geometrically allowed diameters in a certain diameter range. Chirality is not a selective parameter to first order. The concentration of (9,9) or (10,10) tubes is only of the order of 4% as it is read from the Gaussian distribution.

4.3 CONSEQUENCES FOR THE OPTICAL ABSORPTION

The quasi-continuous distribution of tube diameters is expected to shape the optical absorption. The experimental results depicted in Fig. 11 were obtained for a thin film of the same set of unpurified nanotubes as it was analyzed by Raman spectroscopy are. Three characteristic peaks located at 0.7, 1.2, and 1.8 eV are observed. For the evaluation of the spectra the superposition of the joint density of states for the 80 different tubes, weighted with the Gaussian distribution should be a good approximation. This is indeed demonstrated by the noisy line in Fig. 11. The same type of peaks is observed. Their slightly lower position in energy is due to the fact that for γ_0 a value of only 2.7 eV was used in this evaluation. The low energy peaks in the spectra are retained because of the rather narrow distribution of the diameters and the independence of the transition energies from the chirality. For the higher transition the latter statement is not valid any more. Thus for these transitions the peak structure is washed out.

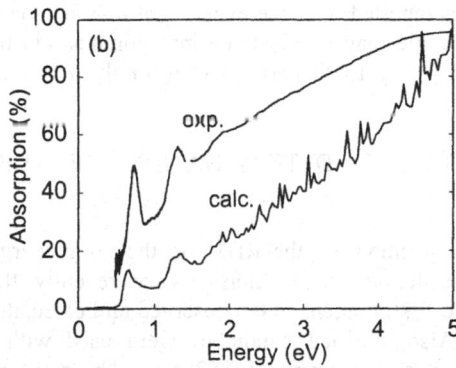

Figure 11. Observed (smooth curve) and calculated (noisy curve) optical absorption for a distribution of SWCNTs.

4.4 CLUSTERING AND TUBE - TUBE INTERACTION

The reason for the appearance of the fine structure in the pattern can be traced back to a clustering of the tube diameters in n,m space. If the 80 values for 1/d or the corresponding frequencies are plotted with equal distances the resulting line is not smooth. The derivative dN/dv of this function versus the frequency v exhibits clear maxima as demonstrated in Fig. 12. The number of clusters observed in the diameter range shown corresponds rather well to the number of oscillators which was required in [33] for a formal fit of the observed fine structure.

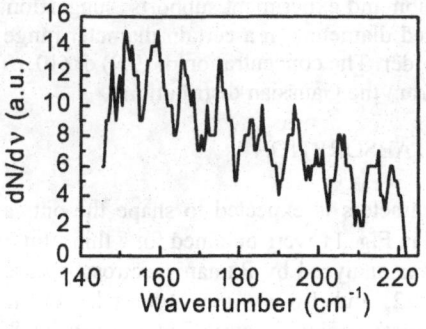

Figure 12. Clustering of Raman lines for the RBM in SWCNTs. dN is the number of tubes which exhibit RBM frequencies within a frequency interval dv.

Finally, the frequency up shift requested for the calculated spectra to match the experiments needs justification. Since the claimed accuracy of the calculation was better than 2% this up shift must be traced back to other sources. One of them is the tube-tube interaction in the nanotube bundles. Calculations were performed for individual tubes whereas the tubes in the experiments were clustered into bundles as described above. The van der Waals interaction between the tubes in the bundles contributes to the force constant of the radial modes. Very recent calculations in which such contributions were included claim an up shift between 8 and 14% for tube diameters in the range from 1 to 1.6 nm [34]. This matches very well to the results obtained above. The slightly lower values reported from the experiment may be due to the fact that not all nanotubes in the sample may be clustered into bundles and the clustering may not be perfect or simply due to the error bar from the *ab initio* calculation.

4.5 A SIMPLIFIED MODEL FOR THE DETERMINATION OF TUBE DIAMETERS

Due to the high sensitivity of the peak position for the RBM on the laser energy, calculated as well as experimental results are rather noisy. Very recently this disadvantage was avoided by plotting the first moment of the observed and calculated spectra versus the excitation energy. Also, purified nanotubes were used with a diameter distribution centered at 1.25 nm and variance of 0.03 nm . The resulting frequencies exhibit the same type of oscillations as they were described above, except for less noise. Results are depicted in Fig.13. The oscillation frequency is about 0.6 eV and one may ask to which property of the electronic structure this frequency is related. A simplified theoretical model which will be discussed in detail elsewhere [35] yields the full drawn line in Fig. 13. An analysis of the model reveals that the oscillations originate from a periodic change in contributions of higher transitions as the excitation energy is up shifted. If higher energy transitions start to contribute they originate from tubes with larger diameter. This behavior is indicated by the dotted lines in Fig. 9. For the area with lower energy only the second lowest electronic transitions from the semiconducting tubes can contribute for the relevant values for d. For the area with the higher energy the second and the third lower transitions are effective but higher tube diameters are involved. Thus, the center of gravity for the response is up shifted.

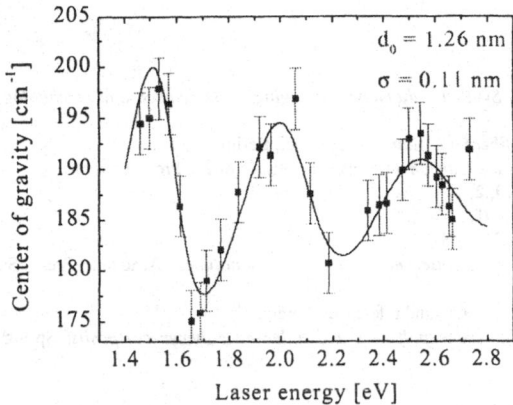

Figure 13. First moments for the Raman response of the RBM in 4" wafer grown SWCNTs from Rice University versus excitation energy (squares). The full drawn line is calculated from a simplified model.

The smooth distribution of the experimental data allows to optimize the fit parameters γ_0 and α simultaneously for a given Gaussian distribution. By plotting the deviation of the calculation from the experiment on a two dimensional parameter diagram a rather sharp minimum for $\gamma_0 = 2.9$ eV but a rather broad minimum for $\alpha = 0.02$ eV is observed.

5. Conclusion

Optical absorption and Raman scattering were demonstrated to be excellent tools to analyze carbon nanostructures. For the case of on cage doped fullerenes a very strong resonance enhancement for the radial modes is observed. The response is very similar to the isostructural and isoelectronic compound obtained from intercalation doping.

With respect to SWCNTs the Raman response from the RBM is particularly useful. There is an overall up shift for the center of gravity of the response with decreasing tube diameter. In addition, for a given material, the center of gravity oscillates with the energy of the exciting laser and so does the peak of the response. This oscillation originates from the oscillating number of resonance transition contributing to the line pattern. From a best fit to a calculation, the π-overlap γ_0 and the life time of the excited state $\tau = 2\pi/\alpha$ were determined to be 2.9 eV and 0.2 ps, respectively. The fine structure of the response is consistent with the contribution of all geometrically allowed tubes. Finally the tube-tube interaction provides an effective stiffening of the RBM of about 8%.

258

Acknowledgements
This work was supported by the Fonds zur Förderung der Wissenschaftlichen Forschung in Austria, Project 12924-TPH

6. References

1. Spear, K.E. and Dismukes, J.P. (1994) *Synthetic diamond: Emerging CVD science and technology,* John Wiley & Sons, New York.
2. Dresselhaus, M.S. et al. (1998) *Graphite fibers and filaments* Springer, Berlin.
3. Saito, R. et al. (1999) *Physical Properties of Carbon nanotubes* Imperial College Press, London.
4   Gruen, D (1999) , Ann. Rev. Mater. Sci. **29**, 211.
5. Grill, A. and Meyerson, B.S. (1994) in [1], 91.
6. Blank, V.D. et al. (1998) Physics Lett. A **205**, 208.
7. Dresselhaus, M.S. et al. (1996) *Science of fullerenes and carbon nanotubes,* Academic Press, San Diego.
8. Krause, M. et al. (1999) J. Chem. Phys. **111**, 7976, and references therein.
9. Eklund, P.C. and Rao, A.M. (2000) *Fullerene polymers and fullerene polymer composits,* Springer Heidelberg.
10. Obraztsova E.D. et al. (1998) Carbon **39**, 821.
11. Rao, A.M. et al. (1993) Science **259**, 955.
12. B. Burger, J. Winter, H. Kuzmany, Z. Phys. B 101, 277 (1996).
13. Hebard, A.F. et al. (1993) Appl. Phys. A **57**, 299.
14. Signorini, S. et al. (1998), in H. Kuzmany, J. Fink, M. Mehring, and S. Roth (eds.), *Molecular Nanostuructures,* World Scientific, p. 519.
15. Sariciftci, N.S. et al. (1992) Science **258**, 1474.
16. deHeer, W.A. et al. (1997) Advanced Materials **9**, 87.
17. Kim, J.M. et al. (2000) International Winterschool on *Electronic Properties of Novel Materials,* Kirchberg, Austria.
18. Haddon R.C. et al. (1991) Nature (London) **350**, 320.
19. Hummelen, J.C. et al. (1995) Science **269**, 1554.
20. Lahamer, et al. (1998) Adv. in Metal and Semicond. Clust. **4**, 179.
21. Stevenson, S. et al. (1999) Nature **401**, 55.
22. Oszlanyi, G. et al. (1997) Phys. Rev. Lett. **78**, 4438.
23. Kuzmany, H. et al (1999) Phys. Rev. B **60**, 1005.
24. Plank, W. et al. (2000) Europ. Phys. J. B, to be published.
25. Haffner, S. et al., (1998) Europ. Phys. J. B **1**, 11.
26. Thess et al. (1996) Science **273**, 483.
27. Journet, C. et al. (1997) Nature **388**, 756.
28. Fan, S. et al. (1999) Science **283**, 512.
29. Kataura, H. et al. (1999) Synthetic Mctals **133**, 2555.
30. Kuerti, J. et al. (1998) Phys. Rev. B **58**, 8869.
31. Kuzmany, H. et al. (1998) Europhys. Lett. **44**, 518.
32. Milnera, M. et al. (2000) Phys. Rev. Lett. **84**, 1324.
33. R.B. Martin, R.B. and Falicov, L.M. (1975) Topics Applied Phys. **8**, 79.
34. Henrard, L. et al. (1999) Phys. Rev. B **60**, 8521.
35. Hulman, M. et al. (2000) unpublished.

ENCAPSULATION OF TRANSITION METALS INTO CARBON NANOCLUSTERS

SUPAPAN SERAPHIN
Department of Materials Science and Engineering
The University of Arizona, Tucson, Arizona 85721, U.S.A.

Fourteen transition metals were co-deposited with carbon in the conventional arc discharge. A composite anode that contained the transition metal or its oxide stuffed into central bores of the graphite rods was used. Both soot sample at the reactor walls and slag at the cathode were characterized using scanning and transmission electron microscopy, and x-ray diffraction. Encapsulation occurs mainly in the slag samples. The arc discharge was modified by having metal in molten form in graphite crucible anode and a helium jet stream flowing perpendicular to the electrodes. The process produces mostly spherical particles. The criteria determining the encapsulation tendency are discussed. A modified arc discharge combined with the decomposition of methane is used to obtain encapsulation of Cu and TiC into carbon cages. The effects of preparation parameters, including the size of the metal pool, flow rate of the gas jet, static pressure, and annealing on the particle morphology and their size distribution were investigated. Ferromagnetism and its dependence on the particle size of Fe, Co, and Ni nanoclusters were identified both in the macroscopic and the microscopic scales.

1. Introduction

Encapsulation of metals into the carbon nanocages provides new aspects of materials science for essentially two reasons. (1) The encapsulant consists of a small number of atoms that is measurably confined by the size of the carbon cage; (2) collective physical properties of these clusters are novel and can be studied as a function of the number of the atoms providing insight into size-dependent properties.

In the nanometer size regime, the number of atoms in the bulk of the particle is close to that in the surface region. By varying the size of the particle, the effects of the bulk vs. surface region can be studied independently. These effects are pronounced in the group of the transition metals where the collective properties vary strongly depending on the number of electrons in d-orbitals [1]. Specifically, the ferromagnetic properties of Fe, Co, and Ni particles inside carbon nanocages will be discussed.

We investigated the encapsulation of fourteen transition metals in the conventional carbon arc discharge process. We hypothesize that the encapsulation tendency is related to the ability of the metal to form carbides [2]. A first group can be

259

L. Pavesi and E. Buzaneva (eds.), Frontiers of Nano-Optoelectronic Systems, 259–273.
© 2000 *Kluwer Academic Publishers. Printed in the Netherlands.*

encapsulated easily in their carbide forms (V, Cr, Mn, Y, Zr, Nb, Mo). A second group cannot be encapsulated at all (Pd, Pt). Third, metals having strong affinity to form carbides (Ti, W) do not allow carbon cages to form, thus no encapsulation occurs. The fourth category is the iron-group: Fe, Co, and Ni which can be encapsulated into carbon cages in its elemental form, not in the carbide form, because each metal acts as catalysts in the carbon deposition process. In addition, single-walled nanotubes and strings of spherical beads were generated when Fe, Co, or Ni were present. The arc discharge was modified to prepare high-yield spherical Fe, Co, Ni particles of uniform size [3,4]. Several preparation parameters including the size of the metal pool [5], flow rate of the gas jet, static pressure [6], and post-deposition annealing were varied to control the size distribution and morphology of particles and the carbon shells. In order to encapsulate Cu and Ti particles into the carbon cages, a combination of modified arc discharge and methane decomposition was used [5]. Detailed preparation processes and morphology of the various products are presented.

We also investigated one of the most prominent collective phenomena of the transition metals, ferromagnetism and its dependence on the particle-size [6,7]. The overall magnetic properties of the encapsulated Fe, Co, and Ni particles were analyzed with a magnetometer. In addition, the magnetic properties of the individual particle was determined using transmission electron holography.

2. Experimental Procedure

2.1. CARBON NANOCLUSTER PREPARATION

The carbon nanoclusters used in this study were prepared by one of the three different methods. A brief description of each method of preparation is given below and the schematic diagram of the experimental setup is shown in Fig. 1.

(a) *Conventional arc discharge* [Fig.1 (a)]: The samples were prepared using the standard arc chamber. The cathode, which is the upper electrode, is a graphitic carbon rod 9-mm in diameter. The anode 6-mm in diameter with a center hole of 3-mm in diameter is filled with one of the transition metals. The arc discharge was generated by a dc current of 75 A at 27 V between the electrodes. The reaction chamber was maintained at either 100 Torr or 550 Torr helium atmosphere. The sample was retrieved from the wall of the reaction chamber in the form of soot.

(b) *Modified arc discharge* [Fig. 1 (b)]: In this method, two electrodes facing each other were lined up in the vertical configuration. A graphite rod 6.5-mm diameter is located on the top as the cathode. The anode, instead of using a graphite rod filled with metal powders like that in conventional arc-discharge, was a graphite crucible with an inner diameter of 25 mm, and filled with molten Fe, Co, or Ni. The thickness of the crucible wall is about 3 mm. A jet of helium gas with a flow rate of approximately 30 m/s was introduced in the direction perpendicular to the centerline of the two electrodes. During the arc discharge at a dc current of (22 V, 175 A) and a helium pressure of 300 Torr, both the metal and the graphite crucible were partially evaporated. The evaporated materials were quenched by the helium jet, and then

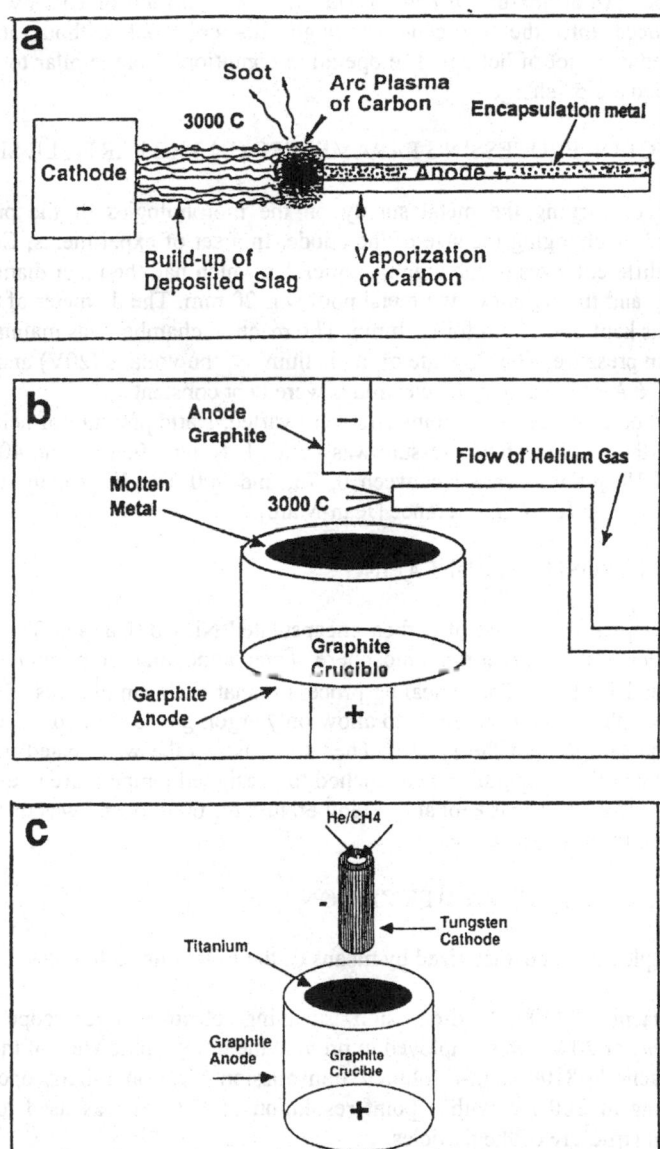

Figure 1. (a) The conventional discharge; (b) the modified; and (c) with methane decomposition deposited on the wall of the reaction chamber. The soot sample was collected from the wall of the reaction chamber.

(c) *Combined modified arc discharge and methane decomposition* [Fig. 1 (c)]: In order to encapsulate Cu and Ti particles into the carbon cages, the graphite crucible was filled by Cu and Ti, respectively. A hole was drilled in the center of the graphite cathode. An admixture of helium and methane in a ratio of 10/1 by volume was introduced into the arc center through this hole but without an additional perpendicular jet of helium. The operation conditions were similar to those of the modified arc discharge.

2.2. EFFECT OF PROCESSING PARAMETERS ON THE PARTICLE SIZES

The effect of varying the metal supply on the morphologies of the products was investigated by changing the size of the anode. In a set of experiments, Co was filled into three different sizes of the graphite crucibles which had the inner diameter of 4, 6, and 13 mm and the height of the metal pool was 20 mm. The diameter of the graphite cathode was kept at 6.5 mm for each run. The reaction chamber was maintained at 200 Torr helium pressure. The flowrate of the helium jet, the voltage (20V) and the current density (~ 0.6 A/mm^2) across the electrodes were kept constant.

Two other processing parameters were varied: static pressure of helium and the velocity of the jet. The static pressure was varied between 40, 150, and 400 Torr. The velocity of He jet was varied between 0, 75, and 300 Torr.l/s. Detail set up of the experiment can be found in reference [Jean-Marc].

2.3. POST-DEPOSITION ANNEALING

A set of samples comprised of carbon encapsulated Ni, Co, Cu, and Ti particles was annealed separately in an argon atmosphere. Three annealing temperatures were used: 600, 900, and 1100°C. The annealing process is that each sample was first put into a quartz tube with two ends confined to allow only argon gas to flow for 30 min in order to purge the sample and the system. Then the quartz tube was heated in an electric furnace. When the temperature had reached the designed temperature under continued argon flow, the sample was kept at this temperature for 60 min, and was then allowed to cool down to room temperature.

2.4. STRUCTURAL CHARACTERIZATION

All the samples were characterized by means of the following techniques:

(1) A Hitachi S-4500 field-emission scanning electron microscope (FE-SEM) operating at 20 kV was employed to provide the topographic view of the samples.
(2) A Hitachi H-8100 high-resolution transmission electron microscope (HRTEM) operating at 200 kV with a point resolution of 0.2 nm was used to reveal the internal structure of the particles.
(3) A Scintag Model 2000 x-ray diffractometer (XRD) using the Cu K_α radiation was used to identify the phase and the crystal structure of each sample.

2.5. MAGNETIC PROPERTY CHARACTERIZATION

The overall magnetic properties of the particles was determined using a vibrating sample magnetometer operating with applied magnetic field up to 1000 Oe.

The ferromagnetism of the individual particle was analyzed using electron holography. The principle is based on the fact that the electron wave transmitted through a magnetic sample changes phase depending on the sample magnetization and its thickness [8-10]. For spherical particles, the phase change due to the sample thickness will be symmetric around the particle center while the phase change due to the magnetization will be asymmetric across the particle. The phase change can be retrieved from a hologram obtained by overlapping the wave scattered by the sample with the (partially) coherent reference wave. The holograms were reconstructed to retrieve both the amplitude and phase of a wave scattered by the sample. The phase image reconstructed from the hologram corresponds to the phase difference, $\Delta\varphi$, between object and reference waves. For a magnetic field, the phase difference is proportional to the integral of the magnetic flux density B over an area between reference beam and object beam along the direction of the optical axis of the optical axis of the microscope. The remnant magnetization, M_r, can be determined when the value of $\Delta\varphi$ is measured from the reconstructed phase image retrieved from the holograms. We determined M_r for particles of various diameters. To illustrate the dependence of the magnetization on the particle size, the M_r values were normalized with respect to the bulk saturation magnetization, M_s (5.1×10^5 A/m for Ni and 1.4×10^6 A/m for Co).

3. Results and Discussion

3.1. ENCAPSULATION TENDENCY

The tendency of encapsulation into carbon nanoclusters can be divided into four categories as follows: (1) Elements that can be encapsulated in the form of their carbides; (2) Elements that are not encapsulated but tolerate the formation of the carbon cages; (3) Elements that form stable carbides, competing with the carbon supply for the graphitic cage formation; (4) The iron-group metals (Fe, Co, Ni) that stimulate the formation of single-walled tubes and strings of nanobeads in the conventional arc discharge condition, and produce the nanometer-size carbon-coated ferromagnetic particles in the modified arc discharge.

3.1.1. Category 1: Encapsulations of the carbides of V, Cr, Mn, Y, Zr, Nb, Mo
Vanadium, an element on the right of Ti in the periodic table, has also been introduced into the carbon arc discharge. In contrast to the encapsulation behavior of boron, the encapsulation of vanadium carbide was observed mainly in the slag. The yield of the carbon tubes is very low. The encapsulation is estimated to be about 5% of the tubes produced. In the soot, numerous vanadium-carbide are embedded in amorphous carbon.

In the slag produced by an anode containing chromium, an element to the right of vanadium in the periodic table, about 5% of the polyhedral graphitic cages are filled by second-phase material. The sizes of the encapsulated particles range from 10 to 50 nm. Most of the hollow cores of the carbon cages are not completely filled [11]. The electron diffraction patterns and the high-resolution images indicate that the dark particles consist of a Cr_3C_2 crystalline phase. Unlike the samples in the slag, the soot contains a high density of Cr_3C_2 particles with their size ranging from 40 to 70 nm. These particles are embedded in amorphous carbon (Fig. 2). The observation of the Mn-related sample shows similar results as those from Cr-related samples with the exception that the fraction of the encapsulation of manganese carbide is slightly higher than in Cr_3C_2.

The results on the encapsulation of yttrium carbide are one of the earliest reports in the field [11]. Among fourteen elements investigated, yttrium carbide is found to be encapsulated at the highest percentage of 30% of the cage production. It should be noted that due to the instability of yttrium in air, Y_2O_3 powder was used in making the composite anode. This is significant because only Y_2C can be encapsulated and Y_2O_3 cannot [12]. We speculate that it may be due to the interfacial compatibility of Y_2C with the graphite structure of the polyhedral carbon cage. It is also found that the encapsulation of yttrium carbide crystals in the particles as well as in the tubes [Fig.3] occurs mostly in the slag and not in the soot since there are very few tubes and polyhedral carbon particles formed in the soot.

ZrO_2 is another oxide stuffed into the anode in order to investigate the encapsulation. Again, only zirconium carbide is encapsulated in the carbon nanoclusters while the unreacted oxide is found to be isolated without any graphitic layers coated, which is found similarly to the yttrium case. Niobium carbide is found to be encapsulated into the hollow cores of about twenty percent of the carbon nanoclusters produced in the slag. Although the density of the NbC particles produced in the slag is not very high and the sizes of the particles are small, almost every NbC particle is encaged by carbon nanoclusters. In the soot portion, the NbC particles are embedded in amorphous carbon. No encapsulation is observed. Molybdenum carbide is encapsulated into 30% of the carbon cages produced in the slag. The encapsulation occurs independent of the particle size. The number of graphitic layers does not correlate directly to the sizes of the metal particles. The small metal particles are sometimes surrounded by graphitic layers thicker than those of larger particles. Most particles contain a gap between the MoC crystal and the carbon cage. This phenomenon may be relevant in tracing the origin of the graphitic formation to either segregated carbon from the supersaturated droplet, or from gas phase reactions at the surface of the particle.

It is interesting to investigate the encapsulation behavior of boron in the carbon arc discharge although it is not a transition metal. Boron is well known for being a dopant for semiconductors and also as an active catalyst for graphitization. TEM observation of the soot prepared from the boron composite anode reveals a morphology never observed before. Normally, soot samples consist of C_{60}, C_{70}, other fullerenes, and amorphous carbon. The soot prepared from the boron composite anode contains numerous polyhedral crystals coated with a few layers of graphitic carbon and networks of graphitic "ribbons". The coated particles have diameters ranging from 10 to 30 nm. Some of them are elongated into a tubular form. The XRD, high-resolution TEM images, and the electron diffraction patterns suggest that the crystal is in the form of B_4C [13]. The slag portion of the sample consists of numerous graphitic networks with very few polyhedral particles of boron carbide. This is in strong contrast to most observations which show that encapsulation occurs predominantly in the slag and not in the soot. With respect to slag runs in a reference arc-discharge with no boron involved, we observe two significant differences: first, the density of the tubes is much lower in the boron run. Second, there are many more graphitic networks in the boron run than in the ones without boron.

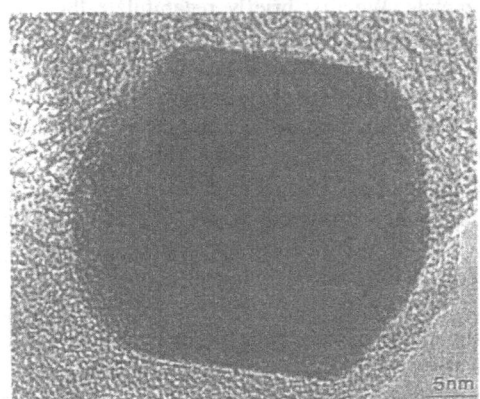
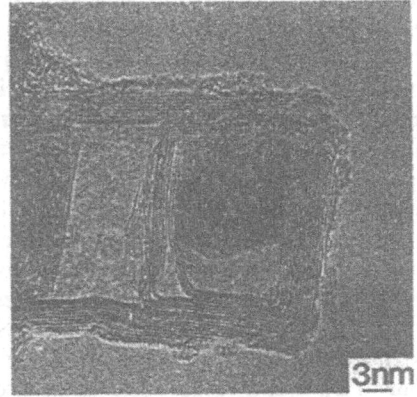

Figure 2. Cr₃O₂ embedded in amorphous C   Figure 3. YC₂ in a carbon tube

3.1.2. Category 2: Elements that are not encapsulated but tolerate the formation of the carbon cages: Pd, Pt

When Pd or Pt was used in the composite anode, there is no encapsulation observed in spite of the existence of numerous carbon cages. We believe that it is due to the fact that these elements do not react readily with carbon to form carbide. It is found that the density of Pd or Pt particles is much higher in the soot portion than in the slag. This may be due to their relatively low melting/boiling points and high vapor pressure.

3.1.3. Category 3: Elements that form molecular or stable carbides: Ti, W

Titanium have a high reactivity with carbon and form very stable carbides. It is found that the titanium-related slag contains numerous titanium carbide crystals embedded in amorphous carbon [12,14]. It is obvious that a strong tendency to form the carbide competes with the formation of carbon cages. A very low density of tungsten particles is observed both in the soot and in the slag. There is no encapsulation observed. Since the boiling point of W is very high, the metal is not easily evaporated during the carbon arc discharge, and the melting tungsten drops down from the anode and forms chunks at the bottom of the reactor. Aluminum and silicon are also strong carbide formers. When aluminum is used in the composite anode, there are no carbon nanoclusters observed in the samples of both slag and soot. Instead, there are numerous aluminum carbide fibers and particles mixed with amorphous carbon and graphitic flakes. The electron diffraction pattern indicates that the aluminum carbide is Al_4C_3. In this aluminum-related sample, it is the first time we have observed that the slag and the soot consist of the same features, suggesting that when aluminum is involved, the product of the carbon arc discharge is independent of the locations of the deposit.

3.1.4. Category 4: The iron-group metals

It was reported earlier that the presence of Fe, Co, or Ni in the arc discharge can stimulate the growth of the single-walled carbon tubes [12,15,16,17] and strings of spherical beads [12,15,18]. In addition, however, and when prepared under modified arc discharge conditions [3,4], the results show that these metal particles can be coated by a few graphitic layers and have a uniform nanometer size. The following results are a summary of our own work related to these metals. We first briefly recapitulate the results obtained with the conventional arc discharge method involving Fe, Co, and Ni.

Long straight single-walled tubes (SWT) of carbon can be produced from a composites anode filled with Fe, Co, or Ni. The size of the tubes ranges from 0.7 to 2.0 nm in diameter and are a few micrometers long. It is found that using a mixture of these metals, such as Fe/Ni, Co/Ni, or Fe/Co to fill in the composite anodes, can enhance significantly the density of the single-walled nanotubes. The tubes form numerous bundles of curved threads and tangle together when observed in the TEM. This phenomenon exhibit itself in the macroscopic scale of the web-like appearance of the soot. It is found that most of the tubes produced by Fe/Ni have a diameter between 1.0 and 1.7 nm, and those from Co/Ni between 1.2 and 1.3 nm.

The strings of spherical beads (SSB) are observed mostly in the soot and the extended slag deposits produced by using Fe, Co, or Ni composite anodes. The strings contain ten to thirty beads. The image also shows that the hollow cores of the beads are mostly empty. Only a few of them are filled with nickel as shown by the darker contrast inside the beads. The diameter of the beads ranges from 10 to 70 nm but in each string the diameters of the nanobeads are almost the same. Further examination of the strings of beads indicates that each carries at one end a Ni particle of slightly larger diameter than that of the chain-beads. The other end is not associated with any metal particle. It is suggested that the formation of the strings begins at the metal particle. Detailed discussions of the growth mechanism of the SSB are presented in the references [12,18].

Under the *modified arc-discharge* conditions described in the experimental procedure, the amount of Fe, Co, or Ni with respect to carbon in the arc discharge process is much higher than the conventional method. The results can be used to elucidate the effect of the ratio of carbon-to-metal supply on the morphology of the products. The results show that the modified arc discharge does not produce any multi-walled, or single-walled tubes, nor strings of carbon nanobeads. Most of the products consist of spherical metals particles of a diameter range 15 to 80 nm (Fig. 4). TEM reveals that the as-made cobalt and nickel particles are covered by 1-2 graphitic layers, while iron particles are surrounded by amorphous carbon. X-ray diffraction analyses indicate that the particles are in the form of elements (Fe, Co, Ni) and not in the form of carbides.

To encapsulate Cu and Ti, the preparation process has to be modified further to enhance the carbon supply. This was carried out by adding the *decomposition of hydrocarbon gas*. Admixture of helium and methane in a ratio of 10/1 by volume was introduced into the arc through a hole of the cathode facing the metal pool in the anode.

This process can coat Cu and TiC particles with 3-4 layers of graphitic cages. It is a significant improvement from using the conventional arc discharge.

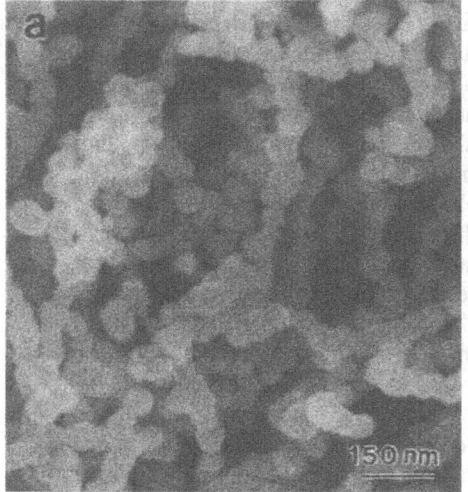

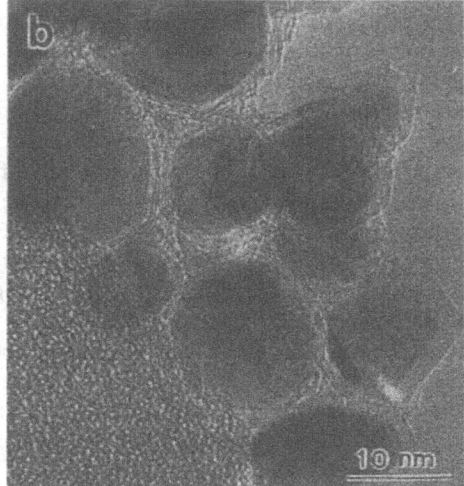

Figure. 4. SEM of spherical Ni particles   *Figure 5. TEM of 900C annealed Ni particle*

Various attempts have been made to identify the encapsulation mechanism. However, as the number of studies grew, it was obvious that the existing hypotheses could explain only some but not all of the case studies. The first group of explanations specifies physical properties of the encapsulated materials to be the determining factor. Saito et al. [15,19] reported that the vapor pressure of the rare-earth elements determines the encapsulation. All the encapsulated elements (Y, La, Ce, Pr, Nd, Gd, Tb, Dy, Ho, Er, and Lu) belong to the group of relatively low vapor pressure metals, and those not encapsulated (Sm, Eu, Tm, and Yb) to the group of high vapor pressure metals, with only one exception, Tm. However, this criterion cannot be used to explain the transition metals in our study. Majetich et al. [20] suggested that encapsulation would not occur if 1) the materials have low melting points; 2) the corresponding carbides have very high melting points. The hypothesis cannot be used to explain our results of metals that do not form carbides e.g. Ag, Cu, Pd. Furthermore, our observation shows that low melting point Mn is encapsulated. Another hypothesis was advanced by Guerret-Piecourt et al. [21], and was based on the electronic configuration of the metal. Incomplete electronic shells in the most stable ionic state of the metal are claimed to be the factor that determines the encapsulation. Here again, the explanation of an incomplete d-shell favoring encapsulation is in disagreement with our and Saito's results. By and large, none of the hypotheses is able to interpret the wide spectrum of encapsulation phenomena without exceptions.

We have postulated the enthalpy of the carbide formation to be responsible for the encapsulation tendency [12,14]. If the carbide does not exist, no encapsulation is observed although cages may be formed. If, on the other hand, the carbide is aggressively formed, no carbon may be left to form a cage. It is obvious that in between these two extreme, the "supply aspect" shifts the balance to either encapsulation or refusal to even form a cage. It is also apparent that some factor in addition to the physical parameters must be investigated to give a basis for a more comprehensive guideline.

This "supply-side mechanisms" can probably be influenced by the configuration and operating conditions at the electrodes, particularly at the anode, and in the rest of the reaction space. Varying anode dimensions and variations of the metal concentration and current densities should influence the encapsulation tendency strongly, much beyond the initial physical properties of the elements and their compounds encapsulated. Variations of results between different studies may then be understood on this basis. Further support comes from the observation of spherical carbon-coated ferromagnetic particles in the modified arc discharge [22,23] operating against a tungsten cathode, a configuration quite different from the conventional arc discharge setup. Elliott et al. [22] and Host et al. [23] suggested that a "local carbon supply" from the area opposite the tungsten cathode vs. the "global carbon supply" existing in the remaining areas of the reactor vessel are factors determining the morphology of the product. This results in entirely different growth phenomena, proving that it is the balance of various processes determining the outcome. We also believe that the interfacial effects are important in the understandings of encapsulation. The compatibility of the carbide with the graphitic walls of the cage plays a role within the constraints of the thermodynamic data. The epitaxial relationship between the carbide crystals of Y_2C, MoC, and NbC, and the carbon cages is evident.

Further studies will elucidate the basic mechanisms, and identify the relative roles of carbon rejected from supersaturated droplets upon cooling vs. gas phase reactions at the surface of such droplets. Our results show evidence of both scenarios. If the carbon cage is formed by the rejected supersaturated droplets upon cooling, there must be a direct correlation of the size of the particles and the numbers of the graphitic layers. MoC encapsulation shows contrasting results. Small size MoC particles are surrounded by thick graphitic layers. The evidence of the partially filled particles and the facetting of the metal carbides may point toward the simultaneous growth of the metal carbides and the carbon cages. The situation is still fluid. Further work in a number of laboratories will better clarify the selection mechanisms which determine the encapsulation phenomena.

3.2. EFFECT OF PROCESSING PARAMETERS ON THE PARTICLE SIZES

3.2.1. Carbon and Metal Supply

Changing the size of the metal pool in the graphite anode in the modified arc discharge affects directly the size of the deposited product. The smaller the metal pool, the smaller the average size of the particles [5]. The average diameters of the particles from the 4-, 6-, and 13-mm Co pool are 10, 15, and 65 nm, respectively. This suggests that the rates of supply of carbon versus metal is one of the parameters that affect the size of the particle formation while other parameters are kept constant. The thickness of the carbon coating is also affect by the amount of the metal and carbon available. The higher the ratio of the carbon to metal is used in the anode, the thicker the carbon coating layers are formed on the surfaces of the particles.

3.2.2. Static Chamber Pressure
The average particle size increased from 5.4 nm to 16.9 nm, (2.5 times) as the static chamber pressure increased from 40, 150, and 400 Torr [6]. This effect is commonly observed in the cluster formation in inert-gas aggregation sources [24].

3.2.3. Velocity of Helium Jet
A faster helium jet resulted in smaller particles as was also observed by Teng et al. for Ni particles [25]. This effect was observed for all considered static pressures above. At 150 Torr for instance, the mean diameter of the particles dropped from 16.9 nm to 7.9 nm and 5.4 nm as the jet velocity increased from 0 to 75 and 300 Torr.l/s.

3.3. ANNEALING EFFECT ON THE STRUCTURAL TRANSFORMATION

To investigate the growth and the dynamic of structural transformation, the encapsulated Fe, Co, Ni, Cu, and TiC particles were annealed at 600, 900 and 1100C. Further graphitization of the carbon coating is observed. Particles only partially coated by carbon are not protected, and are sintered by annealing. Fig. 5 shows morphology of 900C annealed Ni particles having a thicker layer of the graphitic carbon around them. We speculated that the thickening of the graphitic layers may be a result of the precipitation of carbon atoms previously dissolved and trapped in the particles while the particles were quench by the helium jet. The annealing effect occurs at different temperatures for different elements. For instance, the carbon coatings of Co particles were well graphitized when annealing at 600C. When the annealing temperature was increased to 1100C, most Ni and Co particles were sintered to big chunks and left the carbon-coating shells behind as shown in Fig. 6. The size of the shells is generally larger than that of as-made particles. Note a couple of partitions inside the shells. This suggests that several carbon cages may combine together after the metal cores migrated out of the cages.

Figure 6. TEM of 1100C annealed Co sample showing numerous empty graphitic shells with partitions inside

This phenomenon resembles the growth mechanism of "string of graphitic beads" in which at a certain critical stage of formation, the metal particles are pushed out of their graphitic enclosure leaving the empty cage behind [18]. The driving force may involve the surface energy of the particles and the sintering tendency of the metal particles. The dynamic of the structural transformation of these carbon-coated metal particles has been demonstrated earlier by Ugarte [26]. He illustrated that the ejection of the gold particles from the graphitic particles was induced by electron bombardment in a 200-keV TEM.

3.4 MAGNETIC PROPERTIES

Several research teams reported on the magnetic properties of their carbon coated Fe, Co, or Ni particles [27-30]. However, the correlation of the magnetic properties and the particle sizes was not fully established. We analyzed the magnetic properties on macroscopic samples with a magnetometer showing the ferromagnetic character with a ratio of remnant to saturation magnetization, M_r/M_s, of ~0.3 at room temperature. The coercive field of carbon-coated Fe, Co, and Ni particles is 600, 200, and 240 Oe, respectively [4]. The encapsulated particles showed ferromagnetic hysteresis loops with marked size-dependent properties [6]. Fig. 7 illustrates the effects of carbon coated Co particle size on the remnant magnetization (a), and the coercive field (b) at 5K, 80K, and 300K. The magnetization shows a maximum around 10 nm, and increases markedly with decreasing temperature. The coercive field also depends strongly on the particle size. It passes through a maximum when the diameter increases, which shifts to larger sizes with the temperature. At 5K, the coercive field remains nearly constant up to ~10 nm, and decreases afterwards. At 80 and 300K, the coercive field increases first with the diameter to a maximum at ~25 nm. Note that the coercive field is very small for the small particles at room temperature.

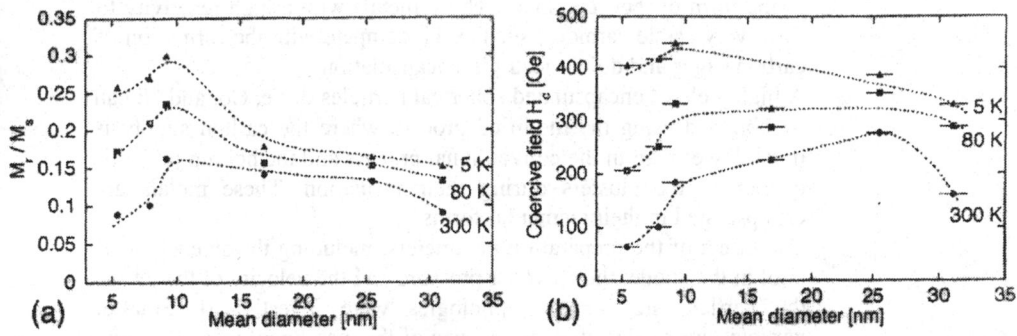

Figure. 7. (a) Remnant magnetization and (b) coercive field as a function of particle diameter and temperature of Co particles

The magnetization of individual particles attached together in a chain was analyzed microscopically by electron holography [7]. It is found that the ratio of remnant magnetization to bulk saturation magnetization M_s (5.1×10^5 A/m for Ni and 1.4×10^6 A/m for Co), M_r/M_s, of Co decreases from 53% to 16% and of Ni decreases from 70% to 30% as the particle diameter increases from 25 to 90 nm (see Fig. 8). It is likely that larger particles contain multiple magnetic domains, which partially compensate each other. After being exposed to a two Tesla external magnetic field, the M_r/M_s of Co increases by 45% from the initial values, with the same dependency on the particle size. The M_r/M_s of Ni particles, on the other hand, increases only 10% from the as-deposited values. The increased magnetization can be attributed to the merging of small domains into larger ones after the exposure to the external magnetic field.

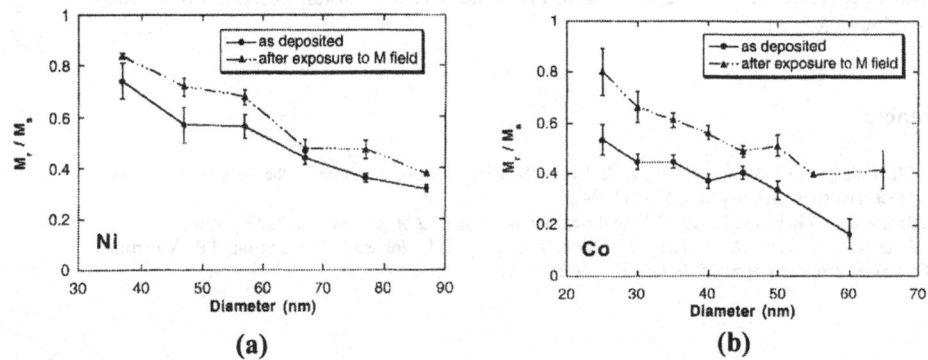

Figure 8. Ratio of M_r/M_s as a function of the particle diameter for (a) carbon coated Ni particles; and (b) Co particles. (Determined by electron holography.)

4. Summary

1. Encapsulation of several transition metals can be achieved using the conventional arc discharge process. Most of metals are encapsulated in the form of their carbides. Those metals with a high reactivity to form very stable carbides, such as Ti, compete with the formation of carbon cages and do not result in encapsulation.

2. A high yield of encapsulated spherical particles of Fe, Co, and Ni can be obtained using the modified process where the carbon supply is much lower than in the conventional process and the helium jet helps quenching the clusters during their formation. These metals are encapsulated in their elemental forms.

3. The effect of the preparation parameters, including the size of metal pool in the anode, the chamber pressure and the velocity of the jet, on the particle size and morphologies were investigated. Smaller particles are produced when the size of the metal pool is smaller, the chamber pressure is lower, and the jet velocity is higher.

4. High temperature annealing induces the structural transformation of the particles. Generally, the graphitic coating of the particles is thicken at 600C to 900C. At 1100C, the metal particles sintered and left numerous empty graphitic shells.

5. Magnetization and the coercive field of Fe, Co, and Ni are strongly dependent on the particle size. The relative magnetization of particle larger than 25 nm decreases approaching the value of the bulk. Large particles tend to form multiple magnetic domains resulting in flux closure inside the particles.

5. Acknowledgements

The author gratefully acknowledges collaborations and discussions with a research team at Materials Electrochemical Research Corp., Tucson, Arizona, and research groups of Profs. Chatelain and Stadelmann at the Swiss Federal University of Technology, Lausanne, Switzerland, and with Drs. Zhou, Jiao, Beeli, and Bonard. Financial support from the U.S. National Science Foundation and the Swiss National Science Foundation is also acknowledged.

References

1. I.M.L. Billas, A. Châtelain, and W.A. de Heer, "Magnetism from the atom to the bulk in iron, cobalt, and nickel clusters" *Science* 265, 1682 (1994).
2. S. Seraphin, D. Zhou, and J. Jiao, "Filling the carbon nanocages" *J. Appl. Phys* 80, 2097 (1996).
3. V.P. Dravid, J.J. Host, M.H. Teng, B. Elliott, J. Hwang, D.L. Johnson, T.O. Mason, J.R. Weertman, "Controlled-size nanocapsules" *Nature* 374, 602 (1995).

273

4.   J. Jiao, S. Seraphin,, X. Wang, and J.C. Withers,"Preparation and properties of ferromagnetic carbon-coated Fe, Co, and Ni nanoparticles." *J. Appl. Phys* 80, 103 (1996).
5.   J. Jiao and S. Seraphin, "Carbon encapsulated nanoparticles of Ni, Co, Cu, and Ti" *J. Appl. Phys.* 83, 2442 (1998).
6.   J.-M. Bonard, S. Seraphin, C. Beeli, J-E. Wegrowe, T. Stöckli, J. Jiao, P.A. Stadelmann, and A. Châtelain, "Production and characterization of carbon encapsulated ferromagnetic cobalt particles", Fullerenes *Proc. 193rd Electrochem. Soc.* , 794 (1998).
7.   S. Seraphin, C.Beeli, J.-M. Bonard, J.Jiao, P.A. Stadelminn, A. Chatelain, "Magnetixation of carbon-coated ferromagnetic nanoclusters determined by electgron holography:, *J. Mater. Res.,* 14, 2861 (1999).
8.   A. Tonomura, *Electron Holography*, Springer Series Optical Sciences 70, (Springer, Berlin) 1993.
9.   A. Tonomura, L.F. Allard, G. Pozzi, D.C. Joy, and Y.A. Ono, *Electron Holography* (Elsevier, Amsterdam) 1995.
10.   C. Beeli, B. Doudin, J-Ph. Ansermet, P.A. Stadelmann, "Study of Co, Ni and Co/Cu nanowires: magnetic flux imaging by off-axis electron holography" *J. Magnetism & Magnetic Mater.* 164, 77 (1996).
11.   S. Seraphin, D. Zhou, J. Jiao, J.C. Withers and R. Loutfy, *Nature* 362, 503 (1993).
12.   S. Seraphin, J. Electrochem. Soc. 142, 290 (1994).
13.   D. Zhou, S. Seraphin, and J.C. Withers, *Chem. Phys. Lett.* 234, 233 (1995).
14.   S. Seraphin, D. Zhou, J. Jiao, J.C. Withers, and R. Loutfy, *Appl. Phys. Lett.* 63, 2073 (1993).
15.   Y. Saito, M. Okuda, T. Yoshikawa, A. Kasuya, and Y. Nishina, "Correlation between volatility of rare-earth metals and encapsulation of their carbides in carbon nanocapsules" *J. Phys. Chem.* 98, 6696 (1994).
16.   S. Iijima and T. Ichihashi, *Nature* 363, 603 (1993).
17.   D.S. Bethune, C.H. Kiang, M.S. de Vries, G. Gorman, R. Savoy, J. Vazquez, and R. Beyers, *Nature* 363, 605 (1993).
18.   S. Seraphin, S. Wang, D. Zhou, and J. Jiao, *Chem. Phys. Lett.* 228, 506 (1994).
19.   Y. Saito, *Carbon* 33, 979 (1995).
20.   S. Majetich, J.H. Scott, E.M. Brunsman, S. Kirkpatrick, M.E. McHenry, and D.C. Winkler, *Electrochem. Soc. Proc.* 95-10, 584 (1995).
21.   C. Guerret-Piecourt, Y. Le Bouar, A. Loiseau, and H. Pascard, *Nature* 372, 761 (1994).
22.   B .R. Elliott, J.J. Host, V.P. David, M.H. Teng, J-H. Hwang, "A descriptive model linking possible formation mechanisms for graphite-encapsulated nanocrystals to processing parameters" *J. Mat. Res.,* 12, 3328 (1997).
23.   J.J. Host, M.H. Teng, B.R. Elliott, J.H. Hwang, T.O. Mason, J.R. Weertman, D.L. Johnson, and V.P. Dravid, " Graphite encapsulated nanocrystals produced using a low carbon: metal ratio*" J. Mat. Res.,* 12, 1268 (1997).
24.   B.G. de Boer and G.D. Stein, *Surf. Sci.* 106, 84 (1981).
25.   M.H. Teng, J.J. Host, J.-H. Hwang, B.R. Elliot, J.R. Weertman, T.O. Mason, V.P. Dravid, and D.L. Johson, *J. Mater. Res.* 10, 233 (1995).
26.   D. Ugarte, *Chem. Phys. Lett.* 209,99 (1993).
27.   S.A.Majetich, J.O.Artman, M.E.McHenry, N.T.Nuhfer, and S.W.Staley, "Preparation and properties of carbon-coated magnetic nanocrystallites" *Phy. Rev. B* 48, 16845 (1993).
28.   K. Lafdi, A. Chin, N. Ali, and J.F. Despres, "Cobalt-doped carbon nanotubes: preparation, texture, and magnetic properties" *J. Appl. Phys.* 79, 6007 (1996).
29.   J.J. Host, J.A. Block, K. Parvin, V.P. Dravid, J.L. Alpers, T. Sezen, and R. LaDuca, "Effect of annealing on the structure and magnetic properties of graphite encapsulated nickel and cobalt nanocrystals," *J. Appl. Phys.,* 83, 793 (1998).
30.   J.A. Block, K. Parvin, J.L. Alpers, T.Sezen, R.Laduca, J.J.Host, V.P. Dravid, "The magnetic properties of annealed graphite-coated Ni and Co nanocrystals" *IEEE Trans. Magn.,* 34, 982 (1998).

DEVELOPMENT OF COATING TECHNOLOGIES USING NANO PARTICLES

YASUO MIHARA

ULVAC-Vacuum Metallurgy Co., LTD.

1. Introduction

We have developed several coating technologies using nano-particles. One of them is the gas deposition method (GDM). Several tens of nanometer sized particles are prepared by the gas evaporation method [1] in an evaporation chamber, and are directly carried by a gas flow into another chamber and ballistically deposited on a substrate to form films. This film formation method is called GDM [2].

An apparatus called Jet Printing System (JPS), based on GDM, has been developed for practical applications. By using JPS, various kinds of metallic films, such as a spot, a line, an area, a coil pattern and their combined pattern are easily prepared. The electrical conductance and the density of the deposited films show the same values as for crystal. The impurities contained in the copper films prepared by JPS were measured by the Glow Discharge Mass Spectrometry. It is proved that the films formed by JPS resulted in higher purity than the copper metal pellets used as source for the evaporations.

In JPS, any one of the heating methods, such as induction, arc or resistance heating has been used to evaporate low melting point metals. For higher melting point metals, such as tungsten, tantalum or molybdenum, a laser heating system has been studied. Nano particles of tungsten were formed by the synchronized irradiation of two pulsed Nd:Yag lasers on a tungsten substrate. The size distribution of nano particles was measured by a low-pressure differential mobility analyzer [3].

A similar deposition technique, called aerosol-JPS, has been developed for ceramic powder [4]. Replacing an evaporation chamber with an aerosol chamber which produces mixture of gas and fine ceramic powders, ceramic films about tens of μ m thick can be easily formed. Piezoelectric lead zirconate titanate (PZT) films were prepared by aerosol-JPS, their piezoelectric properties were measured and an actuator was made with them [5].

L. Pavesi and E. Buzaneva (eds.), Frontiers of Nano-Optoelectronic Systems, 275–290.
© 2000 *Kluwer Academic Publishers. Printed in the Netherlands.*

Another coating technology to make thin films and based on solution of dispersed nano particles has been developed [6]. Solutions with gold, silver or copper nano particles have been manufactured. The particles with a size of less than 10 nm are suspended in an organic solvent without aggregation. These pastes (high viscosity solutions) are painted on a substrate such as stainless steel, glass, alumina plate using the dip-coating method. These painted substrates are baked at 300°C in an air atmosphere for 30 minutes to form metal films.

Two applications of dispersed nano particles films are discussed here. The first one is a repairing system for broken line electrodes in flat panel display. The second is for Cu metallization process in ultralarge scale integrated circuit [7].

2. Deposition of Nano Particles Using a Gas Jet

2.1. GAS EVAPORATION METHOD AND GAS DEPOSITION METHOD

Figure 1 illustrates how nonometer sized particles are formed by the gas evaporation method (GEM) [1]. Source materials, either organic or inorganic, are evaporated in inert gas like He or Ar. Evaporated atoms collide with inert gas atoms and condense into nano particles. Since it is difficult to handle nano particles by conventional ways, the gas deposition method (GDM) [2] has been developed as shown in figure 2. Nano particles and gas are transferred into a chamber that is pumped down and the particles are ballistically deposited on a substrate to form films.

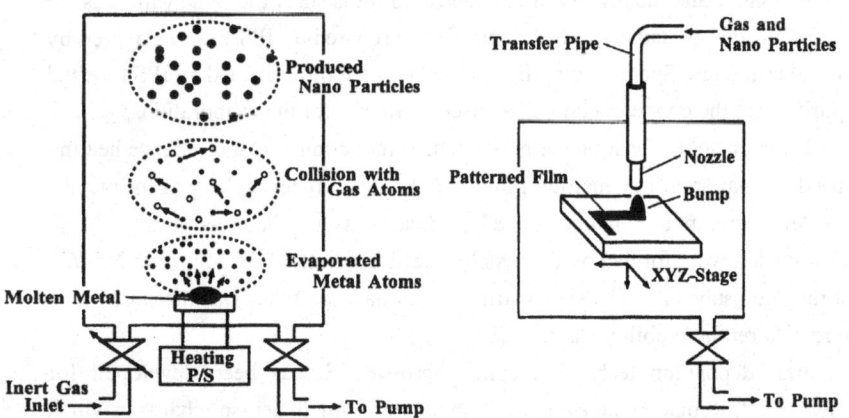

Figure 1. Principles of the gas evaporation method.   *Figure2.* Principles of the gas deposition method.

2.2. JET PRINTING SYSTEM

2.2.1. System for Practical Application

Based on GDM, Jet Printing System (JPS) has been developed for practical applications. Figure 3 shows the schematic diagram of the system for industrial usage. The system has a cassette chamber and a robotic transfer system in a transfer chamber to carry a substrate automatically. A substrate is set on the XY stage and the metal film is formed according to

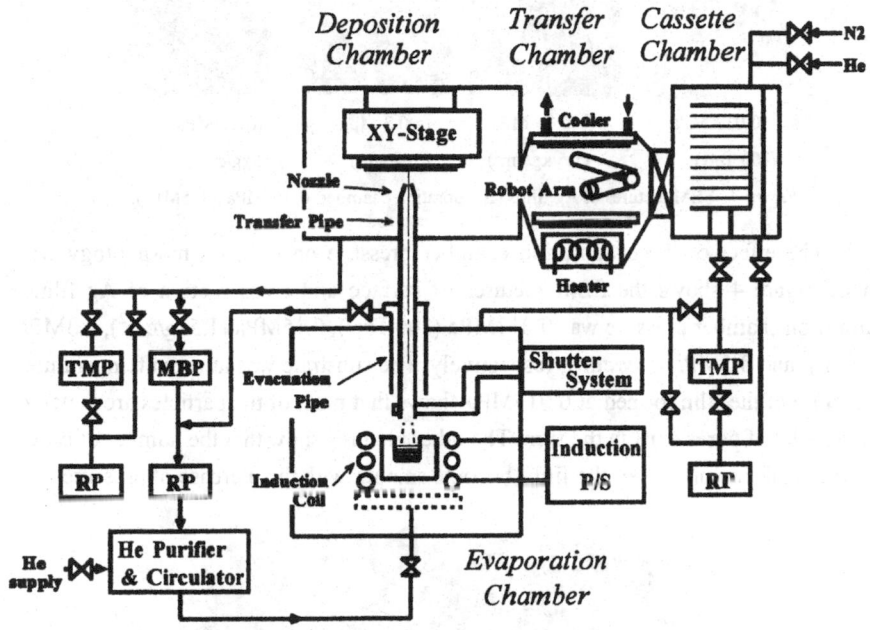

Figure 3. Schematic diagram of Jet Printing System for industrial usage.

a programmed pattern. Nano particles formed in He gas at about 0.067-0.1 MPa (500-760 Torr) are carried through a transfer pipe from an evaporation chamber to a deposition chamber evacuated constantly by a vacuum pump below 1330 Pa (10 Torr). The velocity of the He and nano particles fluxes at the nozzle which is located at the end of the transfer pipe are several hundred meters per second. Nano particles keep their velocity until they collide with the substrate in the deposition chamber while He disperses in random direction immediately after the gas has left the nozzle: This is because of the difference in inertia, momentum, of He compared to the particle. A film is formed continuously by the collision of nano particles to the substrate and patterned by means of moving the stage that holds the substrate.

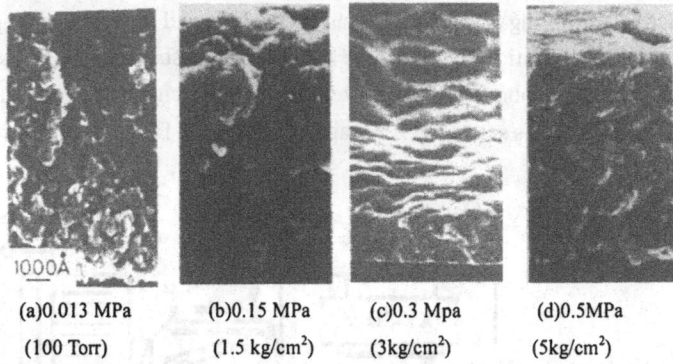

(a)0.013 MPa   (b)0.15 MPa   (c)0.3 Mpa   (d)0.5MPa

(100 Torr)   (1.5 kg/cm^2)   (3kg/cm^2)   (5kg/cm^2)

Figure 4. SEM pictures of Ag films (Evaporation chamber: 0.013MPa – 0.5MPa).

The effect of the evaporation chamber pressure on the film morphology was studied. Figure 4 shows the SEM pictures of surface and cross section of Ag films. Evaporation chamber pressure was 0.013MPa (100 Torr), 0.15MPa (1.5kg/cm^2), 0.3MPa (3kg/cm^2), and 0.5MPa (5kg/cm^2), respectively. The substrate was at room temperature. The picture of the film formed at 0.013MPa shows that most of the particles are sintered however a lot of pores exist in the film. The other pictures show that the number of pores is reduced and the surfaces of the films become smooth with the increase of pressure.

Figure 5. Optical microscope picture of patterned Ag line by using shutter system (line width:40 * m).

An induction-heating source is used as illustrated in figure 3. Stable deposition rate and continuous growth of films without debris are the two important requirements for practical usage. The temperature of the crucible is monitored by an infrared radiation thermometer and also used as a feedback control to the output of the induction power supply. The transfer pipe is heated to avoid adhesion of nano particles. A shutter system is mounted to switch the deposition on and off. Figure 5 shows a demonstration of the shutter system. High purity He gas (higher than 6N) is used to make films with low electrical resistivity; for example, 34 nΩm for aluminum and 21 nΩm for copper were measured by four point probe method. Since high purity He gas is expensive, He gas purifier system has been developed to recycle He gas as shown in figure 3.

Impurity levels of copper films, produced by JPS, were measured by use of Glow Discharge Mass Spectrometry. Two types of copper pellets, 4N (raw material before purification) and 5N (purified products), were tested as source material for the films. Copper pellets were charged in a pyrolytic carbon crucible and heated at 1780℃ in 800 hPa (600 Torr) of He atmosphere for evaporation and the copper films were formed in the deposition chamber at 160 Pa (1.2 Torr). The impurity level is summarized in Table 1. It is easily recognized that the impurity level in films is reduced compared to those in raw material. JPS is usable to prepare various kinds of metallic deposit, such as a spot, a line, an area, a coil pattern and their combined patterns. Figurse 6 and 7 show some examples.

TABLE 1 Impurity contents in raw material and GD films measured by GD-MS

| | 4N Cu | | 5N Cu | |
Element	Raw Material	GD film	Raw material	GD film
P	1.35	0.01	0.88	0.01
S	5.5	0.54	1.11	0.45
Mn	0.16	0.09	0.09	0.07
Fe	0.66	0.2	0.72	0.52
Ni	0.4	0.16	0.42	0.12
As	0.16	0.02	0.18	0,01
Ag	5.34	4.02	1.86	0.34
Sb	0.13	0.01	0.08	0.01
Cr	0.07	0.12	0.06	0.06

(ppm)

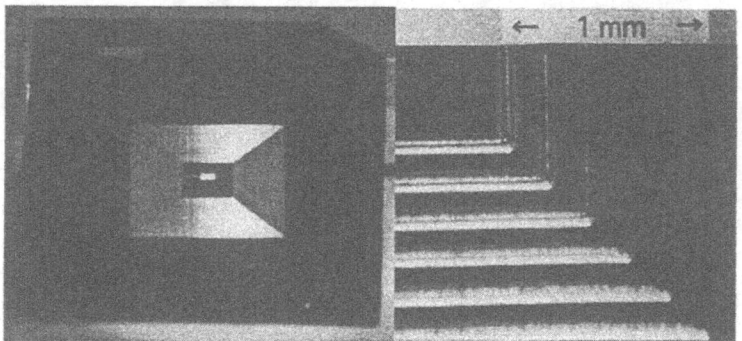

Figure 6. Pattern samples formed by JPS. Cu coil pattern (line/space=130μm /90μm,H30μm).

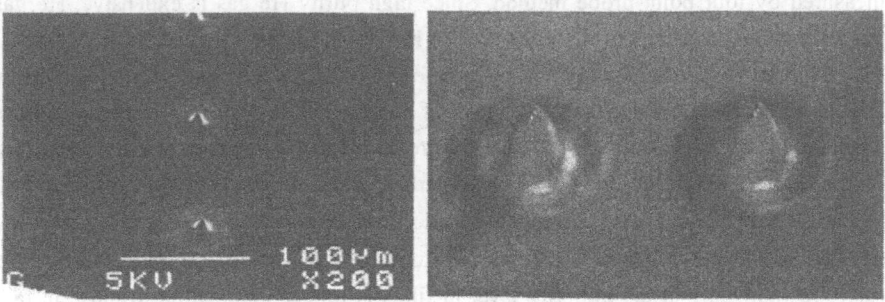

Figure 7. Bumps formed by JPS. Left: SEM photograph of Ni bump (ϕ 30μm × H30μm) on Si. Right: Optical microscope picture of Cu bump (ϕ 0.5mm × H0.5mm) on ϕ 0.8mm Au electrode.

2.2.2. *Nano Particles Formation by Laser Irradiation*

The formation by laser irradiation of nano particles formed by high melting point metals, such as tungsten, tantalum and molybdenum, has been studied [3]. Tungsten nano particles were produced under the irradiation of two synchronized pulses of a Nd:YAG lasers on a tungsten substrate, in low pressure He gas atmosphere, as shown in Fig. 8.

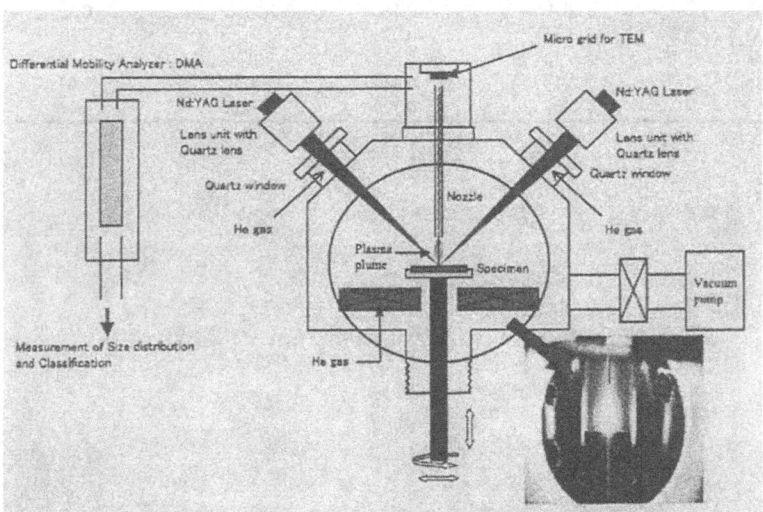

Figure 8. Schematic diagram of experimental apparatus for nano particle formation by laser.

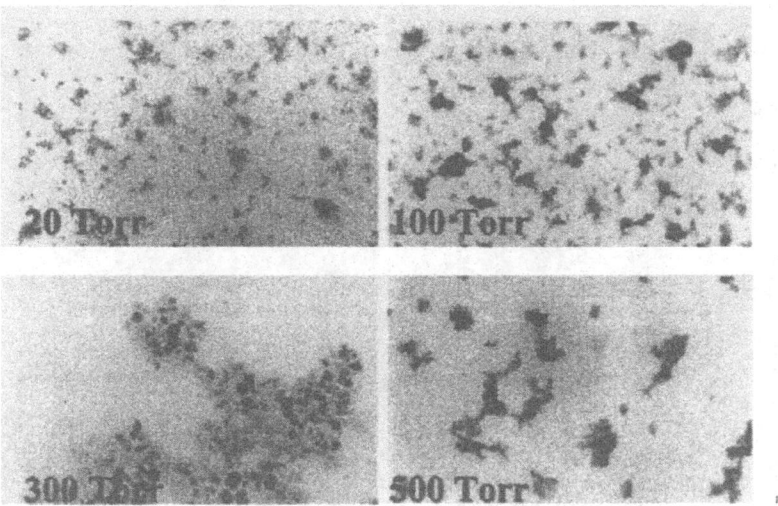

Figure 9. TEM image of tungsten nano particles formed by laser irradiation.

One of the lasers, called assistant laser (max. power: 150 J/Pulse, wavelength: 1064 nm and pulse width: 10 ms), is used for heating the sample. Another laser, called main laser (max. power: 10 J/Pulse, wavelength: 1064 nm and pulse width: 0.2 ms) is for the evaporation of the sample. The main laser irradiation starts 9 ms after the start of the assistant laser irradiation. An high purity tungsten plate was used as the target in order to avoid cracking by thermal shock. The assistant laser power was varied between 46.5-132 J/Pulse to produce efficiently nano particles within 9.7-503 Torr He pressure. The tungsten nano particles formed by using the two lasers were transferred with the He gas to an upper chamber. Some of them were captured on a micro grid for TEM, and the rest were transferred to a low pressure differential mobility analyzer to analyze the distribution of particle sizes.

Figure 9 is a TEM image of tungsten nano particles obtained by laser irradiation. Typical particle size was of several nano meters in this case. Figure 10 shows their size distribution measured by a low pressure differential mobility analyzer. The result shows that by changing the pressure and the laser power one can control the size distribution and shape of UFP. A detailed discussion is found in [3,8-10].

The gas deposition process of high melting point metals will be studied by adding a deposition chamber to the apparatus.

282

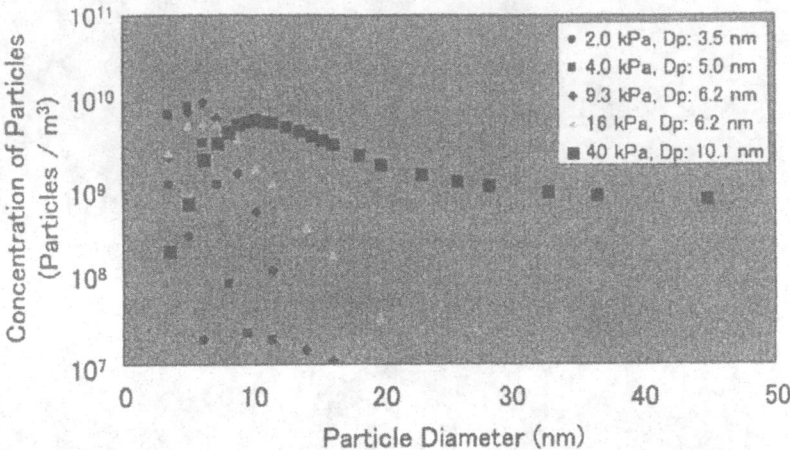

Figure 10. Size distribution of tungsten nano particles formed by laser irradiation.

2.2.3 *Aerosol Jet Printing System*

A similar deposition technique, called aerosol-JPS, has been developed for ceramic powders [4]. Replacing the evaporation chamber with an aerosol chamber which contains mixture of gas and fine ceramic powders, ceramic films about tens of μm thick can be easily formed. Figure 11 shows a schematic diagram of aerosol-JPS. Commercially available PZT powder was placed in the aerosol chamber.

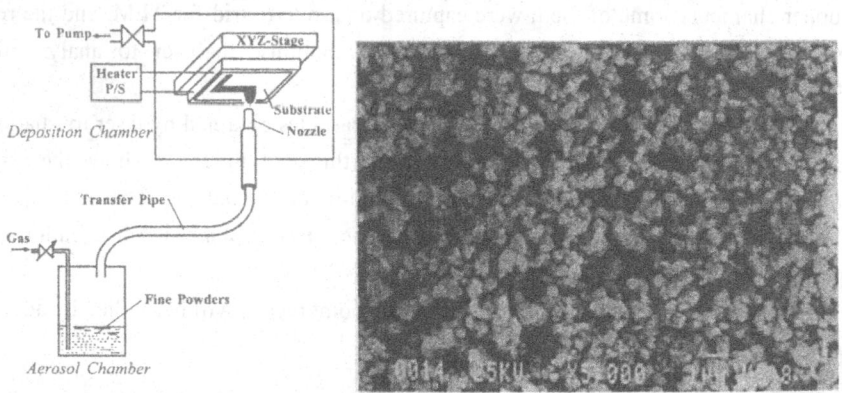

Figure 11. Schematic diagram of aerosol-JPS.   *Figure 12.* SEM photograph of the PZT fine particles.

Figure 12 shows SEM photograph of powder of PZT. The average particle size was 0.3 μm. N2(80%)+O2(20%) gas was introduced in the aerosol chamber to aerosolize the fine powders. The pressure in the aerosol chamber was kept constant to (0.05-0.1Mpa) by controlling the gas flow. Aerosolized fine powder was carried into the deposition chamber and deposited on a substrate in a similar way as described above. Figure 13 shows a SEM photograph of the cross section of PZT film. A dense structure without voids was observed in the film. Figure 14 shows the EPMA analysis of material fine particles and of the film of PZT, respectively. The composition of PZT between raw material and the film is in good accord. Figure 15 shows the diffraction pattern of PZT films of (a) as-deposited film and (b) annealed film at 500°C for 30 minutes. An improvement in crystallization by annealing was observed and the relative dielectric constant was improved from 200 (as-deposited film) to 750 (annealed one).

Figure 13. SEM photograph of PZT film formed by aerosol-JPS.

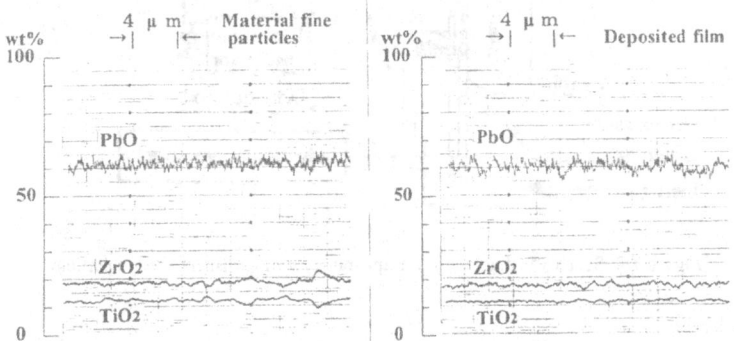

Figure 14. EPMA analysis of material fine powder and PZT film formed by aerosol-JPS.

284

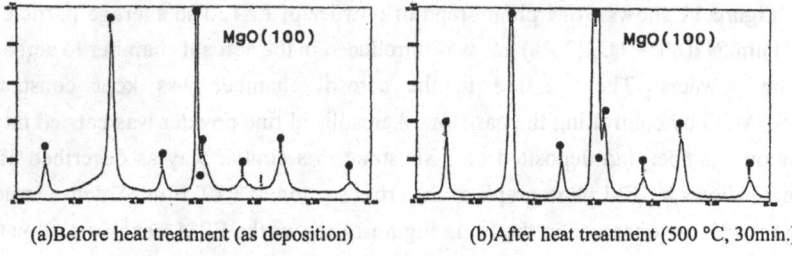

(a)Before heat treatment (as deposition)   (b)After heat treatment (500 °C, 30min.)

Figure 15. X-ray diffraction patterns of PZT film formed by aerosol-JPS.

     J.Akedo employed the aerosol-JPS as a process to create thin film actuator for micro electro mechanical systems (MEMS). He suggested Jet Molding System as a modified form of GDM for three-dimensional fabrication of MEMS-device, and demonstrated the movement of PZT unimorph actuator formed by GDM [5,11-13]. Aerosol-JPS solves the difficulty to get rather thick film, 10-100μm thick, as in other processes like sputtering or sol-gel method. Another advantage of aerosol-JPS is that it is a dry process and that it can form ceramic films without any binder.

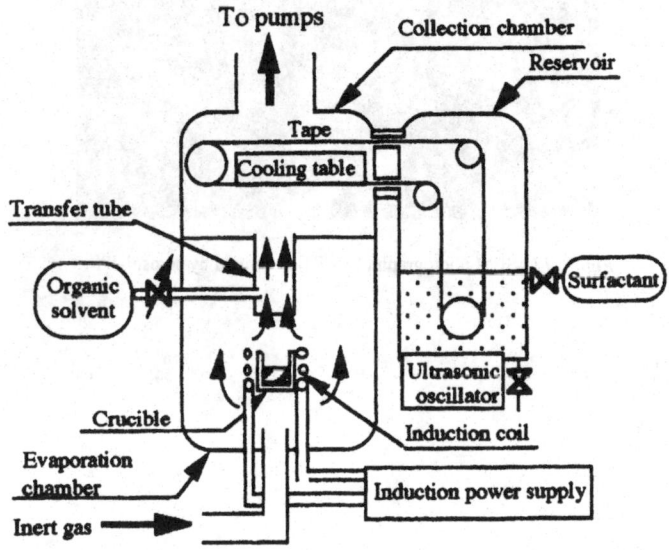

Figure 16. Schematic diagram of the apparatus forming dispersed nano particles.

3. Coating with Nano Particles Solution

3-1. FORMATION PROCESS OF DISPERSED NANO PARTICLES

Solutions with gold, silver and copper dispersed nano particles have been manufactured for development of another coating technology. Schematic diagram of the formation apparatus is shown in Fig.16. It is composed of an evaporation chamber to form nano particles, a transfer tube, a collection chamber, and a reservoir. Metals are evaporated from a crucible located in an induction heating coil. He gas is supplied from the bottom of the evaporation chamber and evacuated through the transfer tube. The evaporated metal atoms collide with inert gas atoms and are cooled down condensing into particles. The particles near the evaporation source are small and in an isolated state. In the region far from the evaporation source, the particles start to form aggregates. Vapor of α-terpineol solvent is supplied to the region where the particles are in an isolated state. The organic solvent vapor is condensed on the surface of particles. The particles covered by the solvent are carried with the He gas flow to the collection chamber. The particles are collected in a form of suspension with condensed organic solvent on a cooled tape. The suspension is stored in the reservoir.

Two kinds of surfactants are added to the suspension to stabilize it. α-Terpineol is usually used as a dispersing organic solvent. The suspension shows colloidal colors such as wine red for Au and yellow for Ag. These suspensions can be condensed to 70 wt.% for Au and 40 wt.% for Ag and become a paste of dispersed nano particles.

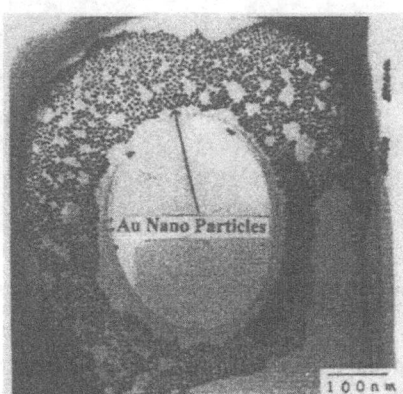

Figure 17. TEM image of Au dispersed nano particles.

A transmission electron microscope (TEM) image of the dispersed Au nano particles is shown in Fig. 17. Au particles are well separated from each other because of the surfactants covering the surfaces of particles. The average particle size is about 8 nm.

3.2. FILM FORMATION BY DISPERSED NANO PARTICLES PASTE

The pastes with dispersed nano particles were painted on a substrate such as stainless steel, glass, alumina or polyimide film using dip coating method. These painted substrates were baked at 300℃ in the air for 30 minutes. A SEM image of the baked film on a glass substrate is shown in Fig.18. It is shown that the particles are grown up to 0.1μm in size but are closely attached to each other resulting in a smooth surface like in vacuum deposition films. The thickness of the films is controlled by changing the metal contents or the coating thickness from 0.05μm up to 2μm.

Adhesion strengths were measured by the scotch tape peeling test. The peeling strength is estimated to be 3 kg/mm^2. The adhesion strength of the film formed on the surface of stainless steel was more than 3 kg/mm^2. Metal additives such as Cu, Bi and B were used to improve adhesion strengths for glass or alumina so that the adhesion strength becomes more than 3 kg/mm^2.

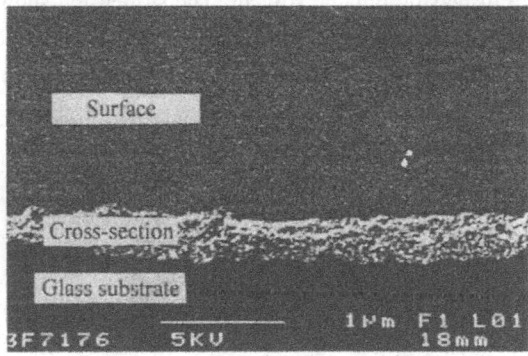

Figure 18. SEM image of Au film formed using dispersed nano particles on glass substrate

The electric conductivity of the films formed by dispersed nano particles were measured by the four point probe method and were 75-150 nΩm for Au , and 42-90 nΩ m for Ag, respectively. The use of metal additives for the improvement of adhesion strength showed negative effect on the electric conductivities.

3.3 APPLICATION

3.3.1 *Repairing System for Flat Panel Display*

The pastes are applied to the repairing system for broken line electrodes in flat panel display. A special apparatus has been developed for the repairing system which has a micro-dispenser unit with a nozzle of 30 μm in diameter to draw a repairing line, a Nd:YAG laser unit to bake the repaired part and a highly accurate XY stage system as shown in figure 19. The system can draw lines with 40 μm in width and the drawing speed is 2 mm/sec. The film thickness can be controlled from 0.3 to 2 μm. The film has the adhesion strength of more than 3kg/mm^2 and was not damaged by the pencil (6H) scratching test. The specific electrical resistance is 188n Ω m.

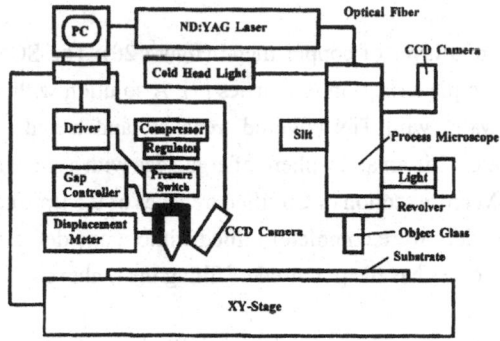

Figure 19. Repairing System for flat panel display.

3.3.2 *Copper Metallization for Ultralarge Scale Integrated Circuit*

Recently, copper has become an attractive material for interconnections in semiconductor devices due to its lower electric resistivity and better electro-migration resistance than that of aluminum alloy. The damascene process is used to form copper interconnects, since the conventional etching method used for aluminum can not be applied easily for copper because of the difficulty of copper etching. The process consists of forming trenches and via-holes in the dielectric layer followed by a filling with copper (metallization). Chemical and mechanical polishing is used to remove excess metal on the surface. The damascene process requires filling a gap with a high-aspect ratio and good flatness. A new process, "Spin on copper Metal (SOM) process", has been developed for this. In the process, a coating solution containing the dispersed particles of copper is spin-coated and filled the trenches and via-holes formed on a silicon wafer. Then a baking

288

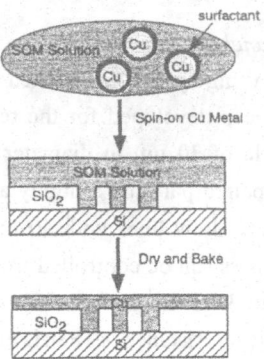

Figure 20. Spin on copper Metal (SOM) process flow.

process follows to form a layer of copper metal (figure 20). The SOM process has the good gap filling and the planarization properties [7]. A solution with 40wt% of Cu was coated on a silicon wafer with TiN covered trenches, and baked at the temperature between 300 °C and 400 °C in an atmosphere of argon gas with a small amount of oxygen. Figure 21 shows a SEM cross section of Cu filled trenches of 0.5 μm and 0.18μm in width. Trenches of 0.5μm in width were completely filled without any defects. The picture also shows that the SOM process has the potential of filling the higher aspect ratio trenches of 0.18μm in width.

The process is simple and inexpensive and does not discharge any wastewater. The specific electric resistance is measured on a flat part on the silicon wafer using the four point probe method and shows as much as the value of copper bulk.

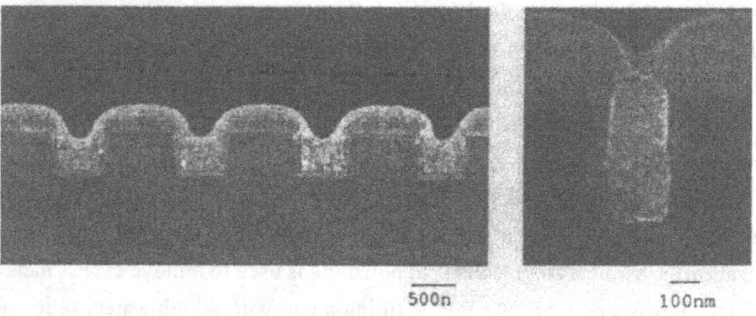

Figure 21. SEM cross section view of Cu filled trenches of 0.5μm and 0.18 μm in width.

4. Summary

Two kinds of coating technologies using nano particles were introduced. One is a deposition of nano particles by the gas deposition method; the other is a coating process with nano particles solution.

Jet Printing System has been successfully developed for industrial usage based on the gas deposition method. Various kinds of metallic films are effectively formed by the system and the properties of metallic film are suitable for practical usage. Formation of nano particles by laser irradiation has been studied. Tungsten nano particles were successfully formed and the distributions of particle size were analyzed by the low-pressure differential mobility analyzer. Further study is now in progress. Aerosol Jet Printing System supplies a new method to prepare easily ceramic film several tenth of μm thick. Jet Printing System is, I) a dry process, ii) a low temperature process, iii) a maskless process, and iv) a clean process.

Dispersed nano particles solution has been successfully developed to make thin films. The particles with a size of less than 10 nm are suspended in an organic solvent without aggregations. The process has the following advantages: (a) the metallization can be done at a low temperature of about 300 °C while general paste needs much higher baking temperature, more than 600 °C; (b) It is possible to fill fine trenches or holes with 0.18 μm in width; (c) The films have a high purity without alkali metal nor sulfur contamination.

Remarks

The gas deposition method was originated in Ultrafine Particle (UFP) Project operated by Research and Development Corporation of Japan from 1981 to 1986. The achievement of UFP project is described in other articles [14,15].

The research of "Nano Particles Formation by Laser Irradiation" is supported by the R&D institute of Photonics Engineering entrusted from the Advanced Photon Processing and Measurement Technology Program of the New Energy and Industrial Technology Development Organization of Japan.

Acknowledgement

I wish to thank Dr. M. Oda, Dr. E. Ozawa, Mr. S. Kashu, Mr. Y. Kawakami, Mr. T. Suzuki, Mr. T. Yoshida, Mr. J. Zheng, Dr. H. Murakami, and Dr. C. Hayashi for all of their generous support. I also wish to thank to Dr. J. Akedo of Mechanical Engineering Laboratory, Ministry of International Trade and Industry for permission of the use of his data.

290

References

1. Kashu, S., Nagase, M., Hayashi, C., Uyeda, R., Wada, N. and Tasaki, A., (1974) Preparation and Properties of Ultra Fine Metal Powders, *Japanese Journal of Applied Physics Supplement* **2,Part 1**,491-493

2. Hayashi, C., Kashu, S., Oda, M., and Naruse, F., (1993) The use of nanoparticles as coatings, *Materials Science and Engineering* **A163**, 157-161

3. Kawakami, Y., Seto, T., and Ozawa, E., (1999) Characteristics of ultrafine tungsten particles produced by Nd:YAG laser irradiation, *Applied Physics* **A69**,S249-S252

4. Kashu, S., Matsuzaki, Y., Kaito, M., Toyokawa, M., Hatanaka, K., and Hayashi, C., (1990) Preparation on Superconducting Thick Films of Bi-Pb-Sr-Ca-Cu-O by Gas Deposition of Fine Poder, in T. Ishiguro and K. Kajimura (eds.), *Advanced in Superconductivity II*, Springer-Verlag, Tokyo, pp.413-418

5. Akedo, J., Ichiki, M., Kikuchi, K., and Maeda, R., (1998) Jet molding system for realization of three-dimensional micro-structures, *Sensors and Actuators* **A69**, 106-112

6. Suzuki, T., Imazeki, N., Yu, G.H., Itou, H., and Oda, M., (1997) Direct drawing system with micro-dispenser using dispersed ultra fine particle paste, *IEMT/IMC proceedings* (IEEE Catalog No.97CH36059), 267-270

7. Murakami, H., Hirakawa, M., Ohtsuka, Y., Yamakawa, H., Imazeki, N., Hayashi, S., Suzuki, T., Oda, M., and Hayashi, C., (1999) Spin-on Cu films for ultralarge scale integrated metallization, *Journal of Vacuum Science & Technology*, **B17**, 2321-2324

8. Kawakami, Y., Ozawa, E., and Sasaki, S., (1999) Coherent array of tungsten ultrafine particles by laser irradiation, *Applied Physics Letters* **74**, 3954-3956

9. Kawakami, Y., Sasaki, S., and Ozawa, E., (1999) Surface Microstructure Effects of Single Crystal Tungsten after Nd:YAG Laser Irradiation, in T. S. Sudarshan, K. A. Khor, and M. Jeandin (eds.), *Surface Modification Technologies* **13**, ASM International, Ohio, pp.345-351

10. Kawakami, Y. and Ozawa, E., (2000) Self-assembled coherent array of ultrafine particles on single-crystal tungsten substrate using SHG Nd:YAG laser, *Applied Physics A*, (in submitted)

11. Akedo, J. and Lebedev, M., (1999) Microstructure and Electro Properties of Lead Zirconate Tianate (Pb(Zr$_{52}$/Ti$_{48}$)O$_3$ Thick Films Deposited by Aerosol Deposition Method, *Japanese Journal of Applied Physics* **38**, 5397-5401

12. Akedo, J., Minami, N., Fukuda, K., Ichiki, M., and Maeda, R, (1999) Electricak Properties of Direct Deposited Piezoelectric Thick Film Formed by Gas Deposition Method -annealing effect of the deposited films-, *Ferroelectrics* **231(7)**, p873-880

13. Akedo, J., (2000) Study on Rapid micro-structuring using Jet Molding −Present status and structureing subject as HARMST-, Journal of Micro-System Technology, (in submitted)

14. Hayashi, C. (1987) Ultrafine Particles, *Journal of Vacuum Science & Technology* **A5(4)**, 1375-1384

15. Hayashi, C. (1987) Ultrafine Particles, *Physics Today* **December**, 1-8

SURFACE ATOMIC SCALE ENGINEERING BY DEPOSITION OF MASS SELECTED CLUSTERS: STM AND HELIUM SCATTERING ANALYSIS

R. SCHAUB, H. JÖDICKE, W. HARBICH, J. BUTTET AND R. MONOT
Institut de Physique Expérimentale
Ecole Polytechnique Fédérale, CH 1015 Lausanne, Switzerland

1. Introduction

Wide effort is actually carried to use metallic clusters for technological applications like nanosized structured new materials, thin films, surface coatings, etc. Prepared in the gas phase, the clusters are generally deposited on surfaces, either as building blocks in a growth process [1], or to function as a tool to modify the characteristics of the surface [2, 3]. No size selection has been applied so far to the clusters in these applications. However, clusters formed and mass selected in the gas phase and subsequently deposited in a controlled way, is a promising alternative in nanostructure formation on surfaces. Clusters are systems containing typically from 2 to 2000 atoms or molecules. They have been studied for their specific properties, which are size dependent and different from both the atoms (or molecules) and the bulk material [4], mostly due to their large surface to volume ratio. Cluster deposition is of fundamental interest as it differs radically from conventional thermal atom deposition, a field which is now well established [5]. The difference arises mainly from the fact that new parameters, such as the cluster size and the deposition energy, may be used to tailor the collision outcome [6]. These input parameters open new perspectives in the controlled growth of such structures, with new phenomena not accessible within conventional atom deposition or atomic manipulation by scanning probe methods. For instance, the size or equivalently the number of atoms constituting the clusters to be deposited can be tuned for a specific inherent property. In that sense, one could prepare the systems as free clusters before deposition and then deposit them. An important issue for future technological applications of cluster deposition is the relation between the size of the incident clusters and the size of the islands obtained on the substrate. Experimental evidence for the importance of the precise definition of the cluster size can be found in the catalytic activity of clusters on oxide substrates [7] and the minimum size of a silver cluster to form an image speck in the photographic process [8].

The controlled deposition of size selected clusters on well prepared surfaces will therefore allow the formation of surface samples with very different desired features, for example with nanosize local alloying, magnetic dots, size selected model catalysts, etc. However these applications request the understanding of the microscopic phenomena involved in the deposition of the clusters and the subsequent evolution of the sample in order to control the parameters of the cluster interaction with the surface. Various events happening in two different time scales are related to the fate of clusters deposited on a surface. The collision itself is a rapid process of the order of picoseconds. It corresponds to the rapid energy transfer of the incident particle energy to the substrate. This phenomenon does not depend significantly on the surface temperature and it can well be modelized by molecular dynamics simulations (MD) [9]. The subsequent evolution of the deposits is governed by surface dynamic processes like the hopping of adatoms, the healing of local defects, etc. It is thus controlled by the surface temperature and the relevant time scale may extend up to hours if the system is cold enough.

291

L. Pavesi and E. Buzaneva (eds.), Frontiers of Nano-Optoelectronic Systems, 291–302.
© 2000 *Kluwer Academic Publishers. Printed in the Netherlands.*

292

We report here about few contributions carried in Lausanne towards a better understanding of the microscopic phenomena involved in the deposition of size selected clusters on a surface.

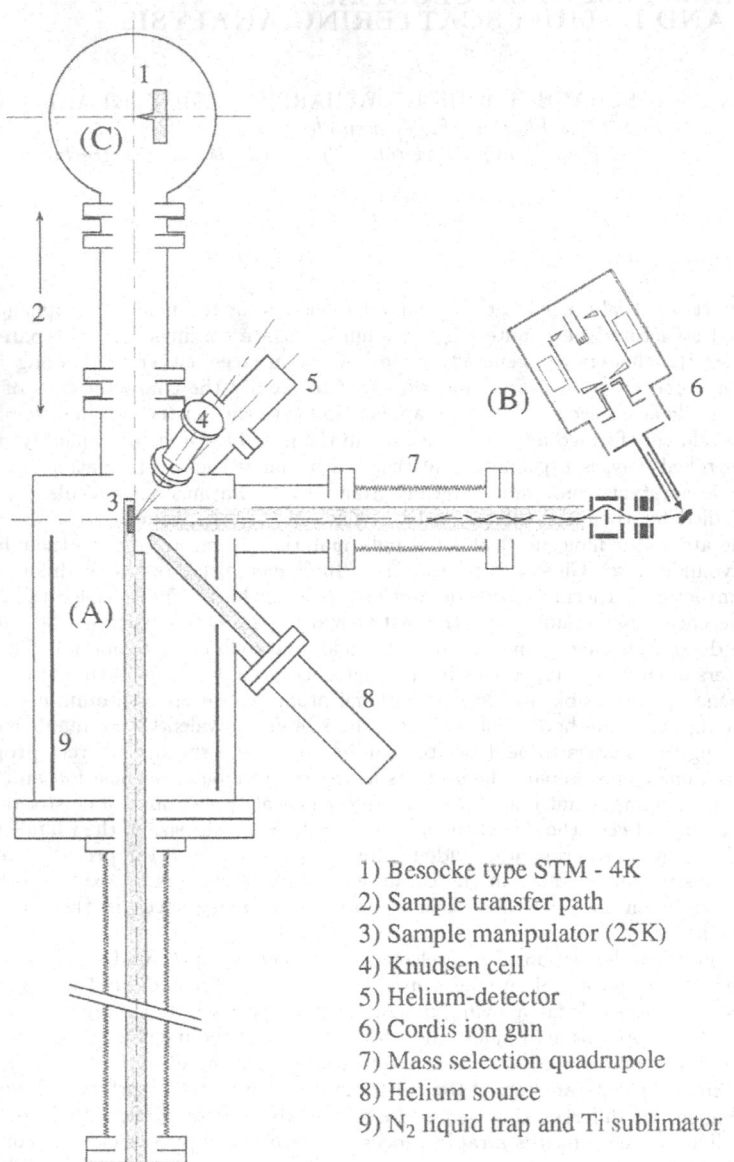

1) Besocke type STM - 4K
2) Sample transfer path
3) Sample manipulator (25K)
4) Knudsen cell
5) Helium-detector
6) Cordis ion gun
7) Mass selection quadrupole
8) Helium source
9) N₂ liquid trap and Ti sublimator

Figure 1. Experimental setup.

The next section presents briefly the experimental setup which has been built recently [10]. The sample is analyzed in situ by two complementary methods: Thermal Energy Atom Scattering (TEAS) and Scanning Tunneling Microscopy (STM). TEAS is used to study the dynamical processes during (or after) the deposition and to gather statistical information about the resulting structures on the surface. Subsequent STM measurements allow to investigate the collision outcome on an atomic scale. Results obtained with the following systems will be successively reported:

– Ag_7 clusters deposition on the bare Pt(111) surface has been studied by TEAS as a function of incoming kinetic energy between 100 eV and 1000 eV and surface temperature in the range 100 K and 400 K. Together with STM measurements at selected surface temperatures and deposition energies, this study reveals that the impact morphologies are of compact shape at low energy, whereas craters with islands of ejected substrate material result from impacts at elevated deposition energies, with a large amount of defects located beneath the surface layer.

– Ag_7 and Ag_{19} clusters deposited in solid Ar and Kr layers condensed on Pt(111). The rare gas acts efficiently as a cushion which dissipates the incoming kinetic energy and thus efficiently softens the landing. TEAS is used to calibrate the coverage of rare gas condensed on the Pt surface and also to probe the sample evolution during heat treatment.

– Ag_7 deposited at 20 eV and 95 eV on the dislocation network formed by two Ag monolayers epitaxially grown on Pt(111). It is observed that the impact energy has a strong influence on the surface morphology. Randomly distributed islands are formed at 95 eV while spatial order is obtained at 20 eV. The structures formed are more stable against surface temperature than the ones produced by thermal deposition of Ag atoms.

2. Experimental

Fig. 1 is a general overview of the experimental setup which is described in detail elsewhere [10]. It consists of three main parts: The deposition chamber (A) allows for surface preparation, thermal atom deposition and cluster deposition. Thermal energy helium scattering (TEAS) is used to monitor in situ the deposition process by recording the helium intensity which is specularly reflected by the sample surface. The temperature of the sample can be varied between 25 K and 1200 K by appropriate cooling and electron beam heating. A minimum base pressure lower than 10^{-11} mbar results from strong turbomolecular and cryopumping (liquid N_2 cooled double-walled Cu jacket) in addition with Ti sublimation. The Pt(111) surface is prepared by 1000 eV Ar^+ sputtering at 500 K and etching in an oxygen atmosphere (10^{-8} mbar) at 1000 K [11]. The cluster source (B) is based on sputtering: secondary ions sputtered from a silver target are formed into an ion beam. A quadrupole is used for size selection and carrying the clusters onto the sample surface. For example, currents of 300 pA of Ag_{19} are typically obtained on a spot of 0.2 cm^2. Deposition times in the order of 5 minutes are necessary to obtain the coverages employed in the experiments reported in section 4. The custom built STM is of the Besocke type [12] and housed in a separate vacuum chamber (C) which can be detached from the deposition chamber for high resolution STM measurements. It can operate between 8 K and 450 K. The STM head is placed in a closed 4 K bath cryostat which allows clean surface conditions to be maintained over days.

3. Deposition of Ag_7 clusters on Pt(111)

TEAS relies on the high reflectivity of low Miller index metal surfaces and the large cross section for He scattering off local defects like adsorbates or vacancies. Practically during the deposition, the intensity of the He beam specularly reflected from the sample surface is recorded. The curve of the relative decrease of this intensity as a function of time contains statistical information of the dynamical outcome of the deposition on the surface. The TEAS measurement of the deposition of thermal Ag atoms is a reference for the deposition of clusters. Fig. 2 shows the specular He signal recorded during the depo-

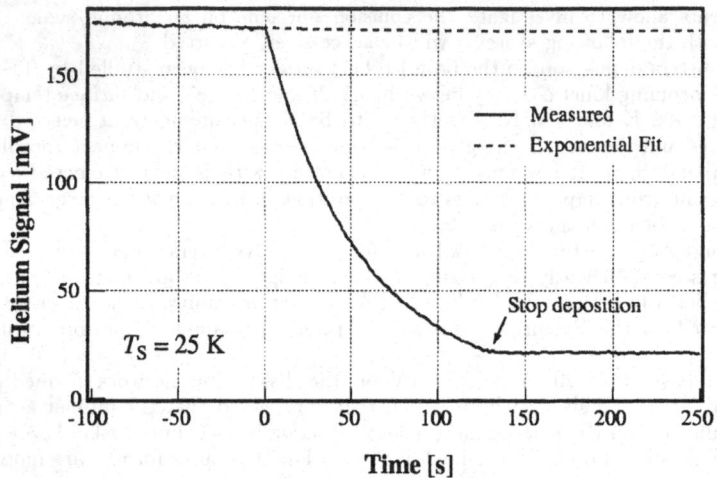

Figure 2. He specular intensity recorded during the deposition of thermal Ag atoms on a Pt(111) surface at 25 K. The flux amounts to $1.35 \cdot 10^{-3}$ ML $\cdot$ s^{-1}. The dashed curve shows the attenuation due to the residual gas contamination.

sition of thermal Ag atoms on the Pt(111) surface at 25 K. This signal is then corrected for the slow signal decrease due to the slight pollution noticeable before the beginning of the deposition at t=0. The initial slope of the intensity decrease gives the effective cross section of the Ag atom for helium scattering. Below 55 K it is found that this cross section is constant indicating that the adatom mobility is frozen. The effective Ag atom cross section is $\Sigma_a = 14.0 \cdot \sigma_{Pt(111)} = 93$ Å^2. Similar He intensity as a function of coverage is recorded for the deposition of clusters at selected incoming kinetic energies and surface temperatures. The effective cross section for diffuse scattering associated to the collision process is also extracted from the initial slope of the He intensity decrease. Fig. 3 shows the effective cross section per Ag atom measured for the deposition of Ag$_7$ clusters at various kinetic energies as a function of surface temperature between 100 K and 400 K. For comparison the value of the cross section for thermal atom deposition measured between 25 K and 500 K is also reported on Fig. 3. Another comparison can be drawn from the effective cross section that would be associated to the Ag$_7$ lying ideally on the Pt(111) surface: for a linear shaped heptamer we calculate $\Sigma_{eff} = 5.8 \cdot \sigma_{Pt(111)} = 39$ Å^2, and for a compact hexagonal shape we get $\Sigma_{eff} = 4.3 \cdot \sigma_{Pt(111)} = 29$ Å^2. The estimation is made by simple geometrical overlap of the atom cross sections. These two values are reported on Fig. 3, labeled as L for the linear chain and C for the compact shape. The cross section for a non-fragmented Ag$_7$ should be between these two values.

The possible scenarios compatible with the behavior of the cross section curves for the different deposition energies, taking into account the fragmentation of the cluster, its implantation and Pt adatom promotion on the surface in a qualitative manner, are obviously numerous. To gain more insight on the morphology of the cluster impact region and hence on the collision processes involved, STM measurements have been performed

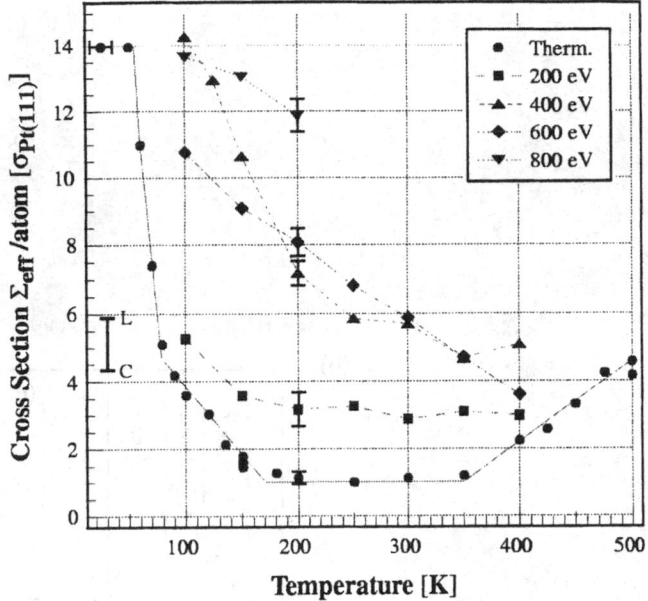

Figure 3. Effective cross section (in units of $\sigma_{Pt(111)} = 6.67$ Å^2) for the deposition of Ag$_7$ clusters and of thermal Ag atoms, as a function of substrate temperature.

at selected impact energies: 95 eV and 1000 eV. These are the upper and lower bounds of the ones measured by TEAS. Fig. 4 and Fig. 5 show STM images of the impact zone resulting from the deposition of Ag$_7^+$ at 95 eV and 1000 eV respectively. Both measurements have been performed on a sample held at 93 K. The coverage is about 0.005 ML.

The deposition at 95 eV: A large number of spots with a diameter of about 15 Å are clearly identified as the impact points. At higher magnification, the atomic structure of some of these impacts is resolved, as shown in Fig. 4. These images are to our knowledge the first high resolution STM images resulting from the impact of metallic clusters on a metallic surface. The structure show particular features. The impact zone can be divided in three distinct parts; zone 1 is probably the central impact region, characterized by a certain disorder of the atoms which are not in registry with the Pt atoms of the surface. The imaged height of these atoms is in the range between 1.8 Å and 3.0 Å and again no order can be found. In the two other zones (2 and 3) the lateral position of the atoms correspond to the one of the Pt substrate lattice. However, the atoms are imaged higher than the surface atoms by up to more than 1 Å. MD simulations for the same cluster with the same impact energy deposited on Pd(100) show that all atoms involved in the collision have adopted lattice sites after a few picoseconds following the impact [9]. In contrast, our images show a certain disorder together with atoms which are imaged on a different height than expected for atoms on regular sites. This is compatible with a local melting of the cluster and surface atoms in the impact zone 1. At the impact, some atoms of the clusters might have fragmented and diffused away, while other were implanted in the surface, in agreement with MD simulation cited above. During the subsequent fast cooling, the material might get quenched in a highly amorphous state.

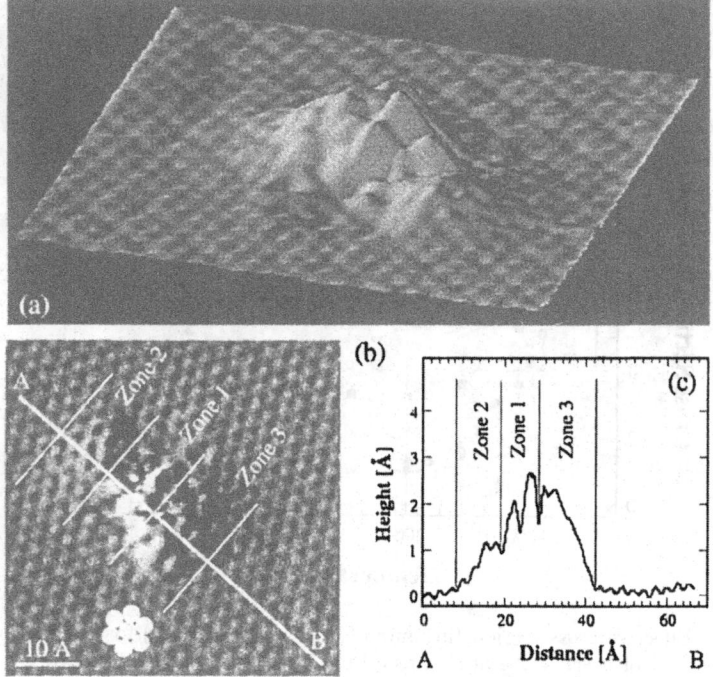

Figure 4. Impact zone after the deposition of Ag_7^+ with 95 eV kinetic energy on Pt(111) at 93 K. (a) shows a 3D projection of the STM image, (b) shows a top view of the same zone and a schematics of a heptamer lying on the surface, (c) is the line profile as marked in (b).

The resulting amorphous region inside the surface could force the surrounding surface atoms to adopt sites higher than their lattice site, or it could alter the LDOS in the impact region, making the surrounding atoms appear higher in the STM image.

The deposition at 1000 eV: The situation changes dramatically when the impact energy of the clusters is increased to 1000 eV. Fig. 5 shows STM images of typical impacts. They consist of a central impact region, a vacancy island (crater) with a diameter of 10 to 15 Å. Small approximately round islands of substrate atoms ejected onto the surface are decorating the rim of the crater. The measured diameter of the impact zone (crater + islands) is 40 ± 10 Å. In general, 3 to 4 islands can be found around a crater. If we admit that every impact leads to a crater, one can estimate the number of atoms belonging to the islands surrounding the craters and calculate the number of adatoms promoted per impact. This adatom yield amounts to 73 ± 20 per impact. The corresponding yield for a single Xe^+ ion at the same energy on Pt(111) is 7.5 [13].

The main observation one can extract from Fig. 3 is the downward trend of the effective cross sections with increasing surface temperature. This signifies that the damages produced on the substrate resulting from the impact are annealed via thermally activated processes. At all investigated deposition energies, the cross sections lie well above the ones associated to the deposition of Ag atoms from the vapor phase. This clearly indi-

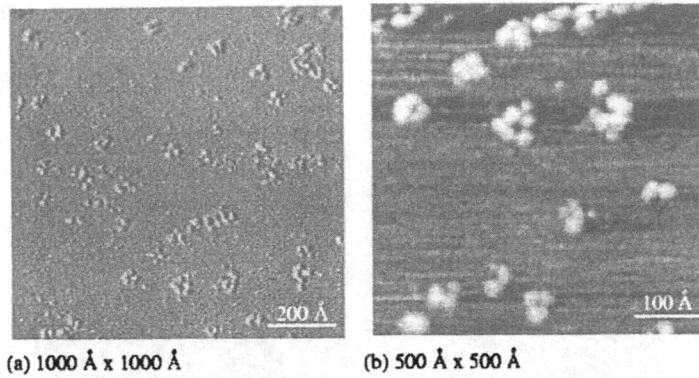

(a) 1000 Å x 1000 Å   **(b) 500 Å x 500 Å**

Figure 5.   STM images at two magnifications of the surface after deposition of Ag_7^+ with 1000 eV at 93 K.

cates that the cluster collision is a much more complicated process, involving phenomena that do not occur during the deposition of thermal atoms. With increasing deposition energy, the defect production increases and spreads away from the impact point. However, in the low temperature range probed, the cross sections for an impact energy of 400 eV are larger than for 600 eV. This striking behavior can be explained by the main conclusion drawn from the STM images for cluster impacts at 1000 eV. There, the analysis has shown that the production of bulk vacancies is a major outcome resulting from the impact process. Therefore, with increasing deposition energy, the defects produced may lie deep under the surface layer. Bulk vacancies, as well as embedded defects, are not detected by He scattering experiments, since the He atoms are strictly sensitive to the topmost surface layer. This holds as long as these defects do not alter in a significant manner the He reflectivity of the surface. The inversion behavior between $\Sigma_{eff}(400 \text{ K})$ and $\Sigma_{eff}(600 \text{ K})$ can thus be understood considering that with increasing impact energy, the damages are located deeper in the bulk.

At the lower collision energy probed by TEAS, 200 eV, the cross section at 100 K is lying in the range defined by the linear and compact Ag_7 (labeled L and C in Fig. 3). With increasing temperature, the cross section decreases and stabilizes around 200 K at a value of $3 \cdot \sigma_{Pt(111)}$ which is smaller than the effective cross section calculated for the ideally compact heptamer. This indicates that the resulting impact structures at 200 eV above 200 K are small compact configurations. At these temperatures they appear to be stable against diffusion towards steps or against dissociation due to effective pinning at the impact points due to implantation. This stabilizing feature could be applied to fabricate samples with families of dots each containing few atoms.

4. Softlanding of Ag clusters on Pt (111)

Obtaining samples with deposited size selected clusters, which keep their number of atoms upon the preparation process, has been a dream for a long time. The first idea which comes to mind is to deposit clusters at zero kinetic energy. However, besides of the technical difficulty of such a method, the large adsorption energy which is suddenly released upon deposition (3.03 eV/atom) results in a process not soft at all. This has been shown by MD simulations of the deposition of Ag clusters at zero kinetic energy

298

on Pd(100) [14]. But mass selected cluster deposition in weakly interacting matrices like the rare gases has shown that softlanding in these substrates is feasible [15].

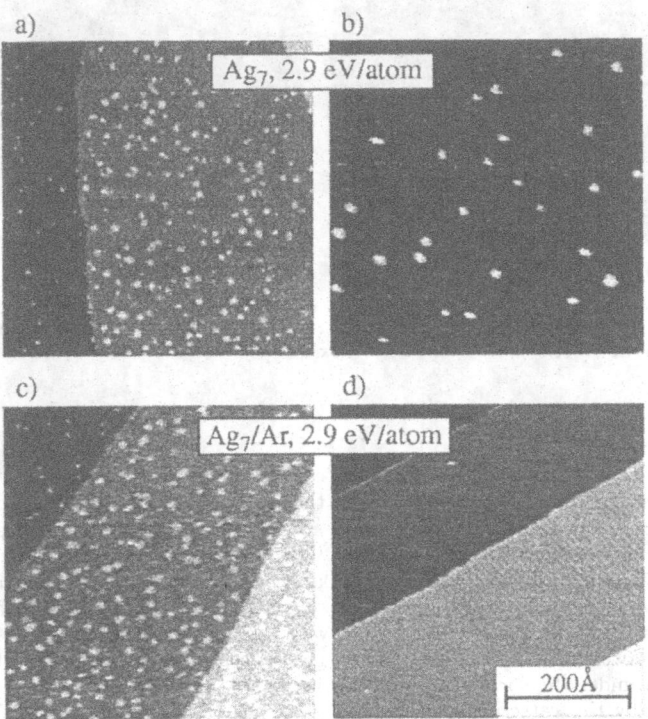

Figure 6. (a) STM image after the deposition of Ag$_7$ clusters on the bare Pt(111) crystal at 80 K. The coverage is about 0.1 ML. (b) After annealing at 300 K still some islands are found on the terraces. Defects created upon the deposition stabilize the islands against thermal decay. (c) Cluster deposition under the same conditions as in (a) but at 25 K into a 10 ML thick Ar buffer layer preadsorbed on the Pt crystal. (d) After annealing at 300 K, no pinning centers are found stabilizing islands, thus demonstrating the softlanding effect of the Ar layer.

Previous results obtained with a simpler experimental setup have demonstrated that few monolayers of rare gas condensed on the Pt(111) act as a cushion which efficiently dissipates the incoming kinetic energy of the clusters [16]. Fig. 6 a) and c) show a comparison of the deposition of Ag$_7$ clusters on the bare surface at 25 K and the deposition in the same conditions on 10 ML of solid Ag condensed previously on the Pt(111) surface. In both cases the incoming kinetic energy was 20 eV and the STM observations were made at 80 K. At this temperature the Ar layer is evaporated and this process evidently did not perturb the sample. No cluster size distribution difference is noticeable between the two samples and the average size is compatible with the selected size. (Unfortunately no atomic resolution was obtainable in this study). But annealing the surfaces at 300 K gives the images shown in Fig. 6 b) and d) which differ considerably. From the studies carried about the growth of Ag particles obtained by vapor deposition on Pt(111) [5] at low temperature and annealed at 300 K, it is known that such particles decay due to

Ostwald ripening: the Ag atoms diffuse on the terraces and condense at the pre-existent Pt step edges. In the case of the deposition of clusters on the bare Pt surface, annealing at 300 K does not lead to the disappearance of particles on the terraces. The remaining islands indicate that surface defects were created during the deposition process which stabilize them against thermal decay. On the contrary, when the clusters are deposited on the solid Ar layer condensed on the Pt surface, annealing at 300 K leads to clean terraces (Fig. 6d) like in the case of Ag particles obtained by thermal atom deposition, demonstrating the soft landing of the clusters into the rare gas cushion. Also one notices the lines of condensed Ag along the surface steps.

With the new experimental setup presented in section 2, the deposition of Ag cluster in rare gas layers has been extensively studied recently. The previous results have been beautifully confirmed and extended. But the lack of space prevents us to present them [11]. Here atomic resolution was routinely obtained with the rare gas surface as well as with the bare Pt(111) surface. All STM measurements were carried at 8 K. In particular Ag_{19} clusters have been deposited on 5 monolayers of condensed Kr at 25 K. An annealing at 60 K evaporates all Kr but the first monolayer on the Pt surface. The clusters are then also covered by one Kr monolayer. The statistics of their height shows that 80 % exhibit a two dimensional structure with an Ag monolayer height and 20 % a three dimensional structure. A higher annealing at 125 K leads to only two dimensional structures for the uncovered Ag clusters. But unfortunately, due to STM tip convolution broadening or to their particular electronic structure, no atomic resolution is observable on the clusters. Luckily, if some Kr atoms are condensed on such sample after the annealing at 125 K, one observes that a ring of Kr atoms surrounds the clusters like a necklace. These Kr atoms can be easily counted and located thank to their high corrugation. Then the geometry of the Ag cluster and their number of atoms can be determined unambiguously. Perfect hexagonal Ag clusters of 19 atoms are observed which proves the softlanding process. The rare gas atoms act then as a probe which allows the precise determination of the structure of the deposited clusters. These results demonstrate for the first time that mass selected clusters can be deposited on a surface while keeping the same number of atoms.

5. Deposition of Ag_7 on a dislocation network

The lattice constant of the Ag crystal is 4.15 % larger than that of Pt. This leads to a number of strain induced phenomena when the first few monolayers of Ag are deposited on the Pt(111) surface [17]. The first Ag monolayer grows pseudomorphic on the Pt surface. When two monolayers are deposited at room temperature, followed by a heat treatment at 800 K, the strain relief is achieved on a small length scale by the creation of a trigonal network of crossing dislocations. Fig. 7 a) and b) show an STM image of the resulting ordered surface and the schematics of its structure. Three types of sites can be distinguished: small triangles (ST) and large triangles (LT) with hcp stacking and polygons (PG) with fcc stacking.

The dislocations form repulsive barriers for diffusing adparticles, hence the condensation of thermal Ag atoms on such a reconstruction network can lead to the formation of an islands superlattice. Indeed at 110 K, the deposition of 0.1 ML of Ag atoms yields an island array with exactly one island in each PG region [18]. The deposition of Ag_7 clusters on such a dislocation network has been studied in order to compare with the deposition of thermal atoms. The TEAS facility was used to precisely calibrate the deposition of two ML of Ag on the Pt(111) surface [12].

Deposition (at 80 K) of 0.1 ML of Ag_7 at 95 eV kinetic energy shows random distributed islands on the surface i.e. the total amount of coverage in the three structure types corresponds approximately to 9 % (ST), 16 % (LT) and 75 % (PG) which is the

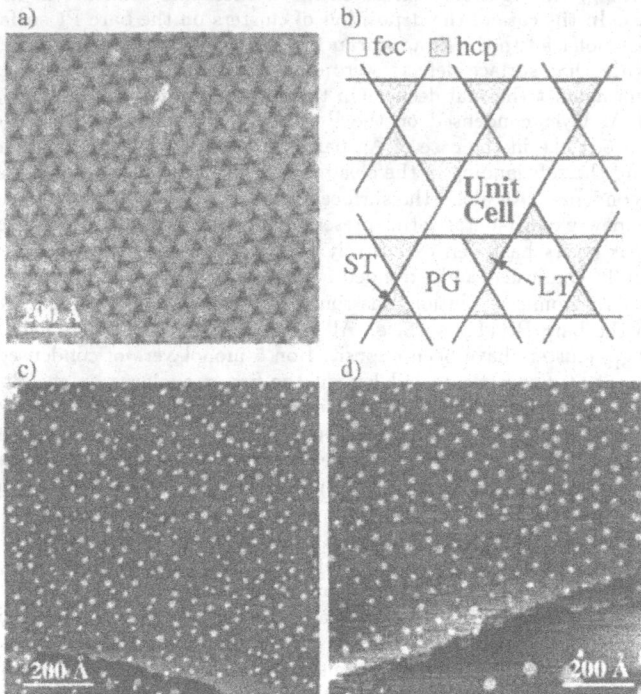

Figure 7. STM images taken at 8 K for: (a) the dislocation network formed by 2 ML of Ag on Pt(111) and (b) the schematics of this structure. (c) 0.1 ML of Ag$_7$ deposited with 20 eV kinetic energy at 200 K. (d) after annealing the sample at 220 K.

surface coverage of these structures. The distribution is stable to about 220 K pointing to the effective pinning of the adclusters.

The situation is quite different at a deposition energy of 20 eV. Fig. 7c) shows the STM picture after the deposition of 0.12 ML at 200 K (the STM measurements are always carried at lower temperature (77 K or 8 K) than the temperature of deposition). The coverage does not correspond to the abundance of ST, LT and PG regions on the surface. A large fraction of the islands is found attached to the dislocation line separating the PG and ST structure. The islands have a hexagonal shape and are much larger than the impacting clusters. The deposition carried at 80 K leads to 9 % of the coverage in the LT regions (compared to 16 % abundance) and the deposition at 200 K results with the LT completely empty. The picture emerging from these observations is the following: At 20 eV deposition energy, an effective pinning of the clusters is only observed at the dislocation line separating the PG and ST structure. Clusters impinging on LT transfer to the PG and attach finally to the pinned structures. This process is valid up to 220 K and the ordering of the islands is clearly noticed in Fig. 7d). However, a drastic reduction in islands density is observed upon annealing at higher temperatures. Energetic cluster deposition can be compared to thermal atom deposition [18]. The confinement effect in the dislocation network leads to the formation of Ag islands in the center of the PG structures at 110 K. But the absence of pinning leads to the destruction of the regular pattern at the surface temperatures employed in our cluster deposition experiments. The importance of the dislocation network to stabilize the silver islands via pinning is shown in Fig. 7 c) and d). The cluster deposition on a dislocation network is then an effective means to produce adsorbates structures with spatial order. The surface morphologies are very different compared to conventional thermal atom deposition. Depending on the impact energy, randomly distributed islands can be formed or preferential pinning can be obtained. The structures are much more stable against temperature than the ones produced by thermal deposition.

6. Conclusion

The results presented briefly in sections 3 to 5 demonstrate how the deposition of clusters is a promising method to create nanostructures on surfaces which could not be obtained by other conventional techniques. Besides of the large choice of elements as substrate or cluster matter, the parameters which are the cluster size and their impact energy open a wide range which can be used to tailor the desired nanostructures. The above results also show how STM and TEAS are powerful complementary experimental approaches to study the outcome of cluster deposition at the atomic as well as the statistical level.

The support of the Swiss National Fond for Scientific Research is thankfully acknowledged.

References

[1] Paillard, V., Mélinon, P., Dupuis, V., Perez, J. P., Perez, A. and Champagnon, B. (1993) Diamondlike Carbon Films Obtained by Low Energy Cluster Beam Deposition: Evidence of a Memory Effect of the Properties of Free Carbon Clusters, *Phys. Rev. Letters* **71**, 4170-4173.

[2] Haberland, H., Karrais, M., Mall, M. and Thurner, Y. (1992) Thin Films From Energetic Cluster Impact: a Feasibility Study, *J. Vac. Sci. Technol.* A **10**, 3266-3271.

[3] Yamada, I. (1999) Novel Materials Processing and Applications by Gas Cluster Ion Beams, *Euro. Phys. J. D* **9**, 55-61.

[4] de Heer, W. (1993) The Physics of Simple Metal Clusters: Experimental Aspects and Simple Models, *Rev. Mod. Phys.* **65**, 611-676.

[5] Brune, H. (1998) Microscopic View of Epitaxial Metal Growth: Nucleation and Aggregation, *Surf. Sci. Reports* **31**, 121-229.

[6] Harbich, W. (2000) *Collision of Clusters with Surfaces: Surface Modification and Scattering*, Springer Series in Cluster Physics, Springer Verlag, Berlin.

[7] Heiz, U., Sanchez, A., Abbet, S. and Schneider, W. D. (1999) Catalytic Oxidation of Carbon Monoxide on Monodispersed Platinum Clusters: Each Atom Counts, *J. Am. Chem. Soc.* **121**, 3214-3217.

[8] Fayet, P., Granzer, F., Hegenbart, G., Moisar, E., Pischel, B. and Wöste, L. (1985) Latent-Image Generation by Deposition of Monodispersed Silver Clusters, *Phys. Rev. Letters* **55**, 3002-3004.

[9] Vandoni, G., Félix, C. and Massobrio, C. (1996) Molecular-Dynamics Study of Collision, Implantation, and Fragmentation of Ag_7 on Pd(100), *Phys. Rev. B* **54**, 1553-1556.

[10] Jödicke, H., Schaub, R., Monot, R., Buttet, J. and Harbich, W. (2000) Deposition of Mass Selected Clusters Studied by Thermal Energy Atom Scattering and Low Temperature Scanning Tunneling Microscopy. An Experimental Set-up, *Rev. Sci. Instrum.*, In Press.

302

[11] Schaub, R. (2000) Controlled Deposition of Mass Selected Silver Clusters on a Pt(111) Surface: Combined Scanning Tunneling Microscopy and Thermal Helium Scattering Study, *PhD thesis* 2183, EPFL, Lausanne.

[12] Jödicke, H. (1999) Development of a New Low Temperature Scanning Tunneling Microscope to Study Mass Selected Cluster Deposition, *PhD thesis* 1986, EPFL, Lausanne.

[13] Michely T. and Teichert, T. (1994) Adatom Yields, Sputtering Yields, and Damage Patterns of Single-Ion Impacts on Pt (111), *Phys. Rev. B* **50**, 11156-11166.

[14] Nacer, B., Massobrio, C. and Félix, C. (1997) Deposition of Metallic Cclusters on a Metallic Surface at Zero Initial Kinetic Energy: Evidence for Implantation and Site Exchanges, *Phys. Rev. B* **56**, 10590-10595.

[15] Fedrigo, S., Harbich, W. and Buttet, J. (1998) Softlanding and Fragmentation of Small Clusters Deposited in Noble-Gas Films, *Phys. Rev. B* **58**, 7428-7433.

[16] Bromann, K., Brune, H., Félix, C., Harbich, W., Monot, R., Buttet, J. and Kern, K. (1997) Hard and Soft Landing of Mass Selected Ag Clusters on Pt(111), *Surf. Sci.* **377-379**, 1051-1055.

[17] Brune, H., Röder, H., Boragno, C. and Kern, K. (1994) Strain Relief at Hexagonal-Close-Packed Interfaces, *Phys. Rev. B* **49**, 2997-3000.

[18] Brune, H., Giovannini, M., Bromann, K. and Kern, K. (1998) Self-Organized Growth of Nanostructure Arrays on Strain-Relief Patterns, *Nature* **394**, 451-453.

VISIBLE AND INFRARED PHOTOLUMINESCENCE FROM DEPOSITED GERMANIUM-OXIDE CLUSTERS AND FROM Ge NANOCRYSTALS

ATSUSHI NAKAJIMA
Department of Chemistry, Keio University,
3-14-1 Hiyoshi, Kohoku-ku, Yokohama 223-8522, Japan
MINORU FUJII, SHINJI HAYASHI
Department of Electrical and Electronics Engineering, Kobe University,
Rokkodai, Nada, Kobe 657-8501, Japan
KOJI KAYA
Institute for Molecular Science, Myodaiji, Okazaki, 444-8585, Japan

1. Introduction

Following Canham's report of visible photoluminescence (PL) from porous silicon,[1] the optical and electronic properties of nano-structures made from silicon (Si) or germanium (Ge) have attracted much attention, because they open a new possibility for photonic applications by the use of group-IV elements. In particular, PL properties of Si nanocrystals (nc-Si) have been widely studied and the relationship between the size of nc-Si and the PL peak energy has been revealed experimentally for at least red and near-infrared (NIR) PL.[2-4] According to these reports, nc-Si with about 4 nm in diameter exhibits a PL peak at about 1.4 eV. As the size decreases further, the PL peak shifts to higher energies and reaches the visible region for nc-Si smaller than 2 nm. In contrast to nc-Si, there have been few reports on the size dependence of the PL spectra for Ge nanocrystals (nc-Ge). The nc-Ge has been prepared by several methods and these samples exhibit strong visible PL at about 2.2 eV independent of the size of nc-Ge (2-15 nm) and the preparation methods.[5-11]

The purpose of this work is to experimentally reveal the size dependence of the PL spectra for clusters in gas phase and nanocrystals in matrices. We will report on: (1) the infrared HOMO-LUMO gap of Ge clusters having 1 nm diameter,[12] (2) the optical properties of deposited germanium-oxide (Ge-O) prepared from vapor depositions of small Ge-O clusters,[13] and (3) the PL of nc-Ge dispersed in SiO_2 films.[14]

The photoelectron spectroscopy (PES) of cluster anions in gas phase[12,15] has been proven to be a powerful technique to study the electronic and geometric structures of atomic clusters in gas phase as a function of size,[16,17] due to its ability to combine size selectivity with quantitative spectral sensitivity. Since photoelectron spectra deliver information about the final states, the anion PES should reveal the electronic properties

303

L. Pavesi and E. Buzaneva (eds.), Frontiers of Nano-Optoelectronic Systems, 303–317.

of the neutral clusters. Together with our developed method of halogen atom (F, Cl) doping, the HOMO-LUMO gap of the neutral Ge_n and Ge_nO_n clusters has been measured. From these gas phase experiments, it has been found that the HOMO-LUMO gaps of Ge_nO_n are 2.9-1.7 eV for n=2-5. And the corresponding visible PL from the deposited Ge-O species on the substrate was observed. Furthermore, the PL properties of nc-Ge in SiO_2, prepared by cosputtering of Ge and SiO_2 and post annealing, were studied, in which the PL spectra strongly depending on the size were observed in the NIR region.

2. Experimental Setup

A couple of experimental methods were employed to characterize the electronic structures and optical properties of Ge and Ge-O samples. In the gas phase, photoelectron spectroscopy (PES) was used for both Ge and Ge-O cluster anions with our developed implement of halogen-atom doping. For the deposited Ge-O samples, measurements of PL and XPS spectra were performed. For Ge nanocrystals in SiO_2, measurements of PL spectra and high-resolution transmission electron microscope (HRTEM) were performed.

2.1. MASS SPECTROMETRY AND PES OF CLUSTER ANIONS

First, the experimental method is explained for photoelectron spectra of the Ge and the Ge-O cluster ions in the gas phase. The apparatus used in this work consists of a cluster source, a time-of-flight (TOF) mass spectrometer, and a magnetic-bottle type electron TOF spectrometer, and most of them have been described in detail previously.[18] Both germanium cluster anions, Ge_n^-, and germanium-chlorine cluster anions, $Ge_nCl_m^-$, were produced by a laser vaporization of a germanium rod in He carrier gas (6 atm) mixed with CCl_4 gas (0.02 % in He) from a pulsed nozzle. Ge-O clusters were generated by a laser vaporization of a GeO_2. The GeO_2 rod was prepared by pressing pure GeO_2 powder. The generated cluster anions generated in a cooled channel (3 cm long, 2 mm diam., and T ~ 100 K) were expanded through a skimmer, and were mass-analyzed by the TOF mass spectrometer with a pulsed electric field (~ 3 keV). The mass resolution of the apparatus, $M/\Box M$, was ~100. After mass-separation, only the target anion selected by a mass gate was allowed to enter the deceleration region. The fourth harmonics of a pulsed Nd^{3+}-YAG laser (266 nm, 4.66 eV) was employed for photodetachment. For the generation of the Ge-O clusters with the doping of fluorine (F) atom, F_2 gas (0.01%) was mixed into the He carrier gas by the pressure ratio of 0.01 % F_2 in He. In the laser plasma on the GeO_2 rod, an F_2 molecule is dissociated into F atoms, which are mixed into the Ge-O cluster. The instrument was calibrated by measuring the photoelectron spectra of Au^- anion, at three different wavelengths of 532 nm, 355 nm, and 266 nm, where the strong line attributed to the $^1S_0 \rightarrow {}^2S_{1/2}$ transition could be observed.[19,20] The repetition rate was 10 Hz and each photoelectron signal was typically accumulated to 10000 - 30000 shots. Laser power of the detachment laser was in the range of 1-3 mJ/cm^2, and no power dependent processes for the spectrum shape were observed.

2.2. PL MEASUREMENT AND XPS OF DEPOSITED GE-O ON SUBSTRATES

The Ge-O clusters, prepared in the laser vaporization source, were deposited onto a substrate in vacuum. The PL was measured under the atmosphere, and the XPS spectrum of the sample exposed to air was measured under vacuum.

While monitoring the mass distribution of germanium-oxide cluster by TOF mass spectrometer, the substrate was inserted downstream after the skimmer, and the Ge-O cluster produced were deposited onto the substrate under the vacuum level of 10^{-5} Torr. The thickness of the deposited materials was roughly estimated to be around 100 - 200 nm by the deposition for 1 hour.

Either a gold or a silicon substrate was used in order to examine the influence of the substrate on the deposited clusters. The gold substrate was prepared by the deposition of the gold vapor on the natural mica at the deposition rate of 0.1 nm/s to the thickness of 100 nm under a ultra-high vacuum.[21] During gold deposition, the substrate was heated up to 620 K for elimination of oxygen from the substrate surface, and after the deposition it is gradually cooled down to room temperature under a vacuum for 24 hours. Thus prepared gold surface is defined as Au (111) by the scanning tunnel microscope. On the other hand, the silicon substrate was used without any cleaning processes, and it was seemingly covered with thin oxide layers.

The PL spectra were measured in air and at room temperature. The excitation source used was the 325 nm (3.82 eV) line from a He-Cd laser, and the collected light was analyzed by a monochromator and was detected by a photomultiplier in a photon energy region from 400 nm to 850 nm (3.1-1.5 eV). The data were acquired with photon-counting electronics and were recorded in a computer.

The deposited samples of Ge-O were analyzed also by an XPS instrument in order to investigate the bond order of Ge-O in the samples. XPS signal from Ge 3d was recorded using a hemispherical analyzer with an Al $K\alpha$. The instrument was calibrated by measuring the Ge 3d XPS spectra of Ge surface (Ge^{0+}) and crystalline GeO_2 powder (Ge^{4+}), which are centered at 29.3 and 32.6 eV, respectively.[22]

2.3. PL MEASUREMENT AND HRTEM OF GE NANOCRYSTALS EMBEDDED IN SIO$_2$

Ge nanocrystals embedded in SiO_2 matrices were prepared by an rf cosputtering method.[23,24] Small pieces of Ge wafers ($2\times2\times0.5$ mm^3, purity 99.9999 %) were placed on a SiO_2 target (10 cm in diameter, purity 99.99 %) and they were cosputtered in Ar gas (2.7 Pa) with a rf power of 200 W, using a magnetron sputtering apparatus. The substrates were fused quartz plates. The substrates were not intentionally heated during the sputtering and kept lower than 100°C by circulating cool water. The thickness of the films was about 3.6 mm. After the cosputtering, in order to grow nc-Ge in SiO_2 matrices, the films were thermally annealed in N_2 gas ambient for 30 min. at 800 centigrade.[23,24] In this method, the size of nc-Ge can be controlled by changing the concentration of Ge in the films. The concentration was controlled by changing the number of Ge targets during the cosputtering. The atomic ratio of Ge to Si in films was determined by electron probe microanalyses (EPMA) for all the samples. The volume fraction of nc-Ge (f_{Ge}) was calculated from the atomic ratio using the densities of bulk Ge crystal (5.33 g/cm^3) and fused quartz (2.20 g/cm^3). The estimated f_{Ge} are distributed from 0.2 % to 7.2 %. After optical measurements, all the samples were studied by cross-sectional HRTEM. The samples for the HRTEM observation were prepared by standard procedures including mechanical and Ar-ion thinning techniques.

The PL spectra were measured at room temperature with a photon counting system in a photon energy region from 0.75 to 2.65 eV using two different set-ups. The excitation source was the 457.9-nm line of an Ar-ion laser. The beam power density was about 1 W/cm^2. The spectral response of the measuring system was calibrated with the aid of a reference spectrum of a standard tungsten lamp. The absorption spectra of the samples from the ultraviolet to visible region were measured by a UV-visible spectrometer.

3. Results and Discussion

3.1. THE INFRARED HOMO- LUMO GAP OF GERMANIUM CLUSTERS

Figure 1 shows the photoelectron spectra of Ge_n^- and Ge_nCl^- (n=4-11) at the 266 nm excitation. In the spectra, the horizontal axis corresponds to the electron binding energy, E_b, defined as $E_b = h\nu - E_k$ where E_k is the kinetic energy of the photoelectron. Arrows indicate threshold energies (E_T), and the E_T value corresponds to the upper limit of the adiabatic electron affinity (EA). The photoelectron spectra of the pure Ge_n^- cluster anions (n=2-15) detached by 193 nm (6.42 eV) photons or by 266 nm (4.66 eV) have already been reported by Cheshnovsky et al.,[25] and by Burton et al.[26] Our photoelectron spectra are almost consistent with theirs. Namely, the obtained EAs in

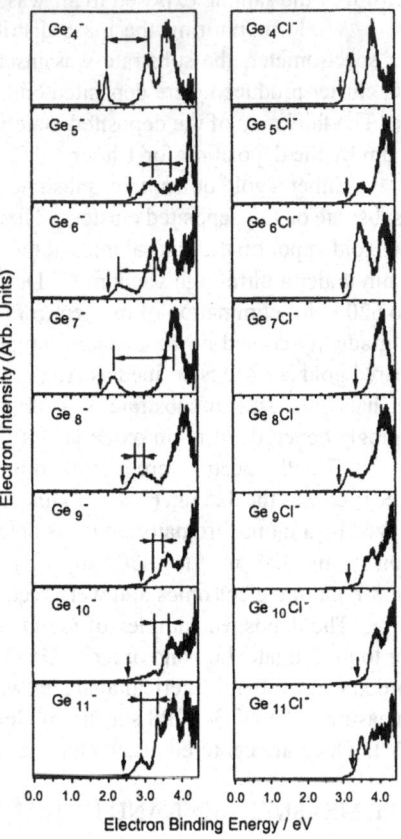

Figure 1. Photoelectron spectra of Ge_n^- and Ge_nCl^- (n=4-11) at 266 nm. Downward arrows indicate thresholds energies (E_T). Comparison between the photoelectron spectra of Ge_n^- and Ge_nCl^- clearly shows that the Cl addition leads to disappearance of the first peaks in the photoelectron spectrum of the corresponding pure Ge_n^-, offering that first peaks is attributed to SOMO and that the Cl atom can be added on the surface of Ge_n framework without any major deformation. The intervals of a sidewise arrow show the HOMO-LUMO gap of the corresponding neutral Ge_n

this work are almost the same with theirs within the experimental uncertainties.

For a closed-shell cluster, the least bound electron in the negative clusters resides alone in what corresponds to the LUMO of the neutral cluster and the next most weakly bound electrons reside in the HOMO of the neutral clusters, because the extra charge produces no large scale changes both in geometry and in electronic states. Without degeneracy in the electronic states, the energy gap should correspond to a HOMO-LUMO gap and this gap originates from the closed-shell electronic structures of the neutral species. In fact, this is precisely what has been predicted by the *ab initio* calculation by Raghavachari on Si_n cluster; all the neutral Si_n clusters are calculated to be singlet,[27] which is applicable to an isovalent Ge_n^- cluster anion because of apparent similarity of the photoelectron spectra between Ge_n^- and Si_n^-. Unless the LUMO degenerates to the HOMO, the first peak in the spectra could be assigned to the singly occupied MO (SOMO) with the excess electron. It is indispensable to develop a method to distinguish the degeneracy of the LUMO with the HOMO in the neutral.

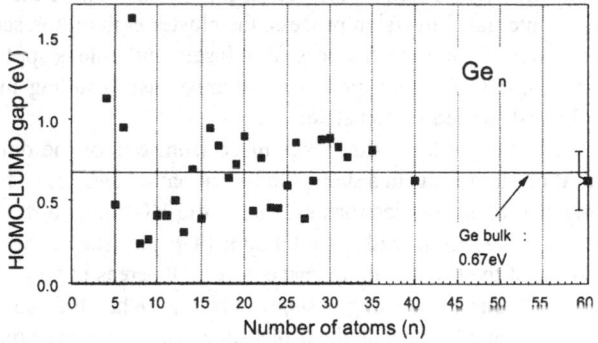

Figure 2. Size dependence of the HOMO-LUMO gaps of Ge_n (n=4-60). A line at 0.67 eV indicates the energy gap of Ge in bulk.

When the photoelectron spectra are compared between Ge_n^- and Ge_nCl^-, a mutual feature can be observed: the first small peak disappears with the Cl atom doping and the other features in the succeeding peaks seem almost the same. From this similarity of the photoelectron spectra, we could be convinced experimentally that (1) the first small bump in the PES of Ge_n^- cluster anions is attributed to the LUMO in the corresponding neutral cluster (the SOMO in the cluster anion) and (2) the Cl atom can be added on the surface of their frameworks without any major geometric deformation. Namely, we can remove the electron in the SOMO from Ge_n^- cluster anion with Cl atom doping and neutralize the Ge_n^- framework as well as the case of Si_nF^- and Ge_nF^-.[15,28] The rigidity of Ge_n/Si_n framework toward the F/Cl atom doping is originated from the covalent character of Ge/Si atoms. As reported elsewhere,[28] indeed, *ab initio* theoretical calculations support the electron removal of the F atom doping to the Si_n^- cluster anion. The doping method of halogen atoms provides us with useful implement to investigate the electronic structures of covalent clusters. Through the distinct assignment on whether the first peak corresponds to the SOMO, the HOMO-LUMO gap in the neutral

Ge$_n$ clusters was determined by the difference in the VDEs of first two peaks. The HOMO-LUMO gap was indicated as a sidewise arrow in fig. 1. Similarly, the PES of Ge$_n^-$ clusters at n=12-60 were measured and the HOMO-LUMO gap was estimated as shown in the fig. 2, by the spectral features on the analogy of those for the smaller clusters, although no PES spectra for the corresponding Ge$_n$Cl$^-$ were measured for n=21-60.

The size dependence of the HOMO-LUMO gap shows that the value oscillates widely in the small cluster, and above n=15 they oscillates around 0.8 eV within 0.2 eV. The HOMO-LUMO gap should qualitatively corresponds to the band gap in bulk. However, the HOMO-LUMO gap obtained corresponds to the energy gap between the ground state and the lowest excited triplet state. Since the triplet state is located below the lowest singlet excited state, the energy of the fluorescence is derived by the addition of the singlet-triplet splitting to the HOMO-LUMO gap. Since the neutral germanium clusters are characterized as singlet species, the cluster having a singlet ground state is photo-excited into the singlet excited state. After rapid relaxation to the lowest singlet excited state via an internal conversion process, the cluster emits fluorescence. Intense photoluminescence would not be seen unless $\Delta S = 0$ selection rule is applicable. As the cluster size increases, the relaxation processes become fast, resulting in fluorescence mainly from the lowest excited singlet state.

When the ejection of a photoelectron takes place from one of the doubly occupied orbitals, the neutral cluster results in a state with two unpaired electrons. Let us consider the system having two unpaired electrons system in the HOMO and the LUMO. In a system containing two electrons with parallel spin (a triplet state), the probability of finding two electrons at the same point in space is zero, whereas in a system containing two electrons with opposite spin (a singlet state), it is not. When the coulomb repulsion between electrons is taken into account, the triplet state is expected to be more stable than the singlet state. The singlet-triplet splitting resulted from an exchange integral, denoted by K$_{hl}$,[29]

$$K_{hl} = \int \varphi_h^*(r_1)\varphi_l(r_1) \cdot r_{12}^{-1} \cdot \varphi_l^*(r_2)\varphi_h(r_1)\, dr_1 dr_2 \qquad (1)$$

In this equation, φ_h and φ_l express the wave function of the HOMO and the LUMO, respectively. The physical interaction between two electrons with parallel spin, as described by the coulomb repulsion term (r_{12}^{-1}) in the Hamiltonian, gradually becomes smaller with the cluster size, because the interaction distance r_{12} becomes large in space. Only for Ge$_4$, the energy of the singlet-triplet splitting is 0.59 eV.[30] Assuming that the interaction distance r_{12} is proportional to the radius of the cluster, the singlet-triplet splitting of G$_{32}$ is calculated to be about 0.3 eV, because the radius of Ge$_{32}$ (the cubic root of n) is simply estimated to be twice longer than that of Ge$_4$. This value is an upper limit for the singlet-triplet splitting at this size. As a matter of fact, when cluster size increases, the overlap between the HOMO and the LUMO decreases along with increasing the interaction radius. Hence, it seems reasonable that the singlet-triplet splitting would be about 0.1 eV around n = 30. Then, the energy of the fluorescence can be predicted only based on the HOMO-LUMO gap especially at larger cluster size. Since the HOMO-LUMO gap is around 0.8 - 1.0 eV around n=30, the wavelength of the fluorescence should be the infrared region of 1500 -1200 nm, where the diameter of Ge_n cluster can be estimated to be 0.7-0.8 nm at n=30 on the assumption of a spherical shape. Even at n~30, the HOMO-LUMO gap obtained is close to the band gap of Ge bulk (0.67 eV).[31] Based on this work, the small energy gap of the Ge_n clusters suggests that reported visible emission could be ascribed not to germanium clusters, but to germanium oxide in the surface layer. Indeed, the PES spectra of germanium oxide cluster anions show very large HOMO-LUMO gap around the composition of 1:1, as described in the following section.

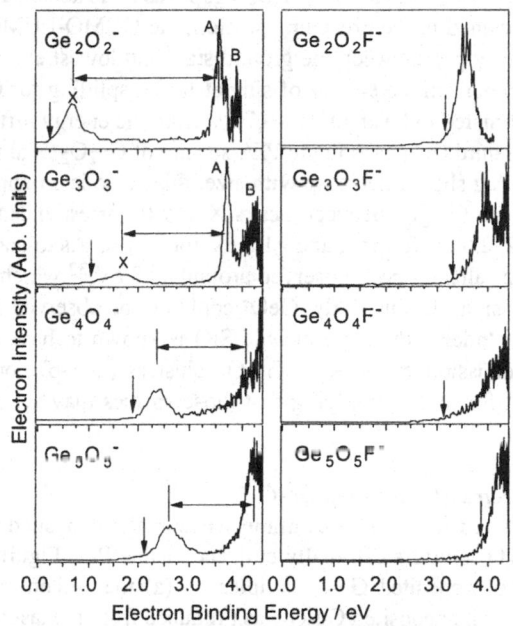

Figure 3. Photoelectron spectra of $Ge_nO_n^-$ and $Ge_nO_nF^-$ (n=2-5) at 266 nm. Downward arrows indicate thresholds energies (E_T; EA). The F-atom addition leads to disappearance the first peaks in the photoelectron spectrum of the corresponding pure $Ge_nO_n^-$, offering that the first peaks is attributed to SOMO and that the F atom can be added on the surface of Ge_nO_n framework without any major deformation. The intervals of sidewise arrow show the corresponding neutral HOMO-LUMO gap of Ge_nO_n.

3.2. VISIBLE PL OF THE DEPOSITED GERMANIUM-OXIDE CLUSTERS

3.2.1. Mass Spectrometry and PES Spectra of Ge-O and Ge-O-F Cluster Anions

For Ge-O, the TOF mass spectra were measured for both $Ge_nO_m^+$ and $Ge_nO_m^-$ cluster ions produced by the laser-vaporization of the GeO_2 rod. The abundant clusters were produced at composition of (n, m) = (n, n; n=1-8) for the cluster cations, while they are at (n, n; n=2-8) and (n, n+2; n=2-8) for the cluster anions. Although the (n, n+2) cluster anions are abundant together with (n, n) anions, the high abundance of (n, n+2) could be attributed to the high electron affinities compared to the other. Namely, both in cations and in anions the (n, n) clusters are abundant in common, implying that the corresponding neutral Ge_nO_n clusters should be more stable compared to the others.

Figure 3 shows the 266 nm photoelectron spectra of abundant $Ge_nO_n^-$ and $Ge_nO_nF^-$ (n=2-5). When the photoelectron spectra are compared between $Ge_nO_n^-$ and $Ge_nO_nF^-$, a common feature can be clearly observed: the first small peak disappears with the F atom doping and the other features in the succeeding peaks seem almost the same. Through the distinct assignment on whether the first peak corresponds to the SOMO, the HOMO-LUMO gap in the neutral Ge_nO_n clusters was determined by the difference in the VDE's of the first two peaks. The HOMO-LUMO gap was indicated by a sidewise arrow in the figure. As mentioned in the preceding section, the HOMO-LUMO gap obtained corresponds to the energy gap between the ground state and lowest excited triplet state. Although there is no report on the energy of singlet-triplet splitting for Ge_nO_n clusters, the splitting can be estimated to be about 0.5-0.7 eV from the energy difference between the second (A) and the third (B) peaks in the PES spectra of $Ge_2O_2^-$ and $Ge_3O_3^-$ (fig. 3).

Since the splitting should decrease with size, this value is an upper limit. From the spectral feature, thus, the gap between peaks X and B corresponds to the emission energy; 3.2 eV for n=2, 2.6 eV for n=3, and >1.8 eV for n=4-6. As to diatomic GeO, the electronic transition has already been observed around 4.68 eV,[32] which corresponds to the energy of UV emission. In this study, GeO$^-$ could not be observed probably due to negative EA of GeO. Indeed, the isoelectronic SiO is known to have negative EA.[33] Then, the estimated emission energies of Ge_nO_n clusters (n=1-6) correspond to the energy of visible light for n=2-6, implying that those species may act as the center of visible emission.

3.2.2. PL and XPS Spectra of Deposited Ge-O

PL spectra were measured in order to examine whether the deposited Ge-O prepared from vapor of the Ge_nO_n clusters can really emit the visible PL. Figures 4(a) and 4(b) show the PL spectra of deposited Ge-O samples on (a) the silicon and (b) the gold substrate, respectively. The deposited Ge-O was produced from the laser vaporization of the GeO_2 rod, where the Ge_nO_m clusters exhibit essentially the same mass distributions as shown in figs. 1 and 2. Comparison between PL spectra of figs. 4(a) and 4(b) clearly indicates that the PL spectra are independent of the substrate. In the PL spectra, sharp and strong peaks were observed at 450 nm (2.9 eV), 480 nm (2.6 eV), and 500 nm (2.5 eV) and no detectable PL was observed from the substrate itself. Interestingly, these energies of the PL peaks roughly correspond to the expected energies for emission of Ge_nO_n clusters, which is estimated by the PES experiment.

Figure 4(c) shows the PL spectrum of deposited Ge_nO_n clusters on the silicon substrate under the exposure of a trace of O_2; O_2 gas (0.07%) was mixed into the He carrier gas. The PL spectra are normalized at their maximum intensities and the scaling factors for the normalization are shown in fig. 4. The PL spectrum exhibits no clear structure and the peak intensity becomes about 50 times weaker than that of the PL spectra of figs. 4(a) and (b). Under the condition of the O_2 mixing, there was no serious difference in the mass distribution of Ge_nO_n precursor clusters for both cations and anions. However, the intensity of the PL is drastically decreased and its spectral feature becomes broad as shown in fig. 4(c). This result indicates that the Ge_nO_n clusters were oxidized by the exposure of O_2 not in the beam but on the substrate. Since the flight time is approximately several hundreds µs from the cluster source to the substrate, it is likely that the oxidation reaction of the Ge_nO_n cluster is so slow that the reaction cannot be observed with mass spectrometry during the flight time. On the substrate, therefore, the deposited Ge_nO_n cluster were gradually oxidized into higher oxidation states. As described later, indeed, the XPS spectra of these samples show that the oxidation states of Ge atoms in the deposited Ge-O become higher with that O_2 exposure.

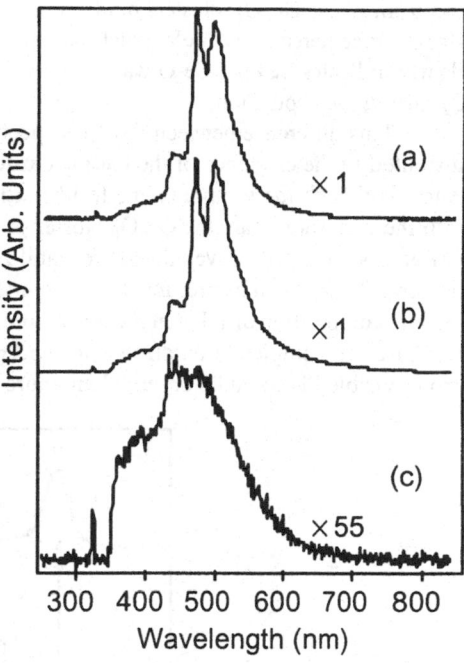

Figure 4. Photoluminescence spectra of deposited Ge-O samples (a) on the silicon substrate, (b) on the gold substrate, and (c) on the silicon substrate under the exposure of a trace of O_2 gas during the deposition. The PL spectra are normalized at their maximum intensities and the scaling factors for the normalization are shown in the figure.

The PL observations were performed both immediately after cluster deposition and after keeping in desiccator for 6 months in order to examine the stability of deposited cluster in air. Comparison between the two PL spectra indicates that there has been no major change both in the PL intensity and peak energies. These results indicate that the Ge_nO_n cluster can survive even in air for a long life.

Figure 5(a) shows the Ge 3d XPS spectra of Ge_nO_m clusters deposited on the silicon substrate from which the visible PL was correspondingly observed as shown in fig 4(a). In figure 5, the Ge 3d XPS spectra of Ge surface (Ge^{0+}) and crystalline GeO_2 powder (Ge^{4+}) are also shown for comparison at top and at bottom, respectively. It has been reported that the XPS spectra of Ge^{0+} and Ge^{4+} are centered at 29.3 and 32.6 eV, respectively, and that the peaks of Ge^{1+}, Ge^{2+}, and Ge^{3+} are observed at 30.1, 31.1, and 32.0 eV.[22] Based on these values along with the band width of 1.8 eV fwhm for pure Ge^{0+} and Ge^{4+}, the observed XPS peak in fig. 5(a) was numerically deconvoluted into 5 components of Ge^{0+}, Ge^{1+}, Ge^{2+}, Ge^{3+} and Ge^{4+}, as shown in the figure; the components occupy 6 % (Ge^{0+}), 8 % (Ge^{1+}), 23 % (Ge^{2+}), 31% (Ge^{3+}), 32% (Ge^{4+}) of the total, respectively. Although the proportion of the Ge_nO_n clusters to the total product was around 30 % in the mass spectra, the corresponding component of Ge^{2+} was decreased

312

into 23 % by the deposition onto the substrate. As shown in fig. 5(b), moreover, the XPS spectrum of the Ge_nO_m clusters oxidized on the substrate with O_2 (PL spectrum in fig. 4(c)), are centered at 32.6 eV which is almost the same with that of Ge^{4+}. This result clearly indicates that the Ge-O was completely changed into GeO_2 with the exposure of O_2 during the deposition.

This difference between the XPS spectra of figs. 5(a) and 5(b) can reasonably be attributed to the existence of the Ge_nO_n clusters, which correspond to the +2 oxidation state. Although the sample seems to be partly oxidized into the higher oxidation state with the exposure to air, the Ge_nO_n cluster produced under the condition of the pure He carrier gas are mostly covered and are stabilized by successive deposition of the Ge_nO_m clusters. Thus, it can be presumed that the internal Ge_nO_n clusters can survive and fairly keep its composition of 1:1. On the other hand, the Ge_nO_n clusters produced under the O_2 exposure are completely oxidized into the Ge_nO_{2n} cluster on the substrate. Since no strong visible PL could be observed anymore from the oxidized sample, it is concluded

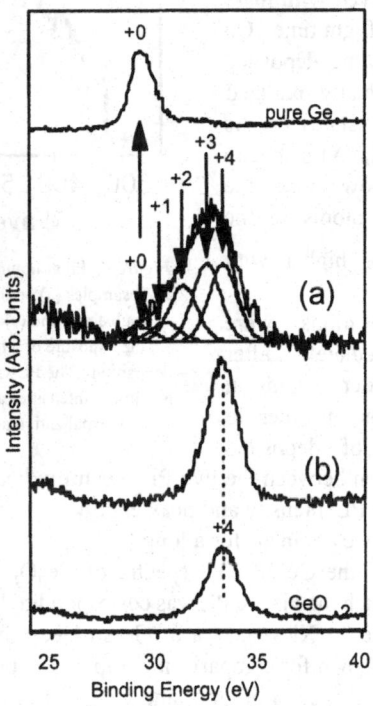

Figure 5. Ge 3d X-ray photoelectron spectra of Ge-O deposited on the silicon substrate; (a) the deposited Ge-O sample which is the same as that in fig. 4(a) and (b) the deposited Ge-O sample with O_2 exposure which is the same as that in fig. 4(c). Ge 3d spectra of Ge surface and the commercial crystalline GeO_2 powder are also shown at top and at bottom. The observed XPS peak in fig. 5(a) was numerically deconvoluted into 5 components of Ge^0, Ge^{1+}, Ge^{2+}, Ge^{3+} and Ge^{4+}, as shown in the figure; the components occupy 6 % (Ge^0), 8 % (Ge^{1+}), 23 % (Ge^{2+}), 31% (Ge^{3+}), 32% (Ge^{4+}) of the total, respectively.

that the oxygen-deficient Ge_nO_n species are indispensable to the visible PL. This conclusion is consistent with recent reports on oxidized Ge nanocrystals,[6] Ge/GeO$_2$ nanocrystals,[34] and Ge (Si) implanted SiO$_2$ film;[35] the neutral oxygen vacancy is responsible for the PL.

Although the study in the gas phase may not be related directly to the electronic structures in the deposited material, these systematic investigations can reveal the role of the Ge_nO_n species. In fact, another result, which is in agreement with ours, has been reported by XPS; germanium oxides, GeO_x (x=1 and 2), are formed at the interface between the visible light-emitting Ge nanoparticles and glassy SiO$_2$ matrices.[8] These results indicate that the Ge_nO_n clusters having O-deficiency can act as the center of the emission. The preparation of O-deficient Ge precursors by laser vaporization would provide a useful implement for the visible luminous device.

3.3. SIZE-DEPENDENT NIR PL FROM NC-GE IN SIO$_2$

In this section, the intrinsic size dependence is revealed for the PL spectra from nc-Ge in matrices. In this experiment, the size of nc-Ge can be controlled by changing the concentration of Ge in the SiO$_2$ films. For all the samples with f_{Ge} 3.6 %, lattice fringes, corresponding to {111} planes of Ge with the diamond structure (0.33 nm), were clearly

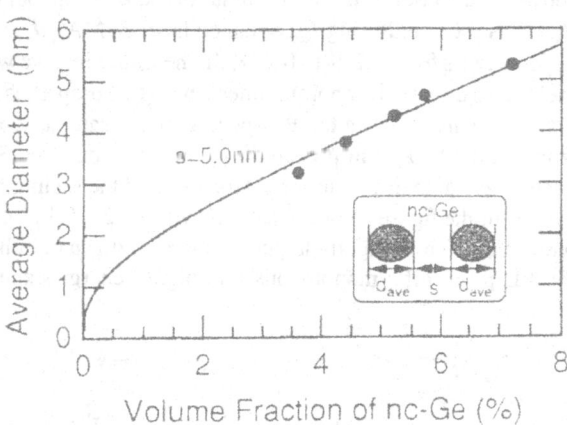

Figure 6 Average diameter of nc-Ge as a function of f_{Ge} (solid circles). Solid curve is the result of the least-squares fitting assuming that nc-Ge are arranged in a simple cubic lattice. The average separation of nc-Ge obtained from the fitting was 5.0 nm.

observed in a cross-sectional HRTEM image. The average diameter (d_{ave}) of the sample determined from several HRTEM images and fig. 6 shows d_{ave} obtained from HRTEM observations as a function of f_{Ge} (solid circles). Solid curve is the result of the least-squares fitting assuming that nc-Ge are arranged in a simple cubic lattice, where the relation ship between f_{Ge} and d_{ave} is expressed as

$$\frac{f_{Ge}}{100} = \frac{\frac{4}{3} \pi \left(\frac{d_{ave}}{2}\right)^3}{(s + d_{ave})^3} \qquad (2)$$

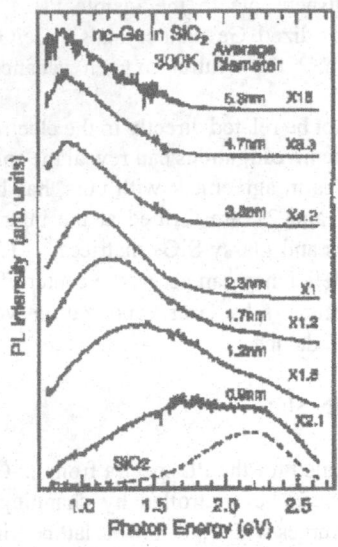

Figure 7. Dependence of PL spectra on the average diameter of nc-Ge. A PL spectrum of a SiO$_2$ film is also shown and a larger factor corresponds to a smaller PL intensity.

where s is the distance between the surfaces of two neighboring nc-Ge and its definition is shown in the inset of fig. 6. By assuming that the model is also applicable to f_{Ge} 1.6 %, d_{ave} was estimated using f_{Ge} obtained from EPMA. d_{ave} decreases from 5.3 to 0.9 nm as f_{Ge} decreases from 7.2 % to 0.2 %. In the following, we will use the size estimated by this method to discuss the photoluminescence (PL) properties of the samples with f_{Ge} 1.6 %.

Figure 7 shows the PL spectra for the samples with various d_{ave}, along with that from a pure SiO$_2$ film prepared by sputtering only the SiO$_2$ target. The PL spectra are normalized at their maximum intensities and the scaling factors for the normalization are shown in the figure. For the sample with d_{ave} = 5.3 nm, a PL peak is observed at about 0.88 eV, which is slightly larger than the band gap of bulk Ge crystal. As d_{ave} decreases, the PL peak shifts monotonously to higher energies and reaches about 1.54 eV as d_{ave}

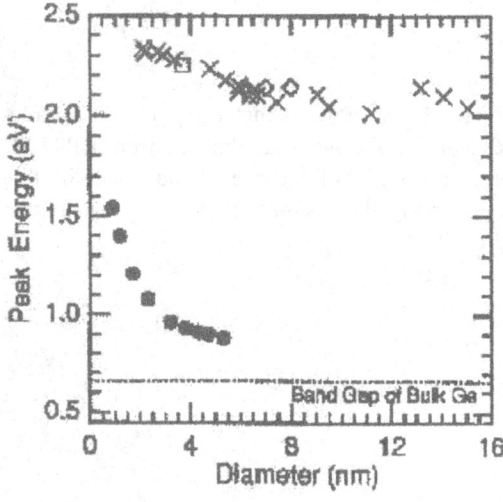

Figure 8. PL peak energy vs. average diameter of nc-Ge. Solid circles represent the present results. Previous experimental results for nc-Ge are taken from Ref.5 (cross), Ref. 6 (square) and Ref. 7 (diamond).

decreases to 0.9 nm. The PL spectra were highly reproducible and stable under the laser ablation. The 2.2-eV peak from the SiO_2 matrix is also observed for the samples with small d_{ave} (d_{ave} 1.7 nm). In the present samples, the size of nc-Ge is controlled by changing f_{Ge}. For the samples with small d_{ave}, f_{Ge} is also very small and the fraction of SiO_2 is large. The large fraction of SiO_2 causes the appearance of the 2.2-eV PL.

In fig. 8, the PL peak energies obtained from fig. 7 are plotted as a function d_{ave} (solid circle) together with the data from previous PL studies.[5-7] The PL peak dependence of the present samples is completely different from those reported in the previous studies; in the present samples, the PL peak shifts monotonously to higher energies as d_{ave} decreases. Moreover, the PL intensity depends strongly on the size. When the intensity is corrected by the amount of nc-Ge obtained in the samples by dividing the raw integrated PL intensity by f_{Ge}, the PL intensity increases about two orders magnitude as the size decreases from 5.3 nm to 0.9 nm.

The band-gap widening is considered to be due to the quantum confinement effects of electrons, holes, and excitons (quantum size effects). Furthermore, the increase in the PL intensity with decreasing the size has also been observed for nc-Si.[4] The increase in the oscillator strength and/or the decrease of the nonradiative Auger recombination processes are considered as the origin of the enhancement of the PL intensity with decreasing the size. The observed size dependence of the PL spectra shown in figs. 7 and 8 is very similar to those of Si and other semiconductor nanocrystal previously reported. This strongly suggests that the PL peak observed in this work originates from the recombination of electron-hole pairs between the widened band gap of nc-Ge.

As described in section 3.1, the HOMO-LUMO gap of Ge_n cluster having about 1 nm diameter is around 0.8 eV, which seemingly corresponds to the peak energy for the PL. For nc-Ge having 1 nm diameter, however, the PL spectrum exhibits a peak at 1.5 eV. This discrepancy can be attributed to the difference in the morphology. In the HRTEM image, the annealed nc-Ge always takes a spherical shape. In contrast, it is reported by using injected ion drift tube techniques that the geometric structure of the Ge_n cluster is a prolate shape around n= 10-50.[36] For the particles having less than 100 atoms, a large fraction of atoms are constituted as a surface atom, so that it is reasonable that the energy gap becomes sensitive to geometric structures.

In summary, these systematic studies have revealed the following three conclusions; (a) Ge nanocrystals themselves emit not visible PL, but near –infrared PL, depending on the size, (b) the comparison between the gas phase and the bulk experiments suggests that there may be a phase transition for nc-Ge around 1 nm diameter which is related to the morphology of the Ge_n clusters, and (c) the Ge_nO_n clusters having O-deficiency can act as the center of the emission.

316

4. Acknowledgements

This work is supported by a program entitled "Research for the Future (RFTF)" of Japan Society for the Promotion of Science (98P01203) and by a Grant-in-Aid for Scientific Research from the Ministry of Education, Science, Sports, and Culture.

References

1. Canham, L. T. (1990) Silicon quantum wire array fabrication by electrochemical and chemical dissolution of wafers. *Appl. Phys. Lett.* **57**, 1046-1048.
2. Takagi, H., Ogawa, H., Yamazaki, Y. Ishizaki, A., and Nakagiri, T. (1990) Quantum size effects on photoluminescence in ultrafine Si particles. *Appl. Phys. Lett.* **56**, 2379-2380.
3. Schuppler, S., Friedeman, S. L., Marcus, M. A., Adler, D. L., Xie, Y. –H., Ross, F. M., Chabal, Y. J., Harris, T. D., Brus, L. E., Brown, W. L., Chaban, E. E., Szajowski, P. F., Christman, S. B., and Citrin, P. H. (1995) Size, shape, and composition of luminescent species in oxidized Si nanocrystals and H-passivated porous Si. *Phys. Rev. B* **52**, 4910-4925.
4. Kanzawa, Y, Kageyama, T., Takeoka, S. Fujii, M., Hayashi, S., and Yamamoto, K. (1997) Size-dependent near-infrared photoluminescence spectra of Si nanocrystals embedded in SiO2 matrices. *Solid State Commun.* **102**, 533-537.
5. Maeda, Y. (1995) Visible photoluminescence from nanocrystallite Ge embedded in a glassy SiO$_2$ matrix: evidence in support of the quantum-confinement mechanism. *Phys. Rev. B* **51**, 1658-1670
6. Okamoto, S. and Kanemitsu, Y. (1996) Photoluminescence properties of surface-oxidized Ge nanocrystals: Surface localization of excitons. *Phys. Rev. B* **54**, 16421-16424.
7. Paine, D. C., Caragianis, C., Kim, T. Y., Shigesato, Y., and Ishahara, T. (1993) Visible photoluminescence from nanocrystalline Ge formed by H$_2$ reduction of Si$_{0.6}$Ge$_{0.4}$O$_2$. *Appl. Phys. Lett.* **62**, 2842-2844.
8. Dutta, A. K. (1996) Visible photoluminescence from Ge nanocrystal embedded into a SiO$_2$ matrix fabricated by atmospheric pressure chemical vapor deposition. *Appl. Phys. Lett.* **68**, 1189-1191.
9. Nogami, M. and Abe, Y. (1994) Sol-gel method for synthesizing visible photoluminescent nanosized Ge-crystal-doped silica glasses. *Appl. Phys. Lett.* **65**, 2545-2547.
10. Saito, A. and Suemoto, T. (1997) Luminescence in selectively excited germanium microcrystallites. *Phys. Rev. B* **56**, R1688-R1691.
11. Craciun, V., Leborgne, C. B., Nicholls, E. J., and Boyd, I. W. (1996) Light emission from germanium nanoparticles formed by ultraviolet assisted oxidation of silicon-germanium. *Appl. Phys. Lett.* **69**, 1506-1508.
12. Negishi, Y., Kawamata, H., Hayakawa, F., Nakajima, A., and Kaya, K. (1998) The infrared HOMO-LUMO gap of germanium clusters. *Chem. Phys. Lett.* **294**, 370-376.
13. Negishi, Y., Nakamura, Y., Nagao, S., Nakajima, A., Kamei, S., and Kaya, K. Visible photoluminescence of the deposited germanium-oxide prepared from clusters in the gas phase. submitted to *J. Appl. Phys.*
14. Takeoka, S., Fujii, M., Hayashi, S, and Yamamoto, K. (1998) Size-dependent near-infrared photoluminescence from Ge nanocrystals embedded in SiO$_2$ matrices. *Phys. Rev. B* **58**, 7921-7925.
15. Negishi, Y., Kawamata, H., Hayase, T., Gomei, M., Kishi, R., Hayakawa, F., Nakajima, A., and Kaya, K. (1997) Photoelectron spectroscopy of germanium-fluorine binary cluster anions: the HOMO-LUMO gap estimation of Ge$_n$ clusters. *Chem. Phys. Lett.* **269**, 199-207.
16. Cheshnovsky, O., Yang, S. H., Pettiette, P. L., Craycraft, M. J., and Smalley, R. E. (1987) Magnetic time-of-flight photoelectron spectrometer for mass-selected negative cluster ions. *Rev. Sci. Instrum.* **58**, 2131-2137.
17. Ganteför, G., Meiwes-Broer, K. H., and Lutz, H. O., (1988) Photodetachment spectroscopy of cold aluminum cluster ions. *Phys. Rev. A* **37**, 2716-2718.
18. Nakajima, A., Taguwa, T., Hoshino, K., Sugioka, T., Naganuma, T., Ono, F., Watanabe, K., Nakao, K., Konishi, Y., Kishi, R., and Kaya, K. (1993) Photoelectron spectroscopy of (C$_6$F$_6$)$_n^-$ and (Au-C$_6$F$_6$)$^-$ clusters. *Chem. Phys. Lett.* **214**, 22-26.

317

19. Hotop, H. and Lineberger, W. C. (1975) Binding energies in atomic negative ions. *J. Phys. Chem. Ref. Data.* **4**, 539-576.

20. Esaulov, V. A. (1986) Electron detachment from atomic negative ions. *Ann. Phys. Fr.* **11**, 493-592.

21. Kubo, K., Kondow, H., and Nishihara, H. *private communication.*

22. Schmeisser, D., Schnell, R. D., Bogen, A., Himpset, F. J., Rieger, D., Landgren, G., and Morar, J. F., (1986) Surface oxidation states of germanium. *Surf. Sci.* **172**, 455-465.

23. Fujii, M., Hayashi, S., and Yamamoto, K., (1990) Raman scattering from quantum dots of Ge embedded in SiO_2 thin films. *Appl. Phys. Lett.* **57**, 2692-2694.

24. Fujii, M., Hayashi, S., and Yamamoto, K., (1991) Growth of Ge microcrystals in SiO_2 thin film matrices: a Raman and electron microscopic study. *Jpn. J. Appl. Phys.* **30**, 687-694.

25. Cheshnovsky, O., Yang, S. H., Pettiette, C. L., Craycraft, M. J., Liu, Y., and Smalley, R. E. (1987) Ultraviolet photoelectron spectroscopy of semiconductor clusters: silicon and germanium. *Chem. Phys., Lett.,* **138**, 119-124.

26. Burton, G. R., Xu, C., Arnold, C. C., and Neumark, D. M. (1996) Photoelectron spectroscopy and zero electron kinetic energy spectroscopy of germanium cluster anions. *J. Chem. Phys.,* **104**, 2757-2764.

27. Raghavachari, K. and Rohlfing, C. M. (1991) Electronic structures of the negative ions Si_2^-—Si_{10}^-: electron affinities of small silicon clusters. *J. Chem. Phys.,* **94**, 3670-3678,

28. Kishi, R., Negishi, Y., Kawamata, H., Iwata, S., Nakajima, A., and Kaya, K. (1998) Geometric and electronic structures of fluorine bound silicon clusters. *J. Chem. Phys.* **108**, 8039-8058.

29. Szabo, A. and Ostlund, N. L. *"Modern Quantum Chemistry"* (Dover Publications, Mineola, 1996), p. 85.

30. Dai, D. and Balasubramanian, K. (1992) Electronic structure of group IV tetramers ($Si_4 - Pb_4$). *J. Chem. Phys.* **96**, 8345-8353.

31. Kittel, C. *"Introduction to Solid State Physics"* 6th Ed. (Willey, New York, 1986).

32. G. Herzberg, *"Spectra of Diatomic Molecule"* (Van Nostrand Reinhold Company Inc., New York, 1950), p. 530.

33. Boldyrev, A. I., Simons, J., Zakrzewski, V. G., and von Niessen, W. (1994) Vertical and adiabatic ionization energies and electron affinities of new Si_nC and Si_nO (n=1-3) molecules. *J. Phys. Chem.* **98**, 1427-1435.

34. von Behren, J., van Bunren, T., Zacharias, M., Chimowits, E. H., and Fauchet, P. M. (1998) Quantum confinement in nanoscale silicon: the correlation of size with bandgap and luminescence. *Solid State Commun.* **105**, 317-322.

35. Rebohle, L., von Borany, J., Yankov, R. A., Skorupa, W., Tyschenko, I. E., Fröb, H., and Leo, K. (1997) Strong blue and violet photoluminescence and electroluminescence from germanium-implanted and silicon-implanted silicon-dioxide layers. *Appl. Phys. Lett.* **71**, 2809-2811.

36. Hunter, J. M., Fye, J. L., Jarrold, M. F., and Bower, J. E., (1994) Structural transitions in size-selected germanium cluster ions. *Phys. Rev. Lett.* **73**, 2063-2066.

New Directions in Nanotechnology – Imprint Techniques

H.- C. SCHEER, H. SCHULZ, D. LYEBYEDYEV
University Wuppertal, Dept. of Electrical and Information Engineering,
Microstructure Engineering
Fuhlrottstr. 10, D-42097 Wuppertal, Germany

1. Introduction

Within the last few years a number of new techniques have emerged for definition of nm-scale patterns. They may be summarised under concepts like bottom-up or top-down, addressing the basic direction of pattern assembly. Bottom-up techniques based on the self assembly offer a very high innovation potential. At the other hand, top-down techniques are somewhat nearer to state of the art technology. These latter may be optimally suited for bridging the gap between conventional techniques and the high end self assembly.

We will report on a class of novel top-down techniques known as nano-imprinting. They use a replication process via imprinting, embossing or molding for definition of nm-scale patterns. The basic principles of these techniques will be shown in contrast to state of the art pattern definition.

In particular, hot embossing, the imprint under pressure and elevated temperature, will be discussed in detail, starting from characteristic issues of this technique. Aspects of processing for definition of 2-dimensional and 3-dimensional surface patterns will be summarised on the basis of imprint results.

2. State of the Art Definition of nm-Scaled Patterns

A number of advanced applications in device physics, optics and optoelectronics as well as sensor technology require patterns with lateral dimensions in the nm range, e.g. metal electrodes.

Whereas for production a number of methods exists to define such patterns in a resist over large areas (lithography via electron beam, ion beam, X-ray or deep UV radiation), research uses almost always electron beam lithography. In most cases, these e-beam writers are upgrades of a secondary electron microscope (SEM).

When nm-scaled electrodes are needed for research applications, electrode beam writing is combined with lift-off. The process sequence is as follows.

First, the substrate is spun with the resist, prebaked and exposed by writing with an electron beam. A common e-beam resist is poly methyl methacrylate (PMMA), which

319

L. Pavesi and E. Buzaneva (eds.), Frontiers of Nano-Optoelectronic Systems, 319–330.

is cracked by e-beam radiation. Exposed areas become dissolvable and are removed in a developer. Then the whole patterned resist surface is covered with a metal, preferably in an evaporation process, featuring minimum coverage of inclined or vertical surface regions. In the last step, the unexposed resist is removed in a wet process, flooding away the metal on top of it. Metal electrodes remain at the substrate replicating the e-beam writing pattern. In order to assure lift-off, the resist profile has to be higher than the envisaged metal electrode thickness. Vertical slopes or even slopes with a negative inclination angle aid in fabrication of sharp electrode line patterns.

An alternative is the electrode definition by dry etching. Therefore the substrate has to be covered with the metal layer before the lithography process. In this case the patterned polymer has to serve as a dry etch mask. This requires good mask selectivity, which is not easily met for a number of e-beam resists.

As a consequence, lift-off is the technique of choice for most research applications. When thickness contrast is not adequate, two or three layer resist systems are often used. In such cases, the e-beam resist comprises the topmost layer. E-beam defined patterns are transferred to a bottom layer resist in a dry etch process via an intermediate thin 'hard mask' layer (e.g. nitride or oxide). Other approaches with two layer resists use different electron beam spreading characteristics resulting in negative pattern slopes.

Novel techniques based on imprint can simplify pattern definition in the nm-scale for a broad range of applications.

3. Introduction to Imprint Techniques

During the last years a number of new techniques have emerged under the heading 'nanoimprinting'. In this paragraph a survey on those applicable for electrode pattern definition on a substrate will be given. 'Volume' techniques like e.g. casting and injection molding related to microsystem technologies like LIGA are not considered.

Common to all nanoimprint techniques is the fact, that they are replication techniques. The patterns envisaged have to be available at an original, the master or stamp, and these originals have to be fabricated following state of the art Si technology, e.g. e-beam writing and dry etch in order to define the pattern in bulk silicon, in quartz or in a poly-Si or oxide layer.

As these originals can be replicated many times, as the replication process is a mechanical one and as it works parallel for the whole sample area, the imprint techniques are expected to be cost and time efficient compared to e-beam.

3.1 MICROCONTACT PRINTING

Microcontact printing was introduced by Whitesides [1]. It is a two step replication process (see Figure 1). In a first step, the original is casted with a monomer and cured. The resulting polymer (poly dimethyle siloxane, PDMS) is elastic and can be pulled off from the original. In step two this elastomeric stamp is used in an imprint-like process. Its surface is 'inked' with a SAM (self-assembing monolayer) of e.g. alkanethiolates [2] or alkylsiloxanes [2, 3] and transfers this ink onto the substrate within the elevated areas of the stamp.

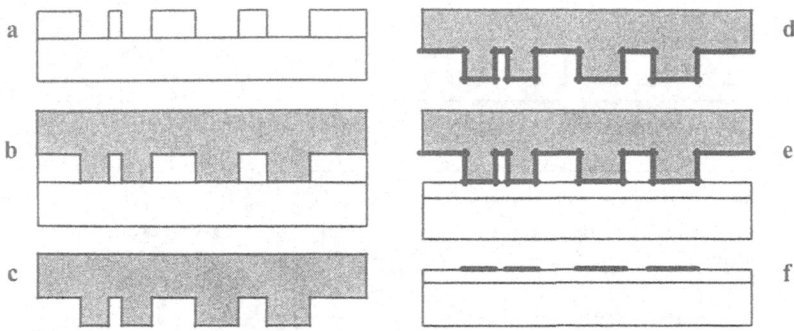

Figure 1. Basic steps during microcontact printing after [1]. Step 1: a - master, b - casted with monomer, c - elastomer after curing and separation. Step 2: d - elstomeric stamp with SAM ink, e - print, f - SAM pattern on Au layer.

Due to the elastic properties, surface contact is not critical even in case of substrate warping. In addition non-flat surfaces can be printed [4, 5], and application of the elastomer to a roller enables large area pattern definition [6].

Minimum feature sizes demonstrated with a PDMS stamp are in the range of 100 nm [2].

SAM pattern definition with alkanethiolates works best on gold surfaces, where they serve as a mask in a subsequent wet etch process in order to define the electrodes [1].

3.2 MOLD LITHOGRAPHY

This technique was introduced by Haisma in 1996 [7] (see Figure 2). He used an original made from quartz. A monomer (e.g. hexanediol diacrylate, HDDA) is spincoated onto the substrate and brought into contact with the quartz mold in a commercial vacuum contact printer. Under vacuum pressure of about 1 bar the viscous monomer fills the pattern relief of the stamp. Curing is done via UV radiation through the quartz mold.

For this type of replication a primer is used at the subtrate and an anti-adhesive on the stamp in order to aid in separation of the sample with the patterned polymer from the original after curing.

Mold lithography results in a thickness contrast in the polymer layer equal to the mold profile. The shrinking of the polymer going along with the curing process helps in separation of the two parts. As the original does not contact the solid sample surface in this process, a residual layer of polymer remains all over the molded area. This residual layer is removed in an anisotropic dry etch process (reactive ion etching, RIE).

Smallest feature sizes demonstrated in mold lithography are 25 nm [7]. Adjustment for multilayer lithography is enabled via the quartz mold.

When molding over an area of 4" (100 mm diameter) a residual layer variation of up to 500 nm was found.

Mold lithography can also be performed with a PDMS elastomeric stamp in contact with the solid substrate of the sample [8].

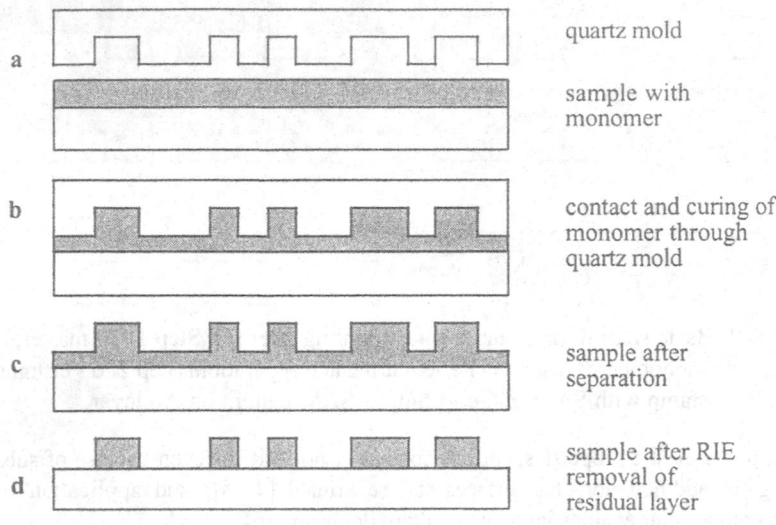

Figure 2. Basic steps during mold lithography after [7].

3.3 HOT EMBOSSING

An imprint technique based on hot embossing was first introduced by Chou [9, 10]. This technique (see Figure 3) uses a thin layer of a thermoplastic polymer spun onto the substrate. The stamp and sample are heated up until a temperature well above the glass transition temperature of the polymer is reached. Then both are brought into contact and a pressure of up to 120 bar is applied in order to emboss the polymer layer. Once again, direct contact between stamp and solid sample surface is avoided, resulting in a residual layer to be removed by RIE. Separation of stamp and sample is done after cooldown to at least glass temperature in order to conserve the plastic deformation obtained.

With this technique the smallest patterns have been shown. A first result gives 25 nm [9], only shortly after 6 nm lines were reported [11].

4. Details of Hot Embossing

4.1 CHARACTERISTIC ISSUES

4.1.1 Positive and negative stamp features
When embossing, a force or pressure is applied from the backside of stamp and sample or the respective stages. But during most time of the imprint, only the elevated area of the stamp is in contact with the polymer layer on the sample, resulting in an increased

local pressure during the process. This pressure depends on the fraction of the elevated area to the overall area [12, 13]. In case of 'positive' stamp features, the local pressure may exceed the external pressure many times, and excellent imprint results are obtained. In case of 'negative' stamp features the initial contact area is high and the local pressure may exceed only slightly the external one. As a consequence, negative stamp features are much harder to imprint [13]. As in general the elevated area to overall area ratio is not given in the literature, comparison and estimation of results is difficult.

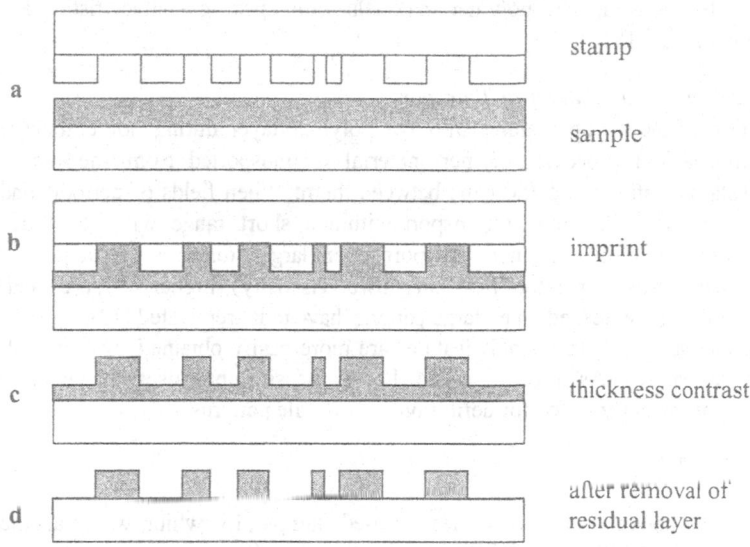

Figure 3. Basic steps during hot embossing after [9]: a - Heating up of stamp and sample, b - contact and pressure application, c - cooldown and separation, d - RIE removal of residual layer.

4.1.2 Suited Polymer Materials

Formation of a thickness relief in a hot embossing process is guided by the viscosity of the thermoplastic polymer at the imprint temperature chosen [14]. Viscosity has to be low in order to allow flowing of the highly viscous medium within acceptable processing times. Viscosity starts to decrease from the glass temperature T_g on. Acceptable values of viscosity for plastic deformation require temperatures of 50 to 100 °C above T_g. The higher the molecular weight of the polymer the higher the temperature necessary.

Thermoplastic polymers are linear polymers without any configuration symmetry. These polymers are 'amorphous' ones, featuring a glassy state below T_g. Their mechanical moduli drop by more than 3 orders of magnitude above T_g and they become viscous in the higher temperature range. Their mechanical response is a result of the entanglements between the long chains, which can slip when high enough temperatures are reached.

Semi-crystalline polymers or duroplastic polymers are not suited for hot embossing. Duroplasts remain stable over their complete temperature range of use and cannot be deformed. This is due to their chemical network - they are highly crosslinked. Semi-crystalline polymers have a regular configuration. Crystalline regions act like crosslink points and hamper polymer flow. Furthermore their crystallinity depends strongly on the preparation conditions, and thus their mechanical response is hardly controlled.

A broad range of thermoplastic polymers suitable for imprint is commercially available. In addition, new polymer materials with specific characteristics are under development [15 - 18]

4.1.3 Pattern Size and Material Transport

Replication of the stamp features into the polymer layer during hot embossing is a purely mechanical process. Polymer material is transported from the areas below elevated stamp features into the gaps between them. When fields of periodic and small patterns are involved, polymer transport within a short range will do. When large patterns are involved, polymer transport over larger distances is required. As a consequence higher temperature (and thus lower viscosity), higher pressures and higher processing times are needed when large patterns have to be replicated [13].

This fact suggests, that small features are more easily obtained by hot embossing than larger features. As a consequence, hot embossing promises to be an imprint technique intrinsically suited for definition of nm-scale patterns.

4.2 RESULTS

In view of the above mentioned issues we used stamps [13] which were patterned over their whole area of 2 x 2 cm^2. Their mean ratio of positive to negative stamp areas is near to 50%. Feature sizes cover the range from several 100 nm up to 100 µm, a typical pattern size for a bond contact. A survey of a typical stamp is given in Figure 4. It is prepared by UV lithography and dry etching. The central band represents the most critical area, where largest features (100 x 100 µm bondpads) and arrays of fine lines (400 nm lines) are in near neighbourhood. The rest of the surface is covered with lines and dashes of different size.

For the embossing experiments we used a simple commercially available hydraulic press, enabling temperatures up to 200 °C and pressures of 200 bar.

4.2.1 Optimum processing conditions

For stamps as described above we found good pattern transfer over the whole area for small patterns up to the largest ones under 100 bar and 100°C above T$_g$ of the polymer [13].

This is demonstrated in Figure 5, where SEMs of increasing magnifications are given from top left to bottom right. Large area pattern replication is demonstrated, largest and finest patterns are defined with good quality.

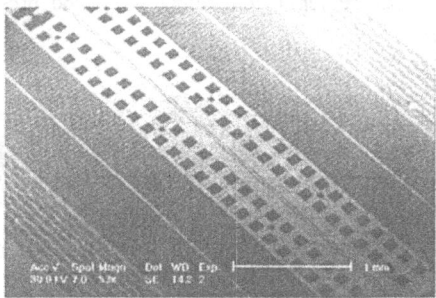

Figure 4. Fully patterned stamp with feature sizes from 400 nm to 100 μm.

Figure 5. Large area imprint results. Large patterns (100 μm) as well as smallest patterns (400 nm) are well resolved.

4.2.2 Processing Conditions and Pattern Size

When the processing conditions above are not met, pattern transfer is inadequate [13], which is demonstrated in Figure 6. At lowest pressure and temperature (left) the fine line array is visible, but the region around the large pad is not filled with polymer. A curved border line of polymer flow is visible around it. When pressure is increased (mid) this border line moves, and at 70°C above T_g and 100 bar the area around the pad is nearly filled. Under optimum conditions, the situation resembles Figure 5 top right.

This result documents clearly, that hot embossing is a technique ideally suited for definition of small patterns, as the fine line field is replicated even under non-ideal procesing conditions. At the other hand, this result has consequences: When only small

326

and periodic patterns have to be replicated, temperature and pressure of the imprint process might be reduced.

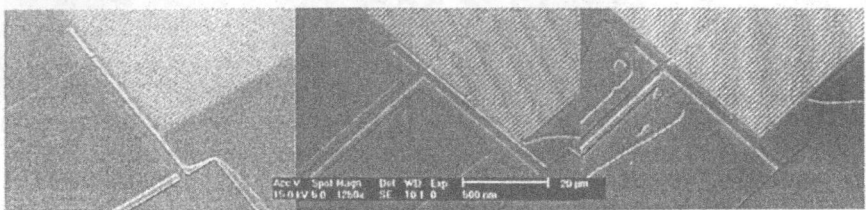

Figure 6. Imprint under non-ideal conditions. Left: T_g + 50 °C and 60 bar, middle: T_g + 50 °C and 100 bar, right: T_g + 70°C and 100 bar.

4.2.3 Imprint Depth

As during hot embossing a residual layer is inevitable, its thickness has to be small compared to the thickness contrast in the polymer so that RIE removal is uncritical.

Figure 7 compares stamp features (left) and imprinted features (right) in case of a positive (top) and a negative (bottom) stamp, achived under optimum imprint conditions [18] with a stamp like the one described. The stamp pattern is reproduced accurately in the polymer. In both cases, negative as well as positive stamp features, the residual layer is about 60 nm and substantially smaller than the thickness contrast in the polymer layer. In order to get such results, the initial polymer film thickness has to be tuned to the stamp feature height and the ratio of elevated to recessed stamp area. In our case, the initial film thickness was about 250 nm.

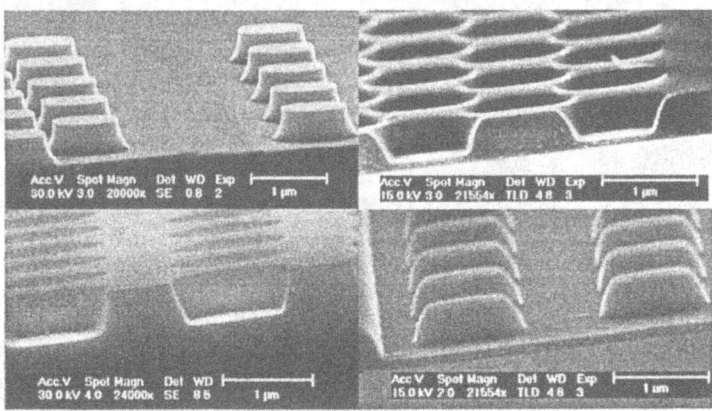

Figure 7. Replication of positive (top) and negative (bottom) stamp features (left) by hot embossing (right). The residual layer thickness is small compared to the thickness contrast in the polymer.

4.2.4 Nanometer Scale Patterns

Figure 8 demonstrates the replication of 50 nm lines [18] within fields of 50 x 50 μm. In this case, the pattern was e-beam written into PMMA. The stamp is a 200 nm thick Ni shim formed in a galvanic process as it is used in general for fabrication of CD masters for injection molding. The patterned area is small.

The feature height is around 150 nm. The imprinted pattern (right) replicates the stamp pattern (left).

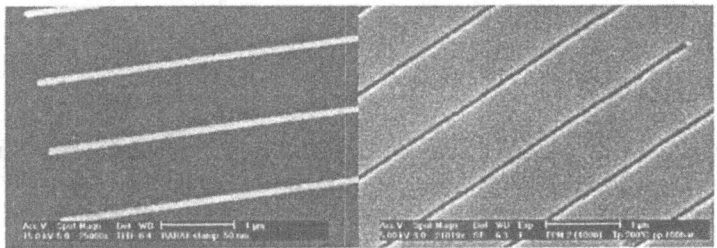

Figure 8. Arrays of 50 nm lines. Left: stamp, right: imprint.

4.2.5 Tree-Dimensional Pattern Replication

Up to now the results referred to 2D-patterns, where only two pattern height levels were relevant. But hot embossing is also applicable to 3D-patterns, where a specific slope or curvature of the master has to be replicated [19, 20].

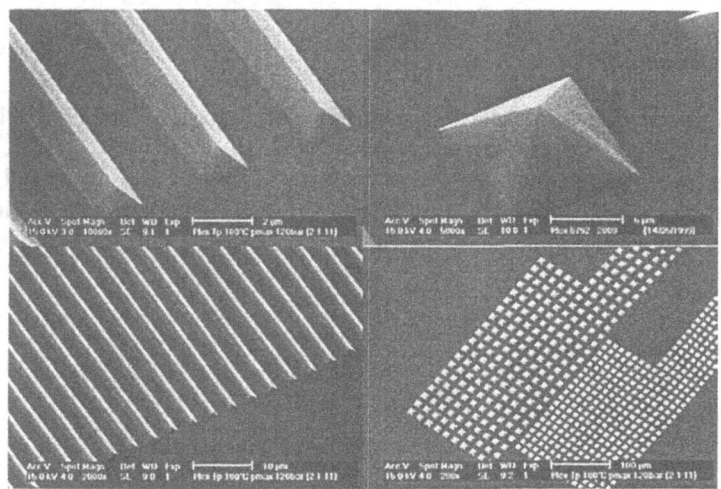

Figure 9. Replication of 3D patterns. Details (top) and large area view (bottom)

Figure 9 shows imprint results using a stamp which is fabricated by anisotropic wet etching in 100-Si. With this technique inverted pyramidal features with rectangular lateral geometries and slope angles of 54.7° can be produced with highest surface quality. When orientation of the mask at the Si wafer is along 110 directions, atomically flat slopes are formed.

Figure 6 demonstrates, that 3D-features are replicated with accuracy over large areas [20]. This is an example, where in general a negative stamp has to be used. In addition, contact area between stamp surface and polymer increases with increasing imprint depth directly after the first contact. As a consequence such patterns are difficult to emboss and require high pressure and temperature.

5. Conclusion

Definition of lateral patterns by imprint methods like hot embossing have demonstrated surprising quality over large areas. As we used a simple commercially available hydraulic press the technique has proven to be simple and cost efficient in the laboratory scale.

At adequate imprint temperature and pressure pattern sizes from several 10 nm to 100 µm have been replicated in parallel, and processing conditions are relaxed when small and periodic patterns are involved. Hot embossing is a technique ideally suited for replication of small lateral patterns.

A large number of thermoplastic polymers are commercially available that can be used in hot embossing. In addition, new polymers with specific characteristics are available recently.

Beyond the definition of 2D-patterns also the replication of 3D-patterns could be demonstrated, offering further benefits for novel nm-scaled devices.

Acknowledgements

The authors acknowledge financial support from the European Commission ESPRIT and from the Deutsche Forschungsgemeinschaft DFG.

For stamps we express our thanks to J. Ahopelto (VTT Espoo), K. Zimmer and K. Otte (IOM Leipzig) and L. Montelius (NSC Lund). Fruitful discussions with T. Hoffmann are highly acknowledged.

6. References

1. Kumar, A. and Whitesides, G.M. (1993) Features of gold having micrometer to centimeter dimensions can be formed through a combination of stamping with an elastomeric stamp and an alkanethiol 'ink' followed by chemical etching, *Appl. Phys. Lett* **63**, 2002-2004

2. Xia, Y., Zhao, X.-M. and Whitesides, G.M. (1996) Pattern Transfer: Self assembled monolayers as ultrathin resists, *Microelectronic Engineering* **32**, 255-268

3. Xia, Y., Mrksich, M., Kim, E. and Whitesides, G.M. (1996) Microcontact printing of octadecylsiloxane on the surface of Silicon dioxide and its application in microfabrication, *J. Am. Chem. Soc.* **117**, 9576-9577

4. Jackman, R.J., Wilbur, J.L. and Whitesides, G.M. (1995) Fabrication of submicrometer features on curved substrates by microcontact printing, *Science* **269**, 664-666

5. Xia, Y., Kim, E., Zhao, X.-M., Rogers, J.A., Prentiss, M. and Whitesides, G.M. (1996) Complex optical surfaces formed by replica molding against elastomeric masters, *Science* **273**, 347-349

6. Xia, Y., Qin, D. and Whitesides, G.M. (1996) Microcontact printing with a cylindrical rolling stanp: A practical step toward automatic manufacturing of patterns with submicrometer sized features, *Adv. Mater.* **8**, 1015-1017

7. Haisma, J., Verheijen, M. and van den Heuvel, K. (1996) Mold-assisted nanolithography: A process for reliable pattern replication, *J. Vac. Sci Technol.* **B14**, 4124-4128

8. Xia, Y., McClelland, J.J., Gupta, R., Qin, D., Zhao, X.-M., Sohn, L.L., Celotta, R.J. and and Whitesides, G.M. (1997) Replica molding using polymeric materials: A practical step towards nanomanufacturing, *Advanced Materials* **9**, 147

9. Chou, S.Y., Krauss, P.R. and Renstrom, P.J. (1996) Imprint lithography with 25-nanometer resolution, *Science* **272**, 85-87

10. Chou, S.Y., Krauss, P.R. and Renstrom, P.J. (1995) Imprint of sub-25 nm vias and trenches in polymers, *Appl. Phys. Lett.* **67**, 3114 - 3116

11. Chou, S.Y., Krauss, P.R., Zhang, W., Guo, L. and Zhuang, L. (1997) Sub-10 nm lithography and applications, *J. Vac. Sci. Technol.* **B15**, 2897-2904

12. Jaszewski, R.W., Schift, H., Gobrecht, J. and Smith, P. (1998) Hot embossing in polymers as a direct way to pattern resist, *Microelectronic Engineering* **41/42**, 575-578

13. Scheer, H.-C., Schulz, H., Hoffmann, T. and Sotomayor Torres, C.M., (1998) Problems of the nanoimprinting technique for nanometer scale pattern definition, *J.Vac. Sci. Technol.* **B16**, 3917-3921

14. Gottschalch, F., Hoffmann, T., Sotomayor Torres, C.M., Schulz, H. and Scheer, H.-C. (1999) Polymer issues in nanoimprinting technique, *Solid State Electronics* **43**, 1079-1083

15. Pfeiffer, K., Bleidiessel, G., Grützner, G., Schulz, H., Hoffmann, T., Scheer, H.-C., Sotomayor Torres, C.M. and Ahopelto, J. (1999) Suitability of new polymer materials with adjustable glass temperature for nanoimprinting, *Microelectronic Engineering* **46** 431-434

16. Gaboriau, F., Peignon, M.C., Turban, G., Cardinaud, Ch., Pfeiffer, K., Bleidiessel, G. and Gruetzner, G. (2000) High density fluorocarbon plasma etching of new resist suitable for nano-imprint lithography, to be published in *Microelectronic Engineering* **53**

17. Pfeiffer, K., Fink, M., Bleidiessel, G., Gruetzner, G., Schulz, H., Scheer, H.-C., Sotomayor Torres, C.M., Hoffmann, T., Cardinaud, Ch. and Gaboriau, F. (2000) Novel linear and crosslinking polymers for Nanoimprinting with high etch resistance, to be published in *Microelectronic Engineering* **53**

18. Schulz, H., Scheer, H.-C., Hoffmann, T., Sotomayor Torres, C.-M., Pfeiffer, K., Gruetzner, G., Bleidiessel, G., Cardinaud, Ch., Gaboriau, F., Peignon, M-C., Ahopelto, J. and Montelius, L. (2000) New Polymer Materials for Nanoimprinting, to be published in *J. Vac. Sci. Technol.* **B18**

19. Schift, H. Jaszewski, R.W., David, C. and Gobrecht, J. (1999) Nanostructuring of polymers and fabrication of interdigitated electrodes by hot embossing lithography, *Microelectronic Engineering* **46**, 121-124

20. Zimmer, K., Otte, K., Braun, A., Rudschuck, St., Friedrich, H., Schulz, H., Scheer, H.-C., Hoffmann, T., Sotomayor Torres C.M., Mehnert, R. and Bigl, F. (1999) Fabrication of 3D micro- and nanostructures by replica molding and imprinting, *Proc. EUSPEN* **1**, 534-537

SELF-ASSEMBLY OF NANOBLOCKS AND MOLECULES IN OPTICAL THIN-FILM NANOSTRUCTURES

N. I. KOVTYUKHOVA[a], E. V. BUZANEVA[b], A. D. GORCHINSKY[b],
P.J. OLLIVIER[c], B. R. MARTIN[c], C. C. WARAKSA[c], T. E. MALLOUK[c]
[a]*Institute of Surface Chemistry, NASU, 31, Pr.Nauky, 03022 Kiev, Ukraine*
[b]*Kiev T.Shevchenko University, 64, Vladimirskaya Str., 01017 Ukraine*
[c]*The Pennsylvania State University, University Park, 16802 PA, USA*

1. Introduction

The growing interest in developing techniques to prepare ultrathin semiconductor nanoparticle films is motivated by the size-dependent electronic and optical properties of semiconductors, which lead to a range of potential applications in electronic and optoelectronic devices, solar cells, photoelectrodes, photocatalysts, and sensors. The wet chemical synthesis of ultrathin semiconductor films represents, in principle, a simple and inexpensive alternative to more technologically demanding chemical vapor deposition (CVD) and physical techniques [1]. However, the realization of practical devices from wet chemical synthesis requires the development of film growth techniques that give similar or better quality films than vapor-phase methods. In particular, precise control of film thickness, crystallinity, and morphology are significant problems to be overcome in wet chemical synthesis.

This paper describes research efforts concentrated on two strategies for wet layer-by-layer deposition of technologically interesting inorganic and composite inorganic/organic nanoparticle films that provide film thickness control to nanometer precision. In general, both methods rely upon layer-by layer (LBL) adsorption of building blocks i.e. molecules, ions or nanoparticles, onto a growing surface, and each adsorption cycle is self-limiting at the extent of a single monolayer. This is achieved by immersing a substrate alternately in solutions of the components. Each adsorption step was followed by rinsing with an appropriate solvent and drying in Ar stream. The objective of the rinsing step is desorption of weakly bound building units forming a second and additional layers, thus providing the deposition of a strongly bound monoparticularly thick layer after each immersing step. An essential condition for a successful LBL process is that the binding of the adsorbed monolayer is irreversible, or sufficiently strong to prevent desorption in the subsequent rinsing/adsorption step. The techniques are applicable to substrates with different surface geometry (planar wafers, spherules high surface area powders, porous solids). The layer-by-layer film synthesis is presented schematically in Fig.1

Our attention will be focused on the design and characterization of ultrathin nanoparticle semiconductor and composite semiconductor/polymer films prepared by:
- LBL self-assembly from inorganic nanoparticles and organic macromolecules,
- direct LBL synthesis from molecular precursors, i.e. the chemical component of the semiconductor.

The most important questions addressed in the course of this work are:
- To study dynamics of film growth on the nanometer length scale;
- To control the quality of inorganic mono- and multilayers;
- To characterize optical properties of the prepared heteronanostructures;

331

L. Pavesi and E. Buzaneva (eds.), Frontiers of Nano-Optoelectronic Systems, 331–346.
© 2000 *Kluwer Academic Publishers. Printed in the Netherlands.*

332

- To extend the process to fabricating useful devices.

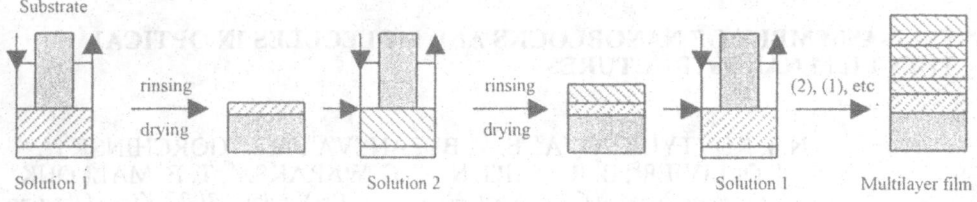

Figure 1. Schematic representation of the principles of wet layer-by-layer film deposition.

2. Layer-by Layer Self-Assembly of Multilayer Inorganic/Organic films

2.1 PRINCIPLES OF THE LBL SELF-ASSEMBLY

These films are grown by a wet chemical layer-by-layer adsorption method, which is similar to that developed earlier for inorganic [2] and organic [3] polyelectrolytes. This method involves spontaneous adsorption (self-assembly) of pre-formed colloidal particles and organic macromolecules in monolayers onto growing surface [4-13]. Alternating these adsorption cycles results in ordered surface heterostructures in which layers of 2D or 3D nanoparticles and polymeric "spacers" alternate along the stacking axis according to the following scheme:

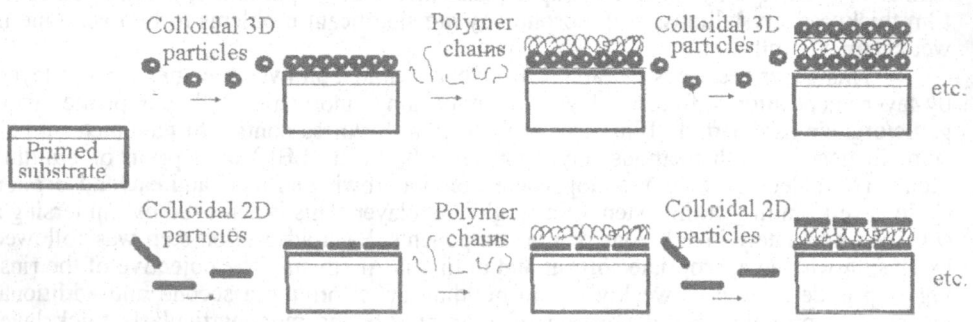

Figure 2. Schematic representation of the sequential adsorption procedure used to grow monolayer and multi-layer films from solutions of 3D and 2D inorganic nanoparticles and organic macromolecules.

The self-assembly of 2D nanoblocks relies on the exfoliation of solids to produce unilamellar colloids of sheets, which typically have nanometer thicknesses and lateral dimensions of tens of nanometers to microns. Exfoliation procedures based on ion exchange and redox reactions have been developed for a variety of lamellar solids.

The advantage of this approach is its versatility arising from a possibility of joining up components with electronic properties ranging from insulating to metallic. By selecting appropriate components, one can prepare nanostructured composite films with the desired optical, electronic, magnetic, mechanical, and thermal properties due to the combination of advantages of the inorganic and organic precursors.

Using this technique, surfaces can be permanently derivatised with films in which layers of semiconductor (e.g. metal chalcogenides [6,7] and oxides [6, 8-10]), insulator (e.g. metal phosphates [4], silicates [6,11], graphite oxide (GO) [6,12]), or metal [13] nanoparticles are interleaved with polymer layers.

Successful self-assembly of the densely packed and well-ordered films depends on the properties of the nanoparticles used, as well as the attachment chemistry. The adsorption-desorption equilibria control nanoparticles and polymers deposition onto the surface as monolayer and multilayer films. Interactions between the substrate surface, nanoparticles and organic macromolecules govern the spontaneous adsorption of the monolayer in each adsorption cycle. Fortunately, relatively weak interactions suffice to hold the structure together, because they act collectively between extended particles and chains. In general, the layer-by-layer self-assembly technique relies on alternate deposition of aqueous oppositely charged nanoparticles and organic macromolecules that are held together by ionic bonds [4, 6-8, 11-13]. However, other types of interactions, such as hydrophobic [11c], hydrogen bonds [14a], and coordinate-covalent bonds [5,10,14b], are also involved.

2.2 FILM GROWTH

Using inorganic building blocks, such as metal chalcogenide [7], clay [11] titanoniobates [8], and GO single sheets [6,12] or metal oxide (e.g. ZnO [10], TiO_2 [9,14a]) nanoparticles, one can achieve complete coverage of the surfaces in four-five adsorption cycles. In all the cases, the quality and morphology of the multilayer films are determined by the coverage in the first adsorbed inorganic monolayer, which in turn can be controlled by the chemical composition of substrate and nanoparticle surface [7,9,11bc,12], the pH of the reaction media [8,12], and the nature of the solvents used in adsorption and rinsing steps [9,10]. To optimize nanoparticle adsorption, the substrate surface is generally primed chemically so that ionizable or other reactive groups could be generated. Typically, alkoxysilyl alkylamines or aminoalkanethiols are used for priming with cationic layer the Si or metal (Au, Ag, Pt) substrates, respectively. Anionic layer, for example, on the Si substrate (Si(OH)) can be created by hydroxylating the surface with H_2O_2-in-H_2SO_4 solution. Also the adsorption of polymers, containing cationic or anionic groups, on bare substrates (including hydrophobic ones [11b,c]) is routinely employed to create positively or negatively charged surfaces. In this case, ionization and molecular weight of the polymer determines surface density of adsorbed inorganic nanoparticles [11c]. In some cases, capping nanoparticles by organic molecules bearing appropriate functions can essentially improve their self-assembly [6, 14].

Morphology of Mono- and Multilayer Films. AFM images in Fig.3 a, b demonstrate effect of the chemical composition of priming layer on the coverage in the first adsorbed nanoparticle layer. One can see that MoS_2 sheets bearing relatively weak negative charge in aqueous solution prefer rather non-polar surface of silicon substrate pretreated with a monolayer of polyethyleneoxide (Si/PEO) (Fig3b) than positively charged surface of silicon substrate primed with 4-((dimethylmethoxy)silyl)butylamine {Si(NH_2)} (Fig.3a). This can be explained by the complex formation between PEO and Li^+ ions that charge-compensate the colloidal MoS_2 sheets. Thus relatively non-polar components, such as MoS_2 and PEO can also be organized in multilayer films [7].

The critical role of pH in controlling electrostatic particle/surface interactions is presented in Fig.4 a, b. At pH 9 (Fig.4 b), the adsorption process selects much larger sheets (900-9000 nm), which cover about 60-65% area of the surface contrary to 30% coverage by smaller sheets at pH 4 (Fig. 4a). The dissociation of the GO hydroxyl groups occurs around pH 9 and significantly increases the negative charge density on the GO sheets, and hence their attraction by the cationic surface [12].

334

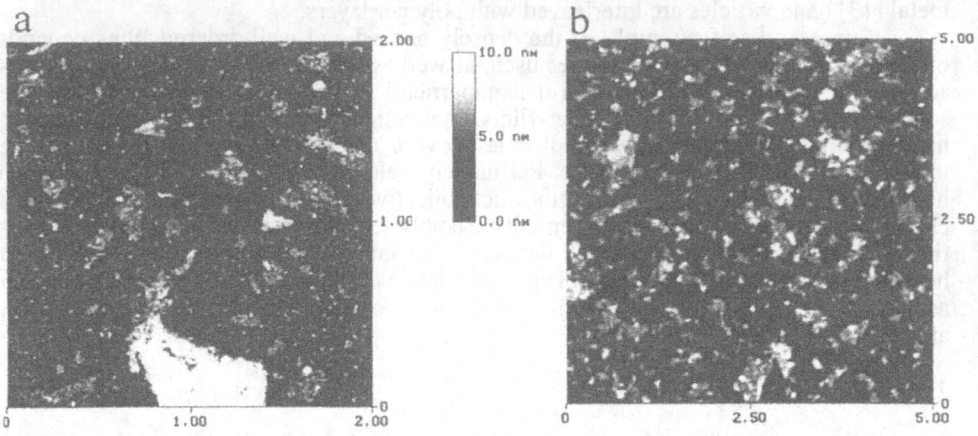

Figure 3. Tapping-mode AFM images of the first adsorbed MoS$_2$ layer on (a) Si(NH$_2$) and (b) Si/PEO substrates.

Also preventing the desorption of nanoparticles during the rinsing step requires careful selection of the rinsing solvent [9,10]. In the case of electrostatic particle/surface interactions, such factors as dielectric constant and pH of the solvent are of crucial importance.

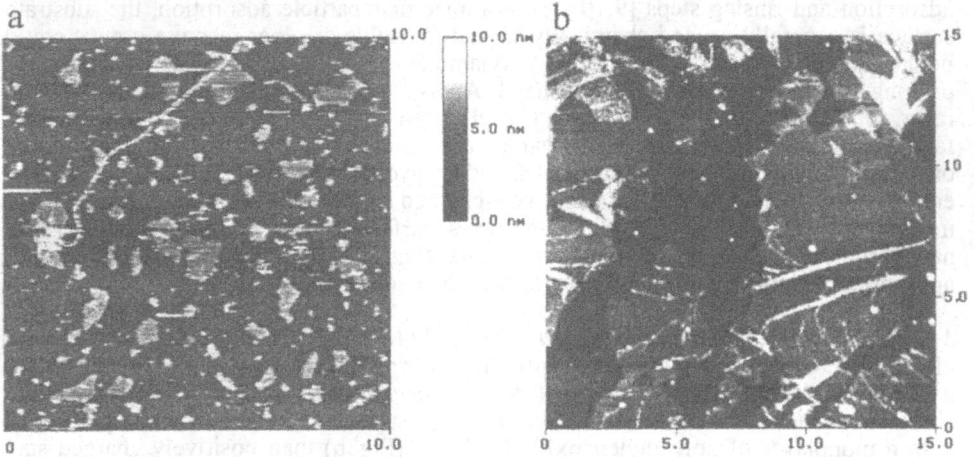

Figure 4. Tapping-mode AFM images of the first adsorbed GO layer on Si(NH$_2$) substrate from (a) aqueous (pH 4) and (b) aqueous ammonia (pH 9) solutions.

It has been shown that the lowest average roughness of the multilayer film surface is observed when the coverage of the first adsorbed layer of nanoparticles is highest [7, 9, 11b, 12]. This was explained by the model proposed by Kleinfeld and Ferguson for the multilayer growth on islands [11b]. In the poorly tiled first nanoparticle layer, widely spaced particles or their agglomerates work as nucleation centres for growing the film, which then covers the surface by vertical and lateral growth of the islands. The formation of individual islands on the top of a densely-packed first nanoparticle layer is less likely, and one can expect the mechanism of film growth close to the ideal one (see Fig 2), i.e. regular saturated-monolayer-at-a-time adsorption of the film components.

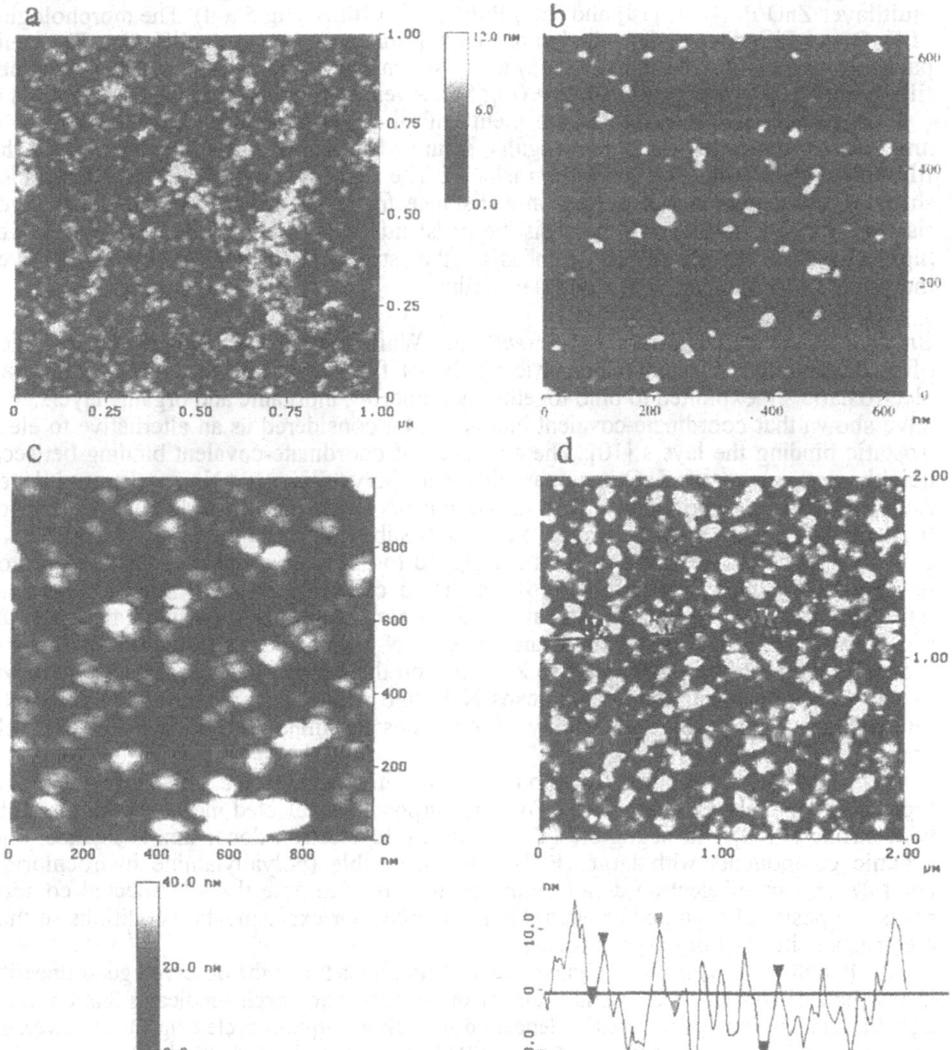

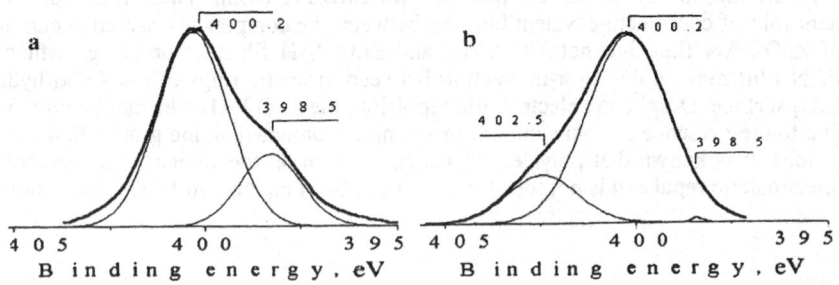

Figure 5. Tapping-mode AFM images of mono- and multilayer films on Si(NH₂) substrate: (a) the first ZnO layer; (b) the first TiO₂ layer; (c) multilayer (ZnO/PAN)₉ZnO film; (d) multilayer TiO₂/PAH)₄TiO₂ film.

Figure 6. N 1s core level spectra of (a) Si-OH/PAN film and (b) Si-OH/(ZnO/PAN)₉ZnO film

This suggestion is confirmed by the comparison of AFM images of monolayer and multilayer ZnO/Polymer [10] and TiO$_2$/Polymer [9] films (Fig.5 a-d). The morphologies of ZnO and TiO$_2$ layers, deposited in one adsorption cycle, are quite different. The well-packed layer of ZnO particles (Fig.5a) leads to densely packed multilayer ZnO/Polymer film completely covering the surface (Fig.5c). Average roughness of the film is about 6 – 8 nm, which is in reasonable agreement with the size of primary ZnO particles (~ 6 nm). Such morphology suggests regular saturated-monolayer-at-a-time growth of the film rather than multilayer growth on islands. The first layer of TiO$_2$ colloidal particles shows separate features disposing on a distance from one another (Fig.5b), that gives rise to a patchy multilayer film consisting of islands 100-150 nm wide and up to 23 nm high (Fig.5d). These results emphasise the strong influence of the first layer morphology on the quality of a multilayer film.

Binding Forces in Multilayer Self-Assemblies. While there have now been many studies of self-assembled inorganic-nanoparticle/polymer films, it is rare that forces other than electrostatic are exploited to bind together neighboring inorganic and organic layers. We have shown that coordinate-covalent bonds can be considered as an alternative to electrostatic binding the layers [10]. The existence of coordinate-covalent binding between neighboring layers of ZnO nanoparticles and polyaniline (PAN) in the multilayer ZnO/PAN film is supported by XPS measurements (Fig.6) [10b]. N 1s core level spectrum of the starting PAN deposited on Si(OH) substrate (Fig.6a) reveals two peaks at 398.5 eV and 400.2 eV, which can be assigned to the neutral imine groups and protonated amine groups, respectively [15]. In the spectrum of PAN incorporated into the multilayer ZnO/PAN film, the peak at 398.5 eV almost disappears, while a new peak at 402.5 eV arises (Fig.6b). This indicates that Zn-N binding forms through coordinating mostly one type of nitrogen atoms to Zn^{2+} ions on the nanoparticles surface. It is known that in transition metal amino complexes N 1s line is high-energy shifted comparing to that of bare amine [16]. The affinity of Zn^{2+} ions to amine and imine ligands is well known.

Using the ellipsometry method, we have investigated the effect of interaction type on the multilayer film growth. For this purpose, we selected inorganic components with strong (ZnO) and negligible (TiO$_2$) affinity to electron-donor amine ligands, and organic components with large (PAN) and negligible (polyallylamine hydrochloride (PAH)) amount of electron-donor amine groups. It is to note that all selected components are positively charged or neutral (PAN) under our experimental conditions so that electrostatic interactions are not involved.

For all multilayer ZnO/Polymer and TiO$_2$/Polymer films, the observed good linearity in the plots of film thickness versus number of the adsorption cycles indicates that on average the same amount of material is deposited in each adsorption cycle (Fig.7). The average increase in the film thickness per each ZnO/PAN adsorption cycle is 63 Å (Fig.7.1), which consistent with average ZnO particles size (~6 nm) estimated by TEM [10], and suggests dense coverage in each deposited monolayer of ZnO particles. The average increase in the film thickness per each TiO$_2$/PAN adsorption cycle, 8 Å (Fig.7.4), is essentially less than that observed for the ZnO/PAN film, 63 Å, despite almost the same average size of the ZnO and TiO$_2$ (~6-7 nm [9]) particles. For ZnO/PAH film (Fig.7.2), average increments in layer thickness, 2.3 nm, is ~2.5 times less than that for ZnO/PAN film. These facts confirm the important role of coordinate-covalent binding between the components, which occurs in the case of ZnO/PAN film, but not TiO$_2$/PAN and ZnO/PAH films. Poor but growth of the TiO$_2$/PAN film may be due to π-interaction between aromatic rings of PAN and hydroxylated TiO$_2$ surface. Despite the electrostatic repulsion, the ZnO/PAH film can be grown, perhaps due to the presence of some amount of the non-protonated amine groups that complex to Zn^{2+} ions. It is known that polyelectrolytes can adsorb on the similarly charged surfaces when electrostatic repulsion is not too strong and can be compensated by total non-ionic free energy of the adsorption. One more example of the self-assembly of similarly charged components in the multilayer film is presented in Fig. 7.3. Multilayer TiO_2/PAH film can be grown with good regularity despite both the components bear positive charge. Thus forces other than electrostatic and covalent also can hold together the self-assembled multilayers.

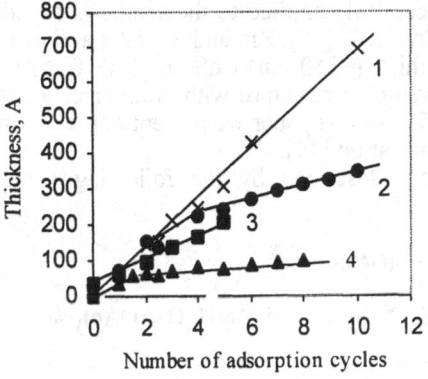

Figure 7. Ellipsometric data for self-assembly of multilayer ZnO/Polymer (1,2) and TiO2/Polymer (3,4) films on Au (1,2,4) and Si(NH₂) (3) substrates:1-(ZnO/PAN)₉ZnO; 2-(ZnO/PAH)₉ZnO; 3-(TiO₂/PAH)₄TiO₂; 4-(TiO₂/PAN)₇TiO₂

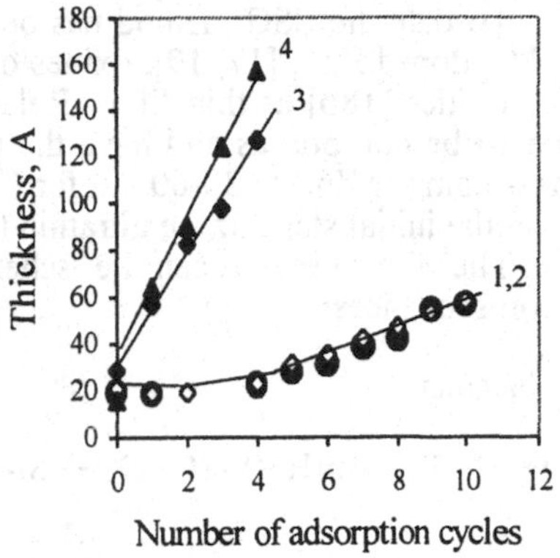

Figure 8. Ellipsometric data for SSG films: 1-Si-OH/ZnS; 2- Si-OH/(Zn, Mn)S; 3- Si(NH₂)/ZnO; 4- Si-OH/SiO₂;

3. Direct Surface Layer-by-Layer Synthesis of Semiconductor Films from Molecular Precursors

3.1 PRINCIPLES OF THE METHOD

This method typically involves a two-step chemisorption/chemical activation cycle [17-19]. One component is adsorbed or reacted chemically with molecules on the surface, but the reaction is self-limiting at the extent of a single monolayer. The chemisorbed monolayer is then activated in the second step, by reaction with an appropriate reagent [17-19] or by redox reactions in the liquid phase. Alternatively, the activation step may occur in a gas phase process, such as UV ozone oxidation or low-temperature plasma treatment.

When the solution-phase activation of the surface is used, the method is called the surface sol-gel (SSG) process [18a]. SSG can be considered as both a surface variant of the bulk sol-gel method and a liquid-phase variant of the vapor-phase atomic layer epitaxy method (ALE) [20]. A technique that is closely related to SSG and ALE is electrochemical atomic layer epitaxy, in which either the adsorption or activation step is a Faradaic redox process. SSG is based on self-limiting surface chemical reactions between the surface and each of the film components. The first examples of SSG involved the synthesis of metal sulfide particles by alternate adsorption of anions and cations from aqueous solutions, and was called SILAR (successive ionic layer adsorption and reaction) [17a]. Later Ichinose et al. generalized the technique to include molecular precursors, such as metal alkoxides, which could be adsorbed and hydrolyzed as monolayer metal oxide films [18a]. SSG combines the advantages of ALE and the bulk sol-gel method: it provides film thickness control at the Ångstrom level, and therefore allows one to tune the band gap and related properties through control of particle size.

Like bulk sol-gel synthesis, SSG does not require high temperatures or expensive high-vacuum equipment. Another strong point of SSG is its compatibility with surface patterning techniques. This has been illustrated by the fabrication of patterned TiO_2 films on Si/SiO_2 substrates bearing microcontact-printed lines of an organic polysiloxane [18b]. A distinct disadvantage is that as a low temperature synthesis technique, SSG may give low-density or incompletely crystallized films. Previously we showed low density surface oxides made by SSG can be thermally annealed to give smooth, adherent, high density thin films [18b].

To date the SSG method has been successfully applied to the synthesis of CdS, ZnS, Mn doped ZnS, [17, 19], oxides of Ti, Zr, Al, B [18], Zn, and Si [19a] and mixed Ti, Ta oxides [18b] as thin films. Relatively thick (~350 nm) CdS and ZnS films was shown to be non-porous and have the polycrystalline structure with grain size ranging from 5-6 nm for ZnS to 30-60 nm for CdS [17a]. In this paper we present our recent results on the initial stage of the ultrathin films formation [19].

The film growth can be schematically described by the following surface reaction sequences:

(I) $Zn(OAc)_2$ S^{2-} $Zn(OAc)_2$

$\}$OH $\rightarrow$ Si-O-**Zn**$(H_2O)_x(OAc)_y$ $\rightarrow$ Si-O-**Zn-S** $\rightarrow$ Si-O-**Zn-S-Zn**$(H_2O)_x(OAc)_y$ etc.

Si

(**II**) $Zn(OAc)_2$ OH^- $Zn(OAc)_2$

$\}$-NH$_2$ $\rightarrow$ Si(NH$_2$)**Zn**$(H_2O)_x(OAc)_y$ $\rightarrow$ Si(NH$_2$)**Zn-OH** $\rightarrow$ Si(NH$_2$)**Zn-O-Zn**$(H_2O)_x(OAc)_y$

Si

(III) $SiCl_4$ H_2O $SiCl_4$ H_2O

$\}$OH $\rightarrow$ Si-O-**Si-**Cl $\rightarrow$ Si-O-**Si-OH** $\rightarrow$ Si-O-**Si-O-Si-**Cl $\rightarrow$ Si-O-**Si-O-Si-OH** etc.

Si

In the aqueous reaction sequences I and II, it is assumed that the weakly coordinating acetate ions are easily displaced as ligands for Zn^{2+} by the more strongly coordinating OH^- or S^{2-} anions. The binding of Zn^{2+} ions to surface Si-OH groups (sequence I) is expected to be quite weak. On the other hand, alkylamines (sequence II) are relatively good ligands for both metal ions. Covalent bond-forming reactions such as those in sequence III are generally assumed to be irreversible in SSG [18a].

3.2. FILM GROWTH

Ellipsometry. The noticeable growth of the sulfide films is observed only after the fourth two-step adsorption cycle, when the films thickness plots become almost linear (Fig.8.1,2). The average increase in film thickness per layer is 6.2 Å and 5.7 Å for the Si-OH/ZnS and Si-OH/(Zn,Mn)S films, respectively. In contrast, for the both oxide films, Si-OH/SiO$_2$ (Fig.8.5) and Si(NH$_2$)/ZnO (Fig.8.4), good linearity in the films thickness plots is observed starting from the first adsorption cycle. The average increase in film thickness per each cycle is 24.5 Å and 35 Å for the Si-NH$_2$/ZnO and Si-OH/SiO$_2$ films, respectively. The average increments per cycle for the oxide films are substantially greater than might be expected for a monolayer-by-monolayer growth process. The ZnO and SiO$_2$ films prepared here have average thicknesses per cycle that are within the range of previously prepared oxide films [18].

It should be noted that no ZnO film formation was achieved on Si-OH substrates during 6 adsorption cycles. This implies that the ZnO film growth requires stronger linking of Zn^{2+} ions to the surface via coordination to NH_2-groups. We have found that proper choice of film growth parameters, such as the chemical composition of the surface and solvent used, is necessary for the successful formation of surface-bound nanoparticles [19]. It was shown previously that the chemical composition of the anion in a metal salt precursor and its concentration affect the rate of film growth [17a]. Similarly, temperature was shown to be an important factor in controlling the kinetics of SSG reactions [18a] and the thermodynamics of metal coordination reactions on surfaces [21].

Morphology of the films. Typical AFM images of the sulfide and oxide films prepared are shown in Fig. 9(a-d). The images of both Si-OH/ZnS and Si-OH/(Zn,Mn)S films grown in ten adsorption cycles (Fig.9 a, b) reveal close-packed layers of well-resolved rounded features about 10 – 30 nm in diameter and 4 – 9 nm in height. The similar morphology was observed for CdS films [17b]. The thickness of the Si-OH/(Zn,Mn)S film is about 4.2 nm, which is in good agreement with the average thickness found by ellipsometry, 4 nm (Fig.8.2). The surface coverage is estimated to be approximately 90% and 75% for the Si-OH/ZnS and Si-OH/(Zn,Mn)S films, respectively. An image of the Si-OH/ZnS film deposited in five adsorption cycles (not shown) also shows evenly distributed rounded features of approximately the same diameter, but their average height is lower (1-7 nm). The surface coverage in the latter case was estimated at about 80%. These data suggest that the sulfide nanoparticles form on the surface from well separated crystal nuclei, rather than from a saturated chemisorbed monolayer, as is normally the case in ALE [20]. The metal sulfide particle growth in the lateral direction is almost completed in about five adsorption cycles, after which growth in vertical direction prevails. The image of the Si-OH/SiO₂ film deposited in four adsorption cycles (Fig. 9d) shows a densely packed particle layer completely covering the surface. Such a morphology is consistent with a high density of nucleation sites for the first layer, which follows from the high density of Si-OH sites and the formation of strong covalent Si-O-Si bonds from the surface reaction of $SiCl_4$. The film surface contains features of about 30 – 60 nm in diameter, which form aggregates. The average roughness of the film is about 3.2 nm. The surface morphology of the $Si(NH_2)/ZnO$ film grown in four adsorption cycles (Fig.9c) is quite different from those described above. No well-resolved rounded features are observed. Instead, the film consists of extended (50 to 500 nm) and rather flat separate islands of different thickness, which cover about 70% of the surface. The average roughness of these islands is about 0.85 nm.

Although the mechanism of the SSG film growth process is still not understood in detail, the significant difference in the morphology of the ZnO film and the other films make us believe in the existence of at least two possible descriptions of the main events of the films growth. The first one, which is consistent with the formation of three-dimensional, rounded features, involves continuous particle growth from nuclei, and approximately follows the Ostwald model for colloids: $(ZnS)_{m+n} + Zn^{2+} + S^{2-} \rightarrow (ZnS)_{m+n+1}$ etc. In this case, the strength of bonds within the particle is greater than that of the bonds anchoring the particle to the substrate. This is the case for ZnS, and also apparently for SiO_2. In the case of the ZnO films, there are two possibilities.

340

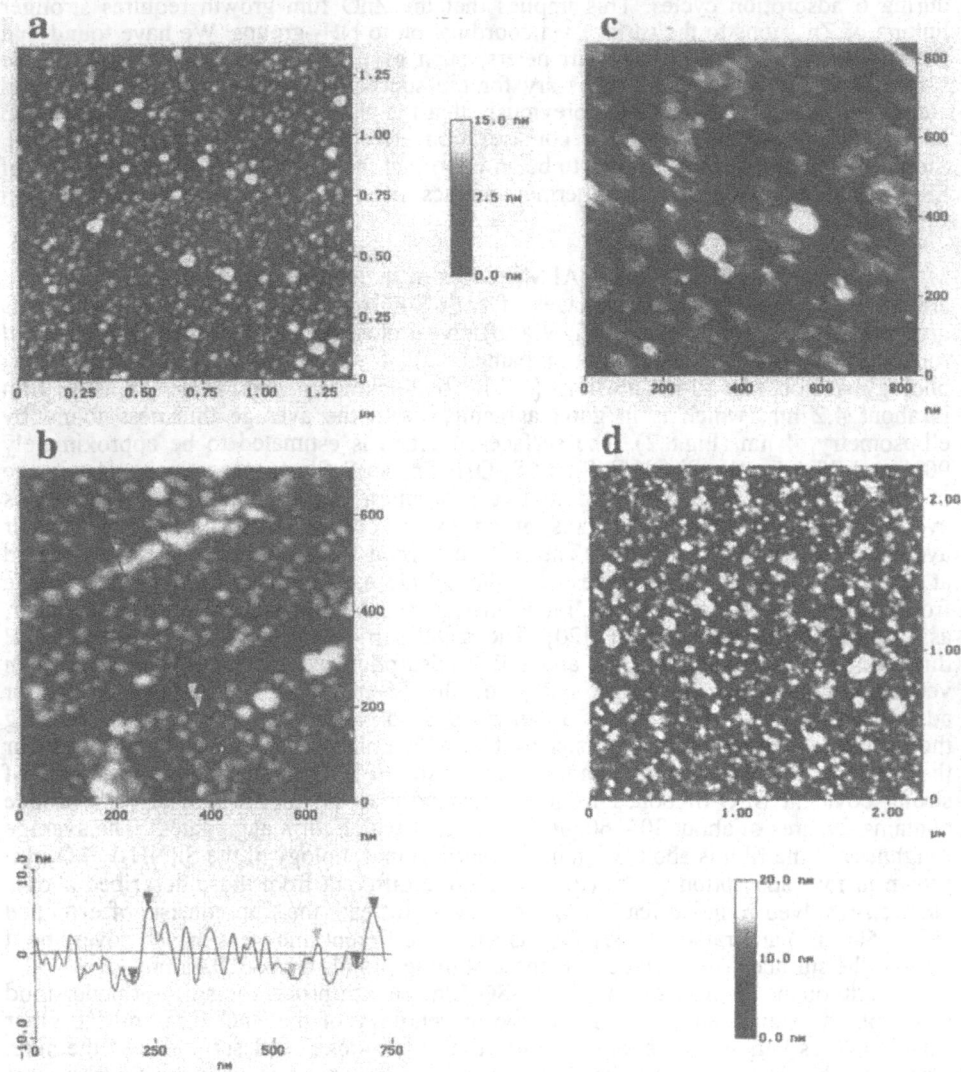

Figure 9. Tapping-mode AFM images of (a) Si-OH/ZnS film deposited in 10 adsorption cycles, (b) Si-OH/(Zn,Mn)S film deposited in 10 adsorption cycles, (c) Si(NH₂)/ZnO film deposited in 4 adsorption cycles, and (d) Si-OH/SiO₂ film deposited in 4 adsorption cycles.

The formation of smooth islands is consistent with the formation of ZnO from islands of Si(NH₂) on an otherwise unreactive surface, or with growth from sparse nucleation sites with the primary growth direction being horizontal. Based on our experience with priming layers of organosilanes, we favor the former explanation. That is, the priming monolayer formed from 4-((dimethylmethoxy)silyl)butylamine is relatively patchy, but it strongly coordinates Zn^{2+} ions. Subsequent reaction with base converts this to a surface Zn-OH film, which coordinates more Zn^{2+} ions in the next adsorption cycle to form a smooth film.

3.3 CHEMICAL COMPOSITION OF THE FILMS

The surface chemical compositions of the $Si(NH_2)/ZnO$, Si-OH/ZnS and SiOH/(Zn,Mn)S films were determined by XPS. The position of the Zn p3/2 line in the spectra of the ZnO film (1022.5 eV) and both of the ZnS-containing films (1021.8 eV) is characteristic of bulk ZnO and ZnS respectively [16].

For Si-OH/(Zn,Mn)S film, the Mn_{2p} XPS spectrum reveals a photoelectron line at 639.8 eV, which is accompanied by two shake-up satellites at 651.3 and 657.1 eV. This spectrum is characteristic of isolated paramagnetic Mn^{2+} [28]. The Zn:Mn:S surface ratio was found to be 1:0.064:0.61 (see Table). It is interesting to note that Zn:Mn ratio in the film is about 3 times higher than in starting solution, consistent with the much lower solubility product of ZnS (4.5×10^{-24}) relative to MnS (3×10^{-13}) [22].

In each of the three XPS spectra, the intense C 1s line from adventitious carbon is asymmetrical and has a shoulder at 288.1-288.3 eV indicating the presence of O-C=O bonds, and hence acetate groups [16]. Residual precursor molecules suggest incomplete sulfidization or hydrolysis. Similar residual ligands have been detected in ZnO colloids [24a] and metal oxide films prepared by SSG [18a].

An Si 2p line, which originates from the uncovered substrate, is observed for all the samples. Its envelope exhibits two distinct features: bulk silicon at 99.1 eV and oxidized silicon at 102.2 - 102.9 eV. The latter peak appears at lower energy than that observed for SiO_2 (103.3-103.7 eV) [30], and is characteristic of the Si(+4) oxidation state in different inorganic environments (e.g. SiO_2, SiOx (x<2), and SiOx(OH)y). The uncertain composition of the oxidized silicon species prevents an accurate determination of the Zn:O ratio in the $Si(NH_2)/ZnO$ film. For the Si-OH/(Zn,Mn)S sample, the total oxygen content is roughly consistent with the oxygen included in the acetate groups and the oxidized silicon, which suggests no significant content of oxide in the semiconductor nanoparticle film.

4. Optical Properties of the Layer-by-Layer Deposited Films

4.1. SSG FILMS

UV-Vis absorption spectra of ZnS and (Zn,Mn)S films deposited in five adsorption cycles on quartz slides (Fig 10a) are quite similar to those reported for ZnS and (Zn,Mn)S colloids [23a]. Band gap energies estimated from the spectra (5.17 eV and 5.07 eV for ZnS and (Zn,Mn)S, respectively) are consistent with quantum size effects, as expected from the small particle si zes observed by AFM.

The photoluminescence (PL) spectrum of the ZnS film excited a 290 nm displays a broad emission centered at 445 nm (Fig.10b.1). Photoluminescence in this spectral region is attributed to the presence of sulfur vacancies in the lattice, as previously found for ZnS colloids [23]. This emission results from the recombination of photogenerated charge carriers in shallow traps [23a,b].

The PL spectrum of the (Zn,Mn)S film (Fig.10b.3) reveals blue-green and yellow emissions at about 438 nm and 580 nm. This Mn^{2+}-based yellow emission has been observed in the photoluminescence of doped ZnS:Mn nanoparticles (see Fig 11c and [23a]) and assigned to the Mn^{2+} $^4T_1$- $^6A_1$ transition. It is known that the Mn^{2+} ion d-electron states act as luminescent centers because of strong interaction with the s-p electronic states of the ZnS nanocrystals, which are excited by bandgap absorption. The yellow emission in the PL spectrum of the (Zn,Mn)S film indicates the existence of an

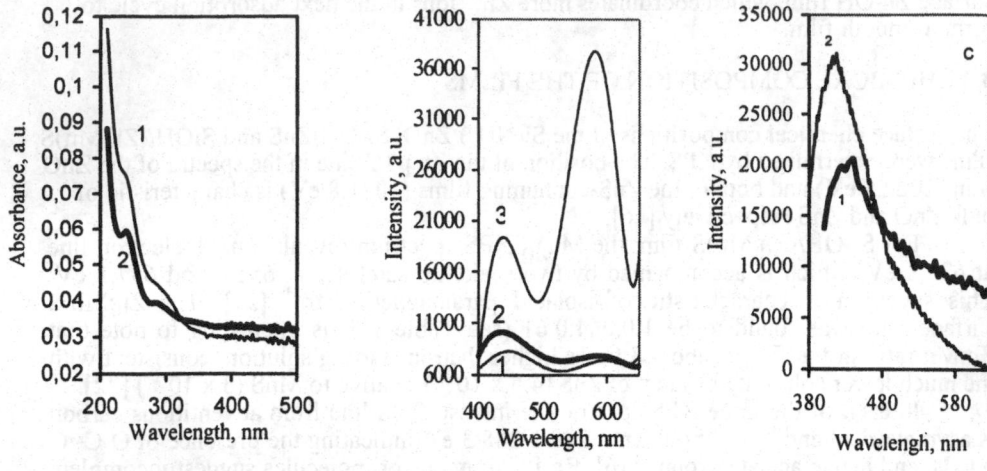

Figure 10. (a) Transmission UV-visible spectra of ZnS (1) and (Zn,Mn)S (2) films deposited in 10 adsorption cycles on quartz slides. Photoluminescence spectra of (b) 1-Si-OH/ZnS, 2-Si-OH/(ZnS)x(MnS)y, 3-Si-OH/(Zn,Mn)S films deposited in 10-12 adsorption cycles and (c) 1-Si(NH$_2$)/ZnO and 2-Si-OH/SiO$_2$ films deposited in 4 adsorption cycles. Excitation wavelengths 290 nm (b) and 340 nm (c).

energy transfer pathway that arises from electronic interaction in the (Zn,Mn)S clusters, and hence the introduction of Mn^{2+} ions into ZnS host lattice. Interestingly, the Mn^{2+} ion introduction into ZnS host lattice is observed only when starting solution contains both Zn^{2+} and Mn^{2+} ions (molar ratio 1:0.02). In case of the film preparation by successive immersing into individual solutions of Zn^{2+} and Mn^{2+} ions in following sequence: Si-OH/(ZnS)$_5$MnS(ZnS)$_3$MnS(ZnS)$_2$ (the sample is referred to as (ZnS)x(MnS)y), no significant Mn-based emission is observed (Fig. 10b.2) [19b].

In the PL spectrum of the ZnO film, a broad emission centered at 440 nm is observed (Fig.4b). As has been previously found for ZnO colloids, the position of the emission peak is strongly dependent on particles size and falls within the range of 420-560 nm [24]. The emission band at 440 nm observed in our experiments implies that the

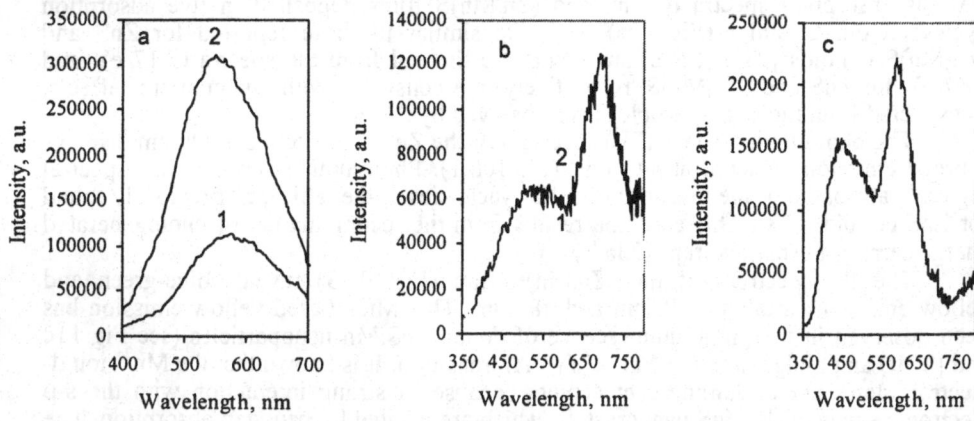

Figure 11. Photoluminescence emission spectra of (a) Si-OH/ZnO (1) and Si-OH/(ZnO/PAN)$_9$ZnO (2); (b) Au/((Zn,Mn)S/PAH)$_9$/(Zn,Mn)S (1) and ITO/((Zn,Mn)S/PAH)$_6$/(Zn,Mn)S (2) films; (c) stock (Zn,Mn)S sol. Excitation wavelengths: (a) 340 nm, (b,c) 280 nm

ZnO film is composed of very small ZnO grains with a size close to that of colloidal particles in freshly prepared sols (ca. 3 nm [24a]). This result is consistent with the smooth, featureless films observed in the AFM image.

The SiO_2 film shows a broad emission band centered at 415 nm when excited at 340 nm (Fig.10c.2). Photoluminescence in this spectral region is due to oxygen vacancies associated with electrons localized on the bridge oxygen atoms of siloxane linkages [25].

4.2. SELF-ASSEMBLED SEMICONDUCTOR/POLYMER FILMS

In the photoluminescence (PL) spectrum of Si-OH/ZnO monolayer a wide band centered at 580 nm is observed (Fig.11a.1). The band position is similar to that observed for the stock ZnO sol suggesting that no increase in the size of ZnO particles take place while they are deposited as one monolayer at a time. This suggestion is in agreement with ellipsometry and AFM results. In contrast, ZnO films deposited by dip-coating revealed an increase in particle size compared to the starting colloid [26].

PL spectrum of the $Si(OH)/(ZnO/PAN)_9ZnO$ film displays a broad emission peaked at 550 nm (Fig.11a). This blue shift in the emission from the composite multilayer ZnO/PAN film in relation to the monolayer ZnO film emission can be due to appearance of new shallow traps for electron-hole pairs. The formation of new surface states may results from chemical binding between the components of the neighbouring layers as was shown above by XPS.

The PL spectrum of the $Au/((Zn,Mn)S/PAH)_9(Zn,Mn)S$ film (Fig.11b.1) reveals blue-green and yellow emissions at about 438 nm and 580 nm. This spectrum is very close to that of stock (Zn,Mn)S sol (Fig.11c), though a blue shift in the yellow emission (~20 nm) suggests the formation of new shallow surface states due to PAH adsorption. Interestingly, the same multilayer (Zn,Mn)S/PAH films deposited onto Au and ITO substrates show quite different PL spectra. In the spectrum of $ITO/((Zn,Mn)S/PAH)_6(Zn,Mn)S$ film (Fig.11b.2), new intensive PL bands centered at around 520 and 690 nm are observed. Since this spectrum substantially differs from spectra of both $Au/((Zn,Mn)S/PAH)_9(Zn,Mn)S$ film and (Zn,Mn)S sol, one can suggest the formation of new shallow and deep surface states at the ITO/film interface. The fact that the same difference is observed in PL spectra (not shown) of bare (Zn,Mn)S films on Au and ITO substrates allow assuming strong interaction between (Zn,Mn)S nanoparticles and ITO surface. It is to note that no significant difference is observed between PL spectra of multilayer ZnO/PAN films deposited on Si(OH) and ITO substrates.

5. Applications of Self-Assembled Multilayer Inorganic/Organic Films

Thin films layer-by-layer self-assembled from inorganic nanobloks and polymer chains tend to show a regularly layered structure [8,11a,13]. Using 2D nanoparticles, for example, in some cases it is possible to prepare well-ordered (along the stacking axis) surface heterostructures that resemble bulk intercalation compounds in their XRD patterns and chemical properties [8]. A great advantage of this synthetic technique is that one can combine judiciously selected components in periodic and aperiodic structures that are not accessible by conventional techniques. Many research efforts have now been focused on making multilayer nanoparticle/polymer films with structure- and sequence-dependent functions. The utility of this approach have been established by several electronic and photonic applications, such as Coulomb blockade devices [27], rectifying diodes [9,10,12,14], electroluminescent devices [28], photon harvesting and charge separation assemblies [29a], multicomponent energy/electron-transfer cascades, which mimics some of the functions of natural photosynthetic assemblies [29b]. We present here briefly one more recent example of photoelectrically active system that composed from layer-by-layer self-assembled ZnO/polyaniline film sandwiched between ITO and Pt electrodes [10]. I-V characteristics of the multilayer ZnO/Polyaniline film were obtained under dark and illumination conditions (Fig.12). The I-V curve of the ~60 nm thick ITO-(ZnO/PAN)$_9$ZnO-Pt device (Fig.12A), which was registered in the dark, reveals rectifying behaviour. (Preliminary experiment has shown ohmic I-V characteristic for Pt/ITO/Pt structure.) The device operates in both forward (ITO "+", Pt"-") and reverse (ITO "-", Pt"+") bias modes at turn-on potentials of 3.2 V and -4 V, respectively. The resistance of the device, calculated for this voltage region, is ~10^{11} Ohm. Close to these electrical properties was previously reported for a ITO-(TiO$_2$/PAN)$_9$ TiO$_2$-Pt device [9].

Upon turning on UV illumination (λ<350 nm), a gradual increase in current with time is observed (see insert in Fig.12, bottom). I-V curve, which was measured after twenty-minute illumination (point B), shows almost ohmic behaviour (Fig.12B). The device resistance has decreased to 10^6 Ohm that is 5 orders of magnitude less than the resistance in the dark. Turning off the illumination results in a gradual current decrease; I-V curve measured in 20 min (point C) is identical to that obtained under dark conditions (point A) (Fig.12 A and C). Repeating the alternate switching between light and dark show good reproducibility ($\pm$ 5%) of current values. The photoelectrical response of the ITO-(ZnO/PAN)$_9$ZnO-Pt device is seen at the excitation wavelengths below 350 nm. One can conclude that the excitation of ZnO nanoparticles initiates the photoelectrical effect while PAN chains enable the field-driven transport of photogenerated charge carriers, which are localized in shallow and deep trap states. The carriers trapped in shallow states are more likely to leave a chain site before recombination and provide current flow. The gradual increase and decrease in the current-time curves (insert in Fig.12) (which follow turning on and turning off the UV-irradiation) may be caused by lasting saturation and emptying the trapping sites, respectively.

Importantly, neither ITO-PAN-Pt nor ITO-ZnO-Pt device does not show any photoelectrical response under our experimental conditions, though ZnO films is known to be photoelectrochemically active [26]. Also ITO-(ZnO+PAN)-Pt device, which was prepared from a mixture (1:1) of ZnO and PAN stock solutions, exhibits no significant change in their I-V characteristics under the UV-illumination. Thus the adjacent ZnO/PAN monolayers organized in the multilayer film via layer-by-layer deposition are responsible for the observed photoelectrical effect. Regularly layered structure, in which every ZnO nanoparticle and PAN macromolecule are in immediate contact with each other, appears to be necessary to provide efficient photoinduced charge injection. Whole multilayer assembly behaves as one electronic unit demonstrating new electronic properties.

6. Conclusions

It has been shown that wet layer-by-layer synthesis is experimentally simple but potentially powerful route to functional surface nanostructures. Photoluminescent ultrathin films can be prepared in 4-10 adsorption cycles by direct surface synthesis, SSG, of semiconductor films from molecular precursors, as well as by self-assembling inorganic/organic multilayers from semiconductor nanoparticles and polymer chains.

Using SSG technique, the coverage and morphology of the films can be controlled by a number of factors including the nature of the surface priming layer and the choice of solvents and solutes. The morphology of the thin films can be understood in terms of nucleation and growth phenomena, which depend on the relative strength of interactions of metal ions with the surface priming layer and with ligands introduced in the following adsorption cycle. Importantly, optical properties of the films were similar to those of sol-gel nanoparticles prepared in the liquid phase.

The quality and morphology of the multilayer inorganic/organic films are determined by the coverage in the first adsorbed inorganic monolayer, which in turn can be controlled by the chemical composition of substrate surface, the pH of the reaction media and the nature of the solvents. Gaining control over attachment chemistry, will allow goal-directed preparation of ultrathin hybrid inorganic/organic films, e.g. densely packed films for electronic applications, or porous films for applications in chemical sensors and catalysts. The latter ones may also be used as matrices for incorporation of nanoparticles and molecules (e.g. dyes). Photoluminescence emission of self-assembled multilayer semiconductor/polymer films strongly depends on the chemical composition of substrate surface and semiconductor nanoparticles. One can expect that judicious selection of the substrate and inorganic film component will enable to tune photoluminescence of the self-assembled heterostructures. The $ITO/(ZnO/PAN)_9ZnO/Pt$ device exhibits an insulator-to-conductor transition under UV irradiation. Importantly, each of the individual components was photoelectrically inactive under these experimental conditions, demonstrating that the whole multilayer assembly behaved as one electronic unit with new electronic properties. The observed electronic properties make inorganic and inorganic/organic films presented here promising candidates for application in nanoscale electronics, optoelectronics, photovoltaics, luminescent display devices and as thin film LED's.

References

1. (a) Fendler J.H., Meldrum F. (1995) The colloid chemical approach to nanostructured materials, *Adv. Mater.* **7**, 607-632; (b) Alivisatos A.P. (1998) Electrical studies of semiconductor nanocrystal colloids, *MRS Bull.* **23**, 2, 18-23; (c) Mallouk T. E., Kim H.-N., Ollivier P. J, Keller S. W. (1996) Ultrathin films based on layered materials, in G. Alberti and T. Bein (eds.), *Comprehensive Supramolecular Chemistry*, vol. 7, Elsevier Science, Oxford, UK, pp. 189-218.

2. Iler R.K (1966) Multilayers of colloidal particles, *J. Colloid Interface Sci.* **21**,569-594.

3. Decher G. (1997) Fuzzy nanoassemblies: toward layered polymeric multicomposites, *Science* **277**, 1232-37.

4. Keller S. W., Kim H.-N., Mallouk T. E. (1994) Layer-by-layer assembly of intercalation compounds and heterostructures on surfaces: towards molecular "Beaker" epitaxy, *J. Am. Chem. Soc.*, **116**, 8817-8818.

5. Colvin V.L., Golstein A.N., Alivisatos A.P. (1992) Semiconductor nanocrystals covalently bound to metal surfaces with self-assembled monolayers, *J. Am. Chem. Soc.* **114**, 5221-5230.

6. Fendler J. (1996) Self-assembled nanostructured materials, *Chem. Mater.* **8**, 1616-1624.

7. Ollivier P.J., Kovtyukhova N.I., Keller S.W., Mallouk T.E. (1998) Self-assembled thin films from lamellar metal disulfides and organic polymers, *Chem. Commun.*, 1563-1564.

8. Fang M., Kim H.-N., Saupe G. B., Miwa T., Fujishima A., and Mallouk T. E. (1999) Layer by Layer Growth and Condensation reactions of Niobate and Titanoniobate Thin Films, *Chem. Mater.* **11**, 1526-32.

9. Kovtyukhova N., Ollivier P.J., Chizhik S., Dubravin A., Buzaneva E., Gorchinskiy A., Marchenko A., Smirnova N. (1999) Self-assembly of ultrathin composite TiO_2/polymer films, *Thin Solid Films* **337**, 166-170.

10. (a) Kovtyukhova N.I., Gorchinskiy A.D., Waraksa C.C. (2000) Self-assembly of nanostructured composite ZnO/polyaniline films, *Mater. Sci&Eng.* **B69-70**, 424-430; (b) Kovtyukhova N.I., Gorchinskiy A.D., Buzaneva E. V. Zankovich S. (2000) Nanocomposite ZnO/polymer films: synthesis, morphology, electrical and optical properties, in *Advanced Materials*, Cambridge Press Publishers, in press.

11. (a) Kleinfeld E. R., Ferguson G. S. (1994) Stepwise formation of multilayered nanostructural films from macromolecular precursors, *Science* **265**, 370-372; (b) Kleinfeld E. R., Ferguson G. S. (1996) Healing of defects in stepwise formation of polymer/silicate multilayer films, *Chem. Mater.* **8**, 1575-1578; (c) Kotov N.A., Magonov S, Tropsha E. (1998) Layer by Layer self-assembly of alumosilicate-polyelectrolyte com-

346

posites: mechanism of deposition, crack resistance, and perspectives for novel membrane materials, *Chem. Mater.* **11**, 886-895.

12. Kovtyukhova N., Ollivier P., Martin B., Mallouk T., Buzaneva E., Gorchinskiy A. (1999) Layer-by-layer assembly of ultrathin composite films from micron-sized graphite oxide sheets and polycations, *Chem. Mater.* **11**, 771-778

13. Schmitt J., Decher G., Dressick W., Brandow S.L., Geer R.E., Shashidhar R., Calvert J.M. (1997) Metal nanoparticle/polymer superlattice films: fabrication and control of layer structure, *Adv. Mater.* **9**, 61-65

14. (a) Cassagneau T., Fendler J.H., Mallouk T. E., Optical and electrical characterization of ultrathin films self-assembled from 11-aminoundecanoic acid capped TiO_2 nanoparticles and polyallylamine hydrochloride, *Langmuir*, in press; (b) Cassagneau T., Mallouk T. E., Fendler J. H. (1998) Layer-by-layer assembly of Zener diodes from conducting polymers and CdSe nanoparticles, *J. Am. Chem. Soc.* **120**, 7848-7859.

15. Monkman A.P., Stevens G.C., Bloor D. (1991) X-ray photoelectron spectroscopic investigations of the chain structure and doping mechanisms in polyaniline, *J. Phys. D: Appl .Phys.* **24**, 738-749.

16. (a) Nicolau Y.F., Menard J.C. (1988) Solution growth of ZnS, CdS and $Zn_{1-x}Cd_xS$ thin films by the successive ionic-layer adsorption and reaction process: growth mechanism, *J. Crystal Growth* **92**, 128-142; (b) Vogel R., Rohl K., Weller H. (1990)Sensitisation of highly porous,polycrystalline TiO_2 electrodes by quantum sized CdS, *Chem.Phys.Lett.* **174**, 241-246.

17. (a)Ichinose I., Senzu H., Kunitake T. (1997) A surface sol-gel process of TiO_2 and other metal oxide films with molecular precision, *Chem.Mater.* **9**, 1296-1298; (b) Fang M., Kim C. H., Martin B.R., Mallouk T.E. (1999) Surface sol-gel synthesis of ultrathin titanium and tantalum oxide films, *J. Nanoparticle Res.* **1**, 43-49.

18. (a) Kovtyukhova N.I., Buzaneva E.V., Waraksa C.C., Martin B.R, Mallouk T.E. (2000), Surface sol-gel synthesis of ultrathin semiconductor films, *Chem. Mater.* **12**, 383-389; (b)) Kovtyukhova N.I., Buzaneva E.V., Waraksa C.C., Mallouk T.E. (2000) Ultrathin nanoparticle ZnS and ZnS:Mn films: synthesis, morphology, photophysical properties, *Mater. Sci&Eng.* **B69-70**, 411-417.

19. Suntola T. (1989) Atomic layer epitaxy, *Mater. Sci. Rep.* **4**, 261.

20. Bell C., Arendt M., Gomez L., Mallouk T.E.(1994) Growth of lamellar Hofmann Clathrate films by sequential ligand exchange Reactions: assembling a coordination solid one layer at a time, *J. Am. Chem. Soc.* **116**, 8374-75.

21. Moulder J.F., Stickle W.F., Sobol P.E., Bomben K.D., *Handbook of X-ray Photoelectron Spectroscopy*, Perkin-Elmer Co, Phys. Electr. Div., Minnesota USA.

22. Meites L. (1963) *Handbook of Analytical Chemistry*, McGraw Hill, New York.

23. (a) Sooklal K., Cullum B. S., Angel S. M., Murphy C. J.(1996) Photophysical properties of ZnS nanoclusters with spatially localized Mn^{2+}, *J.Phys.Chem.* **100**, 4551-4555; (b) Becker W.G., Bard A.J. (1983) Photoluminescence and photoinduced oxygen adsorption of colloidal zinc sulfide dispersions, *J. Phys.Chem.* **87**, 4888-4893; (c) Rabani J. (1989) Sandwich colloids of ZnO and ZnS in aqueous solutions, *J. Phys.Chem.* **93**, 7707-7713.

24. (a) Spanhel L., Anderson M.A. (1991) Semiconductor clusters in the sol-gel process: quantized aggregation, gelation, and crystal growth in concentrated ZnO colloids, *J.Amer.Chem.Soc.* **113**, 2826-2833; (b) Kamat P.V., Patrick B. (1992) Photophysics and photochemistry of quantized ZnO colloids, *J.Phys.Chem.* **96**, 6829-6834.

25. Eremenko A., Smirnova N., Samchuk S., Chuiko A. (1992) Investigation of silica surface chemistry by luminescent probes method, *Colloids and Surfaces* **63**, 83-92.

26. Hotchandani S., Kamat P.V.(1992) Photoelectrochemistry of semiconductor ZnO particulate films, *J .Electrocem. Soc.***139**, 1630-34.

27. Feldheim D. L., Grabar K. C., Natan M. J., Mallouk T. E. (1996) Electron transfer in self-assembled inorganic polyelectrolyte/metal nanoparticle heterostructures, *J. Am. Chem. Soc.* **118**, 7640-7641.

28. Gao M., Richter B., Kirstein S., Mohwald H. (1998) Electroluminescence study on self-assembled films of PPV and CdSe nanoparticles, *J. Phys. Chem.* **102**, 4096-4103.

29. (a) Kaschak D. M. Mallouk T. E. (1996) Inter- and intralayer energy transfer in zirconium phosphate-poly(allylamine hydrochloride) multilayers: an efficient photon antenna and a spectroscopic ruler for self-assembled thin films, *J. Am. Chem. Soc.* **118**, 4222-4224; (b) Kaschak D. M., Lean J. T., Waraksa C. C., Saupe G., Usami H., Mallouk T. E. (1999) Photoinduced energy and electron transfer reactions in lamellar polyanion/polycation thin films: towards an inorganic "Leaf", *J. Am. Chem. Soc.*, **121**, 3435-3445.

SOFT X-RAY SPECTROSCOPY AS A PROBE OF THE ELECTRONIC STRUCTURE OF NANOSTRUCTURED SOLIDS

STEFAN EISEBITT AND WOLFGANG EBERHARDT
IFF, Forschungszentrum Jülich, 52425 Jülich, Germany
S.Eisebitt@fz-juelich.de

Abstract

Soft x-ray absorption and emission spectroscopy is being used in order to investigate the electronic structure of nanostructured materials. Due to the photon-in photon-out nature of the processes, these techniques are well suited for the study of nanostructures and nanocomposite materials. In this overview, results are presented on the local partial density of occupied and unoccupied states in semiconductor systems (porous silicon, CdS nanocrystallites) and molecular nanostructures based on carbon cages (doped single wall carbon nanotubes, $(C_{59}N)_2$). The additional possibility to momentum selective information by resonant inelastic soft x-ray scattering is explained and discussed in the context of nanostructures.

1. Introduction

Nanostructured materials differ in a variety of physical properties from their bulk counterparts. In particular, the *electronic* structure can be altered in nanosized structures. This change is partly due to the increasing influence of the surface. But also the bulk of the nanostructure can posses altered electronic properties when the size of the structure becomes comparable to the electron wavelengths, as the electrons will then begin to feel the effects of *quantum confinement*. While the investigation of this size regime is interesting from a basic science point of view, it is also the prospect for applications which drives materials science into the nano regime. From both the applied and fundamental viewpoint, it is important to understand the *basic* electronic properties of nanostructures. The basic electronic properties as given by the quantum numbers of all the electrons in the system will ultimately determine the positions of the atoms in the structure as well as the macroscopic mechanical, electrical and optical properties, which are usually of interest when nanostructures are designed for application purposes. While it is practically impossible to know the quantum numbers of *all* electrons simply because there are too many electrons even in a chunk of a solid only $(1 \text{ nm})^3$ big, some closely related quantities are accessible. For a broad band solid in which the electrons can be described as Bloch states, the energy-wavevector dispersion $E(\mathbf{k})$ and the resulting den-

347

L. Pavesi and E. Buzaneva (eds.), Frontiers of Nano-Optoelectronic Systems, 347–362.
© *2000 Kluwer Academic Publishers. Printed in the Netherlands.*

sity of electronic states (DOS) are such basic properties, which are in principle accessible experimentally. Most commonly, (angle resolved) valence band photoemission spectroscopy ((AR)PES) is used to obtain this information in an experiment. However, in the investigation of nanostructures one frequently encounters one or several of the following difficulties, which make an experiment based on PES difficult:

- the nanostructure is buried in a matrix or below a cap layer or is part of a more complex nanocomposite material
- the nanostructure constitutes only a small fraction of the entire sample material
- the sample containing the nanostructure does not conduct well and is subject to inhomogeneous or time dependent electrostatic charging
- the individual nanostructure does not come in the form of a single crystal but is polycrystalline or amorphous
- different nanostructures within the nanocomposite have different orientation

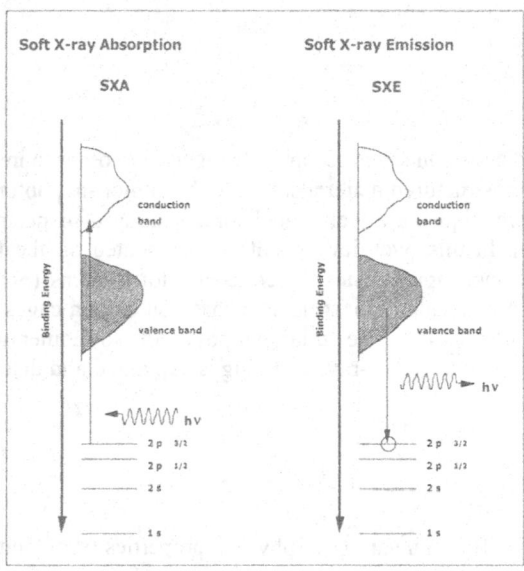

Figure 1. Schematic representation of the core electron absorption process (left) and the soft X-ray emission process (right). Both transitions are described by optical dipole matrix elements.

The difficulties for PES arise due to the fact that it is a surface sensitive, non atom selective technique, which is influenced by electric and magnetic fields. The situations listed above are often caused by the production process of the nanostructure or by the necessary integration of the nanostructure into a larger assembly in order to perform a desired function. It is therefore desirable to have a tool that can probe the electronic structure even under these circumstances. Soft x-ray emission (SXE) spectroscopy [1] and resonant inelastic x-ray scattering (RIXS) are *photon-in photon-out* based techniques which are well suited for the investigation of nanostructures in "real world" arrangements, as they combine atom selectivity and bulk sensitivity. SXE in conjunc-

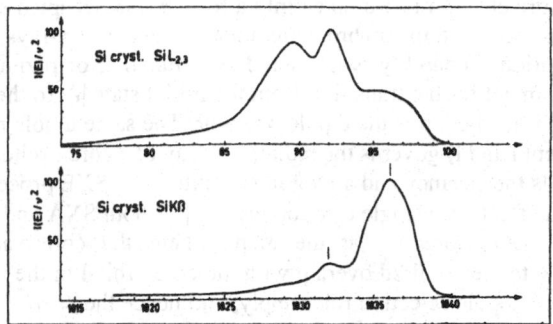

Figure 2. Illustration of the symmetry selectivity of SXE due to the dipole selction rules. Transitions to a Si 1s core level (bottom) probe p-type VB states, while transitions to a 2p level (top) probe (s+d)-type VB states. As a result, SXE probes the *partial* DOS. Reproduced from [2]

conjunction with soft x-ray absorption (SXA) spectroscopy yields information about the DOS, while E(**k**) information can be obtained by RIXS. It is the purpose of this paper to describe these techniques in the context of nanostructure research.

2.   Density of States in Nanostructures

X-ray absorption and emission processes in the vicinity of a core level absorption threshold are well established techniques in order to study the DOS in broad band solids. The processes are depicted schematically in Fig. 1.

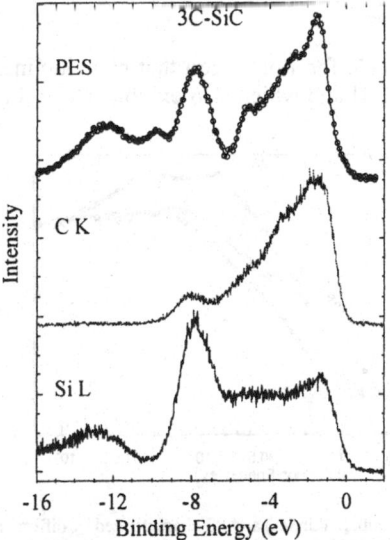

Figure 3. SXE probes the *local* DOS. In a compound material, SXE allows the investigation of the electronic states locally at the different atoms, as the emission energy will be different due to the different core level energies. In this way, the C 1s SXE spectrum measures only the (p-type) states centered at the C atoms, while the Si 2p spectrum measures the ((s+d)-type) DOS at the Si atom. In contrast, a PES measurement will approximate the total DOS, weighted by excitation cross-sections. Reproduced from [5].

In SXA, a core electron is promoted into a formerly unoccupied state by the absorption of a photon. Using a monochromatic, tunable source of x-rays such as a synchrotron, the transition probability is measured as a function of photon energy. The transition matrix element for the transition from the initial state |i> to the final state |f> is given by $< f \mid p \mid i >$, where p is the dipole operator. The same dipole matrix element (except with different i and f) governs the radiative decay of a core excited state, when a valence electron fills the vacancy and a photon is emitted in a SXE process. Neglecting electron correlation effects in a single electron picture [3], both SXA and SXE spectroscopy measure the DOS *locally* at the atom where the core vacancy is created/annihilated due to the required overlap with the core orbital in the matrix element (Fig.3). Furthermore, dipole selection rules apply, and hence the *partial* DOS is measured, e.g. if a s-type core vacancy is prepared, only p-type electrons can contribute to the emission (Fig.2). SXA probes the *unoccupied* local partial DOS (LPDOS), while SXE probes the *occupied* LPDOS. Neither SXA nor SXE alone are capable of probing momentum information such as the bandstructure of a solid. In the remainder of this section, we present several examples for the utilization of these techniques for the investigation of nanostructured solids. SXA and SXE are performed in vacuum chambers at a soft x-ray beamline of synchrotron radiation source. SXE spectroscopy requires a secondary spectrometer in order to energy-analyze the emitted photons. In the 50eV – 1000eV photon energy range, this is typically a spherical reflection grating instrument in a grazing incidence Rowland geometry, which combines the dispersion properties of the grating with the focussing properties of a spherical mirror. Using a 2D position sensitive detector, the entire emission energy spectrum can be recorded in a parallel mode. Some details of SXA and SXE spectroscopy will be discussed when they come up in the context of the examples.

2.1. POROUS SILICON

Porous silicon (PS) is a remarkable form of silicon that photoluminesces efficiently in the visible part of the spectrum.[4] and which also exhibits electroluminescence. In an

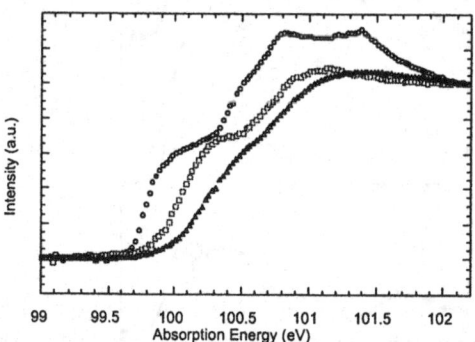

Figure 4. Si $2p_{3/2,1/2}$SXA spectra for bulk Si (circles) and PS characterized by differnt nanoporous structure sizes. A blueshift of the CB minimum can be observed [6].

etching process, silicon structures ranging from the μm to the n m range are created. With a probing depth of about 0.1 μm (in bulk silicon) we have preferentially investigated the nanoporous Si region residing on top of the microporous Si backbone in order to study the quantum size effects on the electronic structure.

Porous silicon (PS) samples were prepared electro-chemically from n-type silicon wafers.[6] During the etching process the samples were illuminated with white light filtered by different long-wavelength-pass filters, in order to provide holes for the electrochemical reaction. As a result, samples with different size distributions were generated. If the band gap of the resulting Si structure is larger than maximum light energy $h\nu_{max}$, the etching can not proceed efficiently. An increasing blue shift of the visible photoluminescence was observed for samples produced using increasing $h\nu_{max}$. Shifts of the CB minimum and the VB maximum as a function of the structure size can be observed in the SXA and SXE spectra [6].

SXA spectra at the Si $L_{3,2}$ edge are presented in Fig 4. For bulk Si (circles) the typical absorption spectrum showing the spin-orbit split L-edge and various DOS features is observed [7]. For the PS samples (squares, triangles) the CB edge is found to be shifted

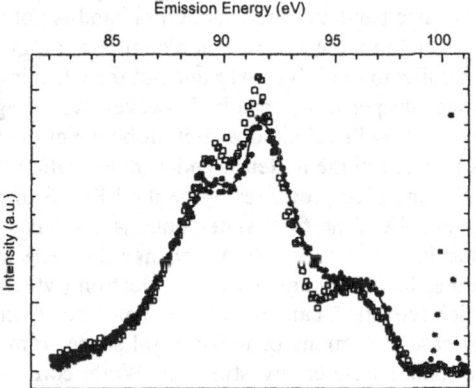

Figure 5. SXE spectrum of the entire VB of bulk Si (open squares) and a PS sample (filled circles). The spectra were excited below the $2p_{1/2}$ absorption threshold. Diffuse reflection of the soft x-rays used for excitation into the Rowland spectrometer can be observed for the PS sample, due to ist surface roughness. Changes of the LPDOS are observable from the VB maximum up to about 1o eV binding energy [6]

towards increasingly higher energy. The spin-orbit features of the absorption edge are washed out due to inhomogeneous broadening, reflecting the size distribution of the generated structures. Changes in the density of occupied states can bee seen in the VB overview in Fig.5.

For the spectra, the excitation energy was chosen to be below the Si L_2 absorption edge as determined from the SXA spectra. Due to this high resolution in the excitation only Si $2p_{3/2}$ core holes are created and thus only transitions from the valence band to the $2p_{3/2}$ component of the spin-orbit split Si 2p core level contribute to the spectra, facilitating the analysis in a VB density of states (DOS) picture. In Fig.6, the correlation

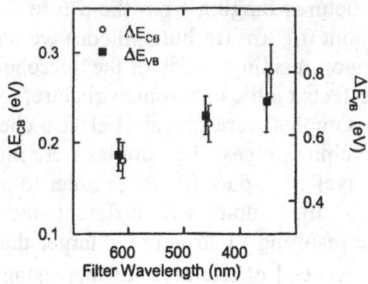

Figure 6. Shifts of the VB maximum (squares, right scale) and the CB minimum (circles, left scale) in PS relative to to bulk Si as a function of the optical filter cut-off wavelength used in the etching process.
[6]

between the synthesis parameters and the observed CB minimum and VB maximum shifts is evident.

This example demonstrates that by combining SXE and SXA, the consequences of quantum confinement for both the occupied and unoccupied states can be studied. The information on the valence band and the conduction band is not convoluted (as in optical absorption) but referred to the core level as a common energy reference. While here the focus was on the shifts of the VB maximum and the CB minimum, one can of course also observe changes deeper in the bands. However, regarding the LPDOS information one has to keep in mind that SXA does not probe the ground state properties, if electron correlation is relevant in the material under investigation. According to the final state rule [8] the SXA and SXE processes probe the LPDOS of the *final* state of the respective transition. For SXE, the final state contains a valence hole, which in a broad band material delocalizes fast and does not change the DOS significantly. The SXA final state on the other hand contains a valence electron (which also delocalizes fast) and a *core hole* which remains localized at the atomic site at which the measurement takes place. In this case, deviations of the spectral shape from the LPDOS may occur, depending on the material under investigation. While core exciton effects are small in silicon, they will become apparent in the investigation of CdS nanocrystallites.

It is a general advantage of the study of a decay process, that a specific state can be prepared in the previous excitation process. In the PS example, we made use of selective excitation to only produce Si $2p_{3/2}$ core vacancies. In a similar fashion, one can selectively excite atoms in a specific chemical state, using the chemical shift experienced by a core level due to the formation of a chemical bond. In this way, one can *e.g.* distinguish between Si and SiO_x, which makes it possible to selectively investigate Si nanostructures in a SiO_2 matrix [9].

2.2. CdS NANOCRYSTALLITES

Although it was possible to study the effects of quantum confinement in porous silicon, the drawback of this material for an investigation of the quantum size effects on the electronic structure is, that one has to deal with a distribution of different sizes within one sample. For a basic study, it is of course desirable to investigate material with perfect size control. Such semiconductor structures in the nm range can now *e.g.* be produced by wet chemistry. One example for this class of materials are CdS nanocrystallites. The nanocrystallites consist of a core of stochiometric CdS which is surrounded by organic ligands. We have investigated [10] a nanocrystallite (NC) of 40 Å diameter [11], and the more molecule-like NCs $Cd_{32}S_{14}(SR)_{36}$ (17.5 Å) [12] and $Cd_{17}S_4(SR)_{26}$ (13.5 Å) [13]. To extend the size range into a truly molecular regime, we also studied a pure Cadmium thiolate whose single crystal structure consists of $Cd_8 (SR)_{16}$ units [14].

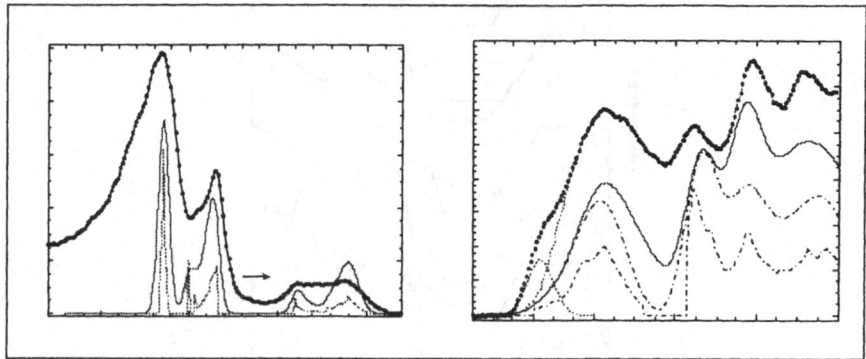

Figure 7. S $p_{3/2}$ SXE (left) and S $p_{3/2, 1/2}$ SXA compared to a calculation of the total DOS at the S atom[Xu]. The S-DOS has been broadened to account for the experimental resolution. In the SXA case, in order to simulate transitions from $2p_{3/2}$ and $2p_{1/2}$, the calculated DOS was superimposed with itself in an intensity ratio of 2:1 and after an energy shift corresponding to the S 2p spin-orbit splitting [5]. The lowest energy feature in the SXA spectrum is not a DOS feature but due to a core exciton [10].

For CdS bulk material, we compare the calculated total DOS at the sulfur atom [15] to the SXE and SXA spectra in Fig.7 [5]. The agreement is very good, when we take into account (a) that SXE and SXA to/from a S 2p level measure selectively the states of s or d symmetry (while the theory calculates the total S DOS. To partially compensate for that, the calculated S-DOS in the upper valence band – which is dominated by p-type states - was divided by a factor of eight), (b) that the SXE spectra experience an increasing final state lifetime broadening for states of larger binding energy, (c) that the SXA spectra are the superposition from the spin-orbit split $2p_{3/2}$ and $2p_{1/2}$ states, and finally (d) the experimental resolution. Only one feature in the SXA spectrum can not be reproduced by the calculation after broadening and spin-orbit superposition: the shoulder at the absorption threshold. This feature is due to a core exciton, which is formed in the absorption process [5]. As we will see below, the core exciton is affected by the quantum confinement as well.

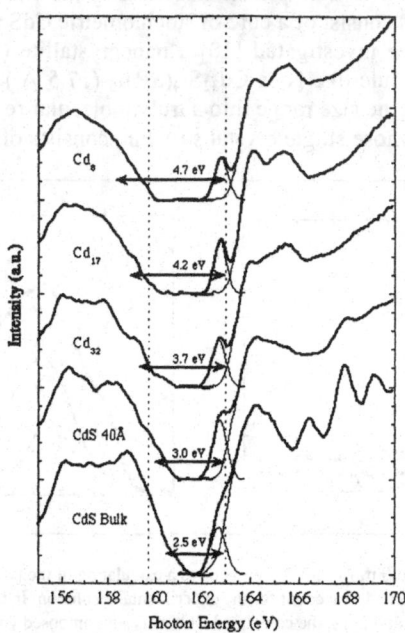

Figure 8. The SXE and SXA spectra in the vicinity of the band gap plotted on a common energy scale(defined by the S $2p_{3/2}$ binding energy) for CdS bulk material (bottom) and CdS nanocrystallites of (from bottom to top) decreasing size. The core excitonic absorption and the band absorption have been separated in the SXA spectra. The superimposed arrows indicate the results from optical absorption spectroscopy [5]. An increase of the band gap with decreasing particle size is observed and can be separated into VB and CB contributions. Furthermore, the core exciton binding energy (as given by the energy separation of the excitonic to the band absorption in SXA) increases for decreasing particle size. The effects can be described quantitatively by a quantum confinement model using a *finite* potential well [10].

The S L_3 SXE spectra and the S $L_{3,2}$ SXA spectra for the different samples (labeled by the number of Cd atoms per cluster) are shown on a common energy scale in the vicinity of the band gap in Fig.8 For decreasing particle size, an increase of the band gap can be observed. This opening of the band gap can be separated into the VB and CB contributions, as both are determined in separate experiments with the S $2p_{3/2}$ core level binding energy as a common energy reference. These shifts can now be compared with experimental predictions. Quantitative agreement is be obtained in a quantum confinement model for an electron in a quantum well, calculated in the effective mass approximation, if the well is not modeled with infinitely high potential walls but with a finite height potential of reasonable depth.[10] The leakage of the electron wavefunctions out of the potential well decreases the confinement, leading to quantitative agreement with the SXA/SXE data. As an additional quantum confinement effect, we observe an increasing binding energy of the *core* exciton with decreasing particle size. The core exciton is present in the final state of the SXA process and is formed by the S $2p_{3/2}$ core hole and the former S $2p_{3/2}$ electron which has been promoted into the CB.

2.3. DOPING IN SINGLE WALL CARBON NANOTUBES

Carbon nanotubes have in recent years attracted increasing interest as a new modification of carbon, related to the fullerenes and graphite. Especially for single wall carbon nanotubes (SWNTs) interesting electronic properties have been predicted early on. [16-18] For these nanotubes, the electronic structure strongly depends on the chirality vector (n,m) defining the type of nanotube: (n,n) tubes ("armchair" type) are predicted to be metallic, while (n,m) tubes with n,m are wide-gap or narrow-gap semiconductors, depending on the particular m and n. If 2n+m or n+2m is an integer multiple of 3, the SWNT is predicted to be a narrow gap semiconductor. This behaviour can be understood in a straight forward quantum confinement approach starting from the bandstructure for a single graphene sheet [19]. Only recently it has become possible to synthesize larger amounts of SWNTs with high purity and narrow distributions in the tube diameter, making more detailed experimental studies of the electronic structure as a function of tube size and/or chirality possible. While the choice of the geometry of the pristine SWNTs by suitable preparation conditions can be used to influence the electronic structure, an additional handle for materials design is doping of SWNTs.

Purified SWNT material ("buckypaper") from the Smalley group [20] was doped interstitially with K using a SAES getter source in ultra high vacuum [21]. The pristine SWNT material is a mixture of room temperature metallic ((49±7)%) and semiconducting SWNTs ((51±7)%), as quantified by scanning tunneling spectroscopy at many different locations on the buckypaper [22]. The individual SWNTs aggregate in bundles and ropes as can be seen in scanning tunneling microscopy images [22]. These metastructures form a disordered, spaghetti-like network which results in a sample that macroscopically is similar to a thin sheet of paper. The doping process was monitored by recording SXA spectra containing the C 1s and K 2p absorption edges. The electronic structure of the SWNT carbon cages has been monitored by the C 1s SXA and SXE. As in the previous examples, the absorption and emission energy scales are referred relative to each other with high accuracy by calibrating the Rowland spectrometer with the elastically scattered light from the synchrotron beamline. In Fig. 9, we can therefore plot both experiments on a common energy scale. In both the SXA and SXE spectra, differences due to the doping can be observed.

For the pristine sample, the most prominent feature in the SXA spectrum in the vicinity of the absorption threshold is the π^* resonance, where the C 1s electron is promoted into unoccupied π states. In the doped sample, we observe two peaks π^* and $\pi^{*'}$. $\pi^{*'}$ is located at 1.25 eV lower absorption energy. During doping process, the π^* peak decreases continuously and the $\pi^{*'}$ peak emerges. We interpret these two peaks as being due to contributions of undoped (π^*) and doped ($\pi^{*'}$) material to the spectrum, the energy difference is a result of the different screening in doped and undoped SWNTs, in agreement with similar observations in K doped C_{60} [23]. The smaller fwhm of $\pi^{*'}$ as compared to π^* suggests that the low energy states are being occupied by the doping electron, as expected. In the occupied states, the influence of doping can be observed by SXE without disturbing core excitonic effects, as there is no core hole present in the final state of the SXE process [21,24,25]. The states close to the Fermi energy (E_F) which have been newly occupied due to the doping are directly visible in the SXE spectra as a peak at around 291 eV emission energy. From the SXE and SXA data one can

356

thus conclude that SWNTs can be doped by potassium in a way similar to the "traditional" doping in bulk semiconductors where the DOS at E_F is increased in the host material, in agreement with recent electron energy loss spectroscopy (EELS) results on the unoccupied states [26].

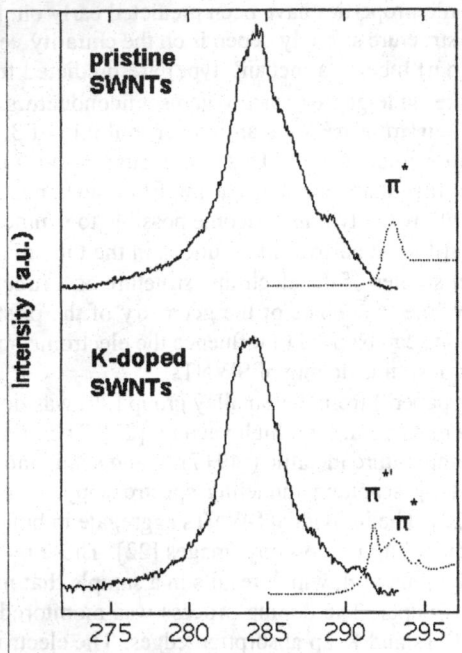

Figure. 9. C 1s SXE (left) and SXA (right) spectra of pristine (top) and K doped (bottom) single wall carbon nanotubes, reflecting the occupied and unoccupied LPDOS on the carbon cage [21].

2.4. $(C_{59}N)_2$ - A MOLECULAR NANOSTRUCTURE

A variety of fullerenes can now be produced with some additional functionalization, where different atoms or entire functional groups are introduced into the carbon network. A simple example is $C_{59}N$, where one carbon atom of a C_{60} molecule has been substituted by a nitrogen atom. As the N atom has one electron more than a carbon atom, this configuration was investigated early on as a means to dope a fullerene on-cage [27]. In the solid, $C_{59}N$ dimerizes to $(C_{59}N)_2$, by forming a C-C bond between C atoms which are adjacent to the N atom.

We have carried out a study of the local electronic structure at the N atoms in $(C_{59}N)_2$ as compared to the electronic structure on the carbon cage using SXA and SXE [28]. For the unoccupied states (not shown), we observe no significant empty DOS at the N atom corresponding in energy to the low-energy carbon cage t_{1u} states, in agreement with the EELS results [29]. The C 1s and N 1s SXE spectra from $(C_{59}N)_2$ are presented in Fig. 10, allowing us to study the occupied electronic states atom selectively. The spectra are referred to a common binding energy scale by subtracting the respective core level binding energies from the SXE emission energies.

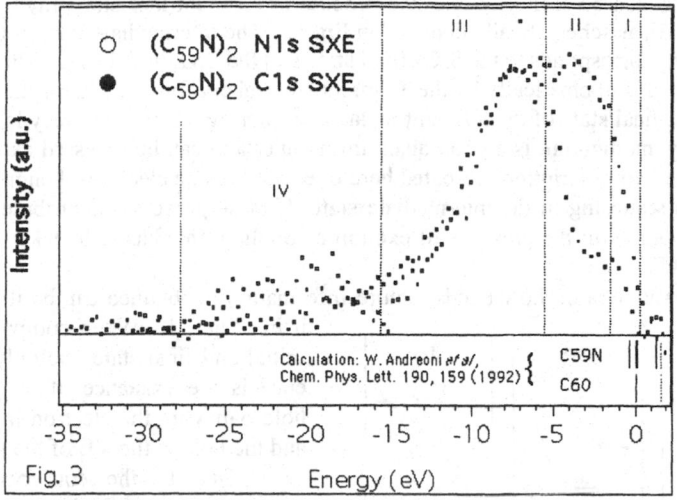

Figure 10. SXE performed atom selectively at the N and C atoms in $(C_{59}N)_2$, representing the local p-DOS at the respective atomic sites. The influence of the „doping" electron can be seen in the vicinity of E_F in region I, in agreement with theoretical results reproduced at the bottom [27].

The core level binding energies were determined by x-ray photoelectron spectroscopy and are in agreement with the values reported in Ref. 29. At the valence band maximum (region I), we observe a clear difference between the C1s and N1s spectra. The DOS locally at the N atom extends to lower binding energies than the C DOS. This small but significant difference is in good agreement with results from electronic structure calculations of $C_{59}N$ and C_{60} [27], which we have reproduced at the bottom of the figure (solid bars: occupied states, dashed bares: empty states. Zero energy in both C_{60} and $C_{59}N$ corresponds to the C_{60} HOMO). The calculations show that the N-induced highest occupied molecular orbital (HOMO) in $C_{59}N$ is 1.35 eV higher in energy than the HOMO in C_{60}. We are therefore directly observing the influence of the doping electron at the N site. Our observations are experimental proof for the statement that the electron is mainly localized at the N site. The "extra" electron introduced by the N atoms can therefore *not* act like a doping electron which can contribute significantly to the conductivity in the system. The intensity differences between the two spectra in regions II and IV are due to N 2p hybridized states and distortion induced C 2s contributions, respectively, which will be discussed elsewhere [28]

3. Wavevector Information

As pointed out above, neither SXA nor SXE alone are capable of probing momentum information such as the bandstructure of a solid. The reason for this is that in both cases, a core vacancy is involved: in the final state of SXA and in the initial state of SXE. As a result, the translational symmetry of the crystal is broken at exactly the point where the measurement takes place and momentum is not a good quantum number any more. It was realized recently, that bandstructure information in solids can be extracted when

SXA and SXE can be seen as a one step process of resonant inelastic x-ray scattering (RIXS) [30-33], as schematically depicted in Fig.11. The intermediate state of the resonant scattering corresponds to the SXA final state and the SXE initial state. The scattering is described mathematically by the Kramers-Heisenberg formula.[32,34] In both the initial and the final state of the resonant inelastic scattering, no core vacancy is present. Consequently, momentum is a good quantum number and can be accessed experimentally. The one step description presented here does not include electron-phonon or electron-electron scattering in the intermediate state. If these processes contribute significantly, for example in the presence of excitonic coupling, this has to be taken into account as well [33].

The basic idea on how bandstructure information is obtained can be illustrated using Fig. 11. When comparing the initial and final state, the only difference is the existance of an electron hole pair with the electron in the CB and the hole in the VB of the solid.

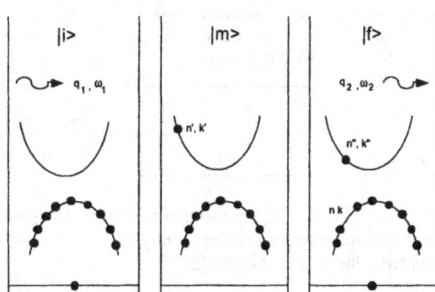

Figure 11. The effects of momentum conservation are observable, if SXA (i→m) and SXE (m→f) have to be described by a coherent one-step process of resonant inelastic x-ray scattering.. As the momentum transfer of the soft x-ray photon is small compared to the electron momenta, vertical excitations are created (modulo a reciprocal lattice vector) [33].

Due to the conservation of momentum, the CB electron and the VB hole have to have a momentum difference corresponding to the momentum transfer of the soft x-ray photon, modulo a reciprocal lattice vector. As the momentum transfer by the soft x-ray photon is small compared to the electron momenta, electron and hole have to have the same crystal momentum. In a simplifying two step way of speaking [33] we can say that the position of the electron in the CB bandstructure after the SXA event will select which VB electrons can participate in the SXE process.

The experimental handle to access bandstructure information is the excitation energy. For a given excitation energy, the core electron can only be efficiently promoted to certain unoccupied E(k) states, such that energy is conserved ('resonance condition'). As illustrated in Fig. 12, this restriction selects the regions in the Brillouin zone from which VB electrons may contribute to the SXE process. Generally, not all the intensity in the experimental RIXS spectra can be accounted for by the RIXS model outlined above. Intensity is found at energies related to occupied states which should not contribute to the spectra, if the excited electron in the intermediate state was to determine the possible crystal momenta for the emission perfectly. In fact, RIXS spectra are usually separated into a component corresponding to the LPDOS k-unselective or incoherent fraction) and the remaining component (k-selective or coherent fraction), which is then analyzed in the momentum conservation framework. Mechanisms giving rise to k-unselective contributions are discussed in detail in Ref. [33]

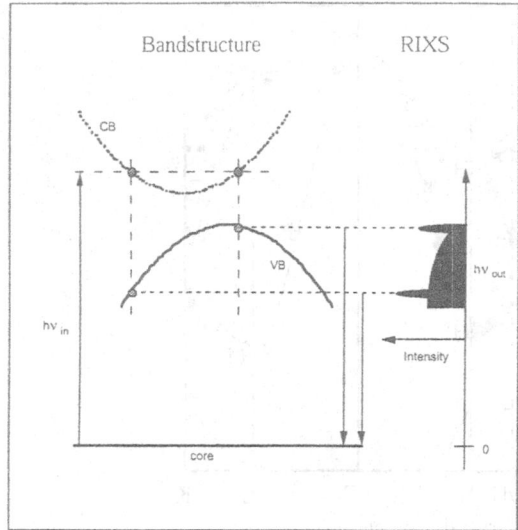

Figure 12. Schematic illustration on how k-information can be obtained in a RIXS experiment. Via the energy of the incident photon an isoenergy plane of unoccupied states is selected. Occupied states with the same wavevectors will contribute preferentially to the scattering [33]

As an example, we reproduce in Fig 13 bandmapping results using RIXS in a bulk crystal of 3C SiC [35]. RIXS allows to trace the E(k) bands while retaining the advantages of SXE spectroscopy, namely bulk sensitivity, atom and symmetry selectivity, and insensitivity to electric and magnetic fields. A drawback is the fact that the momentum selectivity is obtained indirectly via an energy selection, as a result an isoenergy surface of states in the Brillouin zone is probed and input about the dispersion E(k) of the unoccupied states is needed in order to extract information on the unoccupied states. While these are severe drawbacks it is important to note that in the context of nanostructured materials, RIXS might often be the only technique which is applicable, as the main competitor ARPES suffers from the problems outlined in the introduction.

To our knowledge, RIXS has not yet been applied to nanostructured materials for bandmapping purposes. One interesting physics question in the context of nanosized (broad band) material is where between the bulk regime and the atomic regime the description of electronic structure in terms of E(k) breaks down. Even without theoretical input, a deviation from the bulk behavior should be noticable. Furthermore, RIXS can serve as an experimental check of band structure calculations for nanostructures, similarly as SXE and SXA results can be compared to theory on the k-integrated DOS level. For solids formed by regular arrays of nanostructures such as in potential photonic band gap materials, RIXS has the potential to answer the question of wether disperging electronic states which are delocalized in the entire solid (*i.e.* beyond one nanoparticle) exist. Especially due to its applicability in "real world" systems where the nanostructures are part of larger aggregates (matrices, heterostructures, self assembled structures, etc.) without the need of electric conductivity in the sample we expect RIXS to become an important tool in the study of the basic electronic properties of nanostructured materials in the future.

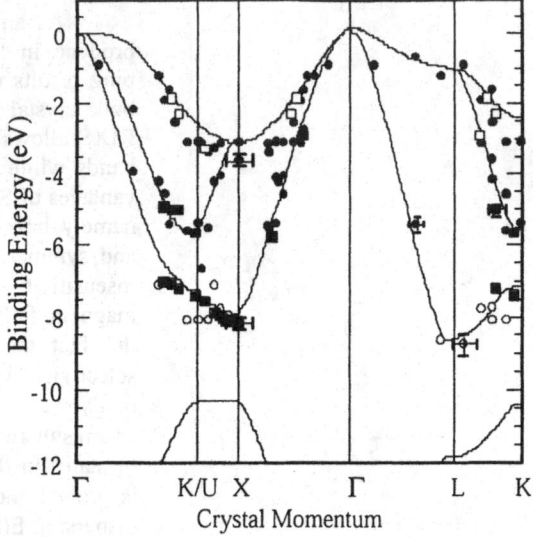

Figure 13. Site and symmetry selective band mapping on 3C-SiC. The dispersion of C p-type states (red circles) and Si s-type states (blue squares) can be followed separately by RIXS [35]. The experimental data is compared to a band structure calculation [36]. The measurement is bulk sensitive and hence not influenced by a possibly different SiC polytype configuration at the sample surface.

4. Outlook

As many nanostructures are being designed with electronic applications in mind, it is important to characterize and understand the basic electronic properties of the ingredients in such structures. SXA, SXA and RIXS will be valuable tools in this endeavor, as they yield very detailed local and atom selective information. Of major importance in the context of nanocomposite materials is the technical aspect, that the experiments are atom selective, bulk sensitive and solely based on a photon-in photon-out process, which is not disturbed by electric or magnetic fields. This combination of capabilities allows the selective investigation of nanosized structures in composite materials. This includes samples containing biological components. Soft x-rays are especially well adapted in this context, as the abundant atoms in biomolecules (C, N, O) contain exclusively core levels with binding energies in the *soft* x-ray energy range. A certain drawback is that SXA, SXE and RIXS spectroscopy rely on a tunable and bright source of soft x-rays, which currently can only be provided by synchrotron radiation. Thus, the experiments can in general not be performed at the home laboratory, but have to be carried out at a synchrotron light facility. The third generation of such facilities, exclusively dedicated to the use of the radiation, is since several years available for scientific (and industrial) community, providing very bright x-ray beams to many users simultaneously. Especially SXE and RIXS require the high brightness of the source and profit directly from the technological improvements in this area. With even brighter sources (several orders of magnitude) such as the free electron laser being under construction right now, it will be possible to address more complex questions in the future, making use of increased energy resolution, temporal resolution or spatial resolution.

What does this general statement imply for the study of nanostructured materials in particular? With more intensity it will be possible to study more dilute systems atom selectively. In addition, it will be feasible to investigate atomic species which are currently inaccessible due to their low absorption cross-section/fluorescence yield. The moderate *time resolution* achievable might be useful in the *in situ* monitoring of synthesis processes, *e.g.* as a monitor of the electronic properties of precipitates in a matrix during Ostwald ripening. It is obvious, that *spatial resolution* is important when dealing with spatially structured samples such as nanocomposites. Currently, the state of the art spatial resolution for images based on a certain spectra feature in SXA is about 10 nm x 10 nm, achieved in a scanning mode. Similar spatial mapping based on SXE rather than SXA might become possible in the future. As SXE measures a decay after a selectively prepared electronic excitation, this will offer new information which can be spatially mapped out. Finally, energy resolution will translate directly into improved *chemical selectivity*. It is already possible to use the core level shift induced by a chemical bond in order to discriminate between the same atoms in a different chemical state, such as Si atoms which have Si neighbors vs. Si atoms which have O neighbors. Such experiments can be extended to atomic neighbors which introduce smaller core level shifts.

Acknowledgement

In this overview article, we have tried to illustrate the use of soft x-ray spectroscopy for the study of the electronic structure of nanostructured solids, using examples of work which has been conducted in our group. While many people have contributed to this work over the years, Annette Karl (doped carbon cages) and Jan Lüning (CdS, SiC) deserve special credit in the context of this article.

References

[1]   Both the terms "x-ray emission" and "x-ray fluorescence" are commonly used for this process.

[2]   J.-E. Rubensson,, 23. IFF Ferienkurs des Forschungszentrums Jülich *Synchrotronstrahlung zur Erforschung Kondensierter Materie* (Jülich, 1992)

[3]   N. Wassdahl, J.-E. Rubensson, G. Bray, P. Glans, P. Bleckert, R. Nyholm, S. Cramm, N. Mårtensson and J. Nordgren, Phys. Rev. Lett **64**, 2807 (1990)

[4]   T. Canham, Appl. Phys. Lett. **57**, 1046 (1990)

[5]   J. Lüning, Dissertation Universität zu Köln and Reports of the Forschungszentrum Jülich, Jül-3544, ISSN 0944-2952, Jülich. (1997)

[6]   S. Eisebitt, J. Lüning, J.-E. Rubensson, T. van Buuren, S. N. Patitsas, T. Tiedje, M. Berger, R. Arens-Fischer, S. Frohnhoff, and W. Eberhardt, Solid State Comm. **97**, 549 (1996)

[7]   Bianconi, R. Del Sole, A. Selloni, P. Chiaradia, M. Fanfoni, and I. Davoli, Solid State Comm. **64**, 1313 (1987)

[8]   U. von Barth and G. Grossman, Phys. Rev. B **25**, 5150 (1982)

[9]   R. Williams, Diplomarbeit Universität zu Köln (1997)

[10]   J. Lüning, J. Rockenberger ,S. Eisebitt, J.-E. Rubensson, A. Karl, A. Kornowski, H. Weller and W. Eberhardt, Solid State Comm. **112**, 5 (1999)

362

[11]   T. Voßmeyer, L. Katsikas, M. Giersig, I.G. Popovic, K. Dies-ner, A. Chemseddine, A. Eychmüller, H. Weller, J. Phys. Chem. **98**, 7665 (1994)
[12]   T. Voßmeyer, G. Reck, B. Schulz, L. Katsikas, H. Weller, J. Am. Chem. Soc. **117** (1995) 12881.
[13]   T. Voßmeyer, G. Reck, L. Katsikas, E.T.K. Haupt, B. Schulz, H. Weller, Science **267**, 1476 (1995)
[14]   T. Voßmeyer, G. Reck, L. Katsikas, E.T.K. Haupt, B. Schulz, H. Weller, Inorg. Chem. **34**, 4926 (1995)
[15]   Y.-N. Xu and W.Y. Ching, Phys. Rev. B **48**, 4335 (1993)
[16]   N. Hamada, S. Sawada, and A. Oshiyama, Phys. Rev. Lett. **68**, 1579 (1992)
[17]   R. Saito, M. Fujita, G. Dresselhaus, and M.S. Dresselhaus, Appl. Phys. Lett. **60**, 2204 (1992)
[18]   *Science of Fullerenes and Carbon Nanotubes* by M.S. Dresselhaus, G. Dresselhaus, and P. Eklund, (Academic Press, San Diego, 1996)
[19]   S.Eisebitt, A. Karl, W. Eberhardt, J.E. Fisher, C. Sathe, A. Agui and J. Nordgren, Appl. Phys. A **67**, 89 (1998)
[20]   A.G. Rinzler, J. Liu, H. Dai, P. Nikolaev, C.B. Huffman, F.J. Rodriguez-Macías, P.J. Boul, A.H. Lu, D. Heymann, D.T. Colbert, R.S. Lee, J.E. Fischer, A.M. Rao, P.C. Eklund, and R.E. Smalley, Appl. Phys. A **67**, 29 (1998)
[21]   S. Eisebitt, A. Karl, A. Zimina, R. Scherer, M. Freiwald, W. Eberhardt, F. Hauke, A. Hirsch, Y. Achiba, Proceedings of the IWEPNM 2000 "Electronic Properties of Novel Materials: molecular nanostructures". To be published at AIP, Ed. H. Kuzmany, J. Fink, M. Mehring, S. Roth (2000)
[22]   S. Eisebitt, I. Wirth, G. Kann und W. Eberhardt, Phys. Rev. B **61**, 5719 (2000)
[23]   C.T. Chen *et al.*, Nature **352**, 603 (1991)
[24]   A. Karl, Doktorarbeit Universität zu Köln (2000)
[25]   A. Karl, S. Eisebitt, A. Zimina, R. Scherer, M. Freiwald, W. Eberhardt, to be published.
[26]   T.Pichler, M. Sing, M. Knupfer, M.S. Golden and J. Fink, Solid State Comm. **109**, 721 (1999)
[27]   W.Andreoni, F. Gygi, M. Parinello, Chem. Phys. Lett. **190**, 159 (1992)
[28]   A. Karl, S. Eisebitt, R. Scherer, M. Freiwald, W. Eberhardt, A. Hirsch, to be published.
[29]   T. Pichler *et al.*, Phys. Rev. Lett. **78**, 4249 (1997)
[30]   F. K. Gelmukhanov, L. N. Mazalov, and N. A. Shklyaeva, Sov. Phys. JETP **44**, 504 (1976)
[31]   Y. Ma, N. Wassdahl, P. Skytt, J. Guo, J. Nordgren, P.D. Johnson, J.-E. Rubensson, T. Böske, W. Eberhardt, and S.D Kevan, Phys. Rev. Lett. **69**, 2598 (1992)
[32]   Y. Ma, Phys. Rev. B **49**, 5799 (1994)
[33]   S. Eisebitt and W. Eberhardt, J. El. Spec. Rel. Phen. □**107** (2000) (accepted), invited review on RIXS.
[34]   J.J. Sakurai, *Advanced Quantum Mechanics*, (Addison-Wesley, London, 1967)
[35]   J. Lüning, J.-E. Rubensson, C. Ellmers, S. Eisebitt und W. Eberhardt, Phys. Rev. B **59**, 10573 (1999)
[36]   S. Logothedis, H.M. Polatoglou, J. Petalas, D. Fuchs, R.L. Johnson, Physica B **185**, 389 (1993).

OPTICAL SPECTROSCOPY OF CARRIER RELAXATION AND TRANSPORT IN III/V SEMICONDUCTOR TUNNELING STRUCTURES

ETIENNE GOOVAERTS[*] AND CHRIS VAN HOOF[#]

[*] Physics Department, University of Antwerp
Universiteitsplein 1, B-2610 Wilrijk, Belgium
[#] Inter-university Micro-Electronics Center (Imec vzw)
Kapeldreef, 75, B-3030 Leuven, Belgium.

1. Introduction

Since the proposal and the demonstration of carrier tunneling in III/V semiconductor heterostructures, there has been a considerable and persistent interest in these phenomena [1,2]. One motivation is the investigation of fundamental physical principles of quantum tunneling, and in a detailed description in the case of its realisations using semiconductor materials. This has direct relevance for the understanding and improvement of related systems, e.g., the intersubband infrared lasers also called *quantum cascade* lasers [3]. On the other side there are perspectives for application of semiconductor tunneling structures in a variety of practical electronics devices including ultrahigh-speed devices oscillators [4], multi-valued logic switches [5], fast A/D converters [6] and special purpose light-emitting diodes [7-9]. The highly nonlinear I-V characteristics with eventual negative differential resistance (NDR) is an essential ingredient for most of the proposed applications.

From the late eighties on, both steady-state and time-resolved photoluminescence were applied to investigate charge build-up near and inside a quantum well (QW) [10-14], the escape of electrons and holes from quantum wells [15,16], including partial results for unipolar devices under applied bias [13,17]. In this paper, we will give an overview of the results we obtained since then by optical spectroscopy in different III/V tunneling structures. Some of the device-oriented aspects have already been discussed previously [18]. Here, we will first consider unipolar double-barrier (DB) and triple-barrier (TB) resonant tunneling structures (RTS), in which tunneling times of minority-carrier have been determined. Then, bipolar devices are discussed with either a RTS or only a single barrier in the junction region, the so-called called RTLED (resonant tunneling LED), and TLED (tunneling LED) devices, respectively. Besides detailed understanding of carrier transport and electrical and optical phenomena, these investigations have also led to the design of electro-optic devices with ultrafast switching characteristics which are presently under further investigation.

363

L. Pavesi and E. Buzaneva (eds.), Frontiers of Nano-Optoelectronic Systems, 363-376.
© 2000 *Kluwer Academic Publishers. Printed in the Netherlands.*

2. Device properties and experimental aspects

The devices which will be discussed here, are composed of epitaxially grown layers of GaAs, AlAs or $Al_xGa_{1-x}As$, in most cases produced by molecular beam epitaxy (MBE) on a p- or n-type GaAs substrate. Bottom and top contact layers are p- or n-type GaAs, while in some instances a doped AlGaAs top layer is added, which has the advantage to be transparent for GaAs emission. Between the contact layers a tunneling structure is grown with barriers of higher band gap layers of typically 3 to 6 nm of AlAs, as in our devices, or alternatively AlGaAs, surrounded by lower band gap GaAs. In double- or triple-barrier resonant tunneling structures (DB RTS or TB RTS), a pair of adjacent barriers is de-fining a quantum well (QW) with a GaAs layer thickness of typically 5 to 8 nm. As an example an AlAs/GaAs DB RTS is depicted in Figure 1 at flat

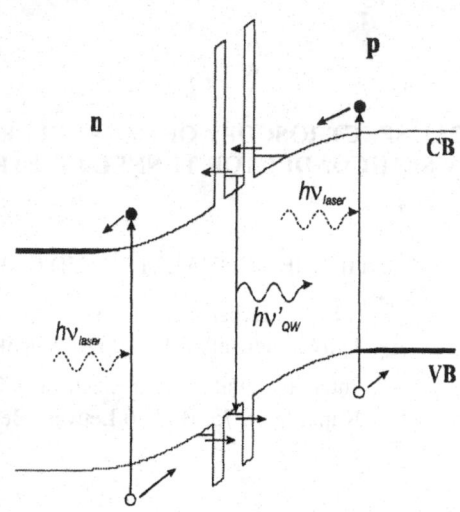

Figure 1.   The band structure diagram of an AlAs/GaAs DB RTS in a *p-i-n* junction under applied forward bias below flat band. The quasi-confined states in the QW are schematically indicated. Also, optical creation and recombination of electron-hole pairs as well as carrier transport processes are depicted.

band, showing also the resonant (or quasi-localised, or confined) states which form in potential wells for the electrons in the conduction band, and for the heavy holes in the valence band. AlAs/GaAs structures yield barrier heights of about 1 eV for electrons (conduction band), and 0.5 eV for holes (valence band) which both is larger than ther-mal excitation energies at room temperature (RT), and even more so at cryogenic tem-peratures. The tunneling devices are defined by lithographic methods to obtain mesa-like structures with metallic contact pads, and finally electrical contacts are obtained by wire bonding (see Figure 2). Under applied external bias the field over the RTS can be varied and carriers can be eventually electrically injected. This results in carrier accu-mulation against the RTS and in the QW-layers and depletion of doped layers. Optical excitation will create minority electrons and holes throughout the device. The carriers will be moving in the device under the influence of the built-in field and of applied ex-ternal bias, and tunneling through or thermally induced hopping over the barriers, as illustrated in Figure 1.

Optical measurements can be performed on devices with an annular (or else grid-type) metal contact, as shown in Figure 2. Alternatively, transparent contact layers (thin layers or ITO) can be employed. In our set-up, a microscope is used for both laser excitation and collection of light. Combined with a CCD-camera, this allows for direct observation of the device in the near infrared.

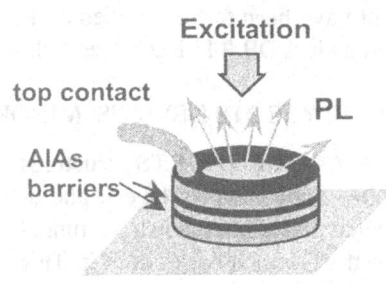

In most of our PL-experiments excitation was performed with a continuous-wave (cw) or a cw mode-locked titanium-sapphire laser (82 MHz repetition rate, $\cong 100$ fs pulse width) for high-resolution/high-sensitivity spectral and for time-resolved measurements, respectively. For the time-resolved measurements the

Figure 2 Schematic device configuration. Laser excitation and collection of PL-light are possible through the GaAs contact layer in the middle of an anular metallic contact. The conventional electrical polarity for forward bias is indicated.

single photon correlation (SPC) technique can be applied, with single channel detection in the spectral domain, and a time resolution of $\cong 250$ ps using a photomultiplier (PM) (a microchannel plate PM would boost this to $\cong 15$ ps). However, detection with a streak camera turns out to be much more powerful, featuring simultaneous (multi-channel) detection over the spectral range of the emission spectra and high time resolution ($\cong 15$ ps in our instrument). We employed a Raman spectrometer (triple 80 cm monochromator, liquid N_2 cooled CCD detector) for high-resolution/high-sensitivity PL-measurements. For low-temperature PL or EL experiments the samples were mounted in vacuum on a cold finger of a liquid N_2 or He cryostat with optical windows and electrical throughputs. As discussed below, impedance matched circuitry was developed for subnanosecond electrical excitation of EL-devices.

3. Carrier tunneling in RTS-devices

In RTS-structures in flat band conditions, the escape time of the carriers photocreated inside the QW could be determined by time-resolved measurements of the PL at the transition energy of the confined excitons [15,16]. However, it was soon realised that a different mechanism becomes important for PL in n-type RTS devices under applied bias: As soon as the applied voltage approaches the first electron resonance, the QW-emission is dominated by the recombination of minority-holes that are created by photoexcitation with electrons electrically injected into the RTS. This process has sometimes been called *electro-photoluminescence*. A strong correlation is indeed found between the intensity of the QW emission and the resonant current in both DB and TB devices [10,14,19-21]. As an example, Figure 3 shows the current and intensity of the emission from both the QW and contact layers as a function of applied bias for an n-type DB RTS [20]. The processes involved in carrier transport under an applied electric field have been further studied by PL-experiments in n-type and p-type RTS devices, as well as in a DB RTLED biased below flat band.

3.1. n-TYPE RTS DEVICES: MINORITY-HOLE TUNNELING

In AlAs/GaAs DB RTS structures with n-doped contact layers, photo-created holes are shown to tunnel into the QW and back out of it. This is observed via the radiative recombination of the holes with the electrically injected majority electrons in the different layers of the device. The bias dependence of the PL linewidth of the QW-transitions shows the effect of electron accumulation before and at the electron tunneling resonances [21], while the quantum confined Stark shift yields the electric field over the structure. From TR measurements of the PL emitted by the QW and the GaAs contact layers, the sequential character of the tunneling of minority holes was demonstrated. For the investigated DB structures with 4 -nm (3-nm) AlAs barriers, a characteristic tunneling time of 2 ns was found, decreasing below 1 ns with increasing electric field over de RTS [20].

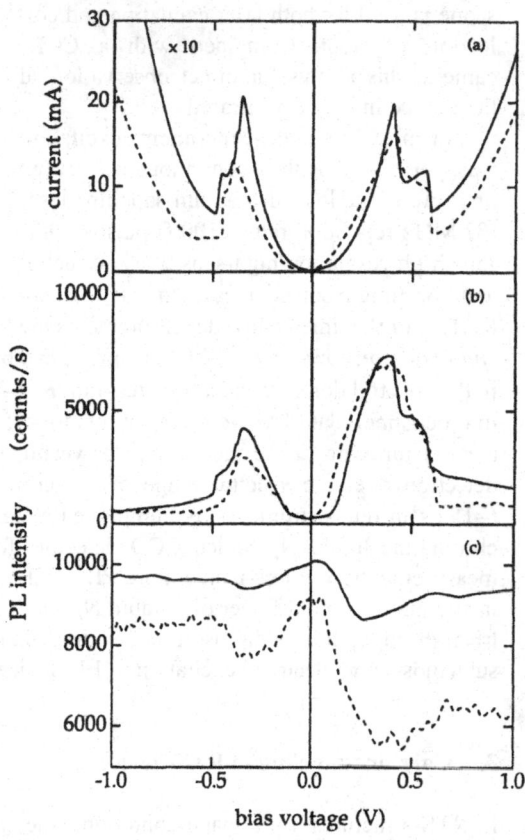

Figure 3. Bias dependence of (a) current, and (b) and (c) PL intensity from the confined exciton (at 1.58 eV) and the n^+-type GaAs (at 1.51 eV), respectively, for a 4-nm (solid line) and 3-nm (dashed) barrier n-type DB RTS (T=80K).

In subsequent investigations it was possible to perform laser excitation with spatial selectivity using different photon energies either below or above the lowest exciton transition in the QW. In an asymmetric TB n-type AlAs/GaAs RTS with 4-nm barriers and two QWs of 5 nm and 8 nm, a more detailed study of the tunneling processes was possible using this method [22], showing the consecutive filling of the narrow and wide QWs upon carrier excitation in- and outside the RTS only. This was analysed in a rate equations model for the population in different layers, assuming sequential (incoherent) tunneling processes.

In the same TB device a charged exciton state could be observed [23], which is called the X^- exciton, and consists of an electron weakly bound to a neutral exciton. It is apparent as an additional PL-transition of the 8-nm QW and is occurring in bias regions where excess electrons are available from current injection, but the carrier density is still low enough to avoid strong many-particle effects. This condition is met just below the electron tunneling resonances and in one case (triple alignment of lowest levels in both QWs with Fermi level in accumulation region), also just beyond the resonance. From the line splitting a binding energy of 2.1 meV was determined for the X^- exciton, comparable to reported values in different QW-structures [24]. Thermal activation with this binding energy was observed in the ratio between the X^- and normal exciton intensities.

Furthermore, high-sensitivity steady-state PL revealed additional exciton transition in this device [25], permitting an accurate modeling of the energy levels and tunneling resonances. Indeed, these transitions involved higher confined states in the QWs as well

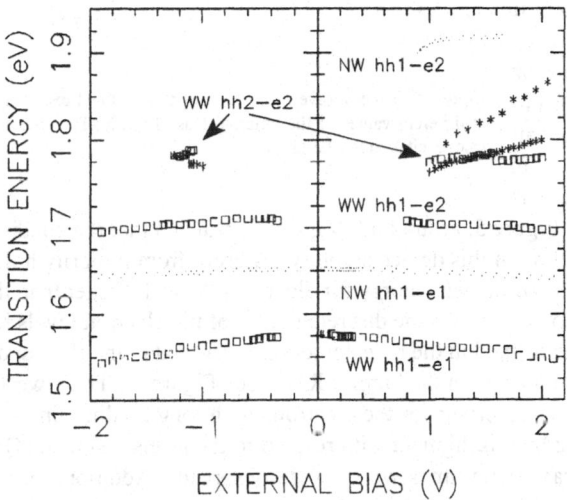

Figure 4. PL-transition energies of higher (squares and crosses) and spatially indirect (stars) excitons measured in an n-type TB RTS as a function of applied bias (T=10 K).

as spatially indirect recombination of carriers inside and outside the QW (Figure 4). These weak transitions become visible because of the injection of carriers at higher bias leading to a high electron accumulation outside the RTS and inside of it in the lowest QW levels (e^1_{WW} and e^1_{NW}). Moreover, at higher bias voltage, injection in the second electron level of the QWs becomes effective and leads to nonthermal population. This extensive spectral information forms a solid basis for modeling of the device as a function of applied bias.

3.2. *p*-TYPE RTS DEVICES: HOLES AND MINORITY-ELECTRONS

In a p-type DB RTS, we have observed the tunneling of heavy and light holes (HH and LH) via the intensity of the QW PL, which is correlated with the resonant tunneling current as a function of applied voltage [26]. Complementary to the case of the n-type devices, the majority holes are now accumulating against the barrier of the RTS and – close to resonance– also inside the QW. One is monitoring the steady-state population

368

of injected holes in the QW while they recombine with the photocreated minority electrons tunneling in and escaping out of it. The tunneling dynamics of the electrons was investigated by TR PL of the QW emission [27]. However, in this case the decay times are strongly dependent on bias voltage, with unexpectedly large values –of more than 10 ns– in specific bias

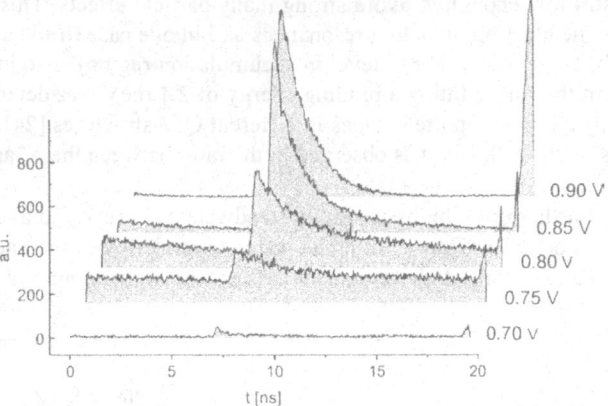

Figure 5. Time-resolved PL-decay of the confined excitons in a p-type RTS at different values of the external bias (T=10 K).The repetition period of the laser excitation is 12.2 ns.

regions (see e.g., Figure 5, around 0.8 V). It was realised that the conditions for minority electron tunneling in this device are very different from minority holes in the n-type RTS, because the confinement energy for electrons is much larger than that for holes. It is then essential to account for the discrete nature of the electron levels in the triangular potential, which is formed in the hole depletion layer next to the RTS under applied bias (a similar situation occurs in the bipolar RTS, see Figure 6). The lowest confined level in this layer forms a reservoir for the electrons with long holding times, whenever tunneling is inhibited by misaligment with respect to the levels inside the QW. This results in the long characteristic times which were observed. Additional information about level alignment could again be obtained from the measurement of higher exciton transitions. This was further supported by modeling of the band structure and energy levels under applied bias.

3.3. BIPOLAR RTS DEVICES BELOW FLAT BAND

Finally, the time dependence of exciton PL was measured in bipolar devices –also called RTLEDs– under external voltage below flat band [28]. The RTS is then located in the middle of two depletion layers at the p- and n-side and there are no carriers injected electrically, into the QW. After direct photoexcitation in the QW, at photon above the confined exciton, the fast escape of electrons is observed with a tunneling time of 600 ps near flat band which decreases at lower bias (i.e., higher electric field over the RTS). At photon energies below the confined exciton, photocreated electrons and holes are accumulating against the barriers on the p- and n- side, respectively, and eventually tunnel through the barriers into and further out of the QW. As in the p-type RTS, very slow PL decay (in the order of 10 ns or longer) was observed in specific bias regions and are related to very long dwell times of electrons (and to some extent of holes) in confined states of the triangular potentials outside of the RTS. At bias voltages far from resonance between these levels and those confined in the QW, only slow nonresonant tunneling is going on.

Also, the PL measurements indicate mixed HH-to-LH injection in the QW leading to a strongly nonthermal population of the confined LH level. A high-energy shoulder on the HH peak appears, with a 22 meV splitting. From the relative intensities of these transitions, a population ratio HH:LH of 5:3 was estimated, at the bias voltage of the hole resonance.

4. Tunneling LEDs

Tunneling LED devices, in which one or more tunneling barriers are grown in the intrinsic region of a *p-i-n* junction, were considered early on [7-9,29,30]. When forward biased, the majority carriers from each of the contact layers are accumulating against and tunneling though the barriers, which for multiple-barrier devices leads to exciton formation and recombination inside the QWs and to efficient EL at the energy of the confined excitons [29-31]. As in the unipolar devices, resonant tunneling of electrons and holes determines the carrier populations in the QW, and in the present case also the EL intensity. It is worth mentioning here that a bipolar resonant tunneling transistor for QW light emission has been demonstrated [32].

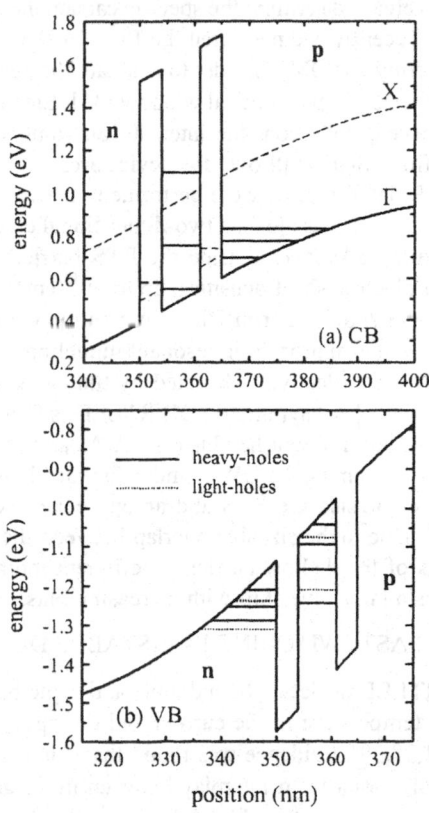

4.1. EMISSION CHARACTERISTICS OF RTLEDs

Efficient EL-emission from carrier recombination in the QW –comparable to or even stronger than the emission from the GaAs contact layers– was first demonstrated in a DB RTLED device [30]. The EL-intensity was shown to correlate directly to the hole and electron resonances and large peak-to-valley ratios were observed (10:1 at 4 K, smaller 1.45:1 but persisting at room temperature). Investigation of the EL-spectrum as a function of applied bias yielded detailed information about the accumulation of carriers outside and inside the

Figure 6. Modelling of the levels in a TB RTLED biased below flat band. Besides the zone-centre band edge, the bottom of the X-point conduction band is shown as well as confined states in the AlAs-layer.

QW (from intensities and line shape) and about the electric field acting on the RTS (from the shift of the QW-transition due to the quantum confined Stark effect). An analogous investigation [33] was performed on two AlAs/GaAs TB RTLEDs which both contained an asymmetrical TB RTS, but with inverted order of the narrow (6 nm) and wide (8 nm) well relative to the p-i-n junction. Depending on the ordering of the wells with respect to injection of electrons and holes, one obtains double or triple resonance conditions, in which either two or three of the levels in the different layers (accumulation layer outside the RTS and confined levels in each of the QWs) are aligned to each other. It was possible to determine the charge redistribution between the two wells as the level alignments were changing by the applied bias, on the basis of the analysis of the EL-spectra and of parallel band structure calculations. This also clarified the feedback mechanisms between charge build-up and level-alignment which are responsible for sustained resonance conditions over a wider range of bias voltage and for the occurrence of intrinsic bistability. Finally, these effects of accumulation of charge and release determine the specific capacitance characteristics of these devices [34].

Recently, we measured the EL of a DB RTLED by scanning near-field optical microscopy (SNOM) in order to evaluate the spatial dependence of the spectra in a device with a $75\times75\ \mu m^2$ optical window [35]. Electric field inhomogeneities could be directly observed, both from the intensity distribution and from the variation of the quantum confined Stark shift over the device area.

Related studies were performed on two RTLED devices with different configurations. In one case [36], a two-dimensional electron gas (2DEG) system was defined in a 6.7 nm GaAs layer outside the RTS barriers confined by an AlGaAs top layer. Very high electron sheet densities, up to 10^{12} cm^{-2} were reached in the 2DEG. The lineshape analysis shows a Fermi-Dirac distribution with carrier heating above the lattice temperature. When approaching resonant tunneling conditions, the high-energy electrons in the 2DEG are selectively depleted by tunneling into the QW. In another variation on the theme [37] an asymmetric DB RTLED is fabricated in which the barriers have the same width, but different heights (one AlAs and one $Al_{0.14}Ga_{0.86}As$ barrier of 3.7 nm enclosing a 4.8 nm GaAs QW). Under forward bias, holes are accumulating against the low barrier outside the RTS and an appreciable electron population is built up inside the QW. Due to the sizeable overlap between the hole and electron wavefunctions on both sides of the shallow barrier, an efficient indirect recombination is obtained which displays a large blue shift with increasing bias.

4.2. FAST SWITCHING IN BISTABLE DEVICES

In RTLED devices, –both double and triple barrier, – bistable operation was observed at low temperature in the current and the optical characteristics as a function of bias [7-9,33,38,39]. Evidence was found that this bistability is an intrinsic effect and does not simply result from interplay between the negative differential resistance (NDR) of the device and an external load. Instead, when approaching resonant tunneling conditions a redistribution of charge between the QWs and accumulation layers outside the RTS can occur which changes the electrical field over the device, influences the alignment between levels, and thus also the carrier tunneling. Approaching the resonance from below

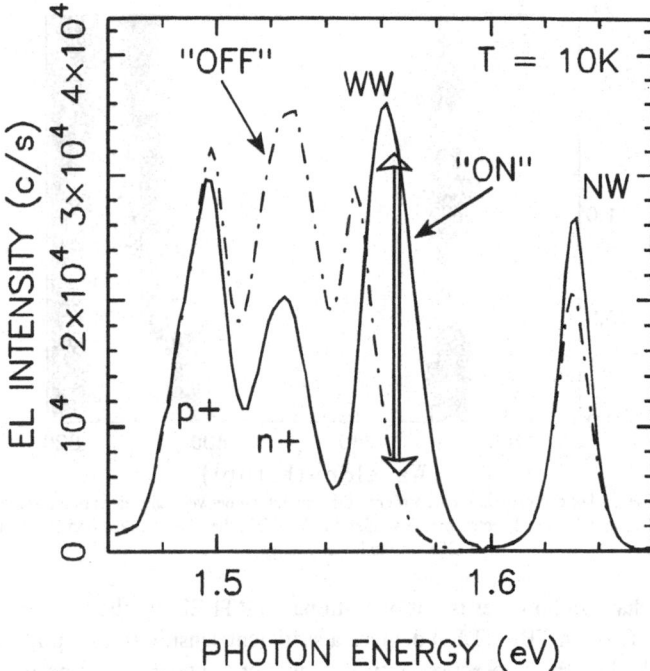

Figure 7. EL-spectrum of a TB RTLED measured (T=10 K) at a voltage in the bistability region of the device in the "ON" (solid line) and "OFF"(broken line) states. The double arrow indicates the intensity change at the monitoring wavelength.

or from above yields different results. This can be observed from the EL-spectra measured in the bistability region, as is illustrated for a TB device in Figure 7. The intensity changes in the wide-well emission, as well as the shift of the transition result from carrier redistribution. This yields the feedback mechanism that is at the origin of bistability in RTLEDs.

While only moderate ON/OFF contrast was reached for DB devices [7], the higher versatility of TB RTLEDs –with additional design parameters and conditions on level alignment– allowed for the demonstration of a giant optical bistability with a contrast of the order of 10^4:1 in optical output at the QW transition [9]. Moreover, it was shown that in TB devices it was possible to influence the bistability region by optical illumination of the device, opening the way for optical switching between the states [8].

Initial measurements of the time reponse of a DB RTLED [7] had put an upper limit of 200 ps (resolution limited) to the characteristic times both for switching on and off, using excitation with a short electrical pulse. The short response times compared to normal *p-i-n* LEDs is related to the tunneling times, an increase of the radiative recombination rates at high carrier densities, and a fast separation of electrons and holes. Earlier results of optical switching of the current bistability in a unipolar RTS-device [36] had yielded about 50 ps for the off- to on-resonance switching event. The latter is induced by photocreation of carriers and the resulting band bending in the depletion re-

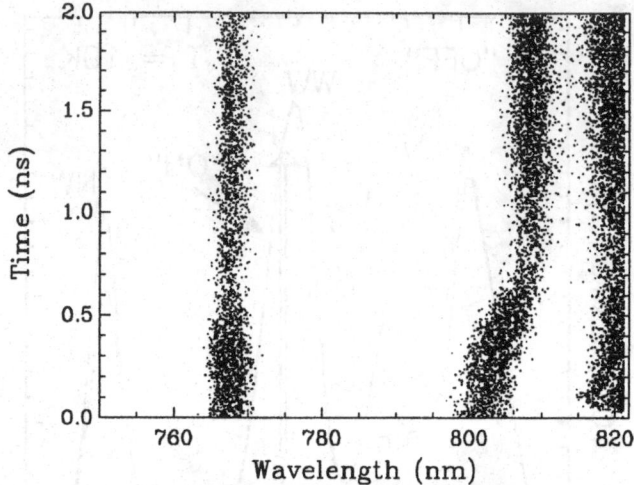

Figure 8. Picosecond streak-camera measurement of the switching between the states of a bistable triple-barrier RTLED. The switching is detected by the sudden Stark-shift of the wide-well exciton emission (around 805 nm) to lower energy.

gion. This mechanism however is not operational in RTLEDs in the absence of any depletion region. Only in TB RTLED devices a sufficient sensitivity for optical excitation was obtained [8,38]. By picosecond optical excitation, optical switching of the device could be induced [38] at pulse energies below 10 pJ as observed from the intensity change of the wide-well emission, and from the quantum confined Stark shift of this transition (see Figure 8). The efficiency and speed of the process is similar for excitation at photon energies above and below the QW transitions, showing that switching results from photocreation of electron-hole pairs outside the RTS. Experiments with excitation below the QW transition were not hampered by PL from direct excitation in the wells, which greatly improved the observation of the switching event.

4.3. ULTRAFAST ELECTRO-OPTIC REPONSE OF TLED DEVICES

Recently, it was found that LEDs with a tunneling structure in the junction region could be profitably applied as high-speed/high-frequency light sources at ambient temperature. It was first demonstrated that DB RTLED devices can be operated at room temperature under electrical pulse excitation (see Figure 9) with reduced switch-on and switch-off times of the order of 200 ps [41,42]. Considering that this was observed for the emission both from the confined QW excitons and from the electron-hole recombination in the doped GaAs regions, it was realised that resonant tunneling is not a key issue in this application. Therefore, tunneling LED (TLED) devices were grown containing only a single AlAs tunneling barrier, showing at least comparable time response of the total emitted intensity [43].

Accurate measurements of the time response with electrical pulses of 250 ps rise and decay times (10% to 90%) necessitated the design of a impedance matched circuitry with coaxial connections to 50 Ω transmission lines on a printed circuit board (PCB) on which the device was directly mounted. This also permits parallel measurements of the current through and voltage over the device using a sampling scope with comparable time resolution. Measurements performed both at room temperature and at T= 80 K in a nitrogen cryostat, showed that thermally activated processes have no influence on the response times.

Deconvolution of the instrumental resolution yields 10-to-90% switching times in the 100 ps range for the fastest devices at relatively high emission efficiencies. This is a major improvement compared to the traditional approach of high doping levels in the junction region of a *p-i-n* LED [44] which reach fast response times at the cost of a strongly reduced quantum efficiency. Very recently [45], an alternative doping scheme with beryllium was demonstrated which does not suffer from this drawback and results in performances comparable to our TLEDs.

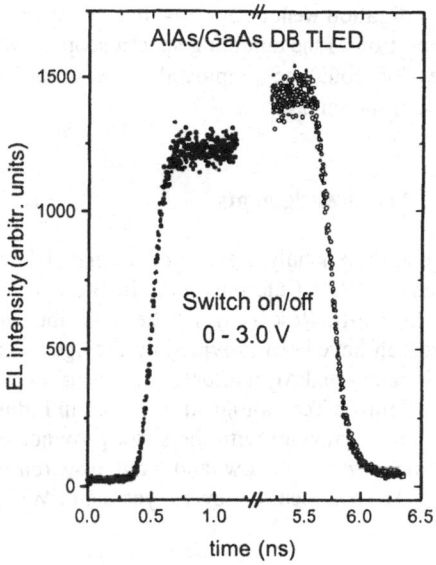

Figure 9. Streak-camera measurement of the switching of the EL-intensity of a DB TLED. The experiment is performed at room temperature under fast electrical excitation with voltage step 0-3 V.

In order to boost the performance of the TLED-device, a microcavity design was explored in which the TLED is placed between a distributed Bragg reflector (DBR) and the GaAs/air interface serving as output mirror [41,46]. Compared to the identical RTLED structure without microcavity, the emission efficiency is drastically increased up to about 100 μW/mA but the speed is reduced to 1.3 Gbit/s (compared to 4.4 Gbit/s without microcavity). However, this reduction is very probably resulting from the higher impedance introduced by the DBR layers below, and by the transparent ITO-contact layer above the TLED.

5. Concluding remarks

In this work, optical spectroscopy has been shown to be a method of choice for the investigation of basic phenomena in III/V tunneling devices, and also for the characterisation of fast electro-optic components based on tunneling structures. Such an approach should be fruitful as well in the study of other nanoscale systems, e.g., quantum wires and quantum dots, in which similar phenomena play an important role. As a closing remark we want to mention the beautiful example of interplay between III/V tunneling structures and quantum dots which was recently obtained in optical and magnetotunneling experiments on tunneling barriers containing self-assembled InAs dots [47]. This investigation went both ways: the spin levels of the dot could be resolved using the high resolution of the tunneling spectroscopy, but also, the delta-function density of states of the dot could be employed as a probe for the density of states in the emitter-accumulation layer.

6. Acknowledgments

The authors kindly acknowledge financial support from the Belgian science supporting agencies IIKW (Interuniversity Institute for Nuclear Sciences) and FWO (National Fund for Scientific Research). Further contributions in the form of scholarships for doctoral research have been provided by the agencies IWONL (Institute for Scientific Research in Industry and Agriculture) and its successor IWT (Flemish Institute for the Promotion of Scientific-Technological Research in Industry), and by the province of Antwerp in its exchange program with the sister province of Shaanxi, China. One of us (E.G.) was a senior research fellow and now research director of the FWO, and the co-author (C.V.H.) has been for several years an FWO postdoctoral research fellow.

References

1. Mizuta, H. and Tanoue, T. (1995) *The Physics and Applications of Resonant Tunneling Diodes* (Cambridge University Press, Cambridge).
2. *Physics of Quantum Electron Devices*, edited by Capasso, F., (1990) *Springer Series in Electronics and Photonics* **28**, Springer-Verlag, Berlin.
3. Faist, J., Capasso, F., Sivco, D.L., Sirtori, C., Hutchinson, A.L. and Cho, A.Y. (1994) *Science* **264**, 553-555; Sirtori, C., Faist, J., Capasso, F., Sivco, D.L., Hutchinson, A.L. and Cho, A.Y. (1996) *Appl. Phys. Lett.* **69**, 2810-2812; Sirtori, C., Capasso, F., Faist, J., Hutchinson, A.L., Sivco, D.L. and Cho, A.Y. (1998) *IEEE J. Quant. Electr.*. **34**, 1722-1729.
4. Goldman, V.J., Tsui, D.C. and Cunningham, J.E. (1987) *Phys. Rev. B* **35**, 9387; Yu, E.T., Jackson, M.K. and McGill (1989) *Appl. Phys. Lett.* **55**, 744.
5. Potter, R.C., Lakhani, A.A., Beyea, D., Hempling, E. and Fathimulla, A. (1988) *Appl. Phys. Lett.* **52**, 2163; Capasso, F., Sen, S., Beltram, F., Lunardi, L.M., Vengurlekar, A.S., Smith, P.R., Shah, N.J., Malik, R.J. and Cho, A.Y. (1989) *IEEE Trans. Electron. Devices* **ED-36**, 2065, and references therein.
6. Fobelets, K., Genoe, J., Vounckx, R. and Borghs, G. (1993) *Semicond. Sc. Techn.* **8**, 2106-2114.
7. Van Hoof, C., Genoe, J., Mertens, R.P., Goovaerts, E. and Borghs, G. (1992) *Electron. Lett.* **28**, 123-124.
8. Raymond, S., Van Hoof, C., Genoe, J., Mertens, R., Borghs, G., Yan, Z.C., and Goovaerts, E. (1993) *Electron. Lett.* **29**, 1301-1302.
9. Van Hoof, C., Genoe, J., Raymond, S. and Borghs, G. (1992) *Appl. Phys. Lett.* **63**, 2390-2392.
10. Young, J.F., Wood, B.M., Aers, G.C., Devine, R.L.S., Liu, H.C., Landheer, D., Buchanan, M., SpringThorpe, A.J. and Mandeville, P. (1988) *Phys. Rev. Lett.* **60**, 2085; Young, J.F., Wood, B.M., Aers, G.C., Devine, R.L.S., Liu, H.C., Buchanan, M., SpringThorpe, A.J. and Mandeville, P. (1989) *Superlatt. Microstruct.* **5**, 411.
11. Eaves, L., Leadbeater, M.L., Hayes, D.G., Alves, E.S., Sheard, F.W., Toombs, G.A., Simmonds, P.E., Skolnick, M.S., Henini, M. and Hughes, O.H. (1989) *Solid-State Electron.* **32**, 1101.
12. Vodjdani, N., Chevoir, F., Thomas, D., Cote, D., Bois, P., Costard, E. and Delaitre, S. (1989) *Appl. Phys. Lett.* **55**, 1528.
13. Vodjdani, N., Cote, D., F., Thomas, D., Sermage, B., Bois, P., Costard, E. and Nagle, J. (1990) *Appl. Phys. Lett.* **56**, 33.
14. Yoshimura, H., Matsusue, T. and Sakaki, H. (1990) *Phys. Rev. Lett.* **64**, 2422.

15. Tsuchiya, M., (1987) *Phys. Rev. Lett.* **59**, 2356; Tsuchiya, M., Matsusue, T. and Sakaki, H. (1988) in: *Ultrafast Phenomena IV*, edited by Yajima, T., *Springer Series in Chemical Physics* **48**, 304 Springer-Verlag, Berlin.

16. Jackson, M.K., Johnson, M.B., Chow, D.H. and McGill T.C. (1989) *Appl. Phys. Lett.* **54**, 552; Yu, E.T., Jackson, M.K. and McGill T.C. (1989) *Appl. Phys. Lett.* **55**, 744.

17. Charbonneau, S., Young, J.F. and SpringThorpe, A.J. and Mandeville, P. (1990) *Appl. Phys. Lett.* **57**, 264.

18. Van Hoof, C., Genoe, J., Brebels, S., Pieters, Ph., Beyne, E. and Borghs, G. (1997) *NATO ASI Series E – Applied Sciences Advanced Study Institute*, pp. 81-96.

19. Van Hoof, C., Goovaerts, E. and Borghs, G. (1990) *Proc. SPIE* **1362**, 291; Goovaerts, E., Van Hoof, C. and Borghs, G. (1990) *Physica B* **175**, 307-310.

20. Van Hoof, C., Goovaerts, E. and Borghs, G. (1992) *Phys. Rev. B* **46**, 6982-6989.

21. Bertram, D., Grahn, H.T., Van Hoof, C., Genoe, J. and Borghs, G. (1990) *Phys. Rev. B* **50**, 17 309-315.

22. Hong, Z. (1992) Ph.D. thesis, University of Antwerp, 'Exciton relaxation and minority-carrier transport in GaAs/AlGaAs heterostructures'.

23. Yan, Z.C., Goovaerts, E., Van Hoof, C., Bouwen, A. and Borghs, G. (1995) *Phys. Rev. B* **52**, 5 907-912.

24. Khen, K., Cox, R., d'Aubigné, Y.M., Bassani, F., Saminadayar, K. and Tatarenko, S. (1993) *Phys. Rev. Lett.* **71**, 1752; Buhmann, H., Mansouri, L., Beton, P., Mori, N., Eaves, L. and Potemski, M. (1995) in *Proc. XXIInd Intern. Conf. on Physics in Semiconductors*, Vancouver 1994, ed. by Lockwood, D.J., publ. World Sientific, Singapore; Finkelstein, G., Shtrikman, H. and Bar-Joseph, I. (1995) *Phys. Rev. Lett.* **74**, 976.

25. Yan, Z.C. (1995) Ph.D. thesis, University of Antwerp, 'Optically-induced switching, tunneling and relaxation processes in AlAs/GaAs triple-barrier resonant tunneling diodes'.

26. Van Hoof, C., Borghs, G. and Goovaerts, E (1991) *Appl. Phys. Lett.* **59**, 2139-2141.

27. Käß, H., Schuddinck, W., Goovaerts, E., Van Hoof, C. and Borghs, G. (1998) *Microelectr. Engin.* **43-44**, 355-361.

28. Romandic, I., Bouwen, A., Goovaerts, E., Van Hoof, C., Mielants, M. and Borghs, G. (1998) *Microelectr. Engin.* **43-44**, 363-369; Romandic, I., Bouwen, A., Goovaerts, E., Van Hoof, C. and Borghs, G. (2000) *Semicond. Sci. Techn.*, **15**, in press.

29. Eaves, L. (1991) Microelectr. Engin. 15, 661-662; Bertram, D., Lage, H., Grahn, H.T. and Ploog, K. (1994) *Appl. Phys. Lett.* **64**, 1012-1014; Buckle, P.D., Cockburne, J.W., Teissier, R.J., Willcox, A.R.K., Whittaker, D.M., Skolnick, M.S., Smith, G.W., Grey, R., Hill, G. and Pate, M.A. (1994) *Semicond. Sci. Technol.* **9**, 533-536; Grahn, H.T., Dertram, D., Lage, H., von Klitzing, K. and Ploog, K. (1994) *Semicond. Sci. Technol.* **9**, 537-539; Maude, D.K., Kuhn, O., Portal, J.C., Henini, M., Eaves, L., Hill, G. and Pate, M.A. (1994) *Semicond. Sci. Technol.* **9**, 540-544: Evans, H.B., Eaves, L. and Henini, M. (1994) *Semicond. Sci. Technol.* **9**, 555-558.

30. Van Hoof, C., Genoe, J., Mertens, R., Borghs, G. and Goovaerts, E. (1992) *Appl. Phys. Lett.* **60**, 77-79; Van Hoof, C., Genoe, J., Bertram, D., Grahn, H.T. and Borghs, G. (1995) *Phys. Rev. B* **51**, 13 491-498.

31. Van Hoof, C., Genoe, J. and Borghs, G. (1996) Phil. Trans. R. Soc. Lond. A **354**, 2447-2462.

32. Genoe, J., Van Hoof, C., Fobelets, K., Mertens, R. and Borghs, G. (1992) Appl. Phys. Lett. **61**, 1051-1053.

33. Van Hoof, C., Genoe, J., Raymond, S., Borghs, G., Yan, Z.C. and Goovaerts, E. (1994) *Proc. SPIE* **2139**, 250-257.

34. Fobelets, K., Van Hoof, C., Genoe, J., Stake, J., Lundgren, L. and Borghs, G. (1996) *J. Appl. Phys.* **79**, 905-910.

35. Mielants, M., Van Hoof, C., Stuer, C., Borghs, G. and Goovaerts, E. (1998) *Mater. Sci. Engin.* B **51**, 9-11.

36. Van Hoof, C., Genoe, J., Portal, J.C. and Borghs, G. (1995) *Phys. Rev. B* **52**, 516-1519.

37. Van Hoof, C., Genoe, J., Portal, J.C. and Borghs, G. (1995) *Phys. Rev. B* **51**, 14 745-748.

38. Yan, Z.C., Goovaerts, E., Van Hoof, C. and Borghs, G. (1994) *Superlatt. Microstruct.* **16**, 239-242; Yan, Z.C., Goovaerts, E., Van Hoof, C. and Borghs, G. (1995) *Lith. J. Phys.* **5-6**, 534-537.

39. Kuhn, O., Genoe, J., Maude, D.K., Portal, J.-C., Eaves, L., Henini, M., Hill, G. and Pate, M. (1998) *Physica E* **2**, 483-488.

40. England, P., Golub, J.E., Mertens, R. and Borghs, G. (1991) *Appl. Phys. Lett.* **58**, 887-889.

41. Van Hoof, C., De Neve, H., Romandic, I., Goovaerts, E., and Borghs, G. (1997) *IEEE Photon. Technol. Lett.* **9**, 1463-1465; *ibid.* (1998) *Mater. Sci. Engin.* B **51**, 72-75.

42. Romandic, I., Zurauskiene, N., Goovaerts, E., Van Hoof, C. and Borghs, G. (1999) *Mater. Sci. Forum* **297-298**, 29-32.

376

43. Van Hoof, C. and Borghs, G. (1998) *IEEE Photon. Techn. Lett.* **10**, 24-26.
44. de Lyon, T.J., Woodall, J.M., McInturff, Kirchner, P.D., Kash, A., D.T., Bates, R.J.S., Hodgson, R.T. and Cardone, F. (1992) *Appl. Phys. Lett.* **59**, 402-404; de Lyon, T.J., Woodall, J.M., McInturff, D.T., Bates, R.J.S., Kash, A., Kirchner, P.D. and Cardone, F. (1992) *ibid.* **60**, 353-355.
45. Chen, C.H., Hargis, M., Woodall, J.M. and Melloch, M.R. (1999) *Appl. Phys. Lett.* **74**, 3140-3142.
46. Van Hoof, C. and Borghs, G., patent pending (97/7), "A device for emitting electromagnetic radiation at a predetermined wavelength and a method of producing such device"
47. Thornton, A., Itskevich, I.E., Ihn, T., Henini, M., Moriarty, P., Nogaret, A., Beton, P.H., Eaves, L., Main, P.C., Middleton, J.R. and Heath, M. (1997) *Superlatt. Microstruct.* **21**, 255-258; Thornton, A.S.G., Ihn, T., Main, P.C., Eaves, L. and Henini, M. (1998) *Appl. Phys. Lett.* **73**, 354-356; Main, P.C., Thornton, A.S.G., Hill, R.J.A., Stoddart, S.T., Ihn, T., Eaves, L., Benedict, K.A. and Henini, M. (2000) *Phys. Rev. Lett.* **84**, 729-732.

NEAR-FIELD SCANNING OPTICAL SPECTROSCOPY OF QUASI-ONE-DIMENSIONAL SEMICONDUCTOR NANOSTRUCTURES

CH. LIENAU,[a] V. EMILIANI,[a] T. GUENTHER,[a] F. INTONTI,[a]
T. ELSAESSER,[a] R. NOTZEL,[b] AND K.H. PLOOG[b]

[a]*Max-Born-Institut für Nichtlineare Optik und Kurzzeitspektroskopie,
Max-Born-Str. 2a, D-12489 Berlin, Germany*
[b]*Paul-Drude-Institut für Festkörperelektronik, D-10117 Berlin, Germany*

Abstract

Carrier dynamics in single GaAs quantum wires are studied in a wide temperature range by near-field scanning optical microscopy with pico- and femtosecond pulses. Luminescence and pump-probe experiments with a spatio-temporal resolution of 250 nm and up to 200 fs allow for a separation of carrier transport along the quantum wire and in the embedding GaAs quantum well from local carrier relaxation phenomena. The drift-diffusive carrier transport from the quantum well into the quantum wire occurs in the pico- to nanosecond regime with diffusion lengths of up to several microns. Diffusive transport along the quantum wire is characterized by carrier motion on a shorter picosecond time scale. In contrast, sub-picosecond relaxation times are found for the redistribution of carriers from high-lying to low-lying quantum wire states. This relaxation is governed by electron-electron and electron-phonon scattering.

1. Introduction

During the last years, the electronic and optical properties of quantum confined systems have been a central topic of experimental and theoretical researches in solid state physics. Semiconductor nanostructures represent important model systems for studying the fundamental physical properties of confined carriers and their optical excitations. In addition, potential device applications of such nanostructures in advanced optoelectronic devices are under intense investigation. Both the dimensionality and the geometrical dimensions of semiconductor nanostructures can be varied in a controlled way using modern epitaxial growth techniques. Their recent progress has allowed for realizing quasi-one-dimensional nanostructures, i.e., quantum wires (QWRs), of high structural quality. Individual quantum wires and arrays of coupled quantum wires have been produced by wet chemical etching, growth on tilted or prestructured substrates and by cleaved-edge overgrowth of quantum wells [1,2,3,4].

377

L. Pavesi and E. Buzaneva (eds.), Frontiers of Nano-Optoelectronic Systems, 377–392.
© *2000 Kluwer Academic Publishers. Printed in the Netherlands.*

The quantization of electronic levels and thus the optical and electronic properties of quantum wires depend sensitively on local variations of the sample composition on a nanometer length scale. As it is difficult to fully avoid compositional disorder on such scales, the properties of ensembles of nanostructures are often dominated by a pronounced inhomogeneous broadening, making the observation of the intrinsic low-dimensional quantum effects difficult. In order to reduce this broadening, experiments on single QWRs using microscopy techniques that provide spatial resolution in the nanometer range are crucial. While a number of experimental techniques offer sufficient spatial resolution, e.g., atomic force microscopy, scanning tunneling microscopy [5] or ballistic electron microscopy [6], these techniques give in general only limited spectroscopic information about subsurface nanostructures [7,6]. Cathodoluminescence spectroscopy has widely been used for monitoring the local emission of nanostructures on sub-micron length scales [8,9]. However, the initial energy distribution of the injected carriers, its relaxation towards quasi-equilibrium and the spatial propagation of carriers inside the nanostructure are difficult to control in such experiments. In contrast, all-optical spectroscopy allows for resonant excitation and thus for generation under well defined conditions. To make use of these advantages, one has to overcome the diffraction limit of conventional optical microscopy, e.g. by using near-field techniques with a spatial resolution of about 100 nm [10,11]. The potential of near-field spectroscopy, in particular at low temperatures [12], for imaging local emission and absorption was demonstrated in experiments on cleaved-edge overgrowth [13,14] and V-groove QWRs [15].

In addition, the combination of near-field microscopy with spectroscopic techniques providing a high temporal resolution in the femto- to picosecond regime offers a unique approach for directly imaging the ultrafast spatio-temporal dynamics of optical excitations in semiconductor nanostructures. It is expected that such experiments will provide considerable insight into the ballistic and diffusive propagation of photogenerated carriers [1617], or the carrier trapping into and carrier relaxation within low-dimensional nanostructures. Here, both nonlinear pump-probe methods with femtosecond time resolution and picosecond luminescence techniques of high detection sensitivity are relevant. Yet, the combination of these techniques with near-field microscopy represents a major challenge for the experimentalists and, so far, only few studies have been reported along this direction [18,19,20,21]. Recently, it was demonstrated that disorder-induced fluctuations of the excitonic resonance energy along the axis of a single QWR can be detected using near-field pump-probe spectroscopy [22].

In this paper, we present near-field optical studies of spatio-temporal carrier dynamics in single GaAs QWRs. Steady-state near-field spectroscopy is applied to analyze the quasi-one-dimensional subband structure and the lateral confinement potential. Processes of carrier and exciton transport are studied by picosecond NSOM, allowing a quantitative measurement of exciton mobilities. Local carrier relaxation, e.g., trapping of carriers into single GaAs QWRs, are investigated in femtosecond near-field pump-probe experiments with a time resolution of about 200 fs. Evidence is found for a transient unipolar transport of electrons along the wire axis on a picosecond time and 100 nanometer length scales.

The paper is organized as follows. After a brief description of experimental techniques (Section 2), results on the electronic structure and the local confinement potential of the QWRs are summarized in Section 3. In Section 4, we discuss picosecond experiments on real-space transfer of carriers and excitons. Femtosecond studies of carrier transport and trapping are presented in Section 5 which is followed by some conclusions (Section 6).

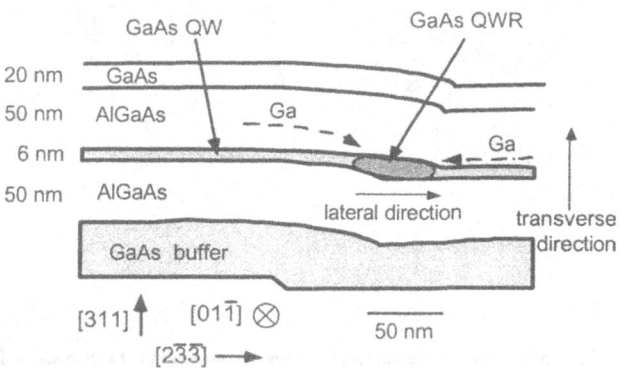

Figure 1. Schematic of the GaAs quantum wire structure. The sample is grown on a patterned GaAs (311)A substrate with 15 nm high mesa stripes. Formation of the quantum wire is due to the lateral migration of Ga atoms during the growth of a nominally 6 nm thick quantum well. At the sidewall, a quantum wire of 13 nm thickness and about 50 nm lateral width is formed. The structure is capped with a 20 nm GaAs layer.

2. Experimental techniques

Fig. 1 shows a schematic of the quantum wire structure investigated in our experiments. The sample was grown by molecular beam epitaxy on patterned GaAs (311)A substrates [4]. Patterning creates 15 to 20 nm high sidewalls of mesa stripes along the [01-1] direction. On this substrate, a nominally 6 nm thick GaAs quantum well layer clad between 50 nm thick $Al_{0.5}Ga_{0.5}As$ barriers was grown. In the growth process, lateral migration of Ga atoms from the flat quantum well areas towards the sidewall occurs. This results in an increase of quantum well thickness up to 13 nm at the sidewall, thus creating a quasi-one-dimensional confinement of carriers over a lateral width of about 50 nm. Such a single quantum wire was investigated in the NSOM experiments.

The NSOM measurements were performed with a near-field microscope designed for operation in the wide temperature range from 10 to 300 K [23]. The sample is mounted on the cold finger of a helium cryostat which is placed in a vacuum chamber. High spatial resolution is achieved by transmitting excitation light through an aperture of about 100 nm diameter at the end of a metal-coated fiber probe. Probe tips with a cone angle of 10° were pulled from single-mode optical fiber puller and then coated with 50-100 nm aluminum. The distance between the fiber probe and the surface of the sample is stabilized to about 10 nm by a shear-force distance regulation with the help of a quartz tuning fork to which the fiber tip is mechanically attached [23,24].

In the experiments reported in Sections 3 and 4, the sample is excited through the fiber probe either by continuous-wave HeNe or Ti:sapphire lasers or by femtosecond

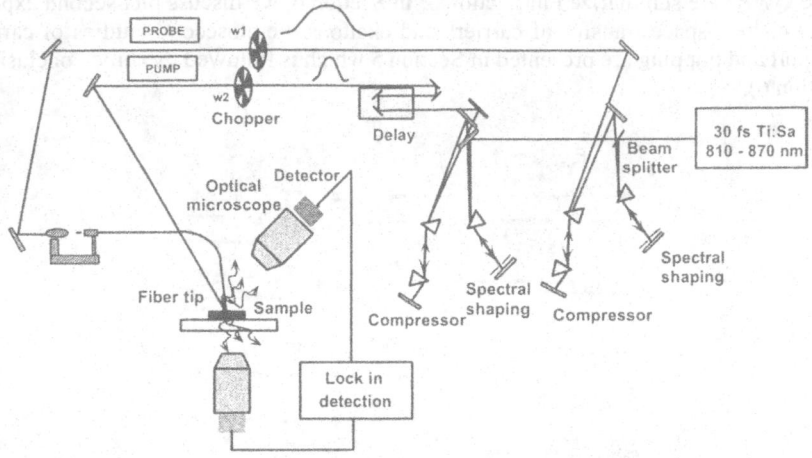

Figure 2. Schematic of the experimental setup for femtosecond near-field spectroscopy. Pump and probe pulses are derived from a mode-locked Ti:sapphire laser using two independent prism arrangements for spectral shaping and compression. In the configuration shown, the pump pulse excites a large area of the sample whereas the probe is transmitted through the near-field probe to detect spatially resolved changes of transmission or reflection. Such nonlinear signals are measured with a lock-in

pulses from a mode-locked Ti:sapphire oscillator. The excitation power on the sample was between 10 and 100 nW, corresponding to very low carrier densities between 10^4 and 10^5 cm^{-1}. The luminescence from the sample was collected with a conventional far-field microscope objective. The luminescence was dispersed in a 0.22 m double monochromator (spectral resolution 1.2 nm) and detected with a silicon avalanche diode, either in steady-state operation or time-resolved with a resolution of 250 ps. Both photoluminescence (PL) and photoluminescence excitation (PLE) spectra of the QWR sample were recorded.

The setup for the femtosecond experiments is shown schematically in Fig. 2. Pump and probe pulses are derived from a modelocked Ti:sapphire oscillator generating 20 to 40 fs pulses which are tunable from 810 to 870 nm (1.53 to 1.425 eV) with an average power of up to 300 mW. The pulse repetition rate is 80 MHz. Pump and probe pulses are derived by splitting the output of the laser and are traveling through separate prism setups for individual spectral and temporal shaping. In particular, group velocity dispersion occurring in the setup is precompensated. The time delay between the two pulses is adjusted with a delay stage in the pump arm. The pump can either be focused onto the sample by a far-field graded index lens (f=8 cm, spot size 30 μm) or be transmitted through the nearfield fiber probe, resulting in a subwavelength excitation spot. The probe pulses are fed into the fiber probe and transmitted onto the sample. Transmitted or reflected probe light is detected through a conventional far-field lens by a photodiode in conjunction with a lock-in amplifier. Background signals are suppressed by spatially and spectrally filtering the probe pulse. In addition, the pump and the probe beam are mechanically chopped at respective frequencies $f_1 = 1.2$ kHz and $f_2 = 2.1$ kHz and the amplified photodiode signal is detected at the sum frequency f_1+f_2. The near-field fiber probes used in the femtosecond experiments were mostly made by chemically etching single mode optical fibers. All femtosecond experiments are taken at a sample temperature of 300 K.

3. Electronic structure and confinement potential of the quantum wire

In the following, we summarize results of spatially resolved PL and PLE experiments under steady state conditions which give insight into the electronic structure and the lateral confinement potential of the QWR [25,26]. In a first series of measurements, the QWR luminescence was recorded as a function of the excitation energy and the position of the excitation tip on the sample. Data for a sample temperature of 10 K were reported in Ref. 25 and give evidence for the quasi-one-dimensional subband structure in the sample. An energy splitting of about 10 meV between the two lowest conduction sub-bands was found, in good agreement with theoretical calculations for a 50 nm wide QWR. The spatial resolution in this measurement was 250 nm, mainly determined by the geometry of the sample with the QWR 50 nm below the sample surface. At 10 K,

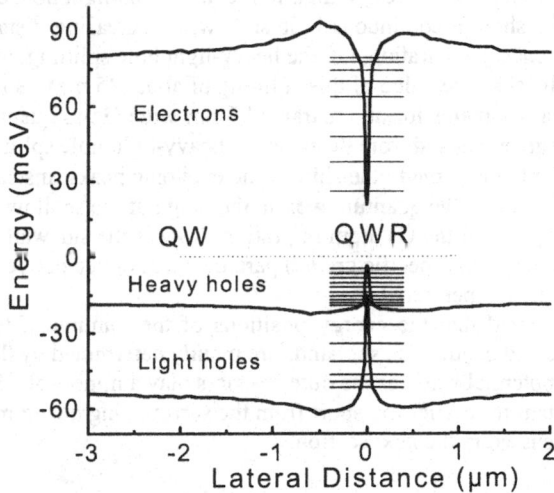

Figure. 3. Lateral confinement potential of the quantum wire structure. The central part was derived from the thickness variation of the GaAs quantum well layer (c.f. Fig. 1), the outer parts from spatially resolved PL and PLE measurements. The lateral potential displays barriers on a submicron scale which strongly affect lateral carrier transfer at low temperatures. The horizontal straight lines indicate the minima of the quasi-one-dimensional valence and conduction subbands.

QWR luminescence only occurs if the excitation tip is located at the QWR position whereas QWR luminescence is absent after excitation of the embedding GaAs quantum well. This points to a very effective suppression of real-space transfer and trapping of carriers from the quantum well into the QWR and thus to the existence of barriers in the lateral confinement potential of the structure. At higher sample temperatures, one finds QWR luminescence originating from carriers initially generated in the quantum well and transferred into the QWR. A spatially resolved PLE study of this behavior gives direct insight into the spatial width and the height of those barriers [26]. The complete lateral confinement potential of the QWR in Fig. 3 was derived from such mea-measurements and - for the central 100 nm around y=0 - from the lateral thickness variation of the quantum well layer which was determined from a cross-sectional TEM image of the structure [4]. The standard ratio of 2:1 for the conduction to valence band offset energy in quasi-two-dimensional $GaAs/Al_{0.5}Ga_{0.5}As$ structures was used in this analysis. The quasi-one-dimensional subband structure and the interband transition energies of the QWR were calculated within the adiabatic approximation for solving the Schrödinger equation, neglecting Coulomb correlation effects. The subband energies and interband transition energies displayed in Fig. 3 (lines) are in reasonable agreement with the low-temperature PLE spectrum reported in Ref. [25].

There are essentially two mechanisms from which the broad barriers in the vicinity of the QWR could originate, (i) a local variation of quantum well thickness due to the growth mechanism which involves lateral transport of Ga atoms, or (ii) strain induced by growing the QWR on a pre-structured mesa-like substrate. A series of excitation spectra of quantum well luminescence was recorded clarify this issue [27]. The PLE spectra display two peaks corresponding to the heavy and light hole exciton transitions. The two peaks show a continuous blue-shift with decreasing distance from the QWR whereas their energy separation, i.e. the heavy-light hole splitting, remains essentially unchanged. The absolute value of this splitting of about 35 meV is in good agreement with theoretical estimates for an unstrained 6 nm wide GaAs quantum well. We conclude from this agreement and from the constant heavy-light hole splitting that strain plays a minor role for the observed blue shift of the excitonic peaks. Instead, the shift is related to a local thinning of the quantum well in the range of the shallow barriers. This thinning is a consequence of the Ga atom migration towards the sidewall in the growth process and determined by the specific growth parameters in molecular beam epitaxy, in particular the substrate temperature.

It should be noted that the energy positions of the minima of the quasi-one-dimensional valence and conduction subbands are mainly determined by the central part of the confinement potential whereas the outer barriers play a minor role. Such barriers, however, are important for carrier transport from the surrounding quantum well into the QWR as will be discussed in the next section.

4. Real-space transfer of excitons

Lateral real-space transfer and trapping of excitons into the QWR was studied in experiments combining high spatial and temporal resolution [28,29]. In those experiments, resonant femtosecond excitation through the near-field probe creates a highly spatially localized distribution of quantum well excitons at a well-defined separation y from the location of the QWR. Real-space transfer of excitons and trapping into the QWR give rise to QWR luminescence. A time-resolved measurement of luminescence intensity reveals the rise time of the QWR emission which is determined by the dynamics of real-space transfer and trapping, and the decay of emission due to radiative recombination.

At a sample temperature of 10 K, the barriers in the vicinity of the QWR suppress a transfer of carriers from the quantum well into the QWR very efficiently. As a result, QWR emission is found only for positions of the excitation tip between the two barriers, as is also evident from the spatially resolved QWR PLE spectrum at 10 K [25] and other steady state PL measurements [30]. In Fig. 4, we present data for a higher sample temperature of 100 K. At this temperature and the low excitation densities in our experiments, transport is dominated by the real-space transfer of excitons [31]. The temporal evolution of the QWR is plotted for different values of y, the lateral distance between the excitation spot and the QWR. The solid line at time zero gives the instrumental response function demonstrating a time resolution of 250 ps. For excitation on the QWR (y=0), the luminescence intensity rises within the time resolution and decays monoexponentially (inset: logarithmic plot of the transients) with a recombination time of 1.5 ns. This result demonstrates that trapping of carriers from the high lying states populated by optical excitation at 1.614 eV to the bottom of the QWR from where the emission originates, occurs much faster than our time resolution. Trapping is connected with the emission of longitudinal optical phonons and characterized by time constants on the order of 1 ps [32].

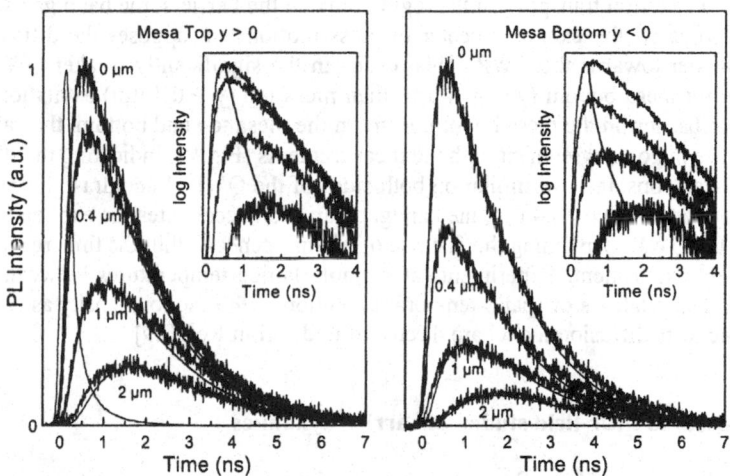

Figure 4. Time evolution of the quantum wire luminescence after resonant excitation of the embedding quantum well on (a) the mesa top and (b) the mesa bottom. The intensity of quantum wire luminescence at 1.54 eV is plotted versus time for different excitation positions y (quantum wire location y=0, sample temperature 100 K). The transients display an increasingly delayed rise for increasing y which is due to the transport of excitons from the excitation spot to the quantum wire: Insets: Data plotted on a logarithmic intensity scale. Solid lines: results of a numerical simulation based on a drift-diffusion model for exciton transport.

With increasing separation of the excitation spot from the QWR location, one observes a continuous increase of the rise time of QWR luminescence, as is evident from the transients for y=0.4, 1.0, and 2.0 μm (Fig. 4). This delay reflects the finite traveling time of excitons from the area of excitation to the QWR, i.e. the dynamics of real-space transfer. The decrease of the overall luminescence intensity for increasing y is due to the fact that a bigger fraction of quantum well excitons undergoes recombination already in the quantum well, i.e. on the way to the QWR. The luminescence recombination time of the quantum well has a value of 1.35 ns. It is interesting to note that this decrease of luminescence intensity is more pronounced for the mesa bottom (y<0) than for the mesa top (y>0). This finding reflects the somewhat higher barrier on the mesa bottom suppressing exciton transfer more effectively.

The time resolved data were analyzed with a drift-diffusion model of exciton transport. There are two relevant current terms, a diffusion term due to the gradient of exciton density, and a drift term caused by the action of the local band gap gradient on the center of mass motion of the excitons. The second term was derived from the lateral potential plotted in Fig. 3. Details of this model have been discussed in Refs.28 and 29. The solid lines in Fig. 4 represent results of the drift-diffusion model for different excitation positions y for T=100 K. For an exciton diffusion coefficient D_{ex} = 13 cm^2/sec, there is very good agreement of the calculations and the experimental results both for y<0 and y>0. This diffusion coefficient corresponds to an excitonic mobility μ_{ex} = 1500 cm^2/Vs in the quantum well, given mainly by the hole mobility and limited in this temperature range by LO phonon scattering.

The model gives a correct description of the influence of the lateral band gap variation on exciton transport. In the region outside the barriers, the band gap variation exerts a force on the excitonic center of mass motion that opposes the diffusive real space transfer towards the QWR. This results in the significantly weaker QWR luminescence for mesa bottom (y < -0.4 µm) than mesa top (y > 0.4 µm) excitation due to the higher barrier on the mesa bottom. Both on the mesa top and bottom, the calculated kinetics are in good agreement with the measurements (Fig. 4), indicating that the exciton diffusion constants are similar on both sides of the QWR. In contrast, in the region inside the barriers, |y| < 0.4 µm, the bandgap variation accelerates the exciton transport towards the QWR, explaining the fast rise of luminescence within the time resolution of the present experiment. Experiments at variable lattice temperatures between 10 and 100 K and simulations of spatio-temporal evolution of the exciton density as calculated within the drift-diffusion model are discussed in detail in Ref. [29]

5. Femtosecond near-field studies of carrier dynamics

The results presented in Section 4 give insight into the transport of quasi-two-dimensional excitons in the QWR structure. In those experiments, the temporal resolution of 250 ps was not high enough to monitor the trapping processes occurring locally in the QWR. For studying ultrafast phenomena, we combined near-field microscopy with femtosecond pump-probe techniques [33]. In such experiments, carrier dynamics in a single QWR is investigated by probing time dependent changes in the nonlinear transmission and/or reflection of the nanostructure which are induced by optical excitation of carriers. In the following, we consider the different contributions to such pump-probe signals and present data on local carrier relaxation and transport phenomena on a 100 fs time scale.

Spatially resolved femtosecond studies were performed with the quasi-two-color near-field pump and probe setup described in section 2. The following optical configurations were used: (i) In a far-field pump/near-field probe geometry, only the probe pulse was sent through the fiber while the pump was focused onto the sample by a far-field graded index lens, resulting in a spot size on the sample of about 30 μm. (ii) In a near-field pump/near-field probe geometry, both pump and probe pulses were transmitted through the same near-field fiber probe. As the substrate of the QWR structure (Fig. 1) is not transparent in the wavelength range of interest, all experiments were performed in a reflection geometry. The probe light reflected from the sample was collected in the far field through a microscope objective, spatially and spectrally filtered and detected with a photodiode. All data were taken at a sample temperature of 300 K.

Far-field pump/near-field probe spectra were measured on the mesa top of the

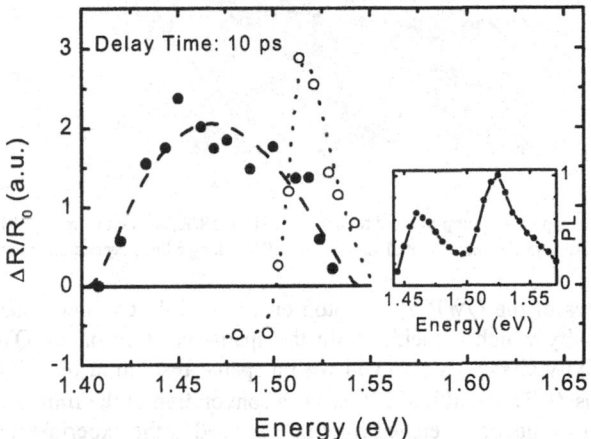

Figure 5. Transient reflectivity spectra of the quantum wire structure for a time delay of 10 ps between pump and probe (T = 300 K). The reflectivity change at the QWR position (solid circles) and on the mesa top quantum well (open circles) are plotted as a function of the photon energy of the probe pulses. Inset: Near-field photoluminescence spectrum at 300 K, displaying the QWR and the quantum well resonance around 1.46 and 1.52 eV, respectively.

sample, i.e. in the quantum well region. The pump pulses with a bandwidth of 40 meV are centered around the quantum well absorption resonance at 1.52 eV. The excitation density was about 3×10^{11} cm^{-2}. The change in reflectivity was recorded with tunable probe pulses of 11 meV bandwidth. For photon energies E_{pr} of the probe between 1.51 eV and 1.55 eV, the creation of electron-hole pairs results in an increase in reflectivity whereas a reflectivity decrease is found for $E_{pr}<1.51$ eV. In Fig. 5, we plot the change of reflectivity $\Delta R/R_0=(R-R_0)/R_0$ for a delay time of 10 ps between pump and probe as a function of E_{pr} (open circles, R,R$_0$: sample reflectivity with and without pump). This spectrum shows a pronounced maximum at $E_{pr}=1.515$ eV and follows – for $E_{pr}\geq1.515$ eV - the shape of the quantum well PL spectrum (inset of Fig. 5). For excitation densities between 10^{10} and 3×10^{11} cm^{-2}, the magnitude of $\Delta R/R_0$ is proportional to the pump intensity. The reflectivity changes rise within the time resolution of the experiment of 200 fs. For $E_{pr}\geq1.515$ eV, the signal decays by carrier recombination on a nanosecond time scale. For $E_{pr}<1.50$ eV, a fast partial decay is found within the first 3 ps, followed by the slow nanosecond decay.

386

Reflectivity changes in the QWR region of the sample were measured with both pump and probe pulses transmitted through the near-field probe. Carriers in high-lying states of the QWR, in the embedding quantum well, in the GaAs cap layer, and – with much lower density - in the GaAs substrate were created by excitation at 1.51 eV. Prob-

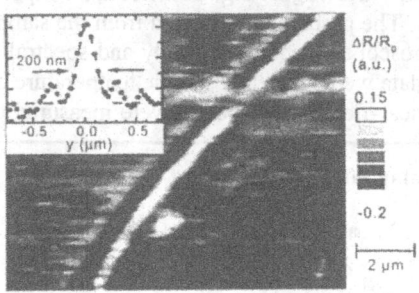

Figure 6. Spatial map of the pump-induced reflectivity change $\Delta R/R_0$ at a delay time t_d= 10 ps. The probe laser is set to 1.45 eV. Inset: Spatial variation of $\Delta R/R_0$ along a line perpendicular to the wire.

ing low-lying states of the QWR at a photon energy of 1.45 eV, one observes a local change of reflectivity which coincides with the spatial position of the QWR (Fig. 6). This local reflectivity change is resolved with a spatial resolution of 200 nm, i.e. less than $\lambda/4$ (Fig. 6 inset). This width of 200 nm is a convolution of the finite spatial resolution obtained with the uncoated etched fiber probes used in the experiment [34] and the lateral spreading of the light pulses within the top layers of the sample. The local change of the QWR reflectivity is superimposed on a background signal from the GaAs cap layer and the embedding quantum well which varies slowly in space. The solid circles in Fig. 5 represent the spectral dependence of $\Delta R_{QWR}/R_0$, the reflectivity change from the QWR minus the background signal for a delay time of 10 ps. The reflectivity spectrum covers the same spectral range as the PL spectrum of the QWR (inset of Fig. 5) and exhibits a similar spectral envelope. The local change of reflectivity is absent for excitation at 1.44 eV where carriers are generated exclusively in the GaAs cap layer and the GaAs substrate. This finding confirms that the local change of reflectivity is due to the QWR.

The time evolution of the reflectivity change due to the QWR gives information on the local carrier dynamics. This was studied with far-field excitation by a 50 fs pulse at 1.52 eV. At this energy, a spatially homogeneous carrier distribution is generated with a diameter of about 30 μm. The excitation density was 3×10^{11} cm^{-2}, corresponding to non-degenerate excitation conditions at T = 300 K. Probe pulses at 1.475 eV, near the QWR resonance, were transmitted through the fiber probe. In Fig. 7 (a), the reflectivity change at this photon energy is plotted as a function of delay time (abscissa) and of the lateral distance y between the QWR position (y=0) and the position of the near-field probe (ordinate). On both sides of the QWR, a spatially homogeneous transient reflectivity decrease is observed that decays on a time scale of several picoseconds (Fig. 7 b, solid circles). This signal is dominated by the carrier-induced reflectivity change of the

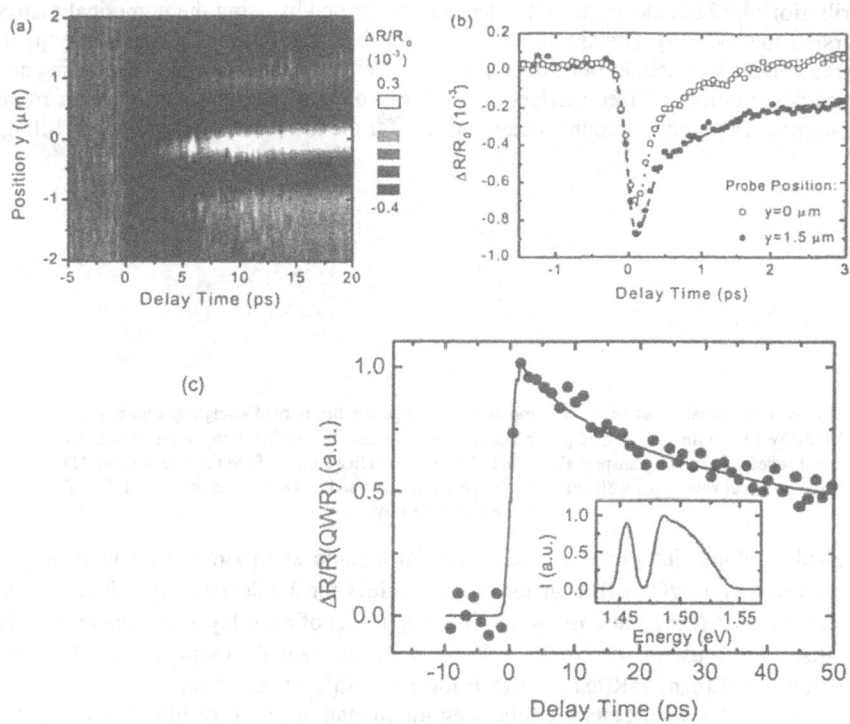

Figure 7. (a) Transient reflectivity change $\Delta R/R_0$ of the QWR structure for far-field excitation at 1.52 eV and near-field probing at 1.475 eV. The change of reflectivity is plotted as a function of time delay between pump and probe (abscissa) and of the lateral distance y between the quantum wire location and the position of the near-field probe (ordinate). One observes a signal enhancement at the quantum wire. (b) Time evolution of $\Delta R/R_0$ at the QWR position (y=0, open circles) and at y=1.5 μm (solid circles). (c) Time dependent reflectivity change of the QWR $\Delta R_{QWR}/R_0 = (\Delta R(y=0) - \Delta R(y=1.5 \mu m))/R_0$.

GaAs cap layer. In contrast, a decrease in reflectivity with a smaller amplitude occurs at the QWR position (y=0, open circles). The QWR contribution, ΔR_{QWR}, is extracted by taking the difference of the two transients in Fig. 7 (b). The time evolution of ΔR_{QWR} (Fig. 7 c) shows a step-like behavior with an ultrafast rise within 200 fs, the time resolution of the experiment, and a constant value up to delay times of 50 ps. Similar transients are measured with pump pulses attentuated by a factor of 3. A similar temporal evolution of ΔR_{QWR} is observed for resonant far-field QWR excitation (E_{pu} = 1.47 eV) and spatially resolved probing at energies of high-lying QWR states around 1.51 eV.

In general, reflectivity spectra of a multilayer semiconductor nanostructure depend sensitively on the details of the layer structure. To interpret our experiments in terms of carrier dynamics, we performed an analysis of the pulse propagation through the multilayer QWR structure (Fig. 1). This structure consists of the GaAs cap layer, the GaAs quantum well in which the QWR is embedded, the GaAs substrate and - in-between - the AlGaAs barriers. For photon energies E_{pump}<1.44 eV, only the cap layer and the substrate are excited, for 1.44 eV<E_{pump}<1.50 eV those layers and the QWR. For higher photon energies, carriers are also created in the quantum well. Carrier generation leads to a modification of the optical susceptibility $\chi(\omega, n_{ex})$ of each layer with carrier density n_{ex}. This results in a change of reflectivity measured at the different probe energies. A numerical simulation of these changes assuming thermalized carrier

388

distributions for both electrons and holes was performed by using the numerical matrix inversion method [35]. The assumption of thermalized distributions is reasonably justified reasonably justified, as our time resolution is less than the sub-100 fs dephasing and thermalization times of free carriers in quasi-two-dimensional quantum wells at room temperature The model accounts quantitatively for the influence of phase-space filling,

 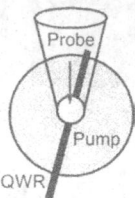

Figure 8. Temporal variation of the pump-induced change in reflectivity of a single quantum wire, $\Delta R/R(QWR)$ on a time scale of 50 ps for localized near-field excitation. The decay of the reflectivity signal reflects the carrier transport along the QWR. The solid line shows a fit to the data within a 1D diffusion model with $D_{QWR} = 80$ cm²/s. Inset: Spectra of pump and probe lasers, centered at 1.50 and 1.46 eV, respectively.

screened Coulomb interaction, here treated in a static approximation, and band-gap renormalization on $\chi(\omega,n_{ex})$. It gives absolute values for the density dependent absorption coefficient $\alpha(\omega,n_{ex})$ and refractive index $n_r(\omega,n_{ex})$ of each layer. A transfer matrix formalism is then used to derive the reflectivity $R(\omega,n_{ex})$ and the carrier-induced change in reflectivity $\Delta R(\omega,n_{ex})=R(\omega,n_{ex})-R(\omega,0)$ for the multilayer structure.

This analysis suggests the following interpretation of our results: For $E_{pump}>1.5$ eV, excitation of the quantum well leads to a decrease of the excitonic quantum well absorption around $E_{pr}=1.52$ eV. The spectral dependence of $\Delta R_{QW}/R$ identifies phase-space filling as the dominant contribution to the excitonic QW nonlinearity, in agreement with earlier studies [36]. This results in an increase in reflectivity (open circles in Fig. 5) with a spectral dependence close to the quantum well PLE spectrum. At probe energies $E_{pr}<1.5$ eV, the reflectivity signal of opposite sign is mainly due to the cap layer. The change in sign occurs because the cap layer signal is most sensitive to the change in refractive index, i.e. the real part of the susceptibility. This effect is due to the large jump in refractive index at the sample surface. The observed decay of the pump-induced reflectivity change on a time scale of a few picoseconds - which is observed in all experiments for probe positions outside the wire region - reflects the trapping of carriers in the cap layer into surface states.

Now we consider the signals that are measured with the fiber probe at the position of the QWR. Here, the additional contribution due to carriers in the QWR results in a smaller amplitude of the change of reflectivity (Fig. 7 c). This contribution is described by the quantity ΔR_{QWR}. The spectrum of ΔR_{QWR} in Fig. 5 resembles the PLE spectrum of the QWR. As for the QW, the pump-induced change in the probe reflectivity reflects the carrier-induced bleaching of the excitonic QWR absorption. This bleaching id due to photoexcited carriers in the QWR and, thus, ΔR_{QWR} is an efficient probe for the temporal evolution of the QWR carrier density. While, at low temperatures ΔR_{QWR} is a probe of the sum of the local concentrations of electrons and holes within the QWR region, the different distribution functions at room temperature makes the signal more sensitive to the electron dynamics [35]. The observation of a reflectivity change around 1.46 eV thus requires carrier redistribution from optically excited high lying QWR and QW continuum states to the bottom of the QWR. The absence of any slower dynamics on the transient in Fig. 7 suggests that this redistribution occurs within the first 200 fs. At room temperature and under our excitation conditions this relaxation involves a complex scattering scenario with contributions from both carrier-LO phonon scattering and carrier-carrier scattering. Experiments performed for different pump and probe wavelengths suggest that a thermalized carrier distribution is formed within the first 200 fs, similar to the thermalization dynamics in quantum wells.

The experiments discussed so far were performed with far-field excitation. Within the excitation spot of 30 μm diameter, a homogeneous carrier distribution is generated. Thus, a diffusive motion of carriers along the QWR is not visible for far-field excitation and the reflectivity signal decays with the electron-hole recombination time of 1.9 ns, identical to the luminescence decay time. A different behavior occurs for near-field excitation through the fiber probe, as is shown in Fig. 8 [37]. For near-field excitation at 1.52 eV and near-field probing at 1.46 eV the reflectivity change of the QWR, $\Delta R_{QWR}/R_0$, displays a fast rise and a subsequent decrease by more than 50 percent on a time scale of 50 ps. The diffusion of carriers out of the spatially confined excitation volume results in a decrease of the local carrier concentration within the probe spot and - thus - in a reduction of the pump-induced reflectivity change which is monitored in this measurement.

Transport of carriers out of the excitation volume can occur both along the QWR and in the lateral direction, i.e. from the QWR into the quantum well. In our experiment, carriers are generated in a 400 nm spot in high-lying QWR and in quantum well states. The data in Fig. 6 (c) suggest that a quasi-equilibrium population of both QWR and quantum well states is formed within the first 200 fs. In such a thermalized distribution consisting of a total about 10^{11} electron-hole pairs per cm^2, the spatial distributions of electrons and holes are different because of the different confinement potentials for electrons (65 meV) and heavy holes (15 meV). The density of electrons in the QWR is estimated to be about four times higher than the QWR hole density. A larger fraction (about 90%) of the photoexcited electrons is confined inside the QWR while most of the holes are populating quasi-continuum states of the embedding QW. Thus, the experiment probes mainly the local dynamics of electrons inside the QWR. Considering the small fraction of electrons that is populating QW continuum states and the ambipolar QW diffusion coefficient $D_{QW} = 12$ cm^2/s, one finds a negligible contribution of the lateral transport to the transient in Fig. 8. Thus the decay of the reflectivity change in Fig. 8 reflects mainly the transport along the QWR axis. Assuming a drift-diffusion model for this transport, the observed decay of the pump-induced change in QWR reflectivity is well described by taking a quasi-one-dimensional diffusion coefficient $D_{QWR} = 80 \pm 20$ cm^2/s. This value is substantially higher than the ambipolar 2D diffusion coefficient $D_{QW} = 12$ cm^2/s. The 2D value corresponds to a mobility of 500 cm^2/Vs which is close to the mobility of holes in GaAs at 300 K. Thus, at room temperature, hole diffusion governs the ambipolar transport through the QW and the hole mobility is limited by scattering with phonons via the optical deformation potential. On the contrary we find experimentally a significantly larger 1D diffusion coefficient D_{QWR} that corresponds to a mobility of 3000 cm^2/Vs, which is close to the RT mobility of electrons in GaAs.

This suggests that, in our experiments, the transport of electrons along the QWR axis is strongly different from a conventional ambipolar diffusive transport regime, with a spatially and temporally correlated motion of electrons and holes and a mobility that is hole limited. Instead, we observe a rapid motion of electrons out of the excitation volume, along the QWR axis, with a mobility that is basically unaffected by the hole distribution within the QWR. This indicates that, because of the excess density of electrons in the QWR that is created with spatially and temporally resolved excitation and the pronounced spatial gradient of the electron density along the QWR, the repulsion between electrons is not compensated by the Coulomb attraction between electrons and holes. Thus electrons propagate mainly independently from the hole distribution and this gives rise to a unipolar transport of electrons along the wire axis and the observed enhancement of the 1D mobility.

6. Conclusions

In conclusion, single quantum well embedded GaAs quantum wires were studied by spatially and temporally resolved near-field spectroscopy in a wide temperature range. The lateral confinement potential of the quantum wire and the quasi-one-dimensional subband structure were determined in steady-state photoluminescence and photoluminescence excitation measurements. We demonstrated the existence of broad shallow barriers in the vicinity of the quantum wire which originate from a thinning of the embedding quantum well in the growth process. Such barriers suppress carrier transport from the quantum well into the quantum wire at low temperatures around 10 K. Exciton transport and trapping was studied in picosecond luminescence experiments at elevated temperatures. An analysis of those results on the basis of a drift-diffusion model highlights the influence of the barriers on the drift-diffusive exciton transport and allows extraction of temperature dependent exciton mobilities. Femtosecond pump-probe experiments with the near-field microscope demonstrate carrier trapping into the quantum wire on a 200 fs time scale at room temperature. Both Coulomb and optical phonon scattering of carriers contribute to this fast local relaxation. First pump-probe studies of carrier transport along the quantum wire suggest an electron mobility substantially higher than expected for ambipolar diffusion of electron-hole pairs and indicate an unipolar transport of electrons along the wire axis. These experiments demonstrate the potential of femtosecond nearfield techniques as a powerful novel tool for studying nonequilibrium carrier dynamics on ultrafast time and nanometer length scales.

7. Acknowledgments

Part of this research was supported by the Deutsche Forschungsgemeinschaft (SFB 296) and by the European Commission through the ULTRAFAST network, the EFRE program and a Marie Curie fellowship for V. Emiliani.

References

1. Kummell, T., Bacher, G., Forchel, A., Nürnberger, J., Faschinger, W., Landwehr, G., Jobst, B., and Hommel, D., (1997) Fabrication of dry etched CdZnSe/ZnSe quantum wires by thermally assisted electron cyclotron resonance etching. *Appl. Phys. Lett.* **71**, 344-346.

2. Goni, A. R., Pfeiffer, L. N., West, K. W., Pinczuk, A., Baranger, H. U., and Stormer, H. L. (1992). Observation of quantum wire formation at intersecting quantum wells. *Appl. Phys. Lett.* **61**, 1956-1958.

3 Kapon, E., Hwang, D. M., and Bhat, R. (1989). Stimulated emission in semiconductor quantum wire heterostructures. *Phys. Rev. Lett.* **63**, 430-433.

4. Notzel, R., Ramsteiner, M., Menniger, J., Trampert, A., Schonherr, H.-P., Daweritz, L., and Ploog, K. H. (1996). Patterned growth on high-index GaAs (n11) substrates: application to sidewall quantum wires. *J. Appl. Phys.* **80**, 4108-4111.

5. Legrand, B., Grandidier, B., Nys, J. P., Stévenard, D., Gérard, J. M., Thierry-Mieg, V. (1998). Scanning tunneling microscopy and scanning tunneling spectrosocopy of self-assembled InAs quantum dots. *Appl. Phys. Lett.* **73**, 96-98.

6. Eder, C., Smoliner, J., and Strasser, G. (1996). Local barrier heights on quantum wires determined by ballistic electron emission microscopy. *Appl. Phys. Lett.* **68**, 2876-2878.

7. Rubin, M. E., Medeirosribeiro, G., Oshea, J. J., Chin, M. A., Lee, E. Y., Petroff, P. M., and Narayanamurti, V. (1996). Imaging and spectroscopy elf single InAs self-assembled quantum dots using ballistic electron emission microscopy. *Phys. Rev. Lett.* **77**, 5268-5271.

8 Walther, M., Kapon, E., Christen, J., Hwang, D. M., and Bhat, R. (1992). Carrier capture and quantum confinement in GaAs/AlGaAs quantum wire lasers grown on V-grooved substrates. *Appl. Phys. Lett.* **60**, 521-523.

9. Grundmann, M., Christen, J., Joschko, M., Stier, O., Bimberg, D., and Kapon, E. (1994). Recombination kinetics and intersubband relaxation in semiconductor quantum wires. *Semicond. Science and Technol.* **9**, 1939-1945.

10. Pohl, D.W., Denk, W., and Lanz, M. (1984). Optical stethoscopy: image recording with resolution $\lambda/20$. *Appl. Phys.. Lett.* **44**, 651-653.

11. Betzig, E., Trautman, J. K., Harris, T. D., Weiner, J. S., and Kostelak, R. L. (1991). Breaking the diffraction barrier: optical microscopy on a nanometric scale. *Science* **251**, 1468-1470.

12. Hess, H. F., Betzig, E., Harris, T. D., Pfeiffer, L. N., and West, K. W. (1994). Near-field spectroscopy of the quantum constituents of a luminescent system. *Science* **264**, 1740-1745.

13. Grober, R. D., Harris, T. D., Trautman, J. K., Betzig, E., Wegscheider, W., Pfeiffer, L., and West, K. (1994). Optical Spectroscopy of a GaAs/AlGaAs Quantum Wire Structure Using Near-Field Scanning Optical Microscopy. *Appl. Phys. Lett.* **64**, 1421-1423.

14. Harris, T. D., Gershoni, D., Grober, R. D., Pfeiffer, L., West, K., and Chand, N. (1996). Near-field optical spectroscopy of single quantum wires. *Appl. Phys. Lett.* **68**, 988-990.

15. Emiliani, V., Lienau, Ch., Hauert, M., Colí, C., DeGiorgi, M., Rinaldi, R., Passaseo, A., and Cingolani, R. (1999). Near-field low temperature photoluminescence spectroscopy of single V-shaped quantum wires. *Phys. Rev. B* **60**, 13335-13338.

16. Steininger, F., Knorr, A., Stroucken, T., Thomas, P., and Koch, S. W. (1996). Dynamic evolution of spatiotemporally localized electronic wave packets in semiconductor quantum wells. *Phys. Rev. Lett.* **77**, 550-553.

17. Hanewinkel, B., Knorr, A., Thomas, P., and Koch, S. W. (1999). Near-field dynamics of excitonic wave packets in semiconductor quantum wells. *Phys. Rev. B* **60**, 8975-8983.

18. Stark, J. B., Mohideen, U., Betzig, E., and Slusher, R. E. (1996). "Time-resolved nonlinear near-field optical microscopy of semiconductor microdisks." In: Ultrafast Phenomena IX (P. F. Barbara, W. H. Knox, G. A. Mourou, and A. H. Zewail, eds.), Springer Series in Chemical Physics, Berlin, pp. 349-350.

19. Levy, J., Nikitin, V., Kikkawa, J. M., Cohen, A., Samarth, N., Garcia, R., and Awschalom, D. D. (1996). Spatiotemporal near-field spin microscopy in patterned magnetic heterostructures. *Phys. Rev. Lett.* **76**, 1948-1951.

20. Nechay, B. A., Siegner, U., Morier-Genoud, F., Schertel, A., and Keller, U. (1999). Femtosecond near-field optical spectroscopy of implantation patterned semiconductors. *Appl. Phys. Lett.* **74**, 61-63.

392

21. Achermann, M., Nechay, B. A., Morier-Genoud, F., Schertel, A., Siegner, U., and Keller, U. (1999) Direct experimental observation of different diffusive transport regimes in semiconductor nanostructures. *Phys. Rev. B* **60**, 2101-2105.

22. Achermann, M., Nechay, B. A., Siegner, U., Hartmann, A., Oberli, D., Kapon, E., and Keller, U. (2000). Quantization energy mapping of single V-groove GaAs quantum wires by femtosecond near-field optics. *Appl. Phys. Lett.* **76**, 2695-2697.

23. Behme, G., Richter, A., Suptitz, M., and Lienau, C. (1997). Vacuum near-field scanning optical microscope for variable cryogenic temperatures. *Rev. Sci. Instrum.* **68**, 3458-3463.

24. Karrai, K., and Grober, R. D. (1995). Piezoelectric tip-sample distance control for near field optical microscopes. *Appl. Phys. Lett.* **66**, 1842-1844.

25. Richter, A., Behme, G., Suptitz, M., Lienau, C., Elsaesser, T., Ramsteiner, M., Notzel, R., and Ploog, K. H. (1997). Real-space transfer and trapping of carriers into single GaAs quantum wires studied by near-field optical spectroscopy. *Phys. Rev. Lett.* **79**, 2145-2148.

26. Lienau, C., Richter, A., Behme, G., Suptitz, M., Heinrich, D., Elsaesser, T., Ramsteiner, M., Notzel, R., and Ploog, K. H. (1998). Nanoscale mapping of confinement potentials in single semiconductor quantum wires by near-field optical spectroscopy. *Phys. Rev. B* **58**, 2045-2049.

27. Emiliani, V., Guenther, T., Intonti, F., Richter, A., Lienau, C., and Elsaesser, T. (1999). Spatially and temporally resolved near-field spectroscopy of single GaAs quantum wires. *J. Phys.: Condens. Matter* **11**, 5889-5900.

28. Richter, A., Suptitz, M., Heinrich, D., Lienau, C., Elsaesser, T., Ramsteiner, M., Notzel, R., and Ploog, K. H. (1998). Exciton transport into a single GaAs quantum wire studied by picosecond near-field optical spectroscopy. *Appl. Phys. Lett.* **73**, 2176-2178.

29. Richter, A., Süptitz, M., Lienau, C., Elsaesser, T. Ramsteiner, M., Nötzel, R., and Ploog, K. H. (1999). Time-resolved near-field optics: Exciton transport in semiconductor nanostructures. *J. Microsc.* **194**, 393-400.

30. Richter, A., Behme, G., Suptitz, M., Lienau, C., Elsaesser, T., Ramsteiner, M., Notzel, R., and Ploog, K. H. (1997). Near-field optical spectroscopy of carrier exchange between quantum wells and single GaAs quantum wires. *phys. stat. sol (b)* **204**, 247-250.

31. Hillmer, H., Forchel, A., and Tu, C. W. (1992). Enhancement of electron-hole pair mobilities in thin GaAs/Al/sub x/Ga/sub 1-x/As quantum wells. *Phys. Rev. B* **45**, 1240-1245.

32. Ryan, J. F., Maciel, A. C., Kiener, C., Rota, L., Turner, K., Freyland, J. M., Marti, U., Martin, D., Moriergemoud, F., and Reinhart, F. K. (1996). Dynamics of electron capture into quantum wires. *Phys. Rev. B* **53**, R4225-R4228.

33. Guenther, T., Emiliani, V., Intonti, F., Lienau, C., Elsaesser, T., Nötzel, R., and Ploog, K. H. (1999). Femtosecond near-field spectroscopy of a single GaAs quantum wire. *Appl. Phys. Lett.* **75**, 3500-3502.

34. Müller, R., and Lienau, C. (2000). Propagation of femtosecond optical pulses through uncoated and metal-coated near-field fiber probes. *Appl. Phys. Lett* **76**, xxxx-xxxx.

35. Schmitt-Rink, S., Ell. C., and Haug, H. (1986). Many-body effects in the absorption, gain, and luminescence spectra of semiconductor quantum-well structures. *Phys. Rev. B* **33**, 1183-1189.

36. Hunsche, S., Leo, K., Kurz, H., and Köhler, K. (1994). Exciton absorption saturation by phase-space filling: influence of carrier temperature and density. *Phys. Rev. B* **49**, 16565-16568.

37. Emiliani, V., Guenther, T., Lienau, C., Nötzel, R, and Ploog, K. H. (2000). Ultrafast near-field spectroscopy of quasi-one-dimensional transport in a single quantum wire. *Phys. Rev. B* **61**, R10583-R10586.

OPTICAL CHARACTERISTICS OF NANOSTRUCTURED III-V COMPOUNDS

I.M. TIGINYANU,[1] C. SCHWAB,[2] A. SARUA,[3] G. IRMER,[3]
J. MONECKE,[3] I. KRAVETSKY,[1] J. SIGMUND,[4] H.L. HARTNAGEL[4]
[1] *Technical University of Moldova, MD-2004 Chisinau, Moldova*
[2] *CNRS / PHASE, BP 20, F-67037 Strasbourg Cedex 2, France*
[3] *Technical University Freiberg, D-09596 Freiberg, Germany*
[4] *Technical University Darmstadt, D-64283 Darmstadt, Germany*

1. Introduction

Over the last years, nanostructuring emerged as an alternative to the search of new materials. It allows one to engineer the material properties just by reducing the dimensions or tailoring the architecture of macroscopic structures on the nanometer scale. Electrochemistry offers an accessible and cost-effective approach for semiconductor nanostructuring. Silicon, for instance, in the form of electrochemically-manufactured nanoporous layers exhibits intense visible luminescence and seems to be very promising for applications in optoelectronics [1].

The possibility for controlling the physical properties just by reducing the dimensions proves to be very attractive, especially when applied to III-V compounds and alloys. One can notice that, compared with porous Si, artificially created porous III-V materials offer potential advantages for applications. Apart from those related to the possibility of changing the chemical composition, the shift from element to compound leads to the occurrence of new physical properties, specific to acentricity, like Fröhlich-type vibrations and optical second harmonic generation (SHG) [2-4]. The Fröhlich modes represent elementary excitations caused by the different polarization of the composite constituents in the electromagnetic field and, consequently, they are expected to occur whenever the wavelength of the incident radiation becomes greater than the average size of the crystallites [5,6]. In connection with strong tendencies to further miniaturization in modern electronics, it is obvious that the properties of many elements in optoelectronic circuits are governed by surface-related excitations. So, the study of surface-related vibrations becomes an important scientific task.

In this work, we present the results of a micro-Raman scattering (RS) study of the first-order modes in porous GaP membranes, with particular emphasis on the surface-related vibrations occurring in honeycomb-like porous GaP structures. The impact of filling the pores with water upon the Fröhlich vibrational modes of porous GaP is studied both analytically and experimentally. Apart from that, we investigate the characteristics of optical SHG in porous membranes of gallium phosphide and deduce the phase matching conditions based on the porosity-induced anisotropy. We show analytically

393

L. Pavesi and E. Buzaneva (eds.), Frontiers of Nano-Optoelectronic Systems, 393–403.
© 2000 *Kluwer Academic Publishers. Printed in the Netherlands.*

that the phase matching conditions can be fulfilled for GaP membranes possessing a degree of porosity higher than 30 %.

2. Experimental Details

The n-GaP substrates used by us were cut from liquid-encapsulation-Czochralsky-grown Te-doped ingots with the free electron concentration of $(0.5-1) \times 10^{18}$ cm^{-3} at 300 K. To fabricate free-standing membranes of porous GaP, (100)- and (111)A-oriented samples were subjected to a 5-MeV Kr$^+$ ion implantation with subsequent anodic etching in an aqueous solution of sulfuric acid under the conditions described in [7,8]. The 2-μm thick porous membranes were detached from the substrate using lateral etching. The formation of the pores was evidenced by images taken with a Scanning Electron Microscope (SEM).

The micro-Raman characterization of porous membranes was carried out at room temperature under excitation provided by the 647.089-nm line of a krypton laser. To avoid local heating of porous structures [2], the laser beam power was limited to 1 mW for a spot diameter at the sample surface of about 2 μm. The scattered light, in a nearly backscattering geometry, was analyzed by a Jobin-Yvon triple monochromator with spectral resolution less than 0.5 cm^{-1}.

Polarized SHG measurements were carried out in a transmission mode. As a fundamental beam, the 1064 nm output of a Q-switched Nd-YAG laser (Spectra Physics GCR-170) with 10-Hz repetition rate and 7-ns pulse width was used. To minimize the influence of the laser output fluctuations, the measured second harmonic (SH) intensity was normalized by the simultaneously monitored laser intensity in the reference channel. The porous membrane was mounted on a step-motorized rotation stage. The direction of the fundamental beam polarization was changed rotating the half-wave plate placed in front of the sample.

3. Morphology of Porous Structures

The formation of the pores in III-V compounds results from the nonuniform dissolution of the material owing to ever present uncontrolled surface defects such as dislocation loops emergencies. In (100)-oriented n-GaP, for instance, after the initial pitting of the surface, further etching proceeds in the directions both perpendicular and parallel to the surface [7,9]. This latter process gives rise to the secondary pores. As the primary pores deepens, more secondary pores are generated and propagate radially away from it. Since the extension of the porous structure occurs underneath the surface, this is termed a 'catacomb'- like porosity [7]. The border between the porous layer and the substrate is hilly rather than flat.

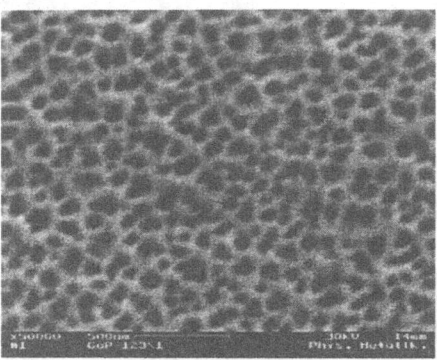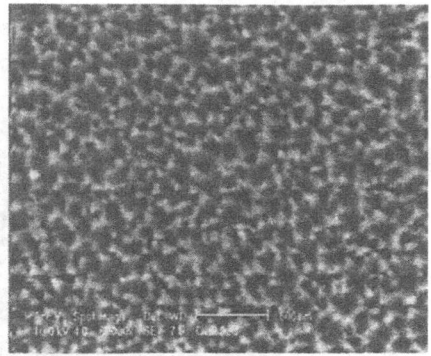

Figure 1. SEM-images of porous GaP membranes prepared on substrates
with (100) and (111)A orientations.

Only two attempts to prepare uniformly sized porous-like structures on III-V materials have been undertaken so far. Takizawa et al [10] fabricated a pillar-like porous structures on (111)A-oriented n-InP using electron beam lithography. Recently [7,8] we have demonstrated that a 5-MeV Kr^+ implantation in n-GaP substrates enables the control of the surface defect density, irrespective of its initial value determined by the crystal growth. The pores left by the dissolved material stretch perpendicularly to the surface and thus leave a network-shaped porous structure whose average lateral dimensions are comparable to the depth of the carrier-depleted surface layer before anodization. Under suitable etching conditions lateral dissolution can be achieved, thus yielding a way to prepare free-standing membranes of porous material. Figure 1 shows SEM images of the top surface for two membranes prepared by Kr^+ ion implantation at the dose $3 \cdot 10^{10}$ cm^{-2} with subsequent anodization of (100)- and (111)A-oriented substrates. One can see that the porous membrane prepared on GaP substrate with a (100) orientation (Figure 1, left-hand image) shows a grid-shaped or honeycomb structure. In contrast, electrochemical etching of a (111)A-oriented GaP surface (Figure 1, right-hand image) leads to a top layer with a pillar structure characterized by lightly interconnected columns stretching perpendicularly to the initial surface.

Further improvement of the morphology of porous GaP was possible to achieve under proper conditions of in-situ illumination. The illumination of samples during anodic etching was found recently to result in a self arrangement of pores [11]. We succeeded to fabricate optically-homogeneous free-standing porous membranes with the thickness up to tens of micrometers. A SEM-image of a porous GaP membrane fabricated on a (111)A-oriented substrate illuminated in-situ is illustrated in Figure 2. One can see that the pores are uniformly distributed and no pronounced fluctuation in their lateral dimensions is observed. The optical transmission spectra of such membranes exhibit pronounced interference fringes at quanta energies lower than the indirect band gap of bulk gallium phosphide, evidencing the optical homogeneity of the porous medium.

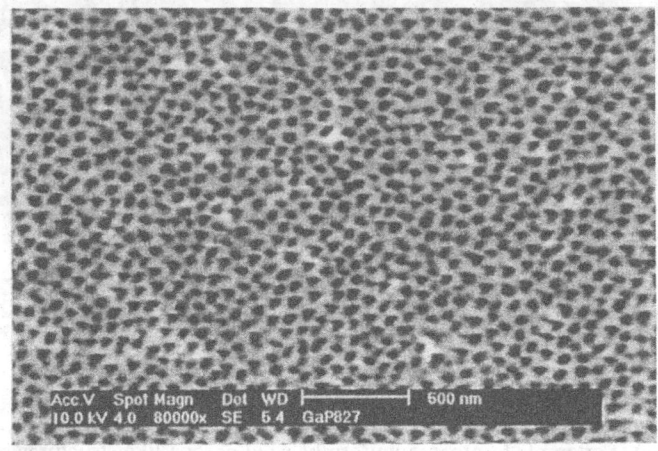

Figure 2. SEM-image of a porous GaP membrane fabricated on a (111)A-oriented substrate illuminated in-situ.

4. Optical Phonon Engineering

As mentioned in the Introduction, nanotexturization of III-V compounds leads to the occurrence of Fröhlich-type vibrational modes. If one considers a single pore in a III-V matrix, than the polarization of the matter in the electromagnetic field gives rise to the appearance of a pore-related dipole which, in its turn, vibrates and emits an infrared radiation. These vibrations of the light-induced polarization, which were predicted by Fröhlich [12], interact with the free carriers like the longitudinal optical (LO) phonons do. Exhibiting a high Raman scattering efficiency, porous III-V layers and membranes can be used as model materials for the purpose of studying the characteristics of the surface-related vibrations as well as the possibilities for their practical use.

Figure 3 illustrates the RS spectra taken from a (100)-oriented bulk GaP sample

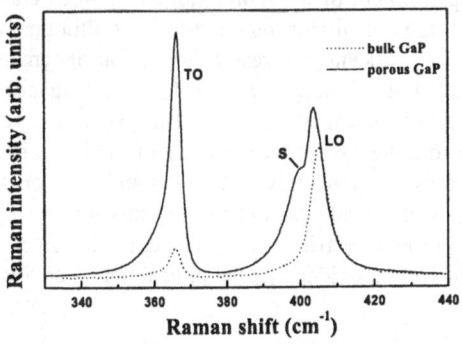

Figure 3. RS spectra from a (100)-oriented bulk GaP substrate (dotted curve) and from a porous membrane (solid curve).

and from a porous membrane obtained by implantation-assisted anodization. The bulk gallium phosphide shows a strong LO phonon at 405.4 cm^{-1}, in agreement with the polarization selection rules. The transverse optical (TO) phonon should be forbidden in (100) backscattering geometry. The observation of a small RS peak at 366 cm^{-1} corresponding to the TO-phonon reflects probably the deviation from a true backscattering geometry. The electrochemical etching results in a strong intensification of Raman scattering at TO-phonons which may be explained taking into account multiple reflections and scattering of light in the porous gallium phosphide network. At the same time a well-defined shoulder appears on the low-energy side of the LO-phonon band attributed earlier to a Fröhlich-type surface-related vibration [2].To demonstrate the Fröhlich character of the RS peak centered in the frequency gap between the TO and LO phonons, a Raman analysis of porous membranes with pores filled in with air and water was undertaken. For this purpose a membrane prepared on a (111)A-oriented sample was immersed in distilled water just after the fabrication and stored there for several hours. The RS spectra of porous membranes with empty and filled pores are shown in Figure 4. One can see that filling in the pores with water has no influence on the position of bulk TO and LO phonons. At the same time, the spectral decomposition evidenced a low-frequency shift of the surface-related phonon from 397 to 388.2 cm^{-1}.

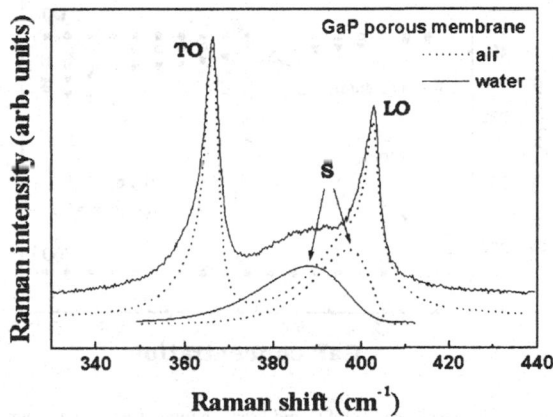

Figure 4. Micro-Raman scattering spectra of a porous GaP membrane before and after filling in the pores with water. In both cases the surface related band was derived from spectral decomposition.

The TO, LO and Fröhlich modes are given by the maxima of Im($\varepsilon(\omega)$) and Im($-1/\varepsilon(\omega)$). On the basis of a one-pole approximation for the calculation of an effective dielectric constant [13] of a heterogeneous material we obtain for a disordered arrangement of parallel pores with circular cross-section interpolated according to [5] to a percolating GaP-skeleton simulated by planes with normals randomly distributed in a plane perpendicular to the pore direction

$$\varepsilon_{eff}(\omega) = \varepsilon_1 \left(1 - \frac{c-\beta}{s} - \frac{\beta}{s-s_0} \right) \qquad (1)$$

where $s = \varepsilon_1 / (\varepsilon_1 - \varepsilon_2)$, $\beta = (1 / 2s_0)^{\cdot} c^{\cdot} (1 - c)$, and $s_0 = \frac{1}{2}(1 - c)$, and c – denotes the GaP concentration, $\varepsilon_2 = \varepsilon_2(\omega)$ is the dielectric function of GaP in the phonon region and ε_1 – is the dielectric constant of material filling the pores ($\varepsilon_1 = 1$ in the case of air and $\varepsilon_1 = 2.3$ in the case of water [14]). From the poles and zeros of $\varepsilon_{eff}(\omega)$ one can deduce un-changed (compared with the bulk value) TO and LO phonon, and a Fröhlich mode which splits into TO-like and LO-like surface related modes for $c \# 1$ (Figure 5). The TO-LO components of the Fröhlich mode show a frequency shift with increasing the degree of porosity, the gradient of this shift being higher for the LO component (low-frequency branch of the Fröhlich mode, see Figure 5). Moreover, the actual frequencies of the Fröhlich mode components prove to depend upon the dielectric constant of the material in the pores. As one can see from Figure 5, the LO-component, expected to prevail in the Raman spectra, show a downward frequency shift of 7-8 cm^{-1} for interme-diate degrees of porosity (40 - 50 %) which is in good agreement with the experimental data (Figure 4).

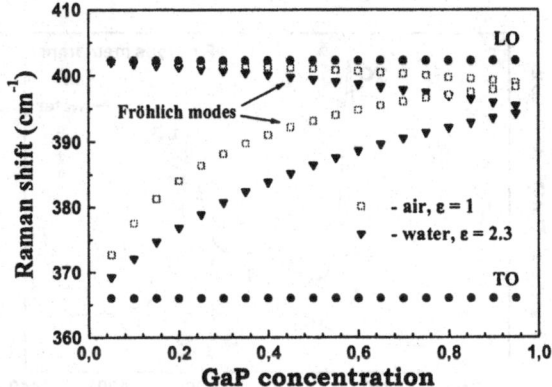

Figure 5. Calculated dependence of TO, LO and Fröhlich mode frequencies upon the GaP concentration in GaP/air and GaP/water nanocomposites.

The dependence of the Fröhlich mode frequencies upon the degree of porosity and the dielectric constant of the surrounding medium is of real practical importance. Since the frequency of a Fröhlich mode can be changed in a controllable manner within the frequency gap of the bulk optical phonons (TO-LO), one can propose completely new design of phonon-assisted optoelectronic devices. In quantum well structure devices [15], for instance, the wells can be made porous. This will allow one to use the Fröhlich-vibration-assisted tunneling as an operating principle, taking the advantage of the unique possibility to control this process by optical means.

5. Optical Second Harmonic Generation

With the development of high-power coherent light sources optical nonlinearity has become a rapidly growing field both in basic research and device applications. Large optical nonlinearities would be required for an all-optical switch expected to provide the building block for optical computers. It is well known that large second-order nonlinear optical coefficients are inherent to III-V compounds. The possibility to introduce anisotropy by anodic etching makes porous III-V materials promising for nonlinear optical applications. Below we present the results of investigation of optical SHG in porous GaP membranes prepared by photon-assisted anodization [8].

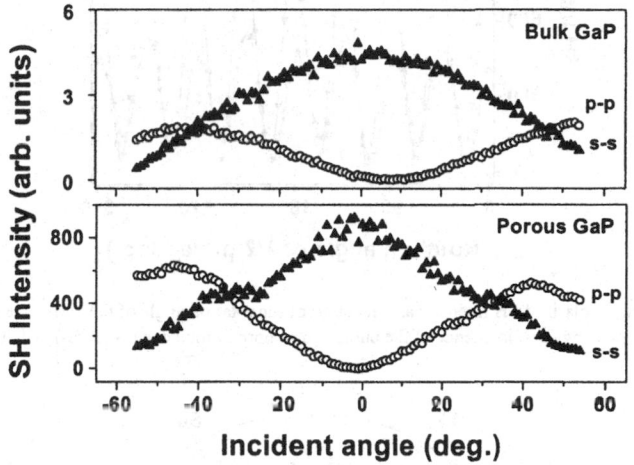

Figure 6. Measured SH intensity as a function of the incident angle of the fundamental beam for bulk and porous GaP. p-p (s-s) denotes p(s)-polarized SH intensity induced by p(s)-polarized fundamental beam, respectively.

Figure 6 illustrates the transmitted p- and s-polarized SH signals ($\lambda_{2\omega} = 532$ nm) from both bulk (111)-oriented GaP and porous membrane as a function of the incident angle of p- and s-polarized fundamental beams. Despite of short coherence length ($L_{coh} \sim$ 1 μm) it is not possible to see Maker fringes in GaP because of a sufficiently high absorption at the SH frequency [16]. The pronounced absorption at 2ω is caused by the fact that the corresponding energy is higher than the indirect band gap of GaP ($2\hbar\omega > E_g$ = 2.24 eV). As one can see from Fig. 1, under identical conditions the porous membranes exhibit a SH intensity of at least two orders of magnitude higher than that inherent to bulk GaP. Another interesting feature is that the fundamental incident angle dependence of the SH intensity for porous membrane measured in s-s polarization geometry shows pronounced shoulders at −35 and +35°. Similar shoulders were found on curve measured in p-s polarization geometry.

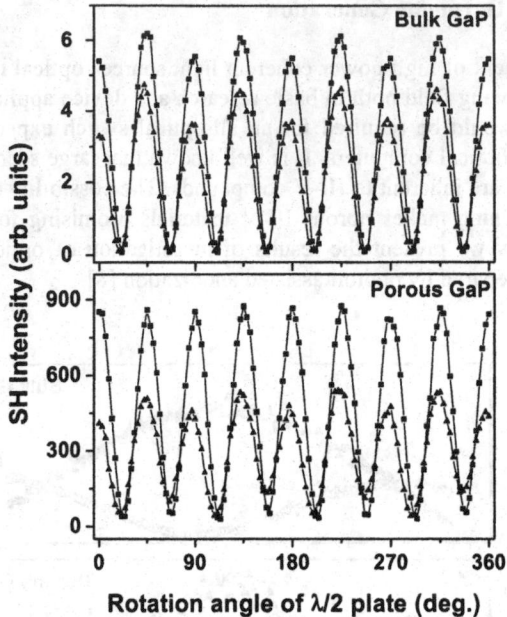

Figure 7. Measured s-polarized SH intensity as a function of the rotation angle of the half-wave plate for bulk and porous GaP at two angles of incidence of the pump beam: normal incidence (squares) and 30° (triangles).

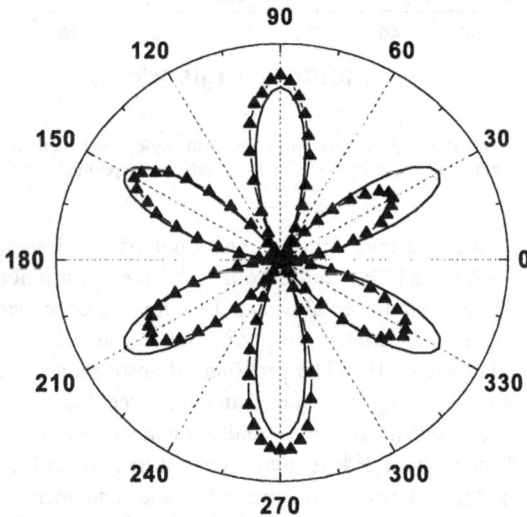

Figure 8. Second harmonic intensity induced by a polarized pump beam at normal incidence as a function of the rotation angle of the porous GaP membrane measured in parallel polarization. Solid triangles represent experimental data, solid line is a fit.

Figure 7 shows the fundamental polarization dependence of the s-polarized SH intensities for bulk and porous GaP at two different angles of incidence of the pump beam. One can notice that the porous membrane exhibits as strong nonlinear optical anisotropy as the bulk GaP. Unlike the case of porous membranes fabricated on (100)-oriented GaP [4], no isotropic contribution induced by the multiple scattering of light in the porous network is observed.

Figure 8 illustrates the rotational dependence of the second harmonic intensity. It reflects perfectly the crystallographic features of (111)-oriented GaP demonstrating the high crystalline quality of the porous skeleton. It is to be noted that in case of strong diffuse scattering any dependence of the SHG signal upon the rotation angle of the porous membrane about the surface normal would be removed.

The absence of an isotropic contribution to the rotational and fundamental polarization dependencies of the SH intensities (Figures 7 and 8) is indicative of the optical homogeneity of the porous medium. In spite of the statistical distribution of pores, the relatively small dimensions of both pores and skeleton entities make the porous medium optically homogeneous and, therefore, the light propagates through it without scattering (since the wavelength of the electromagnetic radiation is much greater than the characteristic dimensions of pores and skeleton, it "sees" a homogeneous medium). The degree of porosity defines the value of the refractive index of the membranes and, in case of preferential orientation of pores along a definite crystallographic direction, the optical anisotropy necessary for phase matching.

According to the effective medium theory [17], in the case of pores aligned along the (111) crystallographic direction, the components of the dielectric tensor of the porous membrane can be written as follows:

$$\varepsilon_\parallel(\omega) = (1-c)\cdot\varepsilon_1 + c\cdot\varepsilon(\omega) \tag{2a}$$

$$\varepsilon_\perp(\omega) = \varepsilon(\omega)\frac{\varepsilon_1\cdot(2-c)+c\cdot\varepsilon(\omega)}{\varepsilon_1\cdot c+\varepsilon(\omega)\cdot(2-c)} \tag{2b}$$

where c is the concentration of GaP, $\varepsilon(\omega)$ is the dielectric function of GaP and ε_1 is the dielectric constant of air. Due to $\varepsilon_\perp(\omega) < \varepsilon_\parallel(\omega)$ for all c, porous GaP is a positive uniaxial material. Type I phase matching is achieved if $n_0(2\omega) = n_{e0}(\omega)$, where

$$n_0(\omega) = \sqrt{\varepsilon_\perp(\omega)}, \quad n_{e0}(\omega) = \sqrt{\varepsilon_{e0}(\omega)} \tag{3a}$$

$$\text{with} \quad \frac{1}{\varepsilon_{e0}(\omega)} = \frac{\cos^2\vartheta}{\varepsilon_\perp(\omega)} + \frac{\sin^2\vartheta}{\varepsilon_\parallel(\omega)} \tag{3b}$$

ϑ is the angle between the exciting laser beam and the optical axis inside the membrane. Taking into account that $\varepsilon_1 = 1$, $\varepsilon(\omega) = 3.1192^2$ and $\varepsilon(2\omega) = 3.4595^2$ [18], one can solve the equation $n_0(2\omega) = n_{e0}(\omega)$ in order to determine ϑ as a function of c. A solution exists for all $c < 0.696$, it means for the degree of porosity $(1 - c) \geq 30$ %. The membranes studied in this work $(1 - c \approx 30$ %) fulfil the phase matching conditions provided that the fundamental and SH beams propagate in directions that are nearly perpendicular to the pores. However, the small thickness of the porous membranes and the relatively high absorption at 2ω make it difficult to study the quantitative characteristics of the SHG under the conditions involved.

6. Conclusion

Nanostructuring changes drastically the optical properties of semiconductor materials. In porous III-V compounds, in particular, it leads to the occurrence of Fröhlich-type vibrational modes with porosity-tunable frequencies as well as to an efficient SHG, at least two orders of magnitude higher than that in bulk gallium phosphide. According to our analysis, the formation of parallel pores stretching perpendicularly to the initial surface gives rise to a pronounced anisotropy in the material properties. The possibility to fulfil the phase matching conditions along with the enhanced nonlinear optical response make nanoporous III-V structures extremely perspective for use in all-optical devices.

This work was partially supported by the NATO Scientific Division under Grant HTECH.LG 961399. I.M.T. and I.V.K. gratefully acknowledge the support by the Alexander von Humboldt Foundation.

References

1.   Cullis, A.G., Canham, L.T., and Calcott, P.D.J. (1997) The structural and luminescence properties of porous silicon, *J. Applied Physics* **82**, 909-965.
2.   Tiginyanu, I.M., Irmer, G., Monecke, J., and Hartnagel, H.L. (1997) Micro-Raman scattering study of surface-related phonon modes in porous GaP, *Physical Review B* **55**, 6739-6742.
3.   Kuriyama, K., Ushiyama, K., Ohbora, K., Miyamoto, Y., and Takeda, S. (1998) Characterization of porous GaP by photoacoustic spectroscopy: The relation between band-gap widening and visible photoluminescence, *Physical Review B* **58**, 1103-1105.
4.   Tiginyanu, I.M., Kravetsky, I.V., Marowsky, G., and Hartnagel, H.L. (1999) Efficient second harmonic generation in porous membranes of GaP, *Physica Status Solidi (a)* **175**, R5-R6.
5.   Monecke, J. (1989) Microstructure dependence of material properties of composites, *Physica Status Solidi (b)* **154**, 805-813.
6.   Monecke, J. (1989) Collective excitations in microcrystal aggregates and composites, *Physica Status Solidi (b)* **155**, 437-444.
7.   Tiginyanu, I.M., Schwab, C., Grob, J.-J., Prevot, B, Hartnagel, H.L., Vogt, A., Irmer, G., and Monecke, J. (1997) Ion implantation as a tool for controlling the morphology of porous gallium phosphide, *Applied Physics Letters* **71**, 3829-3831.
8.   Tiginyanu, I.M. and Hartnagel, H.L. (1999) Nanoporous membranes and heterostructures on III-V compounds for micro- and optoelectronic applications, in *GAAS'99 Conference Proceedings* (Munich, Germany, Oct. 4-5, 1999), Miller Freeman, UK, pp. 194-199.
9.   Erne, B.H., Vanmaekelbergh, D., and Kelly, J.J. (1996) Morphology and strongly enhanced photoresponse of GaP electrodes made porous by anodic etching, *J. Electrochemical Society* **143**, 305-314.
10.   Takizawa, T., Arai, Sh., and Nakahara, M. (1994) Fabrication of vertical and uniform-size porous InP structure by electrochemical anodization, *Japanese J. Applied Physics* **54**, L643-L645.

11.   Tiginyanu, I.M., Hartnagel, H.L., Monecke, J., Kravetsky, I., and Marowsky, G. (1999) Self Arrangement of pores in anodically-etched GaP under in-situ illumination, in *MRS 1999 Fall Meeting Abstracts*, Paper No. G8.3.

12.   Fröhlich, H. (1949) *Theory of dielectrics*, Clarendon Press, Oxford.

13.   Monecke, J. (1994) Bergman spectral representation of a simple expression for the dielectric response of a symmetric two-component composite, *J. Physics: Condensed Matter* 6, 907-912.

14.   Curtnute, B. and Williams, D. (1974) Structure of water and Aqueous solutions, in W.A.P. Luck (ed.), *Proceedings of the Int. Symp. (July 1973, Marburg)*, Verlag Chemie GmbH and Physik Verlag GmbH, p. 207.

15.   Li, S. and Khurgin, J.B. (1993) Feasibility of phonon-assisted electronic devices, *J. Applied Physics* 74, 2562-2564.

16.   Herman, W.N. and Hayden, L.M. (1995) Maker fringes revisited: second harmonic generation from birefringent or absorbing materials, *J. Optical Society of America B* 12, 416-427.

17.   Bergman, D.J. (1978) The dielectric constant of a composite material – a problem in classical physics, *Physics Reports* 43, 377-407.

18.   Singh, S. (1971) Non-linear optical materials, in R.J. Pressley (ed.), *Handbook of lasers with selected data on optical technology*, Chemical Rubber CO, Cleveland, p. 504.

SCANNING PROBE MICROSCOPY (STM, AFM) INVESTIGATION OF CARBON NANOTUBES

L. P. BIRÓ

Research Institute for Technical Physics and Materials Science
H-1525 Budapest, P. O. Box 49, Hungary, e-mail: biro@mfa.kfki.hu

Abstract. The image formation in scanning tunneling microscopy (STM), atomic force microscopy (AFM), and the particularities of imaging supported carbon nanotubes by STM and AFM are discussed. The milestones of STM, STS, and AFM measurements on carbon nanotubes are briefly reviewed. Scanning tunneling spectroscopy (STS) measurements, and atomic resolution images of single-wall and multi-wall carbon nanotubes supported on graphite are compared to typical data for graphite.

1. Introduction

1.1 CARBON NANOTUBES

Carbon nanotubes discovered a decade ago by Iijima [1], are a new member of the carbon allotrope family. A single-wall carbon nanotube (SWCNT) is constituted of a single graphene layer - a graphite-like atomic arrangement of one monolayer thickness - wrapped into a perfect cylinder, Fig. 1.

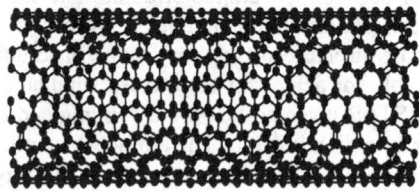

Figure 1. Stick and ball model of a single-wall (9, 11) carbon nanotube

The experimentally found typical diameter values of SWCNTs are in the range of 1 - 2 nm. A multi-wall carbon nanotube (MWCNT) is composed of several coaxial SWCNTs with increasing diameters, in a way that the distance between the walls of two consecutive nanotubes is kept at the value of 0.34 nm, close to the inter-layer distance along the c axis in bulk graphite. The diameter of MWCNTs may range up to 100 nm.

Beyond the beauty of such an arrangement of atoms into a regular, nanoscopic object with cylindric symmetry, the other reasons which produce a still growing interest

405

L. Pavesi and E. Buzaneva (eds.), Frontiers of Nano-Optoelectronic Systems, 405–420.

for these nano-objects reside in their remarkable mechanical and electrical properties [2, 3], which make them candidates for being the basis of a future nanoelectronic industry. Other likely applications of carbon nanotubes range from biology to the manufacturing of new composites with superior stiffness/weight ratio than any presently known material.

Theoretic calculations showed [2, 4] that the electronic structure of a SWCNT is determined by the way in which the graphene sheet is wrapped into a cylinder. The wrapping vector $C = na_1 + ma_2$, Fig. 2, is unambiguously determined by two integers n, and m; the carbon nanotube is "produced" by making the origin and the end of this vector to coincide. In this way, the wrapping vector will become a circle on the surface

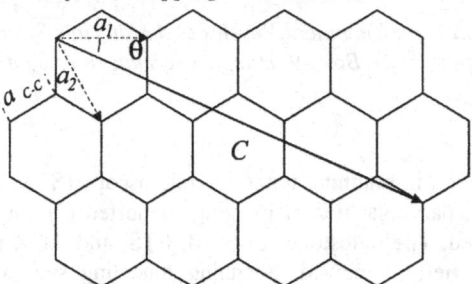

Figure 2. A single-wall (3, 2) carbon nanotube is "produced" by making the end points of the wrapping vector $C = 3a_1 + 2a_2$ to coincide.

of the nanotube, therefore, the tube diameter is given by $d_t = ||C||/\pi = 1/\pi[||a1||(n^2 + m^2 + mn)^{1/2}]$. The vectors a_1 and a_2 are the lattice vectors of graphite; a_{C-C} is the bond length in the c plane, 0.1421 nm. There are two particular directions in the graphene layer: (n, 0), called the "zig-zag" axis (the carbon-carbon bond will be parallel with the tube axis), and the (n, n), called the "armchair" axis (the carbon-carbon bond will be perpendicular to the tube axis). All other directions are called chiral, the chiral angle is θ. On the basis of their electronic structure the SWCNT can be divided in two groups: *semiconducting nanotubes*, these have a vanishing density of states at the Fermi energy, and *metallic nanotubes*, with a finite density of states at the Fermi energy. If the difference (n-m) can be divided by 3, the carbon nanotube will be metallic.

The strong relation between the atomic structure and the electronic structure makes it a necessity to investigate isolated nanotubes, to resolve their atomic structure, and to measure the electronic structure of the very same nanotube. The only experimental tool which was suitable for achieving these tasks simultaneously was the scanning tunneling microscope (STM) invented in 1981 by Binnig and Rohrer [5] rewarded by the Nobel Prize for their invention in 1986. The atomic force microscope (AFM) [6], an offspring which followed very quickly the invention of STM, proved invaluable in manipulating, measuring mechanical properties, and performing electrical measurements on carbon nanotubes on insulating substrates.

1.2. SCANNING PROBE MICROSCOPY

In principle, the concept of an STM is very simple: an atomically sharp, metallic tip is brought within a distance of a few tenths of a nanometer to a conducting surface; due to the quantum mechanical behavior, electrons may tunnel from the tip to the surface and vice versa. A comprehensive overview of the models used to describe the tunneling through a one dimensional potential barrier is given by Wiesendanger [7]. Two major conclusions arise from the various models:

1) The transmission T of electrons through a one dimensional potential barrier, depends exponentially on the width s of the barrier:

$$T \propto e^{-2\chi s}. \tag{1}$$

2) The decay rate χ, characterizing the decrease of the probability to find the electron inside the barrier is:

$$\chi = \frac{2\pi[2m(V_0 - E)]^{1/2}}{h}, \tag{2}$$

where V_0 is the average height of the barrier, E is the energy of the tunneling electron. Independently of the exact shape of the barrier, the strong exponential dependence of T with the barrier width s and the square root of the effective barrier height, $(V_0 - E)^{1/2}$ is typical for tunneling. The exponential dependence makes that the tunneling interface will be very narrow, thus making possible the atomic resolution.

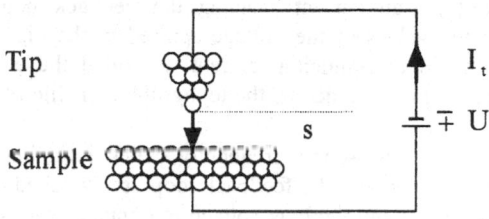

Figure 3. Principle of operation of the STM

Unless a bias is applied between the tip and the surface the two electron fluxes: surface → tip, and tip → surface, will be equal, and will cancel out each other in equilibrium. When an external bias is applied, depending on the polarity of the bias, one of the tunneling directions is made preferential, therefore a net electronic current can be measured in the circuit, Fig. 3. Usually the bias is of the order of 1 V, which yields currents in the 1 nA range.

The practical case of an STM measurement is different from the simple case of tunneling through a one dimensional potential barrier. First of all, in the case of a sharp tip, one has a three dimensional potential barrier instead of a one dimensional one. The real potential barriers may strongly differ from the rectangular shape assumed in deducing the formulas (1), and (2). The most frequently used model for the interpretation of STM experiments is the model given by Tersoff & Hamann [8]. In this theory the tip is treated as a single *s orbital* with a constant density of states (DOS). The tunnel current flowing between the tip and the sample when a bias V is applied, will be:

$$I(V) \propto \int_0^{eV} \rho_s(E)\rho_t(E-eV)T(E)dE, \tag{3}$$

where ρ_s is the sample electronic DOS and ρ_t is the DOS of the tip.

One may note that formula (3) offers the opportunity to acquire information regarding the DOS of the sample. If the tip DOS is flat and the transmission coefficient T may be taken as constant, than Eq. (3) reduces to:

$$I(V) \propto \int_0^{eV} \rho_s(E)dE. \tag{4}$$

Than, the quantity $dI/dV(V)$ α $\rho_s(V)$. However T may be regarded constant only in the limit of small voltages, it gives deviation from this simple dependence at large bias values.

In practical STM instruments, the positioning and scanning of the STM tip is achieved by piezolectric actuators. The width of the STM gap is controlled by a feedback loop which keeps the value of the tunneling current at a value selected by the operator.

An STM can operate in several regimes, the two most important ones are as follows:

- Topographic (constant current) imaging: the feedback loop is on, the image is generated from the values of the voltage applied to the piezo-actuator to maintain a constant value of the tunneling current. Provided the electronic structure at the sample surface is homogeneous, the topographic profile of the surface will be generated.
- Current-voltage spectroscopy, frequently called scanning tunneling spectroscopy (STS). The scanning, and the feedback loop are switched off, the value of the tunneling gap is fixed, and the bias voltage is ramped from -U to +U, and the corresponding current variations are recorded. The function dI/dV gives information about the local DOS of the sample.

Already the first STM experiments produced extremely valuable data. Therefore, the scientific community felt very frustrated by the fact that the STM can be used only on conducting surfaces. Only five years after the invention of STM, again Binnig and coworkers [6], announced in 1986 the birth of the AFM, which was able to image insulator samples too.

In principle, the AFM is a much similar construction to the STM. The major difference is given by the kind of the local probe (the tip), and its interaction with the sample. In order to be able to measure insulators, the tunneling had to be given up. It was replaced by another, well know atomic interaction: the Van der Waals (VDW) interaction. Generally the VDW forces are usually of attractive character and increase rapidly as the distance between the atoms, or molecules is reduced. The force vs. distance dependence is described by:

$$F_{VDW}(s) = -\frac{1}{s^7} \tag{5}$$

This attractive force will increase till the equilibrium distance between the two bodies is reached, after which the global force will decrease very abruptly, turning into a repulsive force arising from Coulomb repulsion of the ion cores [9]. The global force, Fig. 4, will divide the region in the vicinity of the sample into three parts: a) the region dominated by the attractive force, b) the transition region, and c) the region dominated by the repulsive force. The later one is the useful region for contact mode AFM measurements, as one can see from Fig.4, in this region there is a linear correspondence between displacement of the tip and the interaction force.

The detection of the forces of the order of 10^{-7} - 10^{-8} N, acting between a tip with a radius of curvature typically in the range of some tens of nm and the measured sample are possible due to a thin cantilever carrying the tip. The bending of the cantilever pressed against the sample is detected using a laser beam shone onto the back side of the cantilever in combination with optical interference, or a laser beam in combination with a four-quadrant position sensitive detector. A similar feedback loop as in the case of the STM, maintains a constant deformation of the cantilever, i.e., a constant force between the tip and sample. The topographic image is generated from the voltage applied to the tip piezo to maintain the interaction force constant.

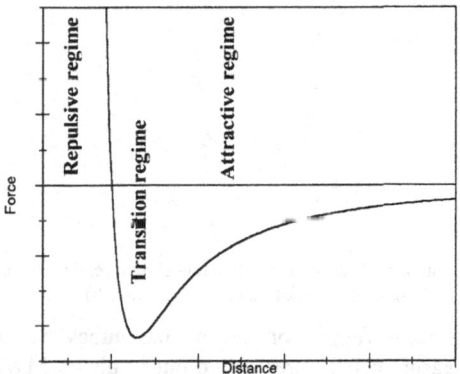

Figure 4. The regions dominated by attractive, or repulsive forces, and the region of transition.

For soft samples, or for samples which may be deformed by the pressure of the AFM tip, and to reduce the frictional contribution to the contact mode AFM images an operation mode based on vibrated cantilevers was introduced. In this "tapping mode" as it is frequently called, the AFM tip is vibrated by a piezo driver at a frequency of several hundreds of kHz with an amplitude of the order of 100 nm. The interaction of the tip with the sample is "switched" from continuous to intermittent, it has a significant magnitude only in one of the maximum elongation positions of the vibrating cantilever. During each interaction period the tip will transfer a certain amount of vibration energy to the sample, therefore the amplitude of the tip in intermittent contact with the sample will be reduced as compared with the amplitude of the free tip. This reduction of the amplitude will be used in the feedback loop to maintain a "constant energy transfer" to the sample during the imaging.

In conclusion, the contact mode AFM image is a constant force image, while the tapping mode AFM image is a constant energy transfer image. These differences may prove important when imaging inhomogeneous samples like carbon nanotubes on a support with different physical properties.

2. STM and AFM on carbon nanotubes

2.1. STM INVESTIGATION OF CARBON NANOTUBES

The first STM experiment proving that atomic resolution is possible on a carbon nanotube was reported by Ge & Sattler [10]. They investigated MWCNTs produced in-situ by the condensation of evaporated carbon on a highly oriented pyrolitic graphite (HOPG) substrate. Superimposed on the atomic lattice, a periodicity of 16 nm was found. This was interpreted as arising from the misorinetation of the two outer layers of the MWCNT, analogously with the generation of Moire patterns known in geometric optic. The STM observation of Moire patterns on HOPG [11] is well known experimentally, however, no clear theoretical description has been given yet.

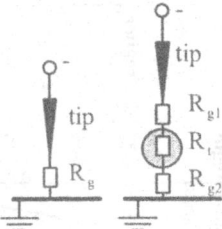

Figure 5. Schematic representation of the STM tip when tunneling directly into the substrate a); and when tunneling through a carbon nanotube b).

The first STS measurement on carbon nanotubes was reported by Olk & Heremans [12]. The measurements were carried out in air on MWCNTs grown by the electric arc method transferred onto an Au substrate by ultrasonication in ethanol. Both semiconductor and metallic carbon nanotubes were found. The comparison of measured gap and diameter values with gap values predicted by theory showed an increasing deviation with decreasing tube diameter. The source of this deviation may be the unavoidable error introduced in the measured diameter by tip/sample convolution effects and by the more complex structure of the tunneling region than in the well known case of a flat, homogeneous sample [13,14]. These effects can be separated in three classes:

- effects arising from the complexity of the system through which the tunneling takes place,
- effects of pure geometric origin,
- effects arising from the different electronic structure of the nanotube and its support.

In Fig. 5 the two cases: a) an STM tip over a flat surface, is compared to b) the case of a tip over a carbon nanotube floating on the Van der Waals potential over a support. While in case a) the current flowing through the STM gap is determined only by

R_g, in case b) there are three resistances, two tunneling gap resistances, R_{g1} and R_{g2}, and the resistance of the nanotube itself R_t. Additionally, in case b) there are two interfaces through which the electron can travel only by tunneling. This situation is frequently called resonant tunneling [15]. The effects arising from the complex structure of the tunneling interface - a nanoscopic object sandwiched between two tunneling gaps – were investigated using a recently developed computer code for the simulation of the tunneling process through a supported carbon nanotube [14], Fig. 6. The figure shows snapshots of the tunneling process, one may note that the time evolution of the tunneling shows significant differences when the tip is situated exactly over the carbon nanotube as compared with the case when the tip is situated over the support. The detailed simulation shows that the nanotube is "charged" during the tunneling, the fraction of the wave packet which tunneled into the tube will form a standing wave pattern and the tunneling from the tube into the substrate will be intermittent.

The effects of geometric origin are the tip/sample convolution effects [14,16]. As a general rule, it can be formulated that in scanning probe microscopy, independently of which object was chosen as tip, always the sharper object (with the smaller radius of curvature) will generate the image. This is illustrated in Fig. 7. For an STM tip, the radius of curvature which has to be taken into account is an effective radius of curvature composed from the geometric radius of curvature, to which one has to add the value of the tunneling gap [16]. The value of the tunneling gap may vary from the support to the nanotube.

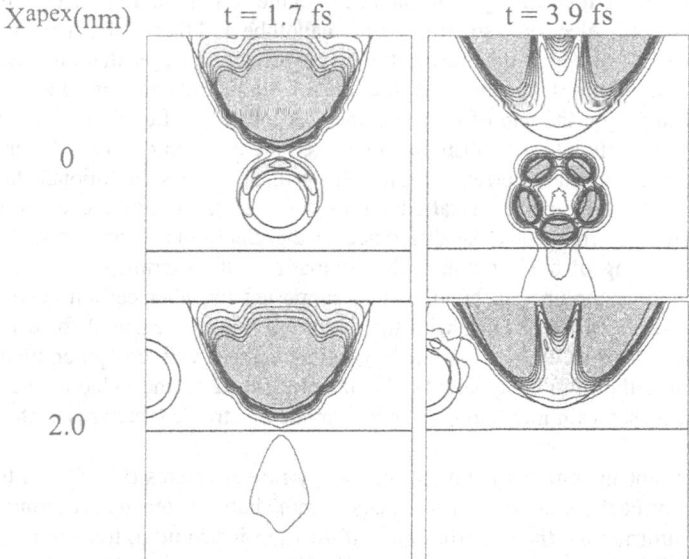

Figure 6. Probability density of the scattered wave packet during the tunneling process through a carbon nanotube (top); and directly into the substrate (bottom). X^{apex} gives the horizontal distance between the apex of the STM tip and the axis of the nanotube, t values indicate the time elapsed since the launching of the wave packet from the bulk of the STM tip [14].

412

Due to the fact that the typical diameter of a MWCNT is in the range of 10 nm, it is extremely difficult to create experimental conditions in which the apex of the tip may be regarded over a length of 10 nm, as having a negligible width ($d_{tip}/2 < d_t/10$) compared with the diameter of the nanotube. Such a tip would be extremely unstable mechanically. The importance of this effect increases with decreasing tube diameter [16].

Figure 7. Illustration of distortion arising due to tip-sample convolution effects for objects with different shapes. When the obejct becomes sharper than the tip, the object will generate the image of the tip.

The effects arising from the difference of the electronic structure of the carbon nanotube and its support will be superimposed on the geometric effects and those arising from the complexity of the tunneling interface. As already discussed, the convolution effects make that the nanotube will have a larger apparent diameter than the geometric one. The existence of the two tunneling gaps will make that the transfer probability of electron through the nanotube will be smaller than in a direct tunneling process from the tip into the substrate, this will yield a smaller tube height than the geometric one. As a result the tube will appear "flattened" in the acquired STM image [16].

When the electronic structure of the nanotube and that of the substrate differ in a way that the DOS of the substrate at the given voltage is larger than that of the nanotube - for example: a semiconductor tube on a metallic substrate imaged at a voltage value situated within the gap of the nanotube – according to Eq. (3), the tunneling current flowing directly into the substrate will exceed the tunneling current flowing through the nanotube. In constant current imaging this will produce an additional flattening of the nanotube. In most experimental situations one prefers to have a good conductor as substrate, so most frequently the difference of the electronic structures will cause the apparent flattening of the nanotube. The comparison of experimental flattening measured for carbon nanotubes in bundles, i.e., supported by other carbon nanotubes with similar electronic structure [17], with the distortion values measured for carbon nanotubes supported on HOPG [13], show in good agreement with computer simulation results [14], that the ratio of apparent half diameter $D_{app}/2$ to the value of the measured height h, increases with increasing difference in the electronic structures of the nanotube and its support.

Current imaging tunneling spectroscopy measurements (CITS) - in this operation mode, for each pixel point of the image, immediately after the acquiring of the topographic information, the spectroscopic information is acquired, too - of carbon nanotubes on HOPG show that in the vicinity of the Fermi level in a voltage range of   -1.5 to 1.5 V, the electronic structure of large diameter MWCNT ($d_t > 50$ nm) is practically identical with that of the HOPG, Fig. 8. In this case the observed flattening is produced only by the tip/sample convolution and by the second tunneling gap. For the larger diameter tube seen in Fig. 8 the ratio $(D_{app}/2)/h$ is 2.81; the geometrically correct ratio, free of convolution effects and neglecting the complexity of the tunneling, should be

0.5. For a SWCNT with identical electronic structure as its support, and a similar radius as the radius of curvature of the STM tip, the simulation [14] gives a ratio of 1.35. This shows that due to the existence of several layers, the tunneling through a MWCNT is more complex.

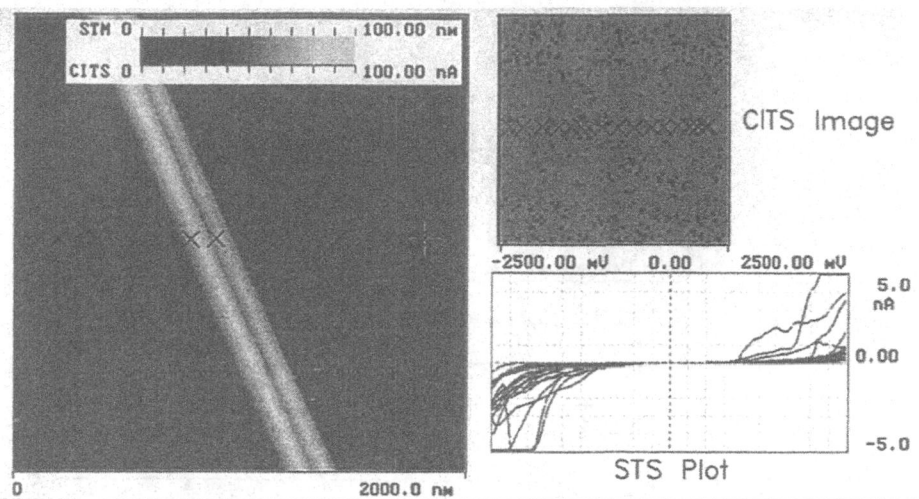

Figure 8. Topographic and CITS image of two multiwall carbon nanotubes on HOPG. The CITS image does not reveal differeneces in electronic structure. In the STS plot the I-V curves recorded in the points marked by crosses are shown, in the range of -1 to 1 V all curves are practically coincident.

As it was pointed out above, the measured flattening has two components: the broadening due to convolution and the smaller apparent height due to the smaller tunneling gap over the nanotube than over the support. The width of the tunneling gap over the nanotube can decrease to zero, in this case one has point contact imaging [18]. In this imaging regime the tube is deformed in its topmost part due to the pressure exerted by the STM tip in contact with the tube, Fig. 9a. However, after reducing slightly the value of the tunneling current from 1.2 nA to 1.0 nA, atomic resolution could be obtained on the same tube, Fig. 9c. The periodicity of the triangular lattice along the line AB is 0.25 nm, while the corrugation amplitude is 0.05 nm, in good agreement with typical values for HOPG. The point contact may have important effects in the STS measurements, too [19].

There are two distinct classes of nanotubes on which atomic resolution was achieved: a) MWCNTs with diameters of several tens of nanometers [10,13], like in Fig. 9c; and b) SWCNTs [20, 21] with diameters typically in the 1 nm range. While the MWCNTs show a similar structure like HOPG, i.e., a triangular lattice composed of tunneling current maxima (light features), and sometimes Moire like superstructures [10]; the SWCNTs show a triangular lattice of minima (dark features), corresponding to the empty centers of the hexagons building up the graphene sheet, Fig 10.

Computer simulation based on a theoretical model using a tight-binding π-electron Hamiltonian [22] was successfully used to calculate STM images of single-wall carbon nanotubes. The comparison of calculated STM images with experimental results

414

allowed the identification of certain effects arising from the curvature of the measured object: i) only the topmost atoms of the nanotube will be "measured" by the STM tip (modelled by an *s orbital*) in their geometrically correct positions, the apparent distance

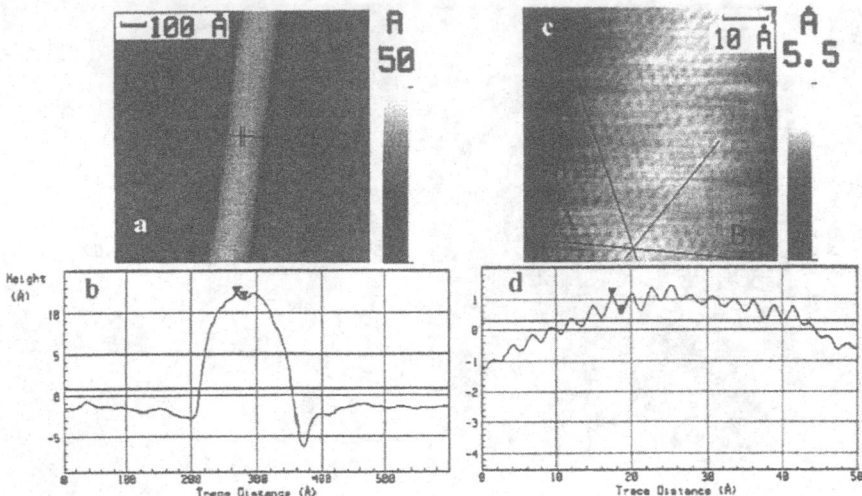

Figure 9. a) Large scale topographic image of a carbon nanotube imaged in point contact regime in the topmost part of the tube, $I_t = 1.2$ nA; $U_b = 100$ mV. b) Line cut along the line marked in a). c) Atomic resolution image on the same tube after reducing the tunneling current, $I_t = 1.0$ nA. d) Line cut along the line marked by AB, the two other lines in the image indicate the other two axes of the triangular lattice, note the corrugation amplitude of 0.05 nm.

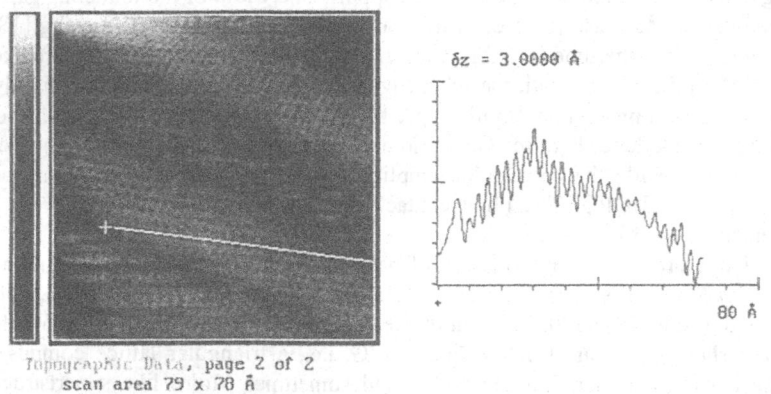

Figure 10. Atomic resolution STM image of single-wall carbon nanotubes. Gray scale at left corresponds to 0.13 nm; $I_t = 1$ nA; $U_b = 100$ mV.

of all other atoms from the topmost one will be inflated by a factor of $1 + s/R$ in the direction transversal to the tube axis, where s is the distance between the STM tip and the nanotube surface, and R is the radius of the nanotube; ii) the atoms all look the same on the nanotube, irrespective of its chirality, but the way in which the bonds between atoms appear in the STM image depends on chirality.

The HOPG-like atomic resolution images of large diameter MWCNTs may be attributed to the reduced curvature of the outer layers in these structures, so that an arrangement like the ABAB stacking of graphite can be achieved. While the SWCNTs behave like a single graphene sheet imaged by STM [23].

The first STS results on SWCNTs were reported simultaneously by two groups [20, 21]. The experimental results are in good agreement with earlier theoretical predictions [2, 4,24]. Metallic and semiconductor carbon nanotubes were found. The typical gap values found for semiconductor nanotubes with diameters in the 1.2 to 1.9 nm range were found to be of the order of 0.5 eV, while the metallic carbon nanotubes with diameters in the range of 1.1 to 2.0 nm had gaps of the order of 1.7 eV [20]. Both groups found that the measured gap values were proportional with $1/d_t$. The differences in the proportionality factors may arise from the differences in taking into account the systematic errors in the determination of the tube diameter due to convolution effects.

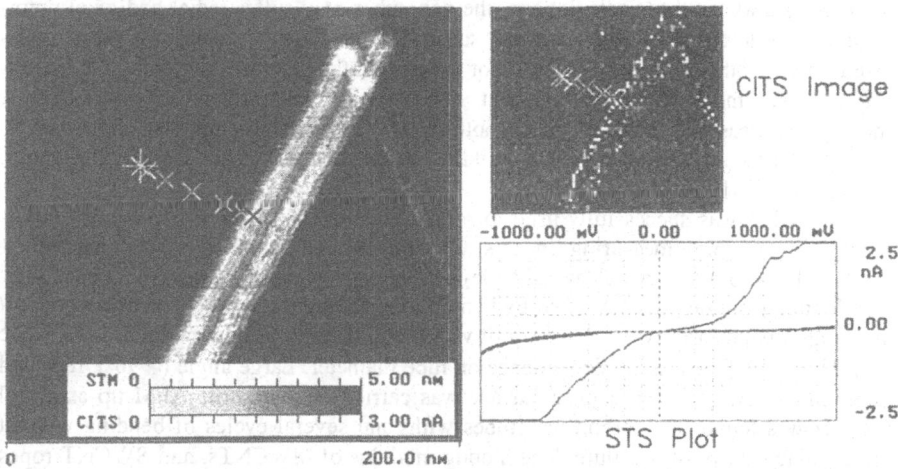

Figure 11. Topographic and CITS STM image of bundles of single and double wall carbon nanotubes. The CITS image is displayed at - 1.0 V. The spectra shown in the STS plot were taken in the points indicated by crosses. One may note the clear difference between the I-V curve of the metallic nanotube and the several, overlapping spectra corresponding to the HOPG.

Spectroscopic measurements carried out on large diameter MWCNTs [25] showed in the central part of the nanotube a differential conductance close that of HOPG, but the carrier mobilities were found to be lower in the tubes as compared to graphite. Close to the tube ends capped by fullerene-like hemispheres a different electronic structure was found, which was attributed to the presence of pentagons. A particularly interesting and relatively less investigated case is the case of nanotubes with a few walls. In Fig. 11, the CITS image of bundles of single and double wall nanotubes grown by the decomposition of C_{60} in the presence of transition metals [26,27] are shown. The typical tube diameters in the bundles are in the range of 1.0 – 2.8 nm. STS measurements indicate both metallic and semiconductor nanotubes, some tubes exhibit a DOS structure that may be interpreted as the combination of the individual DOS corresponding to two tubes. Earlier theoretical calculations showed that in the case of a MWCNT the total DOS will be the sum of the individual DOS of the tubes constituting the multishell structure [28].

Beyond its capabilities to characterize carbon nanotubes, STM is also suitable for being used as a "tool" to modify carbon nanotubes [29] and it is a powerful instrument in searching for new carbon nanostructures like the Y-branching of carbon nanotubes [30] and tightly wound, coiled single-wall carbon nanotubes [31].

2.2. AFM INVESTIGATION OF CARBON NANOTUBES

The first AFM measurements on carbon nanotubes were carried out in order to compare TEM, STM and AFM images of carbon nanotubes [32]. The importance of tip/sample convolution effects in imaging carbon nanotubes by AFM was pointed out soon [33]. However, effects arising from the increasing deformation of SWCNTs and of MWCNTs due to the Van de Waals interaction with the surface [34] were realized much later. As shown by these calculations, the nanotubes may suffer radial and axial deformations due to the interaction with the substrate, or due to the reciprocal pressure exerted on each other by crossing tubes, not to mention the pressure exerted by the tip during a contact mode AFM measurement. These deformation effects may significantly modify the measured height of the nanotube. The larger the diameter of the nanotune and smaller the number of the layers building up the tube, the larger deformations are to be expected [34].

AFM was successfully used to produce the deformation of nanotubes on purpose, for example, measuring in this way the Young modulus of the nanotubes. MWCNTs were fixed on a cleaved $MoSe_2$ surface by depositing a grid of SiO pads [35]. The Young modulus was measured by bending the nanotubes in the plane of the support applying a point load by the AFM tip. It was found that the Young modulus has a value of 1.28 ± 0.59 TPa with no dependence on tube diameter. Large angle (> 90°), repeated bending of MWCNTs on a mica surface was carried out using the AFM tip as a tool [36]. It was found that carbon nanotubes withstand several cycles of bending without undergoing catastrophic failure. The Young modulus of MWCNTs, and SWCNT ropes was measured on a polished alumina ultrafiltration membrane [37]. On such a substrate, nanotubes occasionally lie over the pores suspended over the hollow of the pore. This geometry allows the deformation of the nanotube in a direction perpendicular to the substrate like a clamped beam. For arc grown nanotubes the Young modulus was found to be 810 ± 410 Gpa. Opposite to graphite, the irradiation of the nanotubes with 2 MeV electrons did not caused the reduction of the Young modulus. The along-axis electrical transport through carbon nanotubes was measured using nanotubes fixed under an Au contact using a conductive AFM tip as mobile contact sliding along the tube [38]. Resistivities in the range of 8 – 117 (ohm m) were found, the nanotube diameters used to calculate the resistivities were taken from transversal line cuts taken through the nanotubes.

An overview of various other applications of the AFM in nanotube science is given by Muster and coworkers [39].

A recently developed non conventional way of producing carbon nanotubes is based on the bombardment of a graphite target by high energy (E > 100 MeV), heavy ions [40]. Where dense, nuclear cascades reach the surface of the target sputtering craters are produced with diameters in the 1 μm range, and a depth of tens of nm. Frequently carbon nanotubes emerge from these craters, or are found on the flat bottom of the craters, Fig. 12. A major difference of these samples as compared with the usual nanotube samples for AFM is that when produced by ion irradiation the nanotubes grow in-situ and they do not undergo ultrasonication in some organic solvent. When measured with contact mode AFM, except the distortion arising from convolution, and compression of tubes, no other effects were observed. But when measuring the same samples using a tapping mode AFM, it was frequently found that the nanotubes apparently vibrate in a plane perpendicular to the plane of the substrate. As the other topographic features present in the image were imaged regularly, it can be excluded that the apparent vibration arises due to improper imaging conditions.

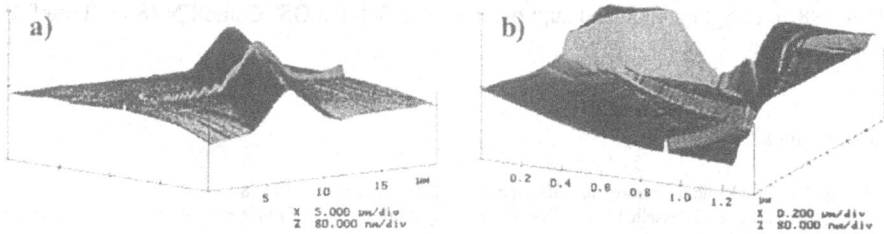

Figure 12. Carbon nanotubes produced by high energy irradiation of graphite. a) Carbon nanotube emerging from a crater and crossing a surface fold produced during the cleavage of HOPG. b) Two carbon nanotubes crossing each other on the flat bottom of the same crater from which emerges the long tube in a). Note the wave-like profile of the tube in a) and of the upper tube in b).

Using a simple computer model to get more insight in the way in which the tapping mode AFM image is generated, and comparing the simulation with the experimental results, it was found that the wave-like profile of the nanotube arises due to the combination of three periodic motions: i) the vertical vibration of the tapping mode AFM tip; ii) the scanning of the AFM tip in the horizontal plane; iii) the oscillation of the nanotube in the (horizontal) plane of the support. The horizontal oscillation is transformed into an apparent vertical vibration, due to the discreteness of the sampling performed by the AFM tip, the convolution effects, and due to the fact that in consecutive scan lines the AFM tip will encounter the nanotube in positions corresponding to different elongations in its horizontal plane of vibration [41].

3. Summary

Scanning probe microscopy: STM and AFM to date are the only tools suitable for imaging individual carbon nanotubes, or related carbon nanostructures and in the same time to perform physical measurements on the imaged nano-object. This makes them ex-

418

tremely attractive, however the way by which the scanning probe microscopy images are generated differs drastically from other imaging techniques like optical or electron microscopy. Therefore cautious image analysis and modeling are most helpful in separating useful information from eventual artifacts and factors that may influence the measurement. The imaging of supported three dimensional objects may strongly differ from the case of flat, homogenous, single crystalline surfaces, effects arising from the complexity of the tunneling interface, or from the Van der Waals, and mechanical interaction of tip and sample cannot be neglected.

Both STM and AFM are essentially three dimensional imaging techniques, therefore, it is strongly recommended that published STM and AFM images to be accompanied by gray scales and/or line cuts, to conserve the full amount of the experimentally acquired three dimensional information.

Acknowledgement

This work has been partly funded by OTKA grants T 030435, T025928, and TéT grant D-44/98 in Hungary, financial support from the Belgian OSTC and FNRS are gratefully acknowledge.

References

[1] Iijima, S (1991) Helical microtubules of graphitic carbon, *Nature* **354**, 56 –58.

[2] Dresselhaus, M. S., Dresselhaus, G., Ecklund P. C. (1996) *Science of Fullerenes and Carbon Nanotubes*, Academic Press, San Diego.

[3] Ebbesen Th. W. (Editor) (1997) *Carbon Nanotubes. Preparation and Properties*, CRC Press, Boca Raton.

[4] Mintmire, J. W., Dunlap, B. I., and White, C. T. (1992) Are fullerene Tubules Metalic?, *Phys. Rev. Lett.* **68**, 631 - 634.

[5] Binnig, G. and Rohrer, H. (1982) Scanning tunneling Microscopy, *Helv. Phys. Acta* **55**, 726 – 735.

[6] Binnig G., Quate, C. F., Gerber Ch. (1986) Atomic Force Microscope, *Phys. Rev. Lett.* **56**, 930 – 933.

[7] Wiesendanger, R. (1994) *Scanning Probe Microscopy and Spectroscopy*, Cambridge University Press, Cambridge.

[8] Tersoff J. and Hamann D. R. (1985) Theory of the scanning tunneling microscope, *Phys, Rev. B* **31**, 805 – 813.

[9] Ciraci, S., Baratoff, A. and Batra, I. P. (1990a) Tip-sample interaction effects in scanning-tunneling and atomic-force microscopy, *Phys. Rev. B* **41**, 2763 – 2775.

[10] Ge, M. and Sattler, K. (1993) Vapor_condenstaion Generation and STM Analysis of Fullerene Tubes, *Science* **260**, 515 – 518.

[11] Jxhie, J., Sattler, K., Ge, M., Verkateswaran, N. Giant and supergiant lattices on graphite, *Phys Rev. B* **47**, 15835 – 15841.

[12] Olk, Ch. H., Heremans, J. P. (1994) Scanning tunneling spectroscopy of carbon nanotubes, *J. Mater. Res.* **9**, 259 – 262.

[13] Biró, L. P., Gyulai. J., Lambin, Ph., B.Nagy, J., Lazraescu, S., Márk, G. I., Fonseca, A., Surján, P., R., Szekeres, Zs., Thiry, P. A., Lucas, A. A. (1998) Scanning tunneling microscopy (STM) imaging of carbon nanotubes, *Carbon* **36**, 689 – 696.

[14] Márk, G. I., Biró, L. P., Gyulai, J. (1998) Simulation of STM images of three-dimensional surfaces and comparison with experimental data: Carbon nanotubes, *Phys. Rev. B* **58**, 12645 – 12648.

[15] Ref . [7], pp. 30 - 34.

[16] Biró, L. P., Lazarescu, S., Lambin, Ph., Thiry, P. A., Fonseca, A., B.Nagy, J., Lucas, A. A., (1997) Scanning tunneling microscope investigation of carbon nanotubes produced by catalytic decomposition of acetylene, *Phys. Rev. B* **56**, 12490 – 12498.

[17] Biró, L. P., B.Nagy, J., Lambin, Ph., Lazarescu, S., Fonseca, A., Thiry, P. A., Lucas, A. A., (1998) Scanning tunneling microscopy of carbon nanoyubes. Beyond the image, in H. Kuzmany, J. Fink, M. Mehring and S. Roth (eds.), *Molecular Nanostructures*, World Scientific, Singapore pp. 419 – 422.

[18] Agrait, N., Rodrigo, J. G., and Vieira, S., (1992) On the transition from tunneling regime to point contact: graphite, *Ultramicroscopy* **42 – 44**, 177 - 183

[19] Márk, G. I., Biró, L. P., Gyulai, J., Thiry, P. A., Lambin, Ph. (1999) The use of computer simulation to investigate tip shape and point contact effects during scanning tunneling microscopy of supported nanostructures, in H. Kuzmany, J. Fink, M. Mehring and S. Roth (eds.), *Electronic Properties of Novel Materials – Science and Technology of Molecular Nanostructures*, American Institute of Physics, Melville, pp.323 – 327.

[20] Wildöer, J. W., Venema, L. C., Rinyler, G. R., Smallez, R. E., Dekker, C. (1998) Electronic structure of atomically resolved carbon nanotubes, *Nature* **391**, 59 - 62

[21] Odom, T. W., Huang, J-L., Kim. Ph., Lieber, Ch. M., (1998) Atomic structure and electronic properties of single-walled carbon nanotubes, *Nature* **391**, 62 - 64

[22] Meunier, V. (1998) Tight-binding computation of the STM image of carbon nanotubes, *Phys. Rev. Lett.* **81**, 5588- 5591

[23] Olk, C. H., Heremans, J., Dresselahaus, M. S., Speck, J. S., Nicholls, J. T. (1990) Scanning tunneling microscopy of a stage-1 CuCll2 graphite intercalation compound, *Phys. Rev. B* **42**, 7524 - 7529

[24] Charlier, J.-C., and Lambin, Ph. (1998) Electronic structure of carbon nanotubes with chiral symmetry, *Phys. Rev. B* **57**, R15037 – R15039

[25] Carroll, D. L., Redlich, P., Ajayan, P. M., Charlier, J. C., Blasé, X., De Vita, A., and Car, R, (1997) Electronic structure and localized states at carbon nanotube tips, *Phys. Rev. Lett.* **78**, 2811 - 2814

[26] Biró, L. P., Ehlich, R., Tellgmann, R., Gromov, A., Krawez, N., Tschaplyguine, M., Pohl, M.-M., Véretsy, Z., Horváth, Z. E., Campbell, E. E. B. (1999) Growth of carbon nanotubes by fullerene decomposition in the presence of transition metals, *Chem. Phys. Lett.* **306**, 155 – 162.

[27] Biró, L. P., Ehlich, R., (submitted to Appl. Phys. Lett.) Room temperature growth of single and multi wall carbon nanotubes by [60]fullerene decomposition in the presence of transition metals.

[28] Lambin, Ph., Charlier, J.-C., Michenaud, J.-P., (1994) Electronic structure of coaxial carbon tubules in H. Kuzmany, J. Fink, M. Mehring, S. Roth (eds.), Progress in Fullerene Research, World Scientific, Singapore, pp. 131 – 134.

[29] Venema, L. C., Wildöer, J. W. G., Temminck Tuinstra, H. L. J., Dekker, C., Rinzler, A. G., Smaller, R., E., (1997) Length control of individual carbon nanotubes by nanostructuring with a scanning tunneling microscope, *Appl. Phys. Lett.* **71**, 2629 - 2631

[30] Nagy, P., Ehlich, R., Diró, L. P., Gyulai, J., (2000) Y-branching of single walled carbon nanotubes, Appl. Phys. A. **70**. 481 - 483

[31] Biró. L. P., Lazarescu, S. D., Thiry, P. A., Fonseca, A., B.Nagy, J., Lucas, A. A., Lambin Ph. (in press) Scanning tunneling microscopy observation of tightly wound, single-wall coiled carbon nanotubes, *Europhys. Lett.*

[32] Gallagher. M. J., Chen, Dong., Jacobsen, B. P., Saris, D., Lamb, L. D., Tinker, F. A., Jiao, J. Huffman, D. R., Seraphin, S., and Zhou, D. (1993) Characterization of carbon nanotubes by scanning probe microscopy *Surf. Sci. Lett.* **281**, L335 – L340

[33] Höper, R., Workman, R. K., Chen, D. Sarid, D., Ydav, T., Withers, J. C., Loufty, R. O. (1994) Single shell carbon nanotubes imaged by atomic force microscopy, *Surface Science* **311**, L731 – L736

[34] Hertel, T., Walkup, R. E., Avouris, Ph., (1998) Deformation of carbon nanotubes by surface van der Waals forces, *Phys. Rev. B* **58**, 13870 - 13873

[35] Wong, E. W., Sheehan, P. E., Lieber, Ch. M. (1997) Nanobeam Mechanics: Elasticity, strength, and toughness of nanorods and nanotubes, *Nature* **227**, 1971 - 1975

[36] Falvo, M. R., Clary, G. J., Taylor II, R. M., Chi, V., Brooks Jr., F. P., Washburn S., Superfine, R. (1997) Bending and buckling of carbon nanotubes under large strain, *Nature* **389**, 582 – 584.

[37] Salvetat, J.-P., Bonard, J.-M., Thomson, N. H., Kulik, A. J., Forró, L., Benoit, W., Zuppiroli, L. (1999) Mechanical properties of carbon nanotubes, *Appl. Phys. A* **69**, 255 – 260.

[38] Dai, H., Wong, E. W., Lieberf, Ch. M., (1996) Probing electrical Transport in Nanomaterials: Conductivity of individual carbon nanotubes, *Science* **271**, 523 - 526

[39] Muster, J., Duesberg, G. S., Roth, S., Burghard, M. (1999) Application of scanning force microscopy in nanotube science, *Appl. Phys. A* **69**, 261 – 167.

[40] Biró, L. P., Szabó, B.,Márk, G. I.,Gyulai. J., Havancsák, K., Kürti, J., Dunlop, A., Frey, L., Ryssel, H. (1999) Carbon nanotubes produced by high energy (E > 100 MeV), heavy ion irradiation of graphite, *Nucl. Instr. and Meth. B.* **148**, 1102 - 1105

420

[41] Biró, L. P., Márk, G. I., Gyulai, J., Rozlosnik, N., Kürti, J., Szabó, B., Frey, L., Ryssel, H. (1999) Scanning probe method investigation of carbon nanotubes produced by high energy ion irradiation of graphite, *Carbon* **37**, 739 – 744.

MOLECULAR SPECTROSCOPY OF NANOPARTICLES

V. YE. POGORELOV, V. P. BUKALO, YU. A. ASTASHKIN,
Department of Physics,
Kyiv National Taras Shevchenko University,
Prospekt Glushkova 6, 03022 Kyiv, Ukraine.

1. Introduction

Molecular spectroscopy is a very useful method to investigate the interaction of nanoparticles with surrounding media. This method is very sensitive to the peculiarities of structure of nanoparticles and nanoparticle systems.

Patrick Bernier and co-authors [1] have shown that by using Raman spectroscopy it is possible to determine the diameter distribution of single-wall nanotubes. Using the resonance Raman effect, [2] it could be observed the overtones of Raman modes that clearly shows the existence of nanotubes armchair.

V. D. Blank and co-authors in [3] have established that the Raman spectra are sensitive to the phase transformations of fullerite in high pressure and high temperature treatment. The phase transitions from *fcc* to orthorhombic and to the rhombohedral phases were accompanied by the splitting of degenerate modes and an activation of silent modes. The temperature increase allows to follow the break-down of C_{60} cages and the formation of cross-linked structure of graphite-like layers which are transformed into graphite at higher temperatures.

H. Kuzmany and J. Winter [4] have shown that molecular spectroscopy could be used to investigate AC_{60} (A=Rb, K) *fcc*-orthorhombic phase transition. Earlier, we have made analogous studies of structural transformations in liquid crystals [5-8]. In addition, we used Raman spectroscopy to study the self-organization of C_{60} in water solution [9]. It was shown that spherical clusters with diameter of 3.56 nm and containing 33 molecules of C_{60} exist in water. We have also investigated the intermolecular interaction between these clusters and mesogenic p-aminoxybenzilidene-p-toluidine (ABT) at T=300 K. The ABT was used in this experiment as the polycrystalline substrate. The interaction of fullerene with ABT could be observed in the splitting of ABT crystal vibrational modes [10]. The Raman spectra of ABT (taken from [10]) are shown in figure 1. In the frequency region 1230-1250 cm^{-1} we can see the corresponding splitting.

L. Pavesi and E. Buzaneva (eds.), Frontiers of Nano-Optoelectronic Systems, 421–429.
© 2000 *Kluwer Academic Publishers. Printed in the Netherlands.*

422

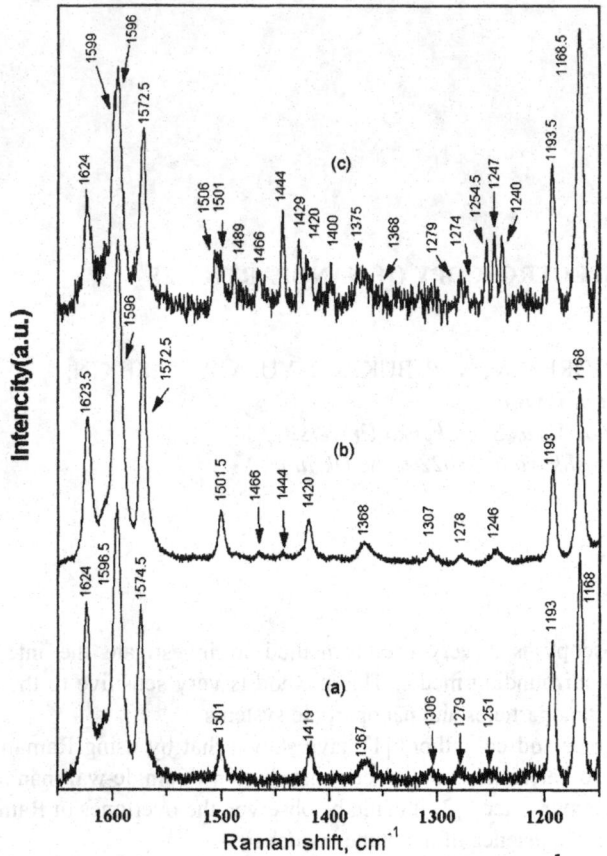

Figure 1. The interaction of the fullerene clusters with ABT substrate.

To interpret the observed splitting, an analysis of the molecular conformations in ABT is carried out in this paper. The investigated object has common features with MBBA— benziliden-aniline core. It consists of two benzene rings connected by plane CH=N-bridges. Here we will discuss the peculiarities of the temperature diagrams of the structural transformation under shock (V>25 K/min) and in slow (V<2 K/min) cooling.

2. Experimental results

A correct model for the mechanism of structural transformations in mesogenic media which can form the liquid crystal (LC) phase is necessary to develop liquid crystal devices as well as to understand a wide-range biophysical phenomena, etc. Optical spectroscopy let us to obtain information about structural transformations in mesogenic media at the molecular level as well as at the atomic level. Such possibility is based on the sensibility of UV and Raman spectroscopy to changes in intramolecular (conformations) and intermolecular structures. It is assumed, that molecular conformations and structural peculiarities are self-congruent factors. So this cause, in our opinion, the need for more precise investigations of molecular conformations under different structural (phase) transformations in mesogenic media, in particularly the liquid crystals (LC). There are many works dedicated to structural transformations in substances which can form a LC-phase. Unfortunately, the methods used in these works (X-ray and neutron diffraction, differential scanning calorimetry) are unsuitable to obtain information about molecular conformations under structural transformations. On the contray, vibrational spectra, particularly Raman spectra, are very sensitive to «order-disorder» transition. Moreover such spectra allow to investigate the influence of the phase transitions on individual parts of investigated molecule. Most of such works are about classical MBBA [11-13]. In ref. [12] the formation of metastable and stable crystal phase in MBBA is connected with the peculiarities of end groups and methoxy-group structure. In ref [13] such phase formation is connected with the difference of equilibrium values of twist-angles between the surface of bridge between benzene rings (C=N-C) and the surface of one of benzene ring.

In our works [5, 7, 14, 15], spectral structural regularities were found which relate the thermo-structural changes in MBBA with corresponding peculiarities of Raman spectra. So we have studied molecular conformations on ABT in order to overcome the contradiction of two different explanations [12,13] concerning the nature of the splitting of the bands in MBBA. The difference between ABT and MBBA is namely in the end groups : CH_3-O- , -C_4H_9 in MBBA and NH_2-O- , -CH_3 in ABT. Such choice allow us to find out a simple relation between molecular conformations and thermo-structural transformations in ABT. Thus, the goal of this work is to investigate the peculiarities of molecular conformations in ABT (NH_2-O-C_6H_4-CH=N-C_6H_4-CH_3) under thermo-structural transformations by using the same experimental techniques and equipment as in ref. [7].

3. Raman spectrum of ABT and its interpretation

ABT belongs to wide range of thermotropic liquid crystals, which molecules includes the bezeliden-anyline (BA) core. The MBBA, EBBA, BA, BT and some other belongs to this class of liquid crystals also. The only difference between such molecules

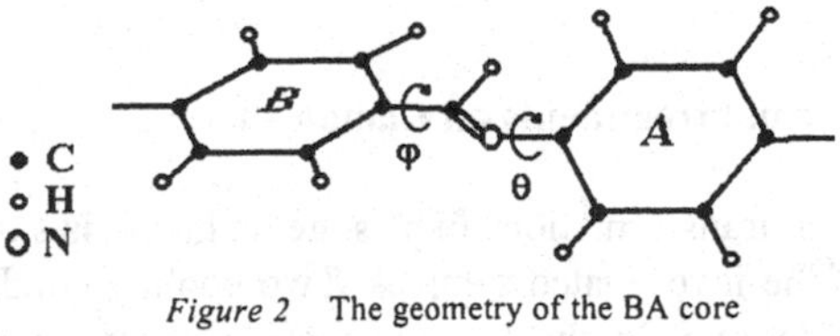

Figure 2   The geometry of the BA core

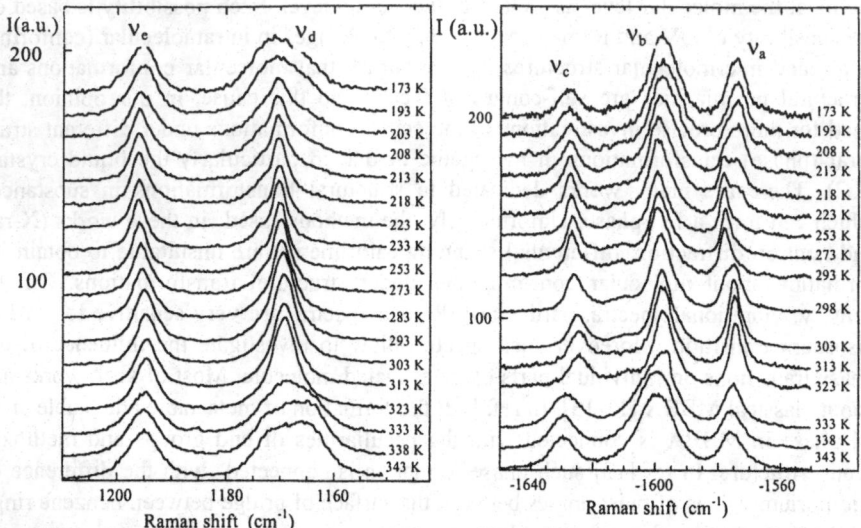

Figure 3 Raman spectra of ABT during the heating cycle in the 1210-1150 cm⁻¹ region

Figure 4 Raman spectra of ABT during the heating cycle, in the 1650-1550 cm⁻¹ region

is in the end groups [13, 16]. The vibrations of the BA core determines the close vibrational spectra of such liquid crystals. At the present, the most investigated spectrum is the MBBA spectrum. The results of computer modeling of that spectrum are reported in [17]. The spectrum of free molecule consists of 117 vibrations. In the high frequency zone, corresponding to intramolecular vibrations, five strong bands are observed for isotropic liquid ABT (T=343 K) at ν_a=1576 cm⁻¹, ν_b=1599 cm⁻¹, ν_c=1629 cm⁻¹, ν_d=1171 cm⁻¹ and ν_e=1197 cm⁻¹. These bands are more often investigated under molecular conformations which accompanies the phase transitions [5, 7, 13-15].

In agreement with ref. [17], the ν_a vibration is interpreted as the valent vibration of benzene ring, the ν_c vibration is the deformation vibration $\beta(CCH)$ of benzene ring, the ν_b band is the first overtone which is magnified by the Fermi resonance between ν_c and ν_b vibrations. In ref. [18] the ν_a vibration is interpreted as the valent vibration of benzene ring also, but ν_c is referred to non-symmetrical vibrations of the methoxy (CH₃-O) group, ν_d is interpreted as the deformation vibrations of the benzene ring and ν_e is the vibration of the CH₃ group. The most detailed interpretation of the vibrational spectrum of free molecules with a BA core was carried out in [17]. We trust it as the most reliable.

4. The influence of thermal treatments on Raman spectra

The nature of the structural transformations in mesogenic media is studied by changing the thermal treatment of the investigated samples. Two cooling conditions were used: shock cooling and slow cooling. In shock cooling cycle, the sample in the isotropic liquid phase is put into liquid nitrogen. After this, the Raman spectra are registered heating the sample. In slow cooling cycle, the Raman spectra are registered both in slow cooling and in heating cycle.

4.1 SHOCK COOLING

A non-oriented nematic ABT sample was used in this experiment. The substance was sealed in a quartz tube with about 3 mm of internal diameter. The sample was annealed for 20 min with T=343K (isotropic phase). Therefore this thermal treatment did not influence its starting structure. The shock cooling was realized by immersing the sample into liquid nitrogen. After that it was annealed for 20 minutes. A reproducible phase composition was checked by Raman spectra. The sample was annealed for 20 minutes at each temperature before the measurement of the Raman spectrum.

The results are shown in Fig. 3 and 4. The temperature dependencies of frequencies $v(T)$ and width $\delta(T)$ of the five more powerful bands are presented in Fig. 5 and 6. A sharp narrowing of the bands was observed at T=213 K. Moreover, at this T the v_c and v_d bands down-shift while the v_e vibration up-shift. At T=223 K, $\delta(T)$ reaches a minimum and remains almost constant up to T=273 K. As observed in Fig. 4, the v_a band splits into two components: $v_{a1}=1573$ cm^{-1} and $v_{a2}=1577$ cm^{-1}. v_{a1} band moves into v_{a2} band in the T range 218K-293K. Moreover, in this T range, the narrowing of the v_d band is accompanied by the appearance of a weak band at around v_d with a frequency of 1171 cm^{-1} and by the disappearance of the band with frequency 1166 cm^{-1}.

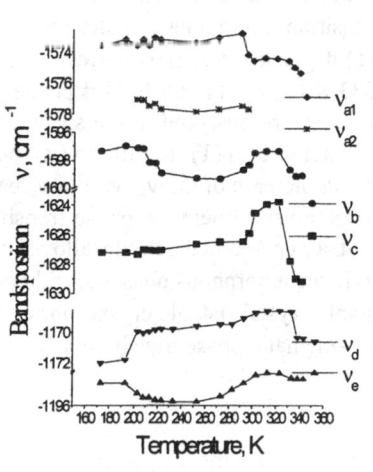

Figure 5 Bands position in shock cooling cycle

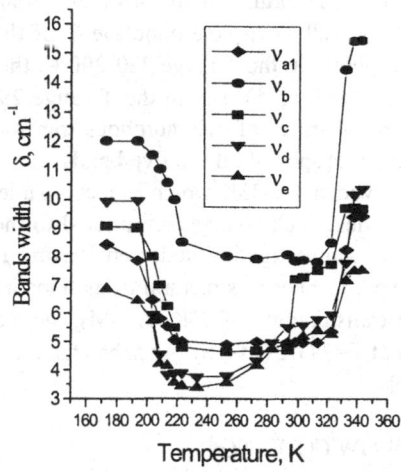

Figure 6 Bands width in shock cooling cycle

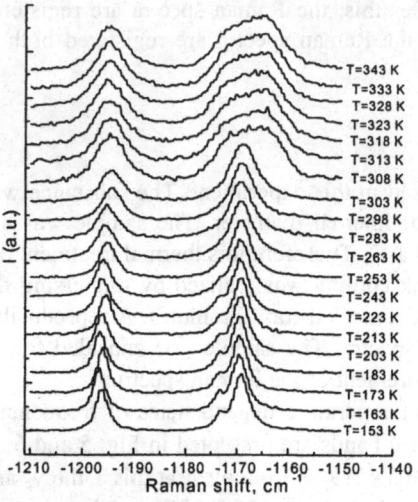

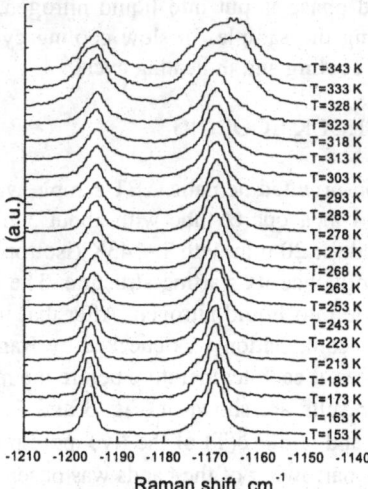

Figure 7 Raman spectra in slow cooling cycle in 1210-1150 cm⁻¹ region

Figure 8 Raman spectra in slow heating cycle in 1210-1150 cm⁻¹ region

Usually, in crystals the raise of T leads to an increase of $\delta(T)$ due to phonon-phonon interaction. The narrowing of $\delta(T)$ shows the removal of non-homogeneous broadening under supercooled nematic→solid crystal transitions. Similar results for MBBA was also obtained in [13]. Thus, the investigation of $\nu(T)$ and $\delta(T)$ allow us to determine the nature of the structural transformations in ABT in the T range 173-213 K as well as to determine the T of the transition amorphous →solid metastable crystal phase. In the T range 220-290 K, the $\delta(T)$ dependence entirely corresponds to a crystal state (Fig. 6). But in the T range 290-333 K both $\nu(T)$ and $\delta(T)$ dependencies show peculiarities of polymorphous transitions which are observed in the spectra as a change of shapes of all the five bands. A sharp increase of $\delta(T)$ for all the five bands is observed at T=333K which is accompanied by an increase of the ν_b, ν_c and ν_d bands frequencies. Such changes are related to the stable crystal→nematic phase transition. Thus, this investigation under an heating rate of a supercooled sample allows us to observe a number of structural transitions in ABT: the amorphous phase→solid crystal phase transition at T=213K (A→M); the metastable crystal→stable crystal phase transition at T=293K (M→S); And the stable crystal→nematic phase transition at T=333K (S→N).

4.2 SLOW COOLING

In slow cooling experiments, the ABT sample was cooled and heated in the T range 343K→153K→343K with a rate of 1 K/min. As it follows from our analysis, a quasi-static thermodynamic conditions takes place under such cooling rate. It means that the

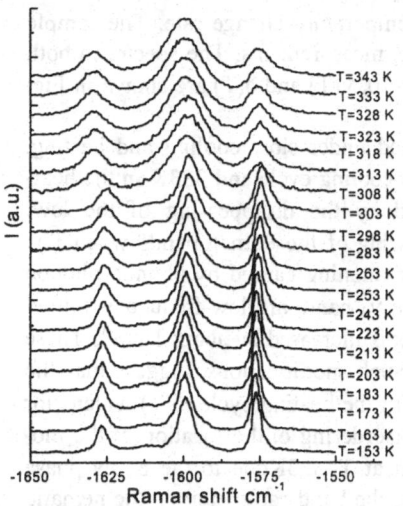

Figure 9 Raman spectra in slow cooling cycle in 1650-1550 cm^{-1} region

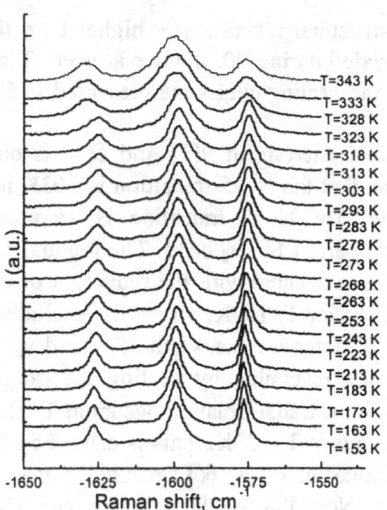

Figure 10 Raman spectra in slow heating cycle in 1650-1550 cm^{-1} region

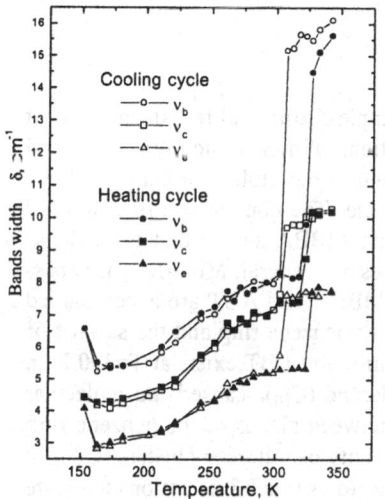

Figure 11 Bands width in slow cooling and heating cycles.

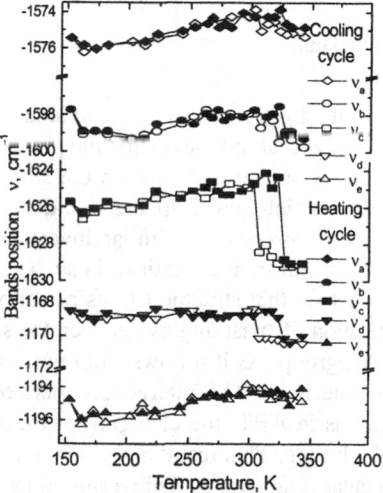

Figure 12 Bands position in slow cooling and heating cycles.

rate of structural relaxation is higher than the temperature change rate. The sample was annealed during 20 minutes at every T before measurements. The spectra in both cooling and heating cycles are presented in Fig. 7-10. $\nu(T)$ and $\delta(T)$ are shown in Fig. 11 and 12.

An hysteresis in $\nu(T)$ and $\delta(T)$ is observed under slow cooling and heating. Thus, the T of the N$\rightarrow$S transition is 303K in the cooling cycle and 333K in the heating cycle. The N$\rightarrow$S transition is accompanied by the disappearing of the low-frequency part of the ν_d band. The gradual narrowing of the Raman bands when T is reduced is connected with the reduction of the broadening caused by phonon-phonon interaction. For T<153K, all bands, excluding the ν_d band, are low-frequency shifted by 1 cm^{-1}. Moreover, the width of ν_b and ν_e bands is increased by about 1 cm^{-1}. These changes are reversible by heating the sample. Such modifications suggest that the polymorphous transformation occurs at T=153K. In the heating cycle, $\delta(T)$ widens for all the bands at T=323K. This is caused by the disordering of the location of the molecular center of mass. $\delta(T)$ reaches a maximum at T=333Kdue to the S$\rightarrow$N phase transition. Note, that on the contrary with MBBA, the band parameters in the nematic phase for ABT are equal before and after cooling cycle of the sample. Thus, the phase transitions in ABT under slow cooling and heating are :

$$I \xrightarrow{338K} N \underset{333K}{\overset{303K}{\longleftrightarrow}} S_2 \xleftrightarrow{153K} S_1$$

5. Conclusions

A spectral investigation was carried to study the complex structural transformations at the molecular and atomic level for the phase transitions of mesogenic media. We find out correlations between changes in the structural and in the molecular conformations of substances which can form the liquid crystal phase. The comparison of results of these investigations and of similar investigations on MBBA shows that the mechanisms of structural transformations in such substances are general. Moreover, it is possible to conclude that structural transformations in MBBA and ABT are accompanied by modification of twist-angles between the surface of benzene ring and the surface of CHN bridge group. As it follows from our investigation the ABT exists at T=300 K in polycrystalline form. The interactions between fullerene $(C_{60})_n$ caused the molecular conformations in ABT: the changing of the angle between planes of the benzene ring and CNH - bridge. This result show a significant activity of fullerene clusters $(C_{60})_n$ in intermolecular interactions. This result allows to hope to use such interactions to create fullerene liquid crystal devices.

References

1.   Patrick Bernier, Wolfgang Maser, Catherine Journet, Annick Loiseau, Marc Lamy de la Chapelle, Serge Lefrant, Roland Lee and Johnn Fischer, *Carbon,* 1998, **36**, No. 5-6, 675.
2.   Pogorelov V.Ye., Kondilenko I.I., Stryzhevskiy V.L., Shinkaryova E.A., *Optica and Spectroscopy,* 1969, **26**, 203.

3. V. D. Blank, S.G. Buga, N. R. Serebryanaya, G. A. Dibutsky, B. N. Mavrin, M. Yu. Popov, R. H. Bagramov, V. M. Prokhorov, S. N. Sulyanov, B. A. Kulnitskiy and Ye. V. Tatyanin, *Carbon,* 1998, **36**, No. 5-6, 665.

4. J. Winter, H. Kuzmany, *Carbon,* 1998, **36**, No. 5-6, 599.

5. Pogorelov V.Ye., Estrela-Llopis I.B., *Mol. Cryst. Liq.Cryst.,* 1995, **265**, 237.

6. Pogorelov V.Ye., Bazhenov V.A., Dekhtyaruk E.S., Pogorelova O.S., *Mechanics of the the composite materials and constructions (Moscow),* 1996, **2**, N. 2, 105.

7. Estrela-Llopis I.B., Pogorelov V.Ye., Bukalo V.P., Astashkin Yu.A., *Mol.Cryst.Liq.Cryst.,* 1998, **320**, 45.

8. V. Pogorelov, Yu. Astashkin, L. Kutulya, G. Semenkova, *Proc. SPIE,* 1999, **4069**, 96.

9. Yu. Prilutski, S. Durov, L. Bulavin, V. Pogorelov, Yu. Astashkin, V. Yashchuk, T. Ogul'chansky, E. Buzaneva, G. Andrievsky, *Mol. Cryst. Liq. Cryst.,* 1998, **324**, 65.

10. V. Pogorelov, E. Buzaneva, Yu. Prilutski, V. Bukalo, S. Zankovych, Yu. Astashkin, A. Benilov, Yu. Kirghisov, P. Scharff, *Synthetic Metals,* 2000. (in print)

11. W.J. Borer., S.S. Mitra, *Phys. Rev. Lett.,* 1971, **27**, 379.

12. J.T.S. Andrews, *Phys.Lett.,* 1974, **46**, 377.

13. M. Mizuno, T. Shinoda *Mol. Cryst. Liq. Cryst,* **69**, 1981, 103.

14. V.Ye. Pogorelov, *Solid State Physics,* 1991, **33**, 1906.

15. V.Ye. Pogorelov, I.P. Pinkevich, I.B. Estrela-Llopis, V.P. Bukalo, *Proceedings SPIE,* 1998, **3488**, 68.

16. B. Ostrowska, A. Tramer, *Acta Physica Polonica,* 1967, **33**, 111.

17. G. Vergoten, Adv. *Raman Spectroscopy,* 1972, **1**, 219.

18. J.P. Heger, R. Mercier, *Rapport Session Societe Suisse de Physique,* 1972, **45**, 886.

EXPLORING THE ULTIMATE LIMITS OF CONTROL: QUANTUM NETWORKS FOR NON-CLASSICAL INFORMATION PROCESSING

G. MAHLER AND A. OTTE

Institute for Theoretical Physics I, University of Stuttgart,
Pfaffenwaldring 57/IV, 70550 Stuttgart, Germany.
email: mahler@theo1.physik.uni-stuttgart.de

Abstract

The down-scaling of electronic devices has been an unprecedented success-story over the last decades. This development has, to a large extent, been based on continuous improvements of structural control. As we approach nanoscopic (atomic) precision, the question arises whether dynamical control (function) will be able to follow down to these ultimate limits, i.e. whether the exploitation of the fundamental rules of quantum dynamics will become possible at large scale. To approach this goal there are at least two different paradigms which could be exploited: "Selective control" and "collective control". The latter should be an interesting alternative for nanoscopic quantum networks.

1. Introduction

"Information is physical"[1]. There is an intimate relationship between mathematics as a powerful description of physics and physics as an implementation of mathematical rules. Usually, a strict distinction is made between the "analog" representation of a number (e.g. as a length, as a current etc.) and its digital variant (based on the elementary distinction "head" or "tail", -1 or $+1$, i.e. 1 bit). In either case information processing is then realized as an appropriate physical process.

The use of some "different physics" may imply either a *different* implementation of the *same* rules or the implementation of *novel* rules (i.e. other basis operations). Along the first route we have witnessed the transition from mechanical to electromechanical and to electronic computers. For the latter, typical examples are analog systems like the Silicon retina[2] and rather speculative ideas based on classical nonlinear dynamics (chaos [3]). A more recent proposal concerns the application of quantum mechanical rules for digital information processing[4, 5, 6].

Analog computing is known for its efficiency but also for its limited flexibility, resolution, and robustness. Digital computation, on the other hand, is universal in the sense that any algorithmic problem can be tackled, while its complexity does not depend on the physical nature of implementation. So one may wonder how quantum computing could, in principle, bypass these constraints.

Appropriate quantum systems will have to be designed for such new applications. A number of options is being investigated right now, gradually turning theoretical models into experimental studies. Semiconductor nanostructures (e.g. electronic quantum dots) might seem to be the first choice. However, the difficulty to isolate the desired electronic degrees of freedom from the rest leads to decoherence times, which are much shorter than can be obtained in other implementations. On the other hand, the perspective to use very large ensembles of quantum dots is certainly attractive: If convenient ways for control could be found (and possibly, some means to reduce damping), the overall system

431

L. Pavesi and E. Buzaneva (eds.), Frontiers of Nano-Optoelectronic Systems, 431–442.

performance might turn out to offer some practical advantages.

Rather than trying to copy the strategies of other prospective physical implementations, it should be more appropriate to adapt the pertinent quantum network architecture to the specific physical needs of nanostructures. One important aspect of such an architecture could be collective control, as will be discussed below.

2. Concrete Quantum Networks

There are two different groups of potential implementations: Systems to support (long distance) quantum communication (primarily photons), and localized quantum networks (effective spins) based on a classical structure (architecture).

The design of quantum networks faces a number of severe problems: The control of geometry and structure will be limited resulting in static variations of the Hamilton-parameters. Furthermore, just as in a classical device only a tiny fraction of the microscopic degrees of freedom will actually be used and interpreted as carrying information. In fact, robust macroscopic or hydrodynamic observables emerge as a consequence of appropriate averaging. However, averaging implies loss of microscopic information with its disastrous effect on quantum coherence. When it comes to the very detailed description of a quantum system, virtually nothing can be neglected, there is "nothing unimportant", a statistical approach cannot be accepted.

The only practical remedy appears to be the design and the exploitation of scale-dependences: Interactions depend on length scales, decoherence depends on timescales. Quantum network states are thus transient and depend on a parameter window.

Options currently under investigation include

- cold ion traps – internal electronic states, collective modes for ion motion in trap [7, 8],

- photons – Fock-states in resonators [9], polarization states or discrete beam directions of free photons,

- NMR on molecular fluids (spin ensembles) [10, 11],

- NMR on contacted spins in solid matrix – gate controlled interactions [12],

- atoms in optical lattices – internal electronic states, collective motion [13, 14],

- Josephson junctions – collective charge states, gate control [15],

- semiconductor quantum dots – charge-states, spin-states [16]).

For these implementations sample networks with less than 10 pseudospins are being realized or under preparation. Whether larger networks will become feasible remains to be seen. Nanoscopic semiconductor systems should turn out to be flexible enough to take up the challenge (cfr.. also Ref.[17]).

3. Abstract Quantum Networks

3.1 FROM BIT TO QUBIT

Suppose that we assign the formal information state "-1" to the ground state wavefunction $| -1 >$ of a semiconductor quantum dot, the information "$+1$" to the first excited state $| +1 >$. Now, contrary to a classical realization, the quantum dot may also be found in the superposition

$$|\Psi >= c_{-1}| -1 > +c_1| +1 >$$

(1)

where the c_i are, in general, complex amplitudes with $|c_{-1}|^2 + |c_{+1}|^2 = 1$. Such super-position states are termed "coherent".

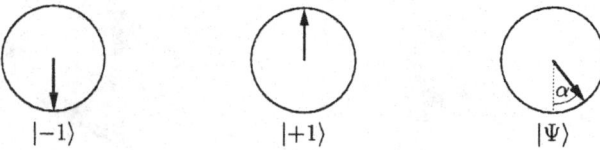

Figure 1. Visualization of a bit and a qubit.

Taking, for simplicity, c_i as real, we may introduce the phase angle α, $\cos\alpha = |c_{-1}|^2 - |c_{+1}|^2$, as the only significant parameter (cfr.. Fig. 1). The superposition state may thus be represented as an arrow, the so-called Bloch-vector. The two points, $\alpha = 0, \pi$ ("south-" and "north-pole") then represent the states for the (classical) bit, while the whole unit circle is a visualization of the qubit.

Allowing for complex amplitudes c_i the Bloch- vector will have 3 components (here we restrict ourselves to two, λ_3, λ_2). We note in passing that the vector-components λ_j can be interpreted as the expectation values of a specific set of operators (observables) $\hat{\lambda}_j$ with $\lambda_j = <\hat{\lambda}_j>$. In this way the concept of a wavefunction can be replaced by a set of expectation values, i.e. directly measurable quantities, reminiscent of classical physics. The latter notation can even be used when wavefunctions are no longer applicable (i.e. for density matrices).

These operators are most conveniently specified by the effects they have on the basis states $|\mp 1>$:

$$\hat{\lambda}_1|\mp 1> = |\pm 1>$$
$$\hat{\lambda}_2|\mp 1> = \mp i|\pm 1>$$
$$\hat{\lambda}_3|\mp 1> = \mp|\mp 1> \qquad (2)$$

The elementary effects thus consist of a state-flip, a state + phase-flip, and a sign-flip, respectively. Any other operator can be composed out of these and the unit operator $\hat{1}$: Rotations by angle α_s with respect to the s-axis, e.g., are generated by [18]

$$\hat{U}_{\alpha_s}^s = \hat{1}\cos\alpha_s + i\hat{\lambda}_s\sin\alpha_s \qquad (3)$$

This "non-classical" instruction combines "do nothing" ($\hat{1}$) with "apply a state-flip" ($\hat{\lambda}_s$ for $s = 1$). So, if the initial state was just one of the basis states, the result would be a superposition.

Note, however, that a superposition of the information $-1, +1$ has no meaning: Does the qubit thus behave like a switch that – erroneously – can also be in between the on- and off-position? Not really, the delicacies of the quantum measurement process guarantee that any measurement result (for $\hat{\lambda}_3$) will either be $\lambda_3 = -1$ (probability $|c_{-1}|^2$ or $\lambda_3 = +1$ (with probability $|c_1|^2$), just the classical points! Nevertheless, the phase is there for intermediate use: In some sense the qubit combines analog features (the phase) with digital features (the measurement outcomes).

This feature of the measurement process also poses a severe constraint on information retrieval: We gain at most one bit of information, the phase enters the detection prob-abilities, not the state *after* measurement. So, in general, the state *before* measurement cannot be reconstructed and thus not copied ("no-cloning" theorem [19]).

The control of the phase is by rotations of the arrow, corresponding to unitary transfor-mations of the state $|\Psi>$. Such a rotation can either be "active" (i.e. by manipulation

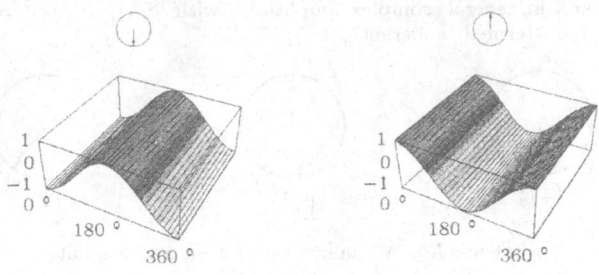

Figure 2.　Coherence pattern, $N = 1$.

of the state) or "passive" (i.e. by manipulating the external reference frame, the "dial" or measurement apparatus). An active rotation can thus always be undone by a corresponding passive rotation; in this sense all the rotated states are "equivalent" (like viewing one given mountain from different directions). A given state can thus alternatively be represented by the "coherence patterns" generated by its projection,

$$\lambda_3(\alpha) = -\cos\alpha \tag{4}$$

on all possible measurement directions (Fig. 2).

3.2 REGISTER

Registers are composite systems, where each subsystem $\mu = 1, 2, \ldots N$ allows the representation of one bit. There are thus 2^N possible states. An implementation could consist of N classical flip-flops or N two-state quantum dots. The latter system will be described by the 2^N product states

$$|\Psi > = |n_1 > \otimes |n_2 > \otimes \cdots |n_N > \equiv |n_1, n_2, \cdots n_N > \tag{5}$$

with $n_\mu = \mp 1$ corresponding to $\lambda_3^\mu = \mp 1$. The set of λ_3^μ uniquely specifies these states: The dimension of the corresponding "classical" configuration space is thus N.

Based on these parameters we can define the correlation functions[18, 20]

$$K_{33\ldots3} = \lambda_3^{(1)} \lambda_3^{(2)} \cdots \lambda_3^{(N)} \tag{6}$$

which are $+1$ if the number of local $|-1>$ states is even, -1 if that number is odd. For $N = 2$ a complete orthogonal state basis would be $|-1, -1>$, $|1, 1>$ with $K_{33} = 1$, denoting strict correlation, and $|-1, +1>$, $|1, -1>$ with $K_{33} = -1$ (strict anti-correlation).

3.3 CLASSICAL NON-SELECTIVITY

If we cannot address the subsystems individually, the pertinent description must avoid reference to specific subsystem indices. The classical way to invoke non-selectivity is a statistical approach and connected with loss of (microscopic) information: Based on partitions into single subsystems, pairs, triples etc. we may introduce the averages,

$$\overline{\lambda_3} = I_1 - I_{-1} \tag{7}$$

$$\overline{K_{33}} = (I_{11} + I_{-1-1}) - (I_{-1+1} + I_{+1-1}) \tag{8}$$

etc., where I_j is the frequency of subsystems in state $|j > (I_1 + I_{-1} = 1)$, I_{jk} is the frequency of pairs in state $|jk >$ etc. I_{jk} can be found by a coincidence measurement. If all subsystems are prepared in the same pure state, the local average is $\overline{\lambda_3} = -\cos\alpha$

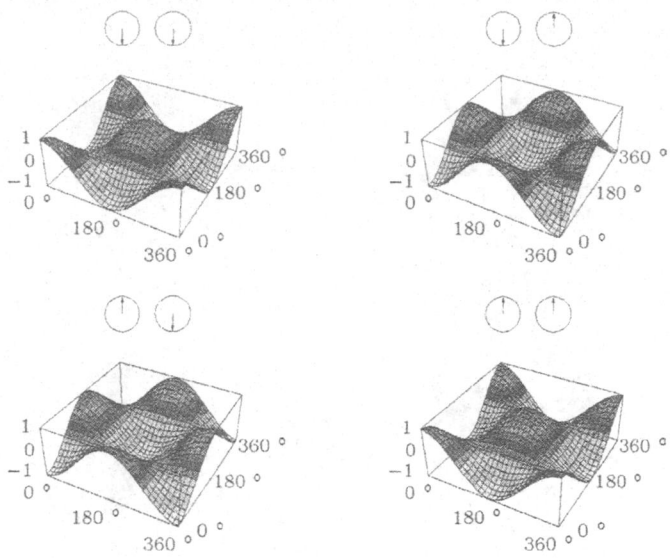

Figure 3. Correlation pattern for product states ($N = 2$).

(cfr.. eq. (4)), which is directly measurable as the well-known (single-particle) interference pattern. Under the same conditions $\overline{K_{33}} = \overline{\lambda_3^{(1)}} \ \overline{\lambda_3^{(2)}}$. Classical pair correlations could imply that $\overline{K_{33}} = \pm 1$ while $\overline{\lambda_3^{(\mu)}} = 0$.

3.4 QUANTUM REGISTER

When for the quantum dot array any superposition is allowed, in addition to the 2^N classical bit-states, the array is said to operate as a quantum register. The corresponding configuration space is now the full Hilbert-space; its dimension is 2^N, which thus grows exponentially with system size N. It should therefore not be too surprising that this huge space has more to offer than its classical counterpart.

In addition to the local coherence pattern we may consider correlation pattern resulting from local rotations with respect to any of the constituents and thus being defined in a multi-dimensional phase-angle space (cfr.. eq. (4)). The product states of a complete orthogonal set have all the same correlation pattern except for a sign (Fig. 3):

$$K_{33...3}(\alpha_1, \alpha_2, ...\alpha_N) = \mp \cos \alpha_1 \cos \alpha_2 \cdots \cos \alpha_N \qquad (9)$$

These are not the most general superpositions, though. In addition to those conserving the product character (e.g., for $N = 2$)

$$|\Psi> = c_1| - 1, -1 > + c_2| - 1, 1 > \qquad (10)$$

there are also states which can no longer be written as a product,

$$|\Psi> = c_1| - 1, 1 > + c_2|1, -1 > \qquad (11)$$

They are called entangled: Entanglement is the concept of coherence extended to composite systems. Specific (maximum) entangled states are "cat-states". They form a

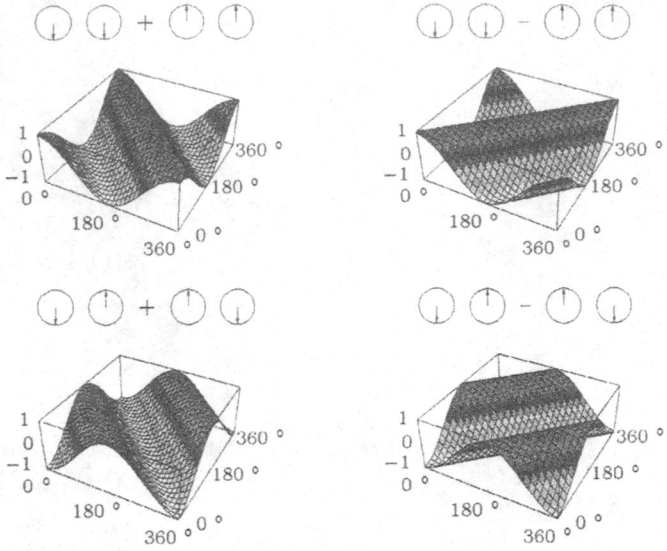

Figure 4. Correlation patterns for cat-states ($N = 2$), cfr.. eqs. (12,13).

complete orthogonal set, just like the product states. For $N = 2$:

$$|\Psi^\pm> = \frac{1}{\sqrt{2}}[|-1,1> \pm |1,-1>]$$ (12)

$$|\Phi^\pm> = \frac{1}{\sqrt{2}}[|-1,-1> \pm |1,1>]$$ (13)

These states have remarkable properties: First of all, their local expectation values are entirely undefined ("undecided"), i.e. $\lambda_j^{(1)} = \lambda_j^{(2)} = 0$. Nevertheless, they are all strictly (anti-) correlated, e.g. for $|\Psi^- >$

$$K_{33} = -1$$ (14)
$$K_{22} = K_{11} = 1$$ (15)

("correlations without correlata"[21]). Note that this is something that could never happen for single pairs in the classical domain! $K_{11} = K_{22} = 1$ indicates that, in addition, there was a strict correlation when the external local reference frames were rotated appropriately. For $N = 2$ the correlation pattern of the 4 cat-states are:

$$K_{33} = \pm\cos(\alpha_1 \pm \alpha_2)$$ (16)

We see from Fig. 4 that each of the 4 cat-states is specified by a unique correlation pattern, while those for the 4 product states are pairwise identical (cfr.. Fig. 3); to uniquely specify those states we would, in addition, need one local coherence pattern. This underlines the special role of quantum correlations.

These considerations can systematically be generalized to $N > 2$. Typical $N = 3$ cat states are

$$|\Phi^\pm> = \frac{1}{\sqrt{3}}[|-1,-1,-1> \pm |1,1,1>]$$ (17)

In all these cases $\lambda_j^{(\mu)} = 0$; $\mu = 1, 2, ...N$. There is a whole hierarchy of correlation tensors and correlation pattern, starting from single subsystems, pairs, triples etc. up to N-particle correlations. The phase patterns with respect to a complete set of basis functions may be viewed as the modes into which any specific state could be decomposed (cfr.. Figs. 2 – 4). This is reminiscent of the Fourier-decomposition in linear wave-theory. For a proper choice of phase angles these correlation functions reduce to the components of the corresponding correlation tensor, $K_{jk...m} = < \hat{\lambda}_j(1) \cdots \hat{\lambda}_m(N) >$ for the state given [18].

4. Quantum Control

4.1 SELECTIVE CONTROL

For Hamiltonian systems the unitary time-evolution is generated by the respective Hamiltonian $\hat{H}$,

$$|\Psi_m(t_f) > = \hat{U}(t_f)|\Psi_m(0) > \qquad (18)$$

$$\hat{U}(t_f) = \exp(-i\hat{H}t_f/\hbar) \qquad (19)$$

By means of this system dynamics, different input states $|\Psi_m(0) >$ are mapped into different output states (for a fixed final time t_f).

Control dynamics, on the other hand, is based on the idea that we start with one and the same initial state $|\Psi_0(0) >$ but modify $\hat{H}$,

$$\hat{H}(t) = \hat{H}_i \qquad t_i \leq t < t_i + \Delta t \qquad (20)$$

where $i = 0, 1, 2, ...M$ and $t_f/\Delta t = M$. This amounts to changing the Hamilton parameters in form of sequential pulses. It has been shown that only two basic types for $\hat{H}_i$ ("quantum gates") are required [22]: local rotations $\hat{U}(\mu)$ (as discussed already, cfr.., eqs.(1,3)) and the quantum-controlled NOT $\hat{U}(\mu, \nu)$ defined by the rules (QCNOT)

$$
\begin{aligned}
\hat{U}(\mu, \nu) | -1, -1 > &= |1, -1 > \\
\hat{U}(\mu, \nu) |1, -1 > &= | -1, -1 > \\
\hat{U}(\mu, \nu) | -1, 1 > &= | -1, 1 > \\
\hat{U}(\mu, \nu) |1, 1 > &= |1, 1 >
\end{aligned}
\qquad (21)
$$

The subsystem ν (right hand side) functions as a control for implementing a π-pulse (NOT-operation) on the left subsystem μ (cfr.. Fig. 5). This is still classical. However, the linearity of quantum mechanics implies that a superposition of inputs generates the corresponding superposition of outputs. This means that, e.g.,

$$\hat{U}(\mu, \nu) \frac{1}{\sqrt{2}}[| -1, -1 > + | -1, 1 >] =$$

$$\frac{1}{\sqrt{2}}[|1, -1 > + | -1, 1 >] \qquad (22)$$

i.e. the superposition of the control (still a product state) leads to the generation of a cat state, with the corresponding change of phase pattern.

The "classical" idea of locally addressable gates can thus be used, together with the superposition principle, to generate non-classical states step by step.

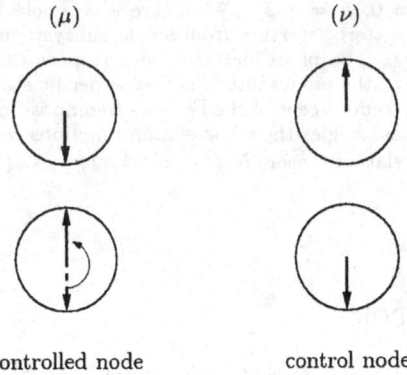

(μ) (ν)

controlled node   control node

Figure 5.   Conditioned state flip.

4.2 IMPLEMENTATION OF QCNOT

The implementation of local rotations is rather simple: In principle, a (frequency-) selective coupling to an external laser-pulse will suffice (assuming that the transition frequency of each subsystem can be selected at will). By construction, this interaction can be switched on and off as desired.

Implementations of selective spin-spin interactions are much more challenging. The simplest way to implement a quantum-controlled-NOT (QCNOT) works for so-called charge transfer-states[18]: In this case the two states of any node do not only refer to different energies, but also to opposite dipole-moments. Dipole-dipole interactions (parameter C_R) thus lead to an energy shift of one system depending on the state of the other (see Fig. 6). The frequency of an external laser pulse can then be chosen such that it acts as a π-pulse on one subsystem only if the neighbor is in the lower (or upper) state. As the neighbor can also be in a superposition state, this two-pseudo-spin array functions as a QCNOT.

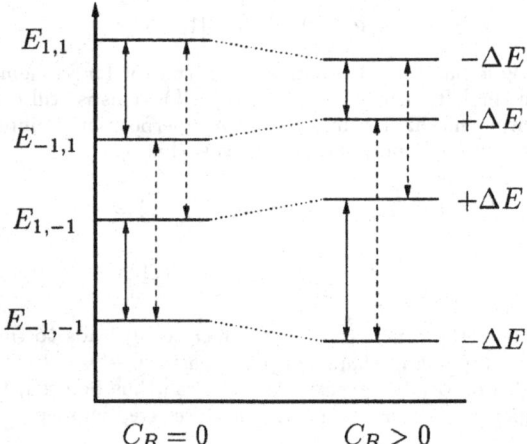

Figure 6.   Energy spectrum of a $N = 2$-network without (left) and with (right) dipole interactions. Solid arrows: transitions within system 1; broken arrows: transitions within subsystem 2. With the interaction the degeneracy of the transition energies is lifted.

4.3 COLLECTIVE CONTROL

Non-selective ("collective") control, on the other hand, has no classical analog: Just like superposition of states it refers to a superposition of "actions" on all members of the quantum register in the sense of an (undecided) "OR" (cfr.. eq. (3)). For example, the $N = 3$ collective operator

$$\hat{E}_{100} = \hat{\lambda}_1(1) + \hat{\lambda}_1(2) + \hat{\lambda}_1(3) \tag{23}$$

when applied to the ground state $| -1, -1, -1 >$ generates the superposition state

$$\hat{E}_{100}| -1, -1, -1 > \sim | +1, -1, -1 > +$$
$$| -1, +1, -1 > +| -1, -1, +1 > \tag{24}$$

The rationale behind this operator is that the underlying external coupling cannot distinguish between the individual subsystems $\mu = 1, 2, 3$ (cfr.. Fig. 7).
While in the classical case non-selectivity leads to a statistical approach (cfr.. Sect. 4) based on concrete though unknown choices, the quantum mechanical dynamics "avoids any decision". Alternatively we may say that the three paths from $| -1, -1, -1 >$ to the state with subsystem $\mu = 1$ or 2 or 3 excited are followed in parallel, i.e. there is no "which-path-information". Only the subsequent measurement would *a posteriori* give us this information and so reproduce the statistical outcome!

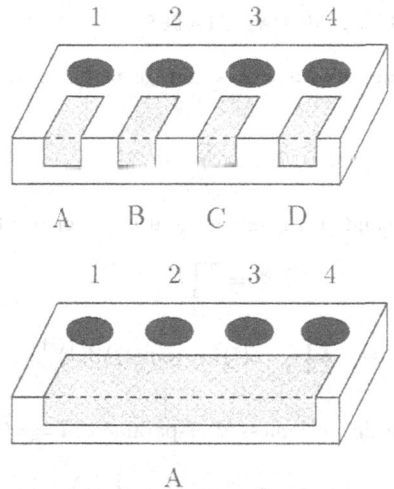

Figure 7. Selective / non-selective control.

Collective operators naturally occur in the description of elementary excitations in solids. A special subset of such operators are permutation-symmetric, like $\hat{E}_{100}$ given above, or

$$\hat{E}_{002} = \hat{\lambda}_3(1)\hat{\lambda}_3(2) + \hat{\lambda}_3(1)\hat{\lambda}_3(3) + \hat{\lambda}_3(2)\hat{\lambda}_3(3) \tag{25}$$

$\hat{E}_{100}$ is a "single-particle"-operator, $\hat{E}_{002}$ a two-particle operator, here for $N = 3$. Only operators of this type (or combinations of them) would be allowed for fundamentally indistinguishable particles (Fermions, Bosons).
In the present case of nanoscopic effective spins bound to a specific position (classical

structure), the subsystems are, in principle, distinguishable by their location. However, the physical couplings will typically have a limited spatial resolution ("wave length") so that external driving [23, 24] (as well as damping [25]) may – at least partly – be of collective type, as described above [26].

Permutation symmetry imposes a structure on the 2^N- dimensional Hilbert-space, which can be characterized by the "total spin" $j = N/2, N/2 - 1, N/2 - 2, \ldots \geq 0$. For $N = 3$ the basis vectors $j = 3/2$ are totally symmetric (Bose-type) and constitute a 1-dimensional representation of the symmetry group, the $j = 1/2$ vectors a 2-dimensional representation of "mixed symmetry". These mixed symmetries do not occur with elementary indistinguishable particles.

Permutation symmetric operators connect states within the same symmetry class only (superselection rule). We thus obtain separate Hilbert-spaces of dimension $(2j + 1)$, i.e. in general a small section of the total space.

So, as long as all the operators are strictly permutation symmetric, the total Hilbert-space decomposes into a direct sum of subspaces which are protected against each other by means of superselection rules. Ironically, networks of indistinguishable subsystems (a typical quantum feature!) have thus Hilbert-spaces which grow only polynomially with N.

The manipulation of symmetry (including, e.g. optical near-field coupling) may allow to prepare the system in subspaces (symmetry classes) which follow neither Fermi- nor Bose-statistics. However, there is a price to pay: The symmetry imposes a constraint (the very superselection rules) that will not allow us to reach any final state from any initial state, contrary to the selective control.

4.4 EXAMPLE: GENERATION OF CAT-STATES

Special (permutation symmetric) cat states emerge based on unitary transformations generated by $\hat{E}_{200}$, where

$$\hat{H} = \hat{E}_{200} = \sum_{j<k} \hat{\lambda}_1^{(j)} \hat{\lambda}_1^{(k)} \tag{26}$$

This is easily seen by expanding the exponential operator function

$$\hat{U} = e^{-it\hat{E}_{200}} = \prod_{j<k} e^{-it\hat{\lambda}_1^{(j)}\hat{\lambda}_1^{(k)}} \tag{27}$$

$$= \prod_{j<k} \left(\cos(t)\hat{1} - i\sin(t)\hat{\lambda}_1^{(j)}\hat{\lambda}_1^{(k)} \right) \tag{28}$$

For $N = 2$ and choosing the dimensionless time-parameter $t = \pi/4$, one finds [23]

$$\hat{U}_{t=\frac{\pi}{4}} = \frac{1}{\sqrt{2}} \left(\hat{1} - i\hat{E}_{200} \right) \tag{29}$$

$$\hat{U}_{t=\frac{\pi}{4}} |00\rangle = \frac{1}{\sqrt{2}} \left(|00\rangle - i|11\rangle \right) \tag{30}$$

Corresponding results for N even are

$$\hat{U}_{t=\frac{\pi}{4}} \propto \hat{1} + e^{i\phi}\hat{E}_{N00} \tag{31}$$

$$\hat{U}_{t=\frac{\pi}{4}} |00\ldots0\rangle \propto |00\ldots0\rangle + e^{i\phi}|11\ldots1\rangle \tag{32}$$

where Φ is a phase angle depending on N. A single non-selective pulse is thus able to produce a cat-state with its dominating N-order correlations!

5. Summary and conclusions

(1) A machine is defined by its functional repertoire and its means of control. For a quantum network the repertoire results from the "designed" structure (constraints) imposed on the underlying Hilbert-space. These constraints result from the external control, which enters as the network-Hamiltonian. This Hamiltonian, in turn, requires the presence of a classical environment, by which its parameters can be manipulated.

(2) State changes are (ideally) induced by unitary transformations; their linearity allows for superpositions (of local states as well as of whole product states), a typical quantum phenomenon. Routes to controlled superpositions can be based on sequential operation of selective gates or on non-selective couplings (exploiting indistinguishability).

(3) Superposition can be specified in terms of certain phase patterns. Any quantum information processing can be seen as a manipulation of such patterns.

(4) Superposition of states may be interpreted as an (undecided) OR, which would be decided by an actual measurement. In this restricted sense many alternative calculation paths are "present" at the same time ("quantum parallelism").

(5) The final read-out (measurement) adds a digital feature to the analog feature (phase) of the network. In this sense the quantum computer may be considered an analog/digital hybrid.

(6) To assess the "efficiency" of the quantum processing a careful accounting of the resources has to be carried out. As real quantum computers of practical size do not yet exist, this accounting has to refer to theoretical models with all their idealizations. Nevertheless, it is believed that certain algorithms (Shor's factoring algorithm[5], Grover's search algorithm[6]) would scale more favorably with problem size than any known classical algorithm on a classical computer.

(7) Quantum computation (just like quantum-simulation) is necessarily a large-scale application, i.e. controlled quantum networks with $N > 10^3$ pseudospins would be required. This is beyond the reach of any known technology. It is hoped that interesting small-scale applications ($N < 10$) like quantum cryptography [27] will be found in the near future.

Financial support by the Deutsche Forschungsgemeinschaft (Grant No. 614/19) is gratefully acknowledged.

References

[1] R. Landauer, "Information is physical", in D. Matzke (ed), *Proc. of the workshop on Physics and Computation*, IEEE Computer Soc., Los Alamitos, p. 1 - 4 (1993)

[2] M. A. Mahowald and C. Mead, "The Silicon Retina", Sci. Am. May 1991, p. 40

[3] J. S. Nicolis, *Chaos and Information Processing*, World Scientific, Singapore 1991

[4] A. Steane, "Quantum computation", Repts. Progr. Phys. 61, 117 (1998)

[5] A. Ekert and R. Jozsa, Rev. mod. Phys. 68, 733 (1996)

[6] L. K. Grover, "Quantum mechanics helps in searching for a needle in a haystack", Phys. Rev. Lett. 79, 325 (1997)

[7] C. Monroe et al., "Demonstration of a fundamental quantum logic gate, Phys. Rev. Lett. 75, 4714 (1995)

[8] B. E. King et al., "Cooling the collective motion of trapped ions to initialize a quantum register", Phys. Rev. Lett. 81, 1525 (1998)

[9] P. Domokos et al., "Simple cavity-QED two-bit universal quantum logic gate", Phys. Rev. A 52, 3554 (1995)

[10] N. A. Gershenfield and I. L. Chuang, "Bulk-spin-resonance quantum computation", Science 275, 350 (1997)

[11] I. Chuang et al., "Experimental realization of a quantum algorithm", Nature 393, 143 (1998)

[12] E. Kane, "A silicon-based nuclear spin quantum computer", Nature 393, 133 (1998)

[13] D. Jaksch et al., "Entanglement of atoms via cold controlled collisions", Phys. Rev. Lett. 82, 1975 (1998)

[14] G. K. Brennen et al., "Quantum logic gates in optical lattices", Phys. Rev. Lett. 82, 1060 (1999)

[15] A. Shnirman, G. Schön, and Z. Hermon, "Quantum manipulations of small Josephson junctions", Phys. Rev. Lett. 79, 2371 (1997)

[16] D. Loss and D. P. DiVincenzo, "Quantum computation with quantum dots", Phys. Rev. A 57, 120 (1998)

[17] E. Buks et al., "Dephasing due to which-path detector", Nature 391, 871 (1998)

[18] G. Mahler and V. A. Weberruss, Quantum Networks, Springer, Berlin, New York 1995, 1998

[19] W. K. Wooters and W. H. Zurek, "A single quantum cannot be cloned", Nature 299, 802 (1982)

[20] G. Mahler, M. Keller, and R. Wawer, "Quantum networks: master equation and local measurements", Z. Phys. B 104, 153 (1997)

[21] N. D. Mermin, "What is quantum mechanics trying to tell us?" Am. J. Phys. 66, 753 (1998)

[22] A. Barenco et al., "Elementary gates for quantum computation", Phys. Rev. A 52, 3457 (1995)

[23] K. Molmer and A. Sorensen, "Multiparticle entanglement with hot trapped ions", Phys. Rev. Lett. 82, 1835 (1999)

[24] A. Sorensen and K. Molmer, "Quantum computation with ions in thermal motion", Phys. Rev. Lett. 82, 1971 (1999)

[25] P. Zanardi and F. Rossi, "Subdecoherent information encoding in a quantum dot array", Phys. Rev. B 59, 8170 (1999)

[26] The implementation of permutation-symmetric 2-particle-interaction is severely constrained by the fact that interactions depend on distance, which, in general, cannot be invariant under permutation.

[27] A. Ekert, "Quantum cryptography based on Bell's theorem", Phys. Rev. Lett. 67, 661 (1991)

CHAOS IN QUANTUM MACHINES

ILKI KIM AND GUENTER MAHLER
Institut für Theoretische Physik I, Universität Stuttgart
Pfaffenwaldring 57, 70550 Stuttgart, Germany

Abstract

We investigate in a quantum Turing architecture the iteration of a sequence of local and pair unitary transformations, which can be interpreted to result from a Turing head (pseudo-spin S) rotating along a closed Turing tape (M additional pseudo-spins). The dynamical evolution of the Bloch-vector of S, which can be decomposed into 2^M primitive pure state Turing-head trajectories, gives rise to fascinating geometrical patterns reflecting the entanglement between head and tape. These machines thus provide intuitive examples for quantum parallelism and, at the same time, means for local testing of quantum network dynamics. Furthermore we show that a single chaotic parameter input $\{\alpha_m\}$ for the local transformations of the Turing head leads to a chaotic dynamics in the entire Hilbert space. The instability of periodic orbits on the Turing head and 'chaos swapping' onto the Turing tape are demonstrated explicitly as well as exponential sensitivity of the state evolution with respect to perturbations in the control.

1. Introduction

In recent years problems of quantum computing (QC) and information processing have received increasing attention. To solve certain classes of problems in a potentially very powerful way, one tries to utilize in QC the quantum-mechanical superposition principle and the (non-classical) entanglement [22], an undertaking which, at the same time, should contribute to our basic understanding of quantum mechanics itself (see e.g. [6]). However, building a large-scale quantum computer remains an extremely difficult task. The major obstacle is the coupling of the quantum computer to the environment, which tends to destroy quantum-mechanical superpositions very rapidly. This effect is usually referred to as decoherence. Present-day technology does not yet support the realization of a practical quantum computer. On the other hand, there might be interesting small-scale physics in a pure quantum regime based on a few pseudo-spins (qubits), which are realizable right now.

Despite the lack of a clear-cut definition, the physics of *complexity* [17] has intrigued physicists for many years. For continuous classical systems with few degrees of freedom the notion of chaos, namely an exponential sensitivity to initial conditions in non-linear systems, has attracted much interest as a sign of uncontrollability. For discrete classical systems in the form of cellular automata the notion of computational irreducibility has been introduced to account for the lack of 'short-cuts', i.e. our inability to predict the respective state evolution without following the detailed dynamics step by step [24]. The deterministic chaos, which occurs in non-dissipative systems, can typically be found

443

L. Pavesi and E. Buzaneva (eds.), Frontiers of Nano-Optoelectronic Systems, 443–454.
© 2000 *Kluwer Academic Publishers. Printed in the Netherlands.*

starting from regular states as a function of some external control parameter. However, there seems to be no direct analogue to chaos in the quantum world: if two quantum states are initially almost identical (that is, their scalar product is very close to 1), they will remain so forever, since the Hamiltonian evolution is a unitary mapping which preserves scalar products. According to this negative result, the semiclassical 'quantum chaology' [3] has been constrained to studying some quantum-mechanical 'fingerprints of chaos' (like spectral properties), and non-trivial transitions from the quantum - to the classical domain and vice versa, following Bohr's correspondence principle (see e.g. [8]). In addition, experimental progress in mesoscopic physics, e.g. the transport of electrons through so-called 'chaotic quantum dots' [16], has attracted a great deal of interest, the results of which give numerical evidence for weak chaos (indicated by level repulsion) [23].

In models of QC based on quantum Turing machines (QTM) [1, 5] the complexity of the computation is characterized by sequences of unitary transformations (or the corresponding Hamiltonians $\hat{H}$ acting during finite time interval steps). Although the above-mentioned linearity of quantum dynamics appears to make the respective evolution "well-behaved" from the start, the limit of control abounds even for modestly large quantum networks [18] due to the, typically, exponentially large Hilbert space, in which the state evolves [7]. It has been shown that if this kind of 'quantum-complexity' could be harnessed, new efficient modes of computation (QC) should become available. However, one will first have to find ways to circumvent that disastrous exponential blow-up.

The study of quantum chaos based on quantum gate networks has so far been proposed e.g. by implementing quantum baker's map on a 3-qubit NMR quantum computer [21], by realizing a quantum mechanical delta-kicked harmonic oscillator or a harmonically driven oscillator in an ion trap [9, 2], and by showing quantum-mechanical localization of an ion in a trap [10]. In all these cases some sort of sensitivity has been predicted with respect to parameters specifying the dynamics (e.g. the respective Hamiltonian).

Here we explicitly describe an iterative map with a few qubits which, though based on standard gates, can be thought to be realized as a QTM architecture: Local transformations of the Turing head controlled by various kinds of sequence $\{\alpha_m\}$ alternate with a quantum-controlled NOT-operation with a second spin on the Turing tape. We will show that the evolution of the Turing head in its reduced space gives rise to geometrical patterns reflecting the entanglement between Turing head and Turing tape. These patterns can be thought to result from the superposition of exponentially many 'primitive' Turing machines, an intuitive example of 'quantum parallelism'. Furthermore the chaotic control e.g. by a Fibonacci-like sequence can generate a chaotic quantum propagation in the 'classical' regime [4], defined here as one with the Turing head being restricted to an entanglement-free state sequence ('primitive') [12]. It will then be shown that chaos on the Turing head can be found also for the quantum-mechanical superposition of those primitives, implying entanglement between head and tape as a genuine quantum feature ('non-classical regime'). Finally, due to this quantum correlation, we find a chaotic propagation even in the reduced subspace of the Turing tape ('chaos swapping'), and as a result also in the total network state $|\psi_n\rangle$. This behavior should be contrasted with that of a regular QTM with its long-time revivals, which are absent in the 'chaotic' QTM.

2. Description of quantum Turing machine

The quantum network [18] to be considered is composed of N $(= M + 1)$ pseudo-spins

$|p)^{(\nu)}$; $p = 0, 1$; $\nu = S, 1, 2, \cdots M$ (Turing head S, Turing tape spins $\mu = 1, 2, \cdots M$) so that its network state $|\psi)$ lives in the 2^{M+1}-dimensional Hilbert space spanned by the product wave-functions $\left|j^{(S)}k^{(1)}\cdots l^{(M)}\right) = |jk\cdots l)$. Correspondingly, any (unitary) network operator can be expanded as a sum of product operators. The latter may be based on the following $SU(2)$-generators

$$\hat{\lambda}_1^{(\nu)} \;=\; \hat{P}_{01}^{(\nu)} + \hat{P}_{10}^{(\nu)}, \quad \hat{\lambda}_2^{(\nu)} = i\hat{P}_{01}^{(\nu)} - i\hat{P}_{10}^{(\nu)}$$

$$\hat{\lambda}_3^{(\nu)} \;=\; \hat{P}_{11}^{(\nu)} - \hat{P}_{00}^{(\nu)}, \quad \hat{\lambda}_0^{(\nu)} = \hat{P}_{11}^{(\nu)} + \hat{P}_{00}^{(\nu)} = \hat{1}^{(\nu)}, \tag{1}$$

where $\hat{P}_{pq}^{(\nu)} = |p)^{(\nu)}{}^{(\nu)}\langle q|$ is a (local) transition operator.

The initial state $|\psi_0)$ will be taken to be a product of the Turing head and tape wave-functions. For the discretized dynamical description of the QTM we identify the unitary operators $\hat{U}_n$, $n = 1, 2, 3, \cdots$ (step number) with the local unitary transformation on the Turing head S, $\hat{U}_{\alpha m}^{(S)}$, and the quantum-controlled-NOT (QCNOT) on (S, μ), $\hat{U}^{(S,\mu)}$, respectively, as follows:

$$\hat{U}_{2m-1} \;=\; \hat{U}_{\alpha m}^{(S)} \;=\; \hat{1}^{(S)} \cos\left(\alpha_m/2\right) - \hat{\lambda}_1^{(S)}\, i \sin\left(\alpha_m/2\right) \tag{2}$$

$$\hat{U}_{2m} \;=\; \hat{U}^{(S,\mu)} \;=\; \hat{P}_{00}^{(S)} \hat{\lambda}_1^{(\mu)} + \hat{P}_{11}^{(S)} \hat{1}^{(\mu)} \;=\; \left(\hat{U}^{(S,\mu)}\right)^\dagger \tag{3}$$

Any such QTM is specified by its tape size M, the external control parameters α_m and the initial state $|\psi_0)$. Here the Turing head is externally driven by $\alpha_m = \alpha$ for all m or by the Fibonacci-like sequence $\alpha_{m+1} = \alpha_m + \alpha_{m-1} \pmod{2\pi}$, $\alpha_0 = 0$. The mth Fibonacci number α_m is then controlled by α_1 via

$$\alpha_m \;=\; \frac{\alpha_1}{\sqrt{5}}\left(\beta^m - \gamma^m\right), \tag{4}$$

where $\beta := \left(1 + \sqrt{5}\right)/2$, $\gamma := \left(1 - \sqrt{5}\right)/2$. The sequence $\{\alpha_m\} \pmod{2\pi}$ acts as a chaotic input (Lyapunov exponent: $\ln\beta > 0$). It is useful for later calculations to note that

$$\beta^{m+1} = \beta^m + \beta^{m-1}, \quad \gamma^{m+1} = \gamma^m + \gamma^{m-1}$$

$$\beta^m = F(m)\cdot\beta + F(m-1), \quad \gamma^m = F(m)\cdot\gamma + F(m-1), \tag{5}$$

where $F(m) := \left(\beta^m - \gamma^m\right)/\sqrt{5}$, the mth Fibonacci number with $F(1) = 1$. The chaotic sequence of Fibonacci-type can be interpreted as a temporal random (chaotic) analogue to 1-dimensional chaotic potentials in real space [15].

First, we restrict ourselves to the reduced state-space dynamics of the head S and tape-spin μ, respectively,

$$\lambda_j^{(S)}(n) \;=\; \left\langle \psi_n \left| \hat{\lambda}_j^{(S)} \otimes \hat{1}^{(1)} \otimes \cdots \otimes \hat{1}^{(\mu)} \otimes \cdots \hat{1}^{(M)} \right| \psi_n \right\rangle$$

$$\lambda_k^{(\mu)}(n) \;=\; \left\langle \psi_n \left| \hat{1}^{(S)} \otimes \hat{1}^{(1)} \otimes \cdots \otimes \hat{\lambda}_k^{(\mu)} \otimes \cdots \hat{1}^{(M)} \right| \psi_n \right\rangle. \tag{6}$$

$|\psi_n)$ is the total network state at step n, $\lambda_j^{(\nu)}(n)$ are the respective Bloch-vectors. We intend to show that these local propagations are chaotic, too. Due to the entanglement between the head and tape, both will, in general, appear to be in a 'mixed-state', which means that the length of the Bloch-vectors in (6) is less than 1. However, for specific initial states $|\psi_0)$ the state of head and tape will remain pure: As

$|\pm\rangle^{(\mu)} := \frac{1}{\sqrt{2}} \left(|0\rangle^{(\mu)} \pm |1\rangle^{(\mu)} \right)$ are the eigenstates of $\hat{\lambda}_1^{(\mu)}$ with $\hat{\lambda}_1^{(\mu)} |\pm\rangle^{(\mu)} = \pm 1 |\pm\rangle^{(\mu)}$, the QCNOT-operation $\hat{U}^{(S,\mu)}$ of equation (3) cannot create any entanglement, irrespective of the head state $|\varphi\rangle^{(S)}$, i.e.

$$\hat{U}^{(S,\mu)} |\varphi\rangle^{(S)} \otimes |+\rangle^{(\mu)} = |\varphi\rangle^{(S)} \otimes |+\rangle^{(\mu)}$$

$$\hat{U}^{(S,\mu)} |\varphi\rangle^{(S)} \otimes |-\rangle^{(\mu)} = \hat{\lambda}_3^{(S)} |\varphi\rangle^{(S)} \otimes |-\rangle^{(\mu)} . \tag{7}$$

As a consequence, the state $|\psi_n\rangle$ remains a product state at any step n for the initial product states $|\psi_0^j\rangle = |\varphi_0\rangle^{(S)} \otimes |P_0^j\rangle$,

$$|P_0^j\rangle \in \left\{ |P_0^{\pm\pm\cdots\pm}\rangle = |\pm\rangle^{(1)} \otimes |\pm\rangle^{(2)} \otimes \cdots \otimes |\pm\rangle^{(M)} \right\} \tag{8}$$

with $|\varphi_0\rangle^{(S)} = \cos(\varphi_0/2) |0\rangle^{(S)} - i \sin(\varphi_0/2) |1\rangle^{(S)}$ and the Turing head then performs a pure-state trajectory ('primitive') on the Bloch-circle

$$|\psi_n^j\rangle = |\varphi_n^j\rangle^{(S)} \otimes |P_0^j\rangle$$

$$\lambda_1^{(S)}(n) = 0, \quad \left(\lambda_2^{(S)} \left(n|P_0^j \right) \right)^2 + \left(\lambda_3^{(S)} \left(n|P_0^j \right) \right)^2 = 1 \quad . \tag{9}$$

Here $\lambda_j^{(S)} \left(n|P_0^j \right)$ denotes the Bloch vector of the Turing head S conditioned by the initial state $|\psi_0^j\rangle$. We show examples for $\alpha_m = \alpha$ and $M = 1, 2$ (Fig 1). From the Fibonacci relation and the property (6) it is found for $|\varphi_n\rangle^{(S)} \otimes |P_0^+\rangle^{(1)}$ ($M = 1$), $n = 2m$ and $\varphi_0^\pm = \alpha_0 = 0$ that

$$\lambda_2^{(S)} \left(2m|P_0^+ \right) = \sin C_{2m} \left(P_0^+ \right), \quad \lambda_3^{(S)} \left(2m|P_0^+ \right) = -\cos C_{2m} \left(P_0^+ \right), \tag{10}$$

where $C_{2m} \left(P_0^+ \right) := \sum_{j=1}^{m} \alpha_j$ and $\lambda_k^{(S)} \left(2m-1|P_0^+ \right) = \lambda_k^{(S)} \left(2m|P_0^+ \right)$. In order to derive the corresponding expression of $\lambda_k^{(S)} \left(n|P_0^- \right)$ for $|\varphi_n\rangle^{(S)} \otimes |P_0^-\rangle^{(1)}$, we utilize the following recursion relations for the cumulative rotation angle $C_n \left(P_0^- \right)$ up to step n

$$C_{2m} \left(P_0^- \right) = -C_{2m-1} \left(P_0^- \right), \quad C_{2m-1} \left(P_0^- \right) = \alpha_m + C_{2m-2} \left(P_0^- \right). \tag{11}$$

Then $C_{2m} \left(P_0^- \right), C_{2m-1} \left(P_0^- \right)$, respectively, satisfy the following expressions:

$$C_{2m} \left(P_0^- \right) = -C_{2m-2} \left(P_0^- \right) - \alpha_m = (-1)^{m-1} \sum_{j=1}^{m} (-1)^j \alpha_j$$

$$C_{2m-1} \left(P_0^- \right) = -C_{2m-3} \left(P_0^- \right) + \alpha_m = (-1)^m \sum_{j=1}^{m} (-1)^j \alpha_j, \tag{12}$$

yielding $\lambda_2^{(S)} \left(n|P_0^- \right) = \sin C_n \left(P_0^- \right), \lambda_3^{(S)} \left(n|P_0^- \right) = -\cos C_n \left(P_0^- \right)$. The Fibonacci property implies that both primitives, $|\varphi_n^+\rangle^{(S)} \otimes |P_0^+\rangle^{(1)}$ and $|\varphi_n^-\rangle^{(S)} \otimes |P_0^-\rangle^{(1)}$, are chaotically driven.

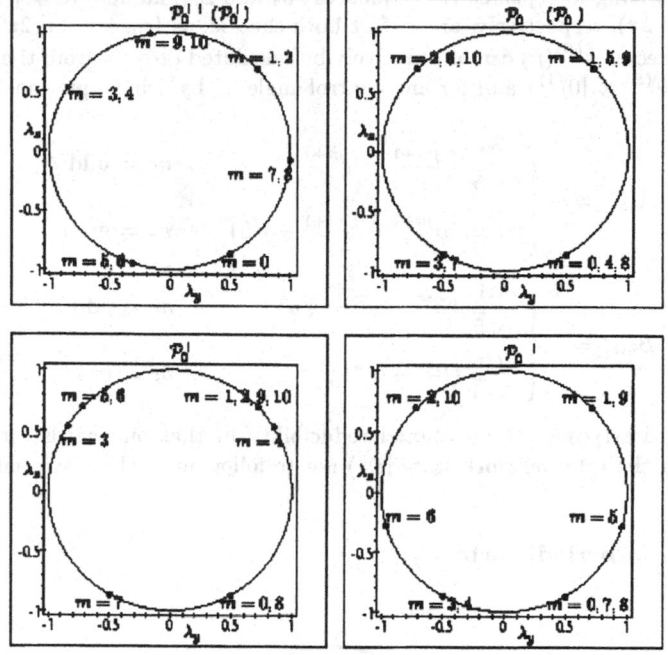

Figure 1 The primitives $\mathcal{P}_0^+$ (aperiodic) and $\mathcal{P}_0^-$ (periodic) for $M = 1$, and $\mathcal{P}_0^{++}$ (aperiodic), $\mathcal{P}_0^{--}$, $\mathcal{P}_0^{+-}$, $\mathcal{P}_0^{-+}$ (periodic) for $M = 2$; $\alpha = \pi/\sqrt{3}$, $\varphi = \pi/6$.

From any initial state, $|\psi_0\rangle = a^{(+)} |\varphi_0^+\rangle^{(S)} \otimes |\mathcal{P}_0^+\rangle^{(1)} + a^{(-)} |\varphi_0^-\rangle^{(S)} \otimes |\mathcal{P}_0^-\rangle^{(1)}$, we then obtain at step n

$$|\psi_n\rangle = a^{(+)} |\varphi_n^+\rangle^{(S)} \otimes |\mathcal{P}_0^+\rangle^{(1)} + a^{(-)} |\varphi_n^-\rangle^{(S)} \otimes |\mathcal{P}_0^-\rangle^{(1)} \tag{13}$$

and, observing the orthogonality of the $|\mathcal{P}_0^\pm\rangle^{(1)}$,

$$\lambda_k^{(S)}(n) = |a^{(+)}|^2 \lambda_k^{(S)}\left(n|\mathcal{P}_0^+\right) + |a^{(-)}|^2 \lambda_k^{(S)}\left(n|\mathcal{P}_0^-\right). \tag{14}$$

This trajectory of the Turing-head S represents a non-orthogonal pure-state decomposition. By using (10), (11), (14) $\left(\text{with } a^{(+)} = a^{(-)} = 1/\sqrt{2}\right)$ we finally have for $|\psi_0\rangle = |0\rangle^{(S)} \otimes |0\rangle^{(1)}$

$$\left(\lambda_2^{(S)}(2m), \lambda_3^{(S)}(2m)\right) = \cos \mathcal{A}_m \cdot (\sin \mathcal{B}_m, -\cos \mathcal{B}_m)$$
$$\left(\lambda_2^{(S)}(2m-1), \lambda_3^{(S)}(2m-1)\right) = \cos \mathcal{B}_m \cdot (\sin \mathcal{A}_m, -\cos \mathcal{A}_m), \tag{15}$$

where $\mathcal{A}_m := \alpha_m + \alpha_{m-2} + \cdots$, $\mathcal{B}_m := \alpha_{m-1} + \alpha_{m-3} + \cdots$. The expression (15) indicates that also in the 'non-classical' regime the local dynamics of the Turing head is controlled by a 'chaotic' driving force, since the sequences $\mathcal{A}_m$ and $\mathcal{B}_m$, namely $\{\alpha_{2m}\}$ (mod 2π) or $\{\alpha_{2m-1}\}$ (mod 2π), respectively, are in fact both chaotic, as $\{\alpha_m\}$ (mod 2π) is.

The Bloch-vector $\vec{\lambda}^{(S)}(n)$ can alternatively be calculated directly from the initial state $\left(\text{here: } |\psi_0\rangle = |0\rangle^{(S)} \otimes |0\rangle^{(1)}\right)$ and for any control angle α_1 by using equation (15) and the relations

$$
\mathcal{A}_m = \begin{cases} \dfrac{\alpha_1}{\sqrt{5}} \left(\beta^{m+1} - \gamma^{m+1}\right) & m = \text{odd} \\[2mm] \dfrac{\alpha_1}{\sqrt{5}} \left(\beta^{m+1} - \gamma^{m+1} - \sqrt{5}\right) & m = \text{even} \end{cases}
$$

$$
\mathcal{B}_m = \begin{cases} \dfrac{\alpha_1}{\sqrt{5}} \left(\beta^{m} - \gamma^{m} - \sqrt{5}\right) & m = \text{odd} \\[2mm] \dfrac{\alpha_1}{\sqrt{5}} \left(\beta^{m} - \gamma^{m}\right) & m = \text{even}, \end{cases} \tag{16}
$$

demonstrating a striking computational reducibility in that one needs for calculating $\vec{\lambda}^{(S)}(n)$ neither the total network state $|\psi_n\rangle$ nor to follow up each individual step n.

3. Instability of periodic orbits

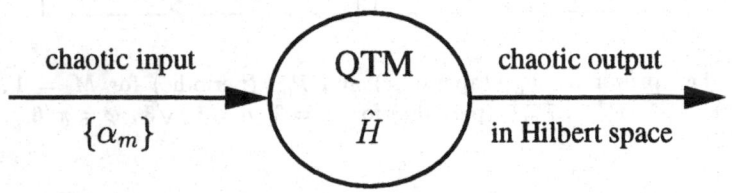

chaotic input $\quad\quad$ **QTM** $\quad\quad$ chaotic output

$\{\alpha_m\}$ $\quad\quad\quad\quad$ $\hat{H}$ $\quad\quad\quad\quad$ in Hilbert space

Figure 2 A input-output scheme of our quantum Turing machine (QTM).

Now we verify that the periodic orbits on the plane $\left\{0, \lambda_2^{(S)}, \lambda_3^{(S)}\right\}$ are unstable, which proves that the dynamics of the Turing head ('output') is indeed chaotic (Fig 2). Because of the alternating character of the dynamics, equations (2), (3), it suffices to check the periodicity only for step $n = 2m$: The periodic orbits for $\lambda_2^{(S)}(0) = 0$, $\lambda_3^{(S)}(0) = -1$ must obey two constraints, $C_{2m}\left(\mathcal{P}_0^+\right) = C_{2m}\left(\mathcal{P}_0^-\right) \overset{!}{=} 2\pi p$, $p \in \mathbf{Z}$ and $\alpha_{m+1} = \alpha_1$ (mod 2π) for $n = 2m + 1$ (one concludes that α_1 must be a rational multiple of π). By using the Fibonacci numbers (4), one finds $C_{2m}^{\text{per}}\left(\mathcal{P}_0^+\right)$ in (10) and $C_{2m}^{\text{per}}\left(\mathcal{P}_0^-\right)$ in (11), respectively, for period $= 2m$ as

$$
C_{2m}^{\text{per}}\left(\mathcal{P}_0^+\right) = \frac{\alpha_1}{\sqrt{5}} \left(\beta^{m+2} - \gamma^{m+2} - \sqrt{5}\right)
$$

$$
C_{2m}^{\text{per}}\left(\mathcal{P}_0^-\right) = \frac{\alpha_1}{\sqrt{5}} \left(-\beta^{m-1} + \gamma^{m-1} + (-1)^m \sqrt{5}\right). \tag{17}
$$

Then let us consider a small perturbation δ of the initial phase angle $\alpha_0 = 0$, implying $|\varphi_0\rangle^{(S)} = \cos(\delta/2)|0\rangle - i\sin(\delta/2)|1\rangle$ and a perturbed Fibonacci-like sequence $\{\alpha_m'\}$

(mod 2π):

$$\alpha_0' = \delta, \ \alpha_1' = \alpha_1, \ \alpha_2' = \alpha_1 + \delta, \ \cdots. \tag{18}$$

Similarly to (16), we obtain for this case

$$C_{2m}'\left(\mathcal{P}_0^\pm\right) = C_{2m}^{\mathrm{per}}\left(\mathcal{P}_0^\pm\right) + \Delta C_{2m}\left(\mathcal{P}_0^\pm\right), \tag{19}$$

where the deviation terms from the periodic orbits read, respectively,

$$\Delta C_{2m}\left(\mathcal{P}_0^+\right) = \frac{\delta}{\sqrt{5}}\left(\beta^{m+1} - \gamma^{m+1}\right)$$

$$\Delta C_{2m}\left(\mathcal{P}_0^-\right) = -\frac{\delta}{\sqrt{5}}\left(\beta^{m-2} - \gamma^{m-2}\right). \tag{20}$$

By using (19), we are able to represent the evolution of the perturbation, $\Delta\lambda_j^{(S)}(n)$, at the $n = 2m$-th step for the Turing-head dynamics as

$$\begin{pmatrix} \Delta\lambda_2^{(S)}(2m) \\ \Delta\lambda_3^{(S)}(2m) \end{pmatrix} = \begin{pmatrix} M_{11} & 0 \\ 0 & M_{22} \end{pmatrix} \begin{pmatrix} \Delta\lambda_2^{(S)}(0) \\ \Delta\lambda_3^{(S)}(0) \end{pmatrix}, \tag{21}$$

where $\Delta\lambda_2^{(S)}(0) = \sin\delta$, $\Delta\lambda_3^{(S)}(0) = -\cos\delta$, $\Delta\lambda_2^{(S)}(2m) = \cos(\delta\alpha_m) \cdot \sin(\delta\alpha_{m-1})$, $\Delta\lambda_3^{(S)}(2m) = -\cos(\delta\alpha_m) \cdot \cos(\delta\alpha_{m-1})$; $M_{11} = \cos(\delta\alpha_m) \cdot \sin(\delta\alpha_{m-1})/\sin\delta$, $M_{22} = \cos(\delta\alpha_m) \cdot \cos(\delta\alpha_{m-1})/\cos\delta$, respectively. One easily shows (cf. equation (5)) that

$$\lim_{\delta\to 0} M_{11} = F(m-1) = \frac{1}{\sqrt{5}}\left(\beta^{m-1} - \gamma^{m-1}\right), \quad \lim_{\delta\to 0} M_{22} = 1, \tag{22}$$

which means that M_{11} grows exponentially (note that $\beta > 1$, $|\gamma| < 1$), and the periodic orbit on the Turing head is thus unstable to any small perturbation δ in the external control.

As an explicit example consider the case of $\alpha_1 = (2/5)\pi$ with $|\psi_0\rangle = |0\rangle^{(S)} \otimes |0\rangle^{(1)}$: By using equation (5) in equation (16) for $C_{2m}^{\mathrm{per}}(+) = 2\pi p$ the periodic orbits with period $= 2m$ have to obey $F(m+2) = 1$ (mod 5). Together with the second condition, $F(m+1) = 1$ (mod 5) for $n = 2m+1$, we obtain $F(m-1) = 1$, $F(m) = 0$ (mod 5). For $C_{2m}^{\mathrm{per}}(-) = 2\pi p$ it follows likewise that $F(m-1) = 1$ (mod 5) for $m = $ even, $F(m-1) = 4$ (mod 5) for $m = $ odd. Considering the common condition for both cases, $F(m-1) = 1$ and $F(m) = 0$ (mod 5), we find the smallest m satisfying both conditions, namely $m = 20$, i.e. $n = 2m = 40$. $\left(\lim_{\delta\to 0} M_{11} = F(19) = 4181 \gg 1, \text{ see Fig 3}\right)$.[1]

Remarkably enough, the local dynamics of the Turing tape also contains some exponential sensitivity to initial conditions $\left(\text{here: } |\psi_0\rangle = |0\rangle^{(S)} \otimes |0\rangle^{(1)}, \lambda_1^{(1)}(n) = \lambda_2^{(1)}(n) = 0\right)$:

$$\lambda_3^{(1)}(n) = \begin{cases} -\cos\left(\alpha_{\left[\frac{n}{2}\right]+1} - \alpha_1 + \delta_{\left[\frac{n}{2}\right]}^{\mathrm{Fib}}\right) & n = 0, 1 \text{ (mod 4)} \\ \cos\left(\alpha_{\left[\frac{n}{2}\right]+1} + \delta_{\left[\frac{n}{2}\right]}^{\mathrm{Fib}}\right) & n = 2, 3 \text{ (mod 4)}, \end{cases} \tag{23}$$

where $\delta_m^{\mathrm{Fib}} := \delta\left(\beta^m - \gamma^m\right)/\sqrt{5}$; $[a] := n$, $a = n + r$, $n \in \mathbb{Z}$, $0 \le r < 1$. One easily confirms that there is no periodic orbit with period $2m = 2$ (mod 4). The following

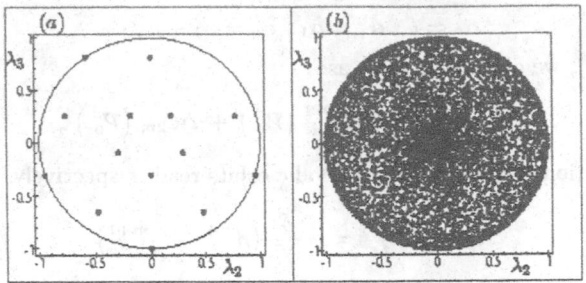

Figure 3 Turing-head patterns $\{0, \lambda_2(n), \lambda_3(n)\}$ under chaotic Fibonacci control for initial state $|\psi_0\rangle = |0\rangle^{(S)} \otimes |0\rangle^{(1)}$. (*a*): $\alpha_1 = \frac{2}{5}\pi$ (periodic); (*b*): $\alpha_1 = \frac{2}{5} \times 3.141592654$ (aperiodic) and total step number $n = 10000$. It is interesting to compare with the cases of $\alpha_m = \alpha$ for all m in Fig. 4 and other non-chaotic sequences in Fig. 5.

expression for the deviation from the periodic orbits with period $2m = 0$ (mod 4) at step $n = 2m + 2$ holds:

$$\Delta\lambda_3^{(1)}(2m+2) = M \cdot \Delta\lambda_3^{(1)}(2), \quad \lim_{\delta \to 0} M = \frac{1}{\sqrt{5}}\left(\beta^{m+1} - \gamma^{m+1}\right)\frac{\sin(\alpha_{m+2})}{\sin(\alpha_1)}, \quad (24)$$

with $\Delta\lambda_3^{(1)}(2) = \cos(\alpha_1 + \delta) - \cos(\alpha_1)$, $\Delta\lambda_3^{(1)}(2m+2) = \cos(\alpha_{m+2} + \delta_{m+1}) - \cos(\alpha_{m+2})$; the perturbation δ in (18) does not enter into the Turing-tape dynamics until $n = 2$! Equation (24) shows the exponential instability of the periodic orbit. Indeed, the Turing tape exhibits chaos only by means of the entanglement with the head ('chaos swapping'), not as a result of a direct chaotic driving force.

4. Exponential parameter sensitivity

The distance between density operators, $\hat{\rho}$ and $\hat{\rho}'$, defined by the so-called Bures metric [11]

$$D^2_{\rho\rho'} := \text{Tr}\left\{(\hat{\rho} - \hat{\rho}')^2\right\}. \quad (25)$$

lies, independent of the dimension of the Liouville space, between 0 and 2 (the maximum (squared) distance of 2 applies to pure orthogonal states, $D^2 = 2(1 - |\langle\psi|\psi'\rangle|^2)$). This metric can be applied to the total-network-state space or any subspace. In any case it is a convenient additional means to characterize a QTM with respect to perturbations δ: For $\alpha_m = \alpha_1$ (Lyapunov exponent = 0) [12] and any δ the distance between the evolving unperturbed and perturbed states remains almost constant (see Fig 6*d*); for the Fibonacci-like driving force, on the other hand, $\left(\alpha_m(\hat{\rho}) = \alpha_m, \alpha_m(\hat{\rho}') = \alpha_m(\hat{\rho}) + \delta^{\text{Fib}}_{m-1}\right)$ we obtain an initial exponential sensitivity, which is eventually constrained, though, by $D^2 \leq 2$ (see Fig 6*a* - *c*) [13]. It is instructive to consider another regular QTM controlled by the rule $\alpha_{m+1} = 2\alpha_m - \alpha_{m-1}$ (Lyapunov exponent = 0): $\alpha_m(\hat{\rho}) = m\alpha_1$, $\alpha_m(\hat{\rho}') = \alpha_m(\hat{\rho}) - (m-1)\delta$. Here we observe a revival in the evolution of D^2, which confirms that periodic orbits are stable (Fig 6*d*), whereas there is no revival for the above chaotic

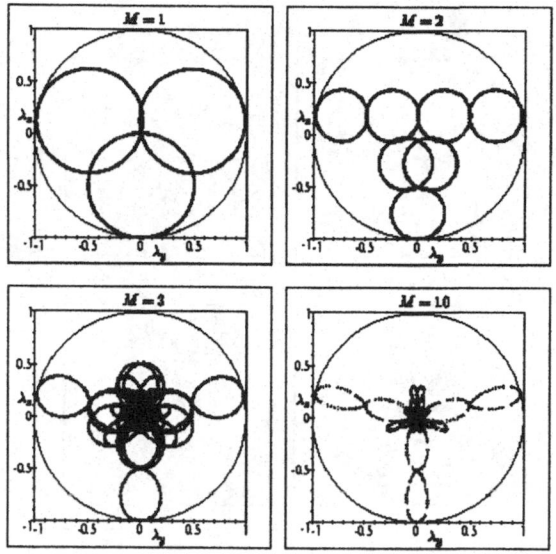

Figure 4 Turing-head-patterns for $|\psi_0\rangle = |00 \cdots 0\rangle$, $M = 1, 2, 3, 10$, and total step number $m = 3000$; $\alpha = \pi/\sqrt{3}$.

system. Finally we display the evolution of D^2 for the total network state $|\psi_n\rangle$, which also shows exponential sensitivity. The respective distances for tape spin 1 are similar to those shown.

The ultimate source of the present chaotic behavior is that any small perturbation δ to the initial state, $|\psi_0(\delta)\rangle$, is connected with a perturbed unitary evolution, $\hat{U}(\delta)$, which implies that the scalar product between different initial states (as a measure of distance) is no longer conserved under these evolutions:

$$O' := |\langle\psi_0(\delta)|\hat{U}^\dagger(\delta)\,\hat{U}(0)|\psi_0(0)\rangle|^2, \quad D^2 = 2(1 - O'). \tag{26}$$

Thus the initial state is directly correlated to its unitary evolution, which can lead to the exponential sensitivity to initial condition, whereas there is no chaos in a generic quantum system evolving by a fixed $\hat{U}$ even if characterized by chaotic input parameters. This O' reminds us immediately of the test function $O = |\langle\psi|\hat{V}^\dagger(t)\,\hat{U}(t)|\psi\rangle|^2$ [19], where $\hat{U}, \hat{V}$ are specified by slightly different external parameters: The corresponding parameter-sensitivity has been proposed as a measure to distinguish quantum chaos from regular quantum dynamics. The origin of chaos in our QTM may thus be alternatively ascribed to a perturbed $\hat{V} = \hat{U}(\delta)$ in the control (*cf.* the comment by R. Schack [20]).

5. Summary

In conclusion, we have studied the quantum dynamics of a small chaotically driven QTM based on a decoherence-free Hamiltonian. Quantum chaos has been shown to occur as an exponential parameter-sensitivity and a cumulative loss of control in a pure quantum regime. This might be contrasted with the usual quantum chaology, which is concerned

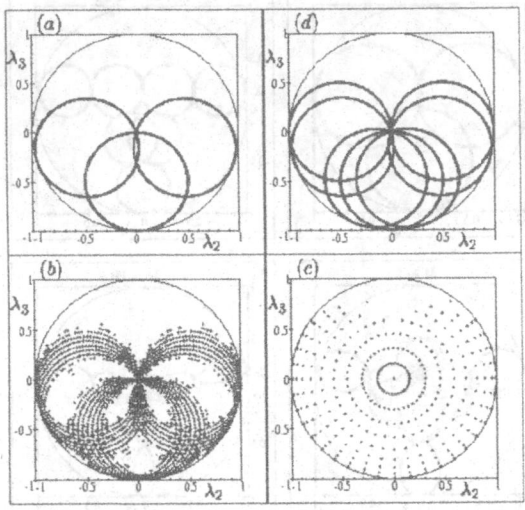

Figure 5 Turing-head patterns $\{0, \lambda_2(n), \lambda_3(n)\}$ for initial state $|\psi_0\rangle = |0\rangle^{(S)} \otimes |0\rangle^{(1)}$ under the control of non-chaotic substitution sequences: (a) quasi-periodic Fibonacci with $\alpha_1 = (2/5)\pi$, $\alpha_2 = \alpha_1 + 0.0005\pi$, (b) as in a) but for $\alpha_2 = \alpha_1 + 0.03\pi$, (c) as in a) but for $\alpha_2 = \alpha_1 + 0.05\pi$; (d) Thue-Morse control with $\alpha_1 = (2/5)\pi$, $\alpha_2 = \alpha_1 + 0.1001\pi$. For each simulation the total step number is $n = 10000$.

essentially with semiclassical spectrum analysis of classically chaotic systems (e.g. level spacing, spectral rigidity). As quantum features we utilized the superposition principle and the physics of entanglement. Our dynamical chaos manifests itself in the superposition and entanglement of a pair of 'classical' (i.e. unentangled) chaotic state-sequences. Due to the entanglement, we can see the chaos in any local Bloch-plane. This indicates that patterns in reduced Bloch-spheres (a quantum version of a Poincaré-cut, Fig 3) should be useful to characterize quantum chaos in a broad class of quantum networks: Here, a periodic orbit would be represented by finite set of fixed points on the plane $\left\{0, \lambda_2^{(S)}, \lambda_3^{(S)}\right\}$. It is noteworthy that this kind of control loss is completely different from the typical control limit of a quantum network resulting from the exponential blow-up of Hilbert-space dimension in which the state evolves [7]. It is expected that a QTM architecture with a larger number of pseudo-spins on the Turing tape would also exhibit chaos under the same type of driving. However, it is just the chaos in small networks which might be interesting for experimental studies, especially in the form of ensembles thereof. The Fibonacci-like sequence should, however, be considered but a special example for chaotic input. Such inputs would, of course, have to be avoided in quantum computation; otherwise the resulting quantum dynamics would easily become chaotic in its entirety!

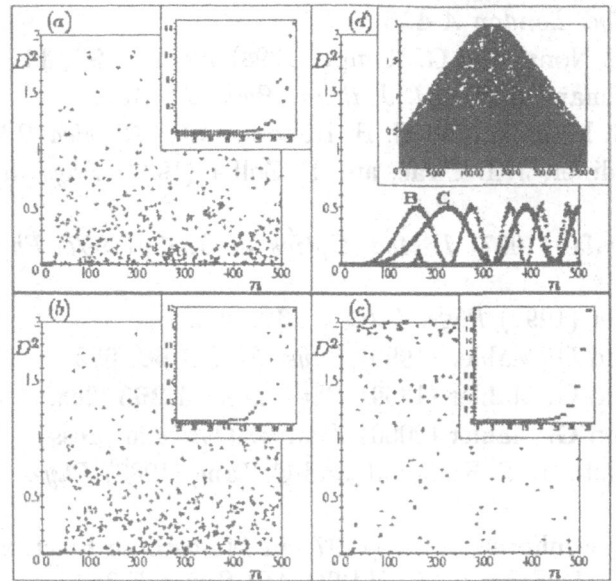

Figure 6 Evolution of the (squared) distance $D^2_{\rho\rho'}$ between QTM state with $(\hat{\rho}')$ and without $(\hat{\rho})$ perturbation δ. $\alpha_1 = \frac{2}{5}\pi$, $|\psi_0\rangle = |0\rangle^{(S)} \otimes |0\rangle^{(1)}$ for $\hat{\rho}$, and $\left(\cos(\delta/2)|0\rangle^{(S)} - i\sin(\delta/2)|1\rangle^{(S)}\right) \otimes |0\rangle^{(1)}$ for $\hat{\rho}'$. (a): chaotic input according to equation (4) (inset shows initial behavior in more detail) for the Turing-head state, $\delta = 0.001$; (b): the same as (a) but for $\delta = 0.00001$; (c): the same as (a) for total network state $|\psi_n\rangle$; (d): $D^2_{\rho\rho'}$ within the Turing head subspace for $\alpha_m = \alpha_1$ ($D^2 \approx 0$, solid line A, $\delta = 0.001, 0.0005$) and $\alpha_{m+1} = 2\alpha_m - \alpha_{m-1}$ (dotted line B, $\delta = 0.001$; boxed line C, $\delta = 0.0005$), inset shows line B on larger scale.

Acknowledgements

We thank J. Gemmer, A. Otte, M. Stollsteimer, F. Tonner, and T. Wahl for fruitful discussions.

Notes

1. For non-chaotic substitution sequences in Fig. 5, see [14].

References

[1] P. Benioff (1982) *Phys. Rev. Lett.* **48**, 1581; (1996) *Phys. Rev. A* **54**, 1106

[2] G. P. Bergman (1999) quant-ph/9903063.

[3] M. V. Berry (1989) *Physica Scripta* **40**, 335.

[4] R. Blümel (1994) *Phys. Rev. Lett.* **73**, 428.

454

[5] D. Deutsch (1985) *Proc. R. Soc. London A* **400**, 97; (1989) *Proc. R. Soc. London A* **425**, 73.

[6] S. Dürr, T. Nonn, and G. Rempe (1998) *Nature* **395**, 33.

[7] R. P. Feynman (1982) *Int. J. theor. Phys.* **21**, 467.

[8] J. Ford, G. Mantica, and G. H. Ristow (1991) *Physica D* **50**, 493.

[9] S. A. Gardiner, J. I. Cirac, and P. Zoller (1997) *Phys. Rev. Lett.* **79**, 4790.

[10] M. E. Ghafar (1997) *J. Mod. Optics* **44**, 1985; (1997) *Phys. Rev. Lett.* **78**, 4181.

[11] M. Hübner (1992) *Phys. Lett. A* **163**, 293.

[12] I. Kim and G. Mahler (1999) *Phys. Rev. A* **60**, 692.

[13] I. Kim and G. Mahler (1999) *Phys. Lett. A* **263**, 268.

[14] I. Kim and G. Mahler (2000) *J. Mod. Optics* in press.

[15] M. Kohmoto, L. P. Kadanoff, and C. Tang (1983) *Phys. Rev. Lett.* **50**, 1870.

[16] L. P. Kouwenhoven *et al.* (1997) in Mesoscopic Electron Transport, edited by L. L. Sohn *et al.*, *NATO ASI Series E* **345**

[17] A. J. Lichtenberg and M. A. Lieberman (1983) *Regular and Stochastic Motion* (Springer, New York).

[18] G. Mahler and V. A. Weberruss (1998) *Quantum Networks: Dynamics of Open Nanostructures* (2nd ed. Springer, New York).

[19] A. Peres (1993) *Quantum Theory: Concepts and Methods* (Kluwer, Dordrecht).

[20] R. Schack (1995) *Phys. Rev. Lett.* **75**, 581.

[21] R. Schack (1998) *Phys. Rev. A* **57**, 1634.

[22] E. Schrödinger (1935) *Naturwissenschaften* **23**, 807, 823 and 844.

[23] D. L. Shepelyansky (1994) *Phys. Rev. Lett.* **73**, 2607.

[24] S. Wolfram (1985) *Phys. Rev. Lett.* **54**, 735.

PROBLEMS AND PERSPECTIVES
IN QUANTUM-DOT BASED COMPUTATION

M. MACUCCI, G. IANNACCONE, S. FRANCAVIGLIA,
M. GOVERNALE, M. GIRLANDA
*Dipartimento di Ingegneria dell'Informazione, Via Diotisalvi, 2,
I-56126 PISA, Italy*

C. UNGARELLI
*School of Mathematical Sciences, University of Portsmouth,
Mercantile House, Hampshire Terrace, Portsmouth PO1 2EG, UK*

Abstract

A review of recent modeling work on quantum cellular automata is presented. Detailed simulations at the single cell level, based on the Configuration-Interaction method, allow to analyze cell bistable behavior for different material systems and geometrical structures. In addition, they allow to evaluate the maximum allowed fabrication tolerances required for correct cell operation. Simulations at the circuit level, based on a simplified physical model of each cell, allow to analyze operation of single logic gates and more complex combinatorial logic networks. A simulation method based on simulated annealing is presented, which allows to compute the ground state and low-lying excited states of the QCA system without the need of exploring the whole configuration space.

1. Introduction

In the last decade many proposals have appeared in the literature for the usage of quantum dots as the building blocks for functional electronic devices, ranging from logic gates to quantum gates. One of the first design concepts was proposed by Bakshi et al. [1], who introduced the idea of quantum dashes: elongated quantum dots, in which an electron was supposed to be shifted either at the top or the bottom of the dash in response to an external electric field or as a consequence of the interaction with the electrons in nearby dashes. Arrays of dashes propagating logic signals and implementing logic functions were envisaged. This idea, however, was faulty. Its proposers did not recognize the fact, later pointed out by Lent et al., that an electron confined in a dash with a width of the order of the De Broglie wavelength of the electron will not behave as a classical particle. Its probability density will be spread all over the length of the dash, unless an unreasonably large electric field is applied, orders of magnitude greater than that which could be produced by electrons in nearby dashes. Confinement in one section of the dash can be obtained by inserting a barrier that divides the dot into two halves: in this case a small electric field suffices to force the electron on either side of the barrier. Tougaw and Lent [2] developed this intuition into a full-fledged proposal for a logic architecture, built around a four dot cell, the Quantum Cellular Automaton (QCA) paradigm. It was shown that a properly arranged two-dimensional array of cells, let to relax down to the ground state, can perform the function of an arbitrary combinatorial logic network. The QCA concept, in spite of its many drawbacks and the severe difficulties in its actual implementation, can still be considered as the most significant and interesting proposal for quantum dot based computation. Other ideas have appeared in the literature, such

455

L. Pavesi and E. Buzaneva (eds.), Frontiers of Nano-Optoelectronic Systems, 455–466.

as the concept by Bandyopadhyay [3], suggesting circuits based on arrays of quantum dots in which the logic information is stored in the orientation of the spin of a single electron. This approach would further compact circuit dimensions and the number of charges involved in the implementation of a logic function, but fabrication challenges are even more prohibitive than for QCA circuits, and no detailed modeling has been performed for the spin interaction, validating the proposed mode of operation.

We have focused on the investigation of the electrostatic QCA concept as originally proposed by Lent *et al.*, developing a numerical model that allows the simulation of QCA cells made up of quantum dots defined by realistic potentials and the evaluation of fabrication tolerances on cell operation. We have then extended our analysis to simple logic circuits made up of arrays of basic cells, determining the error probability as a function of temperature and material parameters by means of a simplified semiclassical model. The semiclassical model has been extended to larger circuits introducing a simulated annealing procedure for the determination of the ground state of the QCA array. Special care has been taken to solve the problem of the system getting stuck into local energy minima during the simulated annealing procedure. Such minima, corresponding to metastable states, could represent a problem also for the actual operation of a QCA network, which, while relaxing down to the ground state, could stop in one of these states, instead.

Finally, we assess the feasibility of QCA logic from a technological point of view, concluding that semiconductor implementations are possible for the purpose of a proof of principle, but cannot be developed into a large scale technology, due to the extreme sensitivity to fabrication tolerances and to the subsequent need for adjustment of each single cell.

2. The QCA concept

The basic building block of QCA logic, as introduced by Lent *et al.*, is represented by a cell with four (or five) quantum dots containing a total of 2 electrons, as shown in Fig. 1. We shall discuss the case of a 4-dot cell, since the fifth dot appearing in some proposals of the Notre Dame group has only the function of improving tunneling properties, while representing a nonnegligible technological complication. If confinement in each of the dots is strong enough, electrons will be localized and behave as classical particles, with the only quantum property being the ability to tunnel through barriers. As a consequence of simple electrostatics, the electrons will tend to align along one of the two cell diagonals, in order to maximize the distance separating them. In the absence of external fields or of nearby cells, alignment along either diagonal will be equally probable. If another cell is placed in the near vicinity, whose polarization state is externally enforced, polarization along the same diagonal will be electrostatically favored (see Fig. 1). We can provide a

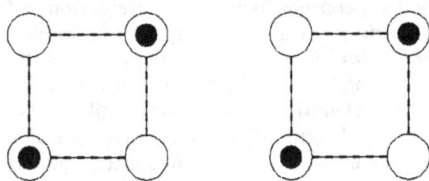

Figure 1. Coupled QCA cells

quantitative definition of the polarization of a cell with the following equation:

$$P = \frac{Q_1 + Q_3 - Q_2 - Q_4}{2q},$$

(1)

where Q_i indicates the charge in the i-th dot (dots are numbered counterclockwise starting from the one in the upper right) and q is the charge of the electron. This expression differs from that given by Lent *et al.* in Ref. [4], and we believe it to be more representative of the actual polarizing action on the following cell [5].

If we encode the two binary logic values with the polarization -1 and $+1$, we observe that logic data can be propagated along a chain of cells (binary wire) in a domino-like fashion. It has been shown by the Notre Dame group [2] that by forming two dimensional arrays of QCA cells it is possible to implement generic combinatorial networks. The majority voting gate is the simplest logic function in QCA logic and is shown in Fig. 2: the logic state of the output cell corresponds to the majority of those at the input cells. This gate can be used also as a programmable gate: holding one of the inputs at the level 0, a two-input AND gate is obtained, while holding one of the inputs at 1, a two-input OR gate results [2]. In general, arbitrarily large logic circuits can be obtained:

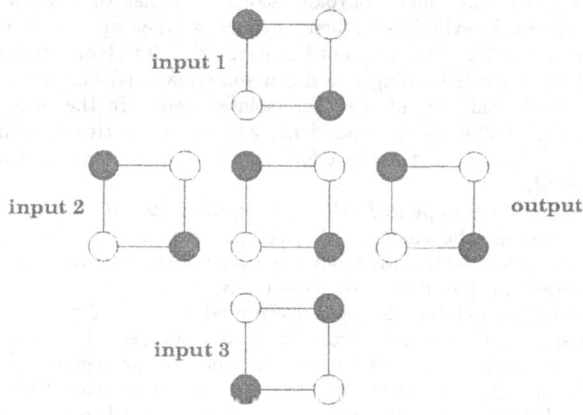

Figure 2. Majority voting gate

the input data will be provided to a certain number of boundary cells, by enforcing the corresponding polarization values. The circuit is then let relax down to the ground state until the polarization of the output cells stabilizes and the output data are read with appropriate noninvasive charge detectors[6]. For this reason, the approach described is usually indicated as "ground state computation," and dissipation plays an essential role in the operation [7]. Alternative QCA based computational approaches have been proposed, with the purpose of preventing the system from getting stuck into metastable states [8] during the relaxation to the ground state. Such approaches [9] are based on the concept of an adiabatic evolution with the system constantly in its instantaneous ground state and involve usage of a clock signal that steers the evolution of the circuit.

3. Calculation techniques

The numerical simulation of a QCA cell with confinement obtained by means of realistic potentials is a rather difficult task, due to the particular interplay between electrostatic interaction and quantum confinement energy. Most of the work existing in the literature on multiple quantum dot systems is based on extremely simplified models exploiting the Hubbard formalism. The detailed electronic structure of the single quantum dot is neglected and the dot and its interaction with the other dots are described by means of a few phenomenological parameters, such as the tunneling energy, the dot confinement energy, the on-site electrostatic interaction. Such description is successful in capturing the overall behavior of the system, and in providing a qualitative understanding of the underlying physics, but does not allow accurate quantitative predictions and the development of effective design tools. Tougaw et al. have studied [2] the cell-to-cell response function by means of a Hubbard-like Hamiltonian and have derived results for a series of idealized cases. Conduction through coupled quantum dots has been studied by several authors within the framework of a Hubbard-like formulation: Stafford and Das Sarma have studied the appearance of collective Coulomb Blockade phenomena in arrays of quantum dots [10], Kotlyar and Das Sarma have investigated transport through coupled double quantum dots [11] and persistent currents in quantum dot arrays [12], while Klimeck et al. have studied the characteristics of conductance peaks in double-dot systems [13].

On the other hand, rather detailed models of single quantum dots either isolated or coupled to leads through tunneling barriers have been developed by a few authors [14, 15], who have been able to treat up to a few tens of electrons, using different types of mean-field approximations. Such approximations, as it has turned out during the first few months of our work, exhibit significant problems when applied to QCA simulation, because convergence of the self-consistent iterative procedure (corresponding to repeated, alternate solutions of the Schrödinger and Poisson equations) has been obtained only in the case of no driver cell present, i.e. for isolated cells. In the presence of a nearby polarizing cell, the computed electron density varies abruptly at each iteration, in a somewhat periodic fashion and every effort to control and quench this oscillation has proven unsuccessful.

Similar problems have appeared with an approach based on the standard Hartree method, which involves the solution of a set of Schrödinger equations (one for each occupied orbital), each of which has a potential term including the contribution from the electrostatic interaction due to the other electrons.

A Hartree-Fock approach has also been attempted, with the application of the Roothan method, but analogous convergence problems have appeared. The reason for the failure of these approaches is represented by the existence of strong degeneracies, due to the presence of coupled quantum dots, and by the relevance of the electrostatic interaction energy in comparison with the quantum confinement energy. Only when the electrostatic energy is very small compared to the confinement energy, the iterative self-consistent procedure converges quickly, with a monotonically decreasing residue. Otherwise, underrelaxation techniques are needed to achieve convergence, and, if the Coulomb interaction becomes comparable to the confinement energy, oscillations appear and no stable solution can be attained.

4. Configuration-interaction method

As a consequence of the above-described convergence problems, we have worked on the development of a "one-shot" method, yielding a direct solution. This method is based on the representation of the many-electron wave function on a basis of Slater determinants, which allows us to recast the computation of the ground state into the solution of an algebraic eigenvalue problem.

Before describing the method, let us introduce the 2-dimensional model that we have used for the cells. We consider two cells next to each other: one has a fixed polarization and will be referred in the following as the "driver cell" and the other, the "driven cell" assumes the configuration corresponding to the minimum energy. Such model is valid as long as the thickness of the dots in the vertical direction, corresponding to the thickness of the 2-dimensional electron gas (2DEG) from which they are obtained by lateral confinement, is small with respect to their other dimensions. The 2DEG is assumed to be obtained by modulation doping next to a GaAs/AlGaAs heterointerface.

The two-dimensional confinement potential in the plane of the 2DEG is computed as the result of the action of the metal gates at the surface of the heterostructure, following the method proposed by Davies et al. [16]. To keep the problem manageable from a computational point of view, the self-consistent rearrangement of mobile charge within the heterostructure, except for that of the two electrons confined in the cell, was not included. In other words, the potential due to the gates is computed with the analytical expressions described in the following. This is used as the bare confinement potential for the definition of a many-body Hamiltonian, whose ground state is then evaluated with the configuration-interaction method.

Let us consider a generic two-particle Hamiltonian with a two-body interaction defined as $g(\vec{r}_j, \vec{r}_i)$:

$$\hat{H} = \hat{H}_1 + \hat{H}_2, \tag{2}$$

$$\hat{H}_1 = \sum_{i=1}^{N} (-\frac{\hbar^2}{2m} \nabla_i^2 + V(\vec{r}_i)) = \sum_{i=1}^{N} \hat{h}(\vec{r}_i), \tag{3}$$

$$\hat{H}_2 = \sum_{i<j} g(\vec{r}_i, \vec{r}_j), \quad g(\vec{r}_i, \vec{r}_j) = g(\vec{r}_j, \vec{r}_i). \tag{4}$$

We consider a complete basis $\{\varphi_i(q)\}$ of spin-orbitals, where q indicates both spatial and spin coordinates, onto which the single-particle wave function can be expanded, and, from this basis, we construct all possible Slater determinants:

$$\Phi_k = \frac{1}{\sqrt{N!}} \begin{vmatrix} \varphi_{n_{1k}}(q_1) & \varphi_{n_{2k}}(q_1) & \cdots & \varphi_{n_{Nk}}(q_1) \\ \varphi_{n_{1k}}(q_2) & \varphi_{n_{2k}}(q_2) & \cdots & \varphi_{n_{Nk}}(q_2) \\ \cdots & \cdots & \cdots & \cdots \\ \varphi_{n_{1k}}(q_N) & \varphi_{n_{2k}}(q_N) & \cdots & \varphi_{n_{Nk}}(q_N) \end{vmatrix}, \tag{5}$$

where k labels the Slater determinants and n_{ik} indicates which spin-orbital appears in the i-th column of the $k-$th Slater determinant. The set $\{\Phi_k\}$ is a complete orthonormal basis for the $N-$electron eigenfunctions Ψ_i:

$$\Psi_i = \sum_{k=1}^{\infty} c_{ik} \Phi_k. \tag{6}$$

The eigenfunctions of $\hat{H}$ can be found solving the equation

$$\mathcal{H} c_i = E_i c_i. \tag{7}$$

The elements of the matrix $\mathcal{H}$ are represented by the matrix elements of the Hamiltonian operator $\hat{H}$ between the Slater determinants with indices k and k'. In practice, we are forced to truncate the series of Slater determinants down to a finite number of elements built from a finite basis of spin-orbitals. For a cell with just two electrons good results can already be obtained with only 12 spin-orbitals, while for cells with 6 electrons (that we have proven to work as effectively as those with 2 electrons [5]) more spin-orbitals are needed.

5. Fabrication tolerances

We have applied the configuration-interaction method to the investigation of the effect of fabrication tolerances, and we have determined the maximum allowed geometrical errors for the gates defining the quantum dots. We have considered a gate layout, sketched

460

in Fig. 3 with circular holes, each of which defines a quantum dot. The holes have a diameter of 90 nm and separation between the centers of 110 nm. The gate voltage has been set at -0.5 V, in order to obtain large enough interdot barriers: cell bistability is strongly dependent on dot confinement, as discussed in Ref. [17]. If the diameter of one

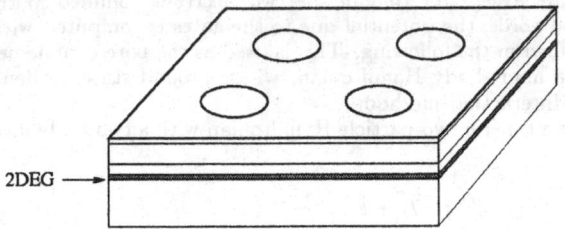

Figure 3. Gate layout for a GaAs QCA cell.

of the dots is smaller than that of the others, the potential minimum defining the corresponding dot will be somewhat shallower and, therefore, the confinement energy will turn out to be slightly larger than those of the other dots. If this increase in the confinement energy overwhelms the electrostatic splitting, i.e. the difference between the electrostatic energies corresponding to alignment of the electrons along the two diagonals, an incorrect polarization state will result. Considering that for cell sizes in the 100 nm range the electrostatic splitting is of the order of 0.02 meV in Gallium Arsenide, an extremely small fluctuation in dot diameter is sufficient to disrupt cell operation. In Fig. 4, we report the cell-to-cell response function, i.e. the polarization of the driven cell as a function of that of the driver cell for a gate with three 90 nm holes and a fourth hole with a diameter of 89.99 nm. This unphysically small error is already large enough to almost prevent cell operation for a separation between cell centers of about 280 nm. An error of a tenth of a nanometer suffices to completely destroy the QCA effect. If cell sizes are reduced, the electrostatic splitting increases, however this increase is linear in $1/R$ (where R is the hole radius), while the confinement energy increases quadratically in $1/R$, thereby leading to even stricter relative precision requirements. On the basis of these results, we

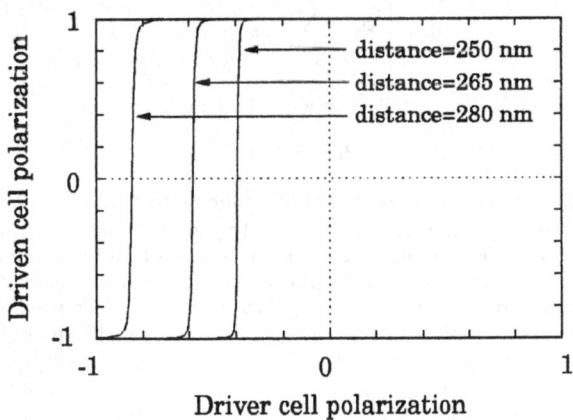

Figure 4. Cell-to-cell response function for a QCA cell with three 90 nm dots and one 89.99 nm dot.

have devised a gate layout (Fig.5) with multiple independent electrodes, which can be independently tuned to compensate for geometrical errors. We have demonstrated [17] that this compensation is possible for errors within normal fabrication tolerances.

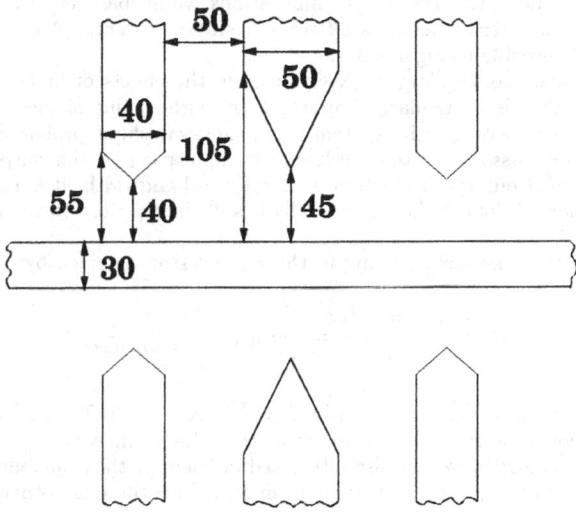

Figure 5.   Layout with independent gates for a QCA cell.

6. Circuit simulation

In order to evaluate the perspectives of QCA architectures it is important to simulate systems with logic functionalities, such as elementary logic gates, and combinatorial logic networks. Since even elementary circuits comprise at least a few cells, it is necessary to adopt a model much simpler than that used for the simulation of a single cell, but still capable of capturing the relevant physics involved in QCA behavior. Otherwise, the computational effort required for the simulation of QCA circuits would be prohibitive.

In our simplified QCA model, we treat electrons in a cell as semiclassical particles, which can tunnel between two adjacent dots. The electron charge is represented by a point charge $-e$ positioned in the center of one dot. In addition, we assume that in the center of each dot there is positive background charge $e/2$, which keeps the cell overall neutral, and therefore takes into account the charge induced on the gates, for a GaAs-AlGaAs split-gate cell, or the neutralizing charge in the doped layer, for a Si-SiO$_2$ cell.

In a semiclassical model, the ground state corresponds to the configuration of minimum electrostatic energy. Being q_i the charge contained in the i-th dot, the electrostatic energy of the configuration is

$$E = \sum_{i \neq j} \frac{q_i q_j}{4\pi\epsilon_0\epsilon_r r_{ij}}, \tag{8}$$

where r_{ij} is the distance between dot i and dot j, ϵ_0 is the vacuum permittivity, and ϵ_r is the relative permittivity of the medium (we consider the case of GaAs/AlGaAs, assuming a uniform relative permittivity of 11.9).

By computing the electrostatic energy of all allowed configurations, we can obtain the energy spectrum and the ground state of the system. We assume that only six configurations are allowed for each cell, i.e., those corresponding to electrons occupying two different dots: two of these configurations are those associated to logical "1" and "0", while we label the other four as "X" states. We do not consider the case of two electrons in the same dot, since such configurations would be energetically unfavoured, and the case of a number of electrons different from two in a cell. A circuit with N cells has therefore 6^N possible configurations.

The semiclassical model allows to easily include the effects of finite temperatures in the simulation. This is particularly important from the point of view of applications. Indeed, at finite temperature the system has a non-vanishing probability to be in an excited state, likely associated to a different configuration of the output cells. Since the outputs are read out with detectors with a finite bandwidth, it is required that the average occupation of dots of the output cells is sufficiently close to the correct value (0 or 1).

The probability of the system being in the ground state is given by

$$P_{gs} = \frac{e^{-E_{gs}/(kT)}}{Z} = \frac{1}{1 + \sum_{i \neq gs} e^{-\Delta E_i/(kT)}}, \qquad (9)$$

where Z is the partition function, given by $Z = \sum_i \exp -E_i/(kT)$, $\Delta E_i = E_i - E_{gs}$, k is the Boltzmann constant and T is the temperature. The summation is performed over all excited states. Besides P_{gs}, we are also interested in knowing the probability of having the correct logical output P_{clo}, which is larger than P_{gs}, since there are other configurations, in addition to that corresponding to the ground state, which share the correct output value. In order to obtain P_{clo} we need to sum over the probabilities corresponding to such configurations:

$$P_{clo} = \frac{\sum_j e^{-E_j/(kT)}}{Z} \qquad (10)$$

In the following we consider systems in which the distance between the centers of adjacent cells is $d = 100$ nm and the distance between adjacent dots in the same cell is $a = 40$ nm. We assume the dielectric permittivity of GaAs. In Fig. 6 we plot P_{gs} and P_{clo} as a function of temperature for a binary wire made of six cells. As expected, both probabilities decrease with increasing temperature: P_{gs} tends to an extremely small value $(1/6^5)$, since all states become equally probable in the limit $T \to \infty$; P_{clo} tends to 1/6 since the six possible states of the output cell become equally probable for $T \to \infty$. Once a maximum acceptable error probability is given, we can obtain the maximum operating temperature directly from Fig. 6.

We can extend our investigation of the maximum error probability to more complex systems, such as the one shown in Fig. 7(c), performing the logic function $AB + CD$. The circuit consists of 18 cells, of which 7 have enforced logic values, so that 6^{11} configurations are allowed. In Fig. 7(a) results are plotted for the input state $A = B = C = D = 1$, while in Fig. 7(b) results are plotted for the input state $A = 0, B = C = D = 1$. The difference between the two curves is due to the different energy splitting between the ground state and the first excited state for the two input configurations: in the case of Fig. 7(a) such energy splitting is 0.159 meV, while in the other case is 0.086 meV.

The computation of the whole energy spectrum imposes a practical limit of 12-13 to the maximum number of cells that can be considered. However, in most cases we only need to know the ground state and the low-lying excited states in order to evaluate circuit

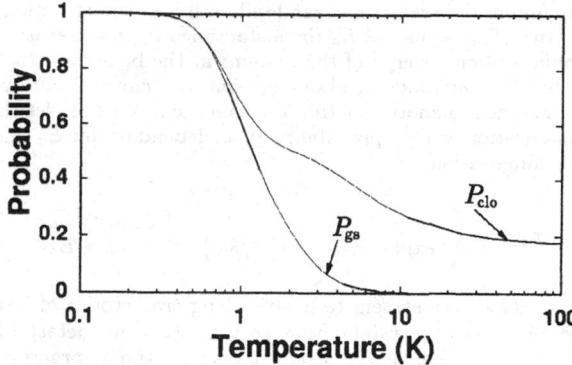

Figure 6. Plot of the probabilities of correct logical output and of being in the ground state as a function of temperature.

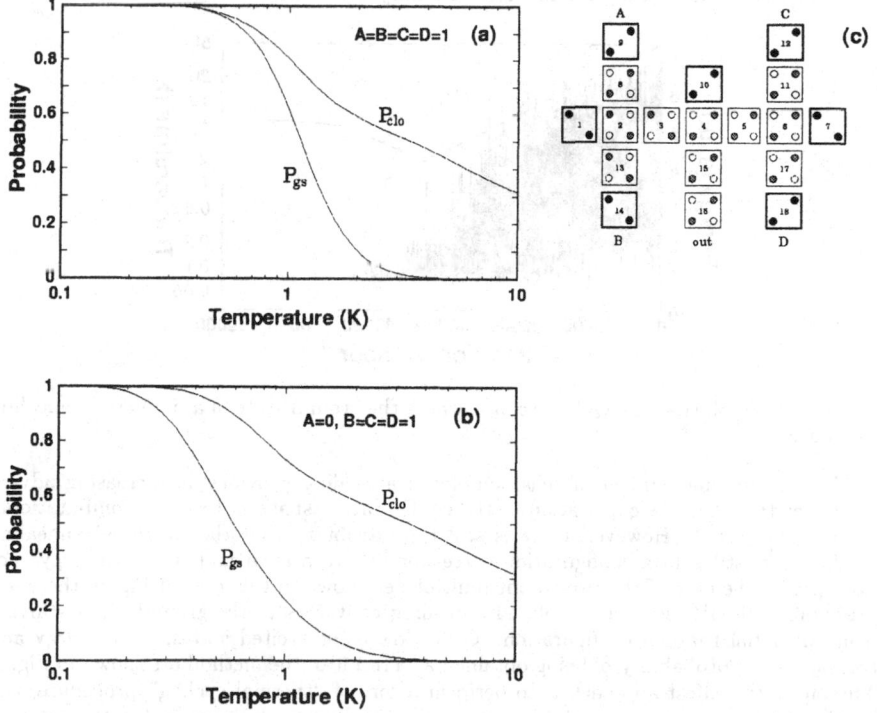

Figure 7. a) and b) Plots of the probabilities of correct logical output and of being in the ground state for the logic circuit shown in c), for two different input vectors.

behavior. Such states can be obtained with a simulated annealing procedure without exploring the whole configuration space. Here, we briefly show how the procedure works.

We start by assuming the system in a randomly selected configuration of energy E_0 at an initial temperature T_0. Let us call E_a the instantaneous (in the sense of corresponding to the current configuration) energy of the system: at the beginning $E_a = E_0$. Then, we pick a new configuration at random, obtained from the current one by simply moving one electron from one dot to another within the same cell. We then let the system evolve into the new configuration with a probability p_{new} depending on E_a and on the energy E_{new} of the new configuration:

$$p_{new} = \begin{cases} 1 & \text{if } E_{new} \leq E_a \\ \exp[-(E_{new} - E_a)/kT] & \text{if } E_{new} > E_a \end{cases} \qquad (11)$$

The first condition allows the system to evolve along trajectories of descending energy, while the second prevents the system from getting stuck in metastable states, corresponding to local minima. This cycle is repeated many times, progressively decreasing the temperature from T_0 to a value at which a stable configuration is reached. If the system is cooled down at a sufficiently slow rate, the stable state is the ground state.

An example of the evolution of the state of the system to the ground state in shown in Fig. 8, where the instantaneous configuration energy and the temperature are plotted as a function of the iteration number for a binary wire with six cells.

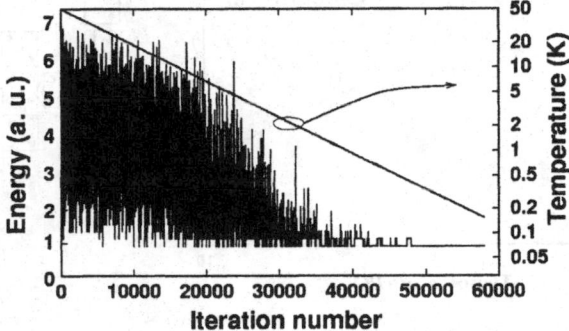

Figure 8. Evolution of a QCA circuit towards the ground state in a simulated annealing calculation.

With increasing number of cells, simulated annealing provides an increasing advantage over the complete exploration of the configuration space in terms of computational resources required. However, there is still the possibility that the simulated annealing method gets stuck in a configuration corresponding to a local minimum of energy. For example, in the case of the two-to-one multiplexer shown in the inset of Fig. 9, there is a probability of 54% that the simulated annealing converges to the ground state, starting from an initial random configuration. Other low-lying excited states, their energy and corresponding probability of being obtained as a result of the method are shown in Fig. 9. Therefore, the safest approach is to perform a sort of "thermal cycling" procedure, i.e., let the system converge to a stable state, and then iteratively increase the temperature T to drive the system to an excited state, and reduce T until a stable state is reached. Among all stable states obtained, the probability that the state of minimum energy is the ground state is $1 - (1 - P_1)^m$ where m is the number of thermal cycles and P_1 is the probability of obtaining the ground state without thermal cycling. For example, in the example considered ($P_1 = 0.54$), the probability of finding the ground state with eight thermal cycles would be 0.998.

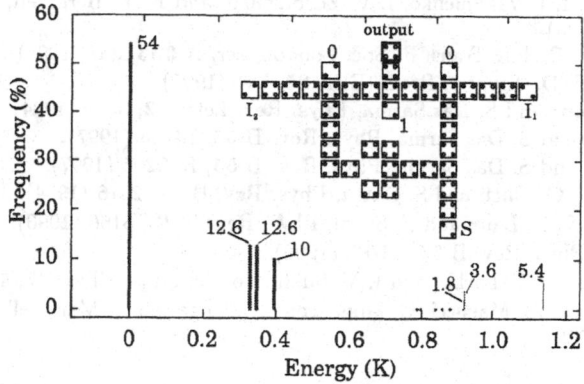

Figure 9. Relative frequency of the outcomes of simulated annealing calculations for the circuit shown in the inset.

7. Conclusions

We have presented some results of our research on the feasibility of logic circuits based on the QCA paradigm. A computational method for the determination of the ground state of a QCA cell has been developed and applied to the investigation of the sensitivity to fabrication tolerances, which has turned out to be extremely large, mainly because of the smallness of the electrostatic interaction. This affects significantly also the maximum operating temperature, which is currently below 1 K. We have devised cell layouts that allow the demonstration of the QCA principle, but we do not expect that it will be possible to develop a large scale integration technology based on the QCA paradigm in the near future. We have also considered alternative implementations, in particular those based on structures at the molecular level, but, at the current state of the art, their implementation still poses formidable technological challenges.

Acknowledgments

This work has been supported by the European Commission through the ANSWERS project (n. 28667), funded within the MEL-ARI and NID-FET initiatives, and by the Italian National Research Council through the MADESS II initiative (research theme on low-dimensional systems)

References

[1] P. Bakshi, D. Broido and K. Kempa, J. Appl. Phys. **70**, 5150 (1991).

[2] P. D. Tougaw, C. S. Lent, and W. Porod, J. Appl. Phys. **74**, 3558 (1993).

[3] S. Bandyopadhyay, B. Das and A. E. Miller, Nanotechnology **5**, 113 (1994).

[4] C. S. Lent, P. D. Tougaw, and W. Porod, Appl. Phys. Lett. **62**, 714 (1993).

[5] M. Girlanda, M. Governale, M. Macucci, and G. Iannaccone, Appl. Phys. Lett. **75**, 3198 (1999).

[6] G. Iannaccone, C. Ungarelli, M. Macucci, E. Amirante, M. Governale, *Thin solid Films* **336**, 145 (1998).

[7] C. K. Wang, I. I. Yakymenko, I. V. Zozoulenko, and K. F. Berggren, J. Appl. Phys. **84**, 2684 (1998).

[8] R. Landauer, Philos. Trans. R. Soc. London, ser. A **353**, 367 (1995).

[9] C. S. Lent, P. D. Tougaw, Proc. IEEE **85**, 541 (1997).

[10] C. A. Stafford and S. Das Sarma, Phys. Rev. Lett. **72**, 3590 (1994).

[11] R. Kotlyar and S. Das Sarma, Phys. Rev. B **56**, 13235 (1997).

[12] R. Kotlyar and S. Das Sarma, Phys. Rev. B **55**, R10205 (1997).

[13] G. Klimeck, G. Chen and S. Datta, Phys. Rev. B **50**, 2316 (1994).

[14] A. Kumar, S. E. Laux and F. Stern, Phys. Rev. B **42**, 5166 (1990).

[15] M. Stopa, Phys. Rev. B **54**, 13767 (1996).

[16] J. H. Davies, I. A. Larkin and I. V. Sukhourov, J. Appl. Phys. **77**, 4504 (1995).

[17] M. Governale, M. Macucci, G. Iannaccone, C. Ungarelli, J. Martorell, J. Appl. Phys. **85**, 2962 (1999).

List of Participants

Yuriy Astashkin
Physics Department
Kyiv National Taras Shevchenko University,
6, Academica Glushkova Ave.,
03022, Kyiv,
UKRAINE
e-mail: ashchuk@expphys.ups.kiev.ua

Anca Angelescu
National Institute for Research and Development
in Microtechnologies (IMT)
P.O. Box 38-160, 72225 Bucharest,
ROMANIA

Ekaterina Astrova
Ioffe Physical Technical Institute,
Polytekhnicheskaya 26,
194021, St.Ptersburg,
RUSSIA
e-mail: east@pop.ioffe.rssi.ru

Olexandr Barabanov
Radiophysical Faculty
Kyiv National Taras Shevchenko University
6 Glushkov St., 03022, Kyiv,
UKRAINE
e-mail: tov@mail.univ.kiev.ua

Sue Bayliss
Solid State Research Centre
De Montfort University
Leicester LE1 9BH
UK
e-mail: sbayliss@dmu.ac.uk

Igor V.Belousov,
Institute for Metal Physics, Ukrainian National Academy of Sciences
Vernadskii Ave. 36,
01680, Kyiv
UKRAINE
e-mail: belousov@scat.kiev.ua

Francois Bessueille
IFOS – Ecole Centrale de Lyon
Ave. Guy de Collongue, BP163
69131 Ecully Cedex
FRANCE
e-mail: bessueil@ec-lyon.fr

Mikola M. Bilyi
6 Akademika Glushkova prosp.,
03022, Kyiv,
UKRAINE
e-mail: anaumenko@mail.univ.kiev.ua

Laszlo P. Biro
Research Institute for Technical Physics
and Materials Science
H-1525 Budapest, P.O. Box 49, Hugary
Street: H-1121 Budapest,
Konkoly-Thege ut 29-33
HUNGARY
e-mail: biro@mfa.kfki.hu

Sergiy Bulavenko
Radiophysical Faculty
Kyiv National Taras Shevchenko University
6 Glushkov St.,
03022, Kyiv,
UKRAINE
e-mail: bulav@mail.univ.kiev.ua

Eugenia V.Buzaneva
Radiophysical Faculty
Kyiv National Taras Shevchenko University
6 Glushkov St.,
03022, Kyiv,
UKRAINE
e-mail: evb@mail.univ.kiev.ua

Leigh T.Canham
Nanotechnology Department
Defense Research Agency
St. Andrews Rd., Malvern
Worcestershire, WR143PS,
UK

Thomas Dittrich
Physical Department E16,
Technical University Muenchen,
85747 Garching,
GERMANY

467

468

e-mail: Thomas Dittrich @physik.TU-Muenchen.de

Alexsandr Doroshenko
Radiophysical Faculty
Kyiv National Taras Shevchenko University
6 Glushkov St.,
03022, Kyiv,
UKRAINE
e-mail: evb@mail.univ.kiev.ua

Volodimyr Duzhko
TUM, Physic Department E16
85748 Garching bei Muenchen
GERMANY
e-mail: vduzhko@physik.TU-Muenchen.de

Stefan Eisebitt
Forschungszentrum JuelichInstitut fuer Festkoerperforschung
52425 Juelich
GERMANY
e-mail: S.Eisebitt@fz-juelich.de

Philippe Fauchet
Department of Electrical
and Computer Engineering
Computer Studies Building Room 518
University of Rochester
p.o.box 270231,
Rochester NY 14627-0231,
USA
e-mail: fauchet@ece.rochester.edu

Jorg Fink
Institute of Solid State Physics
Dresden Postfach 210016
D-01171 Dresden
GERMANY
email: J.Fink@ifw-dresden.de

Sergei V.Gaponenko
Institute of Molecular and Atomic Physic,
National Academy of Sciences of Belarus
F.Skaryna Ave. 70
220072 Minsk
BELARUS
email: gaponen@imaph.bas-net.by

Etienne Goovaerts

Department of Physics,University of Antwerpen – UIA Universiteitsplein 1,
B-2610 Wilrijk-Antwerpen
BELGIUM
email: etienne.goovaerts@uia.ua.ac.be

Kirill.O. Gorchinsky
Radiophysical Faculty
Kyiv National Taras Shevchenko University
6 Glushkov st,
03022, Kyiv,
UKRAINE
e-mail: Lab227@mail.univ.kiev.ua

Olexander.D. Gorchinsky
Radiophysical Faculty
Kyiv National Taras Shevchenko University
6 Glushkov st,
03022, Kyiv,
UKRAINE
e-mail: Lab227@mail.univ.kiev.ua

Ludmila A.. Karachevtseva
Institute of Semiconductor Physics
NAS of Ukraine, Prosp. Nauki, 45,
03650, Kyiv,
UKRAINE
e-mail: kartel@mail.kar.net

Eli Kapon
Institute of Micro and Optoelectronics
Department of Physics
Swiss Federal Institute of Technology
CH-1015 Lausanne, Suisse
SWITZERLAND
e-mail:kapon@dpmail.epfl.ch

Anna Karlash
Radiophysical Faculty
Kyiv National Taras Shevchenko University
6 Glushkov st,
03022, Kyiv,
UKRAINE
e-mail: evb@mail.univ.kiev.ua

Ilki Kim
Institut fur Theoretische Physik,
Universitat Stuttgart
Pfaffenwaldring 57,
D-70550 Stuttgart,

GERMANY
e-mail: ikim@theo.physik.uni-stuttgart.de
Dr.Nina Kovtyukhova
Institute of Surface Chemistry,
31, Nauki Prosp,
01022 Kyiv,
UKRAINE
e-mail: ninok@ukrpack.net

Valeriy F.Kovalenko
Radiophysical Faculty
Kyiv National Taras Shevchenko University
6 Glushkov St.,
03022, Kyiv,
UKRAINE
svp@boy.univ.kiev.ua

Vlad Kudrua
Physics Department
Kyiv National Taras Shevchenko University,
6, Academica Glushkova St.,
03022, Kyiv,
UKRAINE
e-mail: yashchuk@expphys.ups.kiev.ua

Hans Kuzmany
Institut fuer Materialphysik
Universitaet Wien, Strudlhofgasse 4
A-1090 Wien
AUSTRIA
e-mail: kuzman@ap.univie.ac.at

Genagiy Kuznetsov
Radiophysical Faculty
Kyiv National Taras Shevchenko University
6 Glushkov st.,
03022, Kyiv,
UKRAINE
e-mail: vai@rpd.univ.kiev.ua

Serguei Lazarouk
Belorusian State University of Informatics and Radioelectronics Microelectronics Dept.
6 P.Browky St.
220027 Minsk,
BELARUS
e-mail: serg@cit.org.by

Isabelle Ledoux-Rak

Laboratoire de Photonique Quantique et Moleculaire – Ecole Normale Superieure de Cachan – 61 avenue du President Wilson – 94235
Cachan Cedex –
FRANCE
e-mail: ledoux@lpqm.ens-cachan.fr

Valeriy Lozovski
Institute of Semiconductor Physics
NAS of Ukraine,Prosp. Nauki, 45,
03650, Kiev,
UKRAINE

Dmitri Lyebyedyev
Bergische Universitaet Wuppertal
FB Elektrotechnik und
Informationstechnik
Lehrstuhl fuer Mikrostrukturtechnik
Fuhlrottstr.10,
D-42119 Wuppertal
GERMANY
e-mail: dmitri@uni-wuppertal.de

Christoph Lienau
Max-Born-Institut fur Nichtlineare Optik und Kurzzeitspektroskopie
Abteilung C3, Gbd. 19.8
Rudower Chaussee 6
D-12489 Berlin
GERMANY
e-mail: lienau@mbi-berlin.de

Yu. Losytskyy
Physics Department
Kyiv National Taras Shevchenko University,
6, Academica Glushkova Ave.,
03022, Kyiv,
UKRAINE
e-mail: yashchuk@expphys.ups.kiev.ua

Massimo Macucci
Dipartimento di Ingegneria dell'Informazione
Universita` di Pisa, Via Diotisalvi, 2
I-56126 PISA
ITALY
e-mail:massimo@mercurio.iet.unipi.it

Gunter E. Mahler
Institute for Theoretical Physics I
University of Stuttgart

470

Pfaffenwaldring 57
D-70550 Stuttgart
GERMANY
mahler@theo.physik.uni-stuttgart.de

Olga P. Matyshevska
Biology Faculty
Kyiv National Taras Shevchenko University
6 Glushkov st.,
03022, Kyiv,
UKRAINE

Yasuo Mihara
Ulvac-Vacuum Metallurgical Co., Ltd.
516
Yokota,Sanbu-machi,Sanbu-gun,
Chiba,289-1297
JAPAN
e-mail:hayashi@ulvac.co.jp

Salvatore Mirabella
INFM - Universita di Catania
Dipartimento di Fisica
Corso Italia 57
I-95129 Catania
ITALY

Rene Monot
Ecole Polytechnique Federale de
Lausanne
Institut de Physique Experimentale (IPE)
DP-EPFL, PHB-Ecublens,
CH-1015 Lausanne
SWITZERLAND
e-mail: rene.monot@epfl.ch

Vasyl Motsnyi
IMEC
MCP/NMC Division
Kapeldreef 75
B-3001 Leuven
BELGIUM
mailto:motsnyi@imec.be

Viviana Mulloni
INFM and Dipartimento di Fisica
Universita' di Trento via Sommarive 14I-
38050
Povo (Trento)
ITALY
e-mail: mulloni@science.unitn.it
Atsushi Nakajima

Department of Chemistry, Faculty of
Science and Technology, Keio University
3-14-1 Hiyoshi, Kohoku-ku,
Yokohama, 223-8522,
JAPAN
E-mail:nakajima@iw.chem.keio.ac.jp

Androula Nassiopoulou.
Inst. of Microelectronics
IMEL/NCSR Demokritos,P.O. Box
60228,
153 10 Aghia Paraskevi Attikis,
Athens
GREECE.
email: Nassiopoulou@imel.demokritos.gr

Antonina P.Naumenko
Faculty of Physics,
Kyiv National Taras Shevchenko
University
6 Akademika Glushkova prosp.,
03022 Kyiv,
UKRAINE
e-mail: a_naumenko@mail.univ.kiev.ua

Yuri Nazarok
Radiophysical Faculty
Kyiv National Taras Shevchenko University
6 Glushkov st.,
03022, Kyiv,
UKRAINE

Tymish Ogul'chansky
Physics Department
Kyiv National Taras Shevchenko University,
6, Academica Glushkova Ave.,
03022, Kyiv,
UKRAINE
e-mail: yashchuk@expphys.ups.kiev.ua

Stefano Ossicini
Department of Physics
University of Modena
Via Campi 213/A
41100 Modena
ITALY
e-mail: ossicini@imoax1.unimo.it

Thierry Ouisse
Inst. of Microelectronics
IMEL/NCSR Demokritos,

P.O. Box 60228, 153 10 Aghia Paraskevi
Attikis,
Athens
GREECE.
e-mail:.T.Ouisse@imel.demokritos.

Domenico Pacifici
INFM - Universita di Catania
Dipartimento di Fisica
Corso Italia 57I-95129
Catania
ITALY

Lorenzo Pavesi
INFM and Dipartimento di Fisica
Universita' di Trento
via Sommarive 14
I-38050 Povo (Trento)
ITALY
e-mail: pavesi@science.unitn.it

Tatiana Perova
Department of Photophysics
Vavilov State Optical Institute
199034, St.-Petersburg,
RUSSIA
e-mail: perovat@tcd.ie

Valerij Pogorelov
Physical Department
Kyiv National Tara Shevchenko University,
64, Volodimirskaya St.
01033 Kyiv
UKRAINE.
e-mail: pogorelv@expphys.ups.kiev.ua
        pogorelv@mail.univ.kiev.ua

Galina D.Popova
Radiophysical Faculty
Kyiv National Taras Shevchenko University
6 Glushkov st,
03022, Kyiv,
UKRAINE
e-mail: evb@mail.univ.kiev.ua

Paras. N. Prasad
The Institute for Lasers, Photonics and
Biophotonics
Photonics Research Laboratory
428 NSM Complex
SUNY/at Buffalo

USA
e-mail: pnprasad@acsu.buffalo.edu

Francesco Priolo
INFM - Unita di Catania
Dipartimento di Fisica
Universita di Catania
Corso Italia 57
I-95129 Catania
ITALY
e-mail: priolo@ct.infn.it

Yuri.I. Prylutskyy
Biological Department
Kyiv National Taras Shevchenko University,
64, Volodimirskaya str.
01033 Kyiv
UKRAINE
e-mail: prilut@hd.ups.kiev.ua

Sergiy Putselyk
Radiophysical Faculty
Kyiv National Taras Shevchenko University
6 Glushkov st.,
03022, Kyiv,
UKRAINE
e-mail: evb@mail.univ.kiev.ua

Hella.-C. Scheer
Universitaet Wuppertal
Fachbereich Elektrotechnik
Lehrstuhl fuer Mikrostrukturtechnik
Gauss-Strasse 20
42097 Wuppertal
GERMANY
e-mail: scheer@ween01.elektro.uni-wuppertal.de

Heinz Schweizer
4.Phys.Inst. University Stuttgart, Pfaffen-waldring 57,
70550 Stuttgart
GERMANY
E-mail:h.schweizer@physik.uni-stuttgart.de

Nadrian C. Seeman
Department of Chemistry, Mail Code 5180
New York University
New York, NY 10003-6688,

472

U.S.A.
E-mail: ned.seeman@nyu.edu

Supapan Seraphin
University of Arizona
Materials Science and Engineering
The University of Arizona Tucson,
Arizona 85721
USA
e-mail: seraphin@u.arizona.edu

Viktoriya B. Shevchenko
Physical Faculty
Kyiv National Taras Shevchenko University,
6, Academica Glushkova Ave.,
03022, Kyiv,
UKRAINE

Yaroslav Shtogun
Radiophysical Faculty
Kyiv National Taras Shevchenko University
6 Glushkov st.,
03022, Kyiv,
UKRAINE
e-mail: Lab227@mail.univ.kiev.ua

Uri Sivan
Department of Physics
Technion-Israel Institute of Technology
Haifa 32000
ISRAEL
e-mail: phsivan@techunix.technion.ac.il

Valerij A.Skryshevsky
Radiophysical Faculty
Kiev National Taras Shevchenko University
6 Glushkov st.,
03022, Kyiv,
UKRAINE
e-mail: skrysh@uninet.kiev.ua

Vladimir P. Sohatsky
Radiophysical Faculty
Kyiv National Taras Shevchenko University
6 Glushkov st.6,
03022, Kyiv,
UKRAINE
svp@boy.univ.kiev.ua
George I. Stegeman

School of Optics/CREOL
University of Central Florida
4000 Central Florida Blvd.
Orlando, FL 32816-2700,
USA.
e-mail: george@creol.ucf.edu

Genrikh V.Stepanov
Institute of RadioEngineering and Electronics
Russian Academy of Sciences
Moscow, Center, Mohovaya 11,
RF, RUSSIA
e-mail:gvs198@ire216.msk.ru
gvs198@mail.cplire.ru

Ivan M. Tiginyanu
Technical University of Moldova
MD-2004 Chisinau
MOLDOVA
e-mail: tiginyanu@mail.md

Clivia M Sotomayor Torres
Institute of Materials Science
& Dept of Electrical Engineering
University of Wuppertal
Gauss-Strasse 20
42097 Wuppertal
GERMANY
e-mail: clivia@uni-wuppertal.de

Vladimir Tolmachev
Ioffe Physical Technical Institute,
Polytekhnicheskaya 26,
194021, St.Ptersburg,
RUSSIA
c-mail: east@pop.ioffe.rssi.ru

Valeri P. Tolstoy
Department of Chemistry,
St.Petersburg State University.University
st. 2, St.Petersburg, 198904,
RUSSIA
Valeri.Tolstoy@pobox.spbu.r

Tatiana S.Veblaya
Radiophysical Department
Kyiv NationalTaras Shevchenko University
64 Volodymirskaya Str.,
01033 Kyiv,
UKRAINE
Gor@mail.univ.kiev.ua

Victor A. Vikulov
Radiophysical Department
Kyiv NationalTaras Shevchenko University
64 Volodymirskaya Str.,
01033 Kyiv,
UKRAINE
e-mail: skrysh@uninet.kiev.ua

Kostyantyn I. Yakovkin
Radiophisical Department
Kyiv National Taras Shevchenko University,
6, Academica Glushkova Ave.,
03022, Kyiv,
UKRAINE
e-mail: Ya-kos@mail.ru

Valerij N. Yashchuk
Physics Department
Kyiv National Taras Shevchenko University,
6, Academica Glushkova Ave.,
03022, Kyiv,
UKRAINE
e-mail: yashchuk@expphys.ups.kiev.ua

Sergiy Zankovych
Institute of Materials Science
& Dept of Electrical and Information
Engineering
University of Wuppertal
Gauss-Strasse 20
42097 Wuppertal
GERMANY
e-mail:sergiy@uni-wuppertal.de

Andriy Ya. Zhugayevych
Institute of Physics,
Department of Theoretical Physics
46 Prospect Nauky,
03650 Kyiv,
UKRAINE
e-mail: zhugayev@iop.kiev.ua

SUBJECT INDEX